AMNIOTE ORIGINS

AMNIOTE ORIGINS

Completing the Transition to Land

Edited by

Stuart S. Sumida
California State University
San Bernardino, California

Karen L. M. Martin
Pepperdine University
Malibu, California

ACADEMIC PRESS
San Diego London Boston New York Sydney Tokyo Toronto

Cover photograph: Holotype of *Aerosaurus wellesi*. Photo by Stuart S. Sumida.

This book is printed on acid-free paper. ♾

Academic Press, Inc.
525 B Street, Suite 1900, San Diego, California 92101-4495, USA
http://www.apnet.com

Academic Press Limited
24-28 Oval Road, London NW1 7DX, UK
http://www.hbuk.co.uk/ap/

Library of Congress Cataloging-in-Publication Data

Amniote origins : completing the transition to land / edited by Stuart S. Sumida, Karen L. M. Martin
p. cm.
Includes bibliographical references and index.
ISBN 0-12-676460-3 (alk. paper)
1. Amniotes, fossil--Evolution. 2. Amniotes--Evolution.
I. Sumida, Stuart Shigeo. II. Martin, Karen L. M.
QE847.A46 1996
596'.038dc20 96-41836
CIP

PRINTED IN THE UNITED STATES OF AMERICA
96 97 98 99 00 01 BB 9 8 7 6 5 4 3 2 1

CONTENTS

ACKNOWLEDGMENTS AND DEDICATION

The genesis of this book required far more than the ideas of the editors alone. The original impetus was the invitation to organize a symposium, "Paleontological and Neontological Approaches to the Origin of Amniotes", for the Fourth International Congress of Vertebrate Morphology. We thank speakers and the audience for their enthusiastic response. We are most grateful to Susan Herring and James Hanken for their support, interest, and suggestions, and to Susan Abrams and R. Eric Lombard for their counsel when the project was in its embryonic stages. We are extremely grateful to Dr. Chuck Crumly and Mr. David Berl-Hahn for shepherding us through the process of organization, editing, and publication of this work. Their keen interest has been a driving force behind this production.

We are grateful for many perceptive comments and suggestions from the following people who reviewed parts of this book: Donald G. Buth, Chuck Crumly, James Hanken, James Hicks, Victor H. Hutchison, Lee Kats, Sean Modesto, Kevin Padian, David Polcyn, Elizabeth Rega, Bob Roberts, David Sever, Vaughan H. Shoemaker, Richard E. Strauss, Hans-Dieter Sues, and Thomas Vandergon. *Dankeshön* and *cheers* to Dr. Elizabeth Rega, *mercí* to Charles Solomon, and *muchas gracias* to Therese Whitney who provided thorough advice on the vagaries of accents and diacritical symbols in German and British English, French, and Spanish, respectively. Advisors and colleagues have played an important role in the development of the conceptual direction of this collection. We are particularly grateful to David Berman, Eric Lombard, Peter Vaughn and the UCLA Physiological Ecology Group for providing stimulating input and inspiration.

Support for production portions of this volume has been provided by the California State University San Bernardino Division of Graduate Studies and Research, the National Geographic Society (grant 5182-94), the National Science Foundation (BIR-9925034), the North Atlantic Treaty Organization (grant CRG.940779), and the University Research Council of Pepperdine University.

Friends and family have been alternately patient and impatient as the needs arose. For keeping us focused, sane, and thoughtful, we are thankful to: Alex Martin, Doug Martin, Greg Martin, David Polcyn, Elizabeth Rega, Charles Solomon, and Katherine Thorne.

Finally, it must be noted that Dr. Everett C. Olson is a coauthor of one of the chapters in this volume. He has been a driving force in vertebrate paleontology and was a profound influence on both of the editors during the formative years of our professional careers. We are proud that he saw fit to contribute to this project, and we are even more proud that we may provide a vehicle for one of his final publications. Everett C. Olson has passed on, but his influence continues to refresh vertebrate paleontology. It is with great pleasure that we note the enduring nature of his influence. Ole was throughout his life a teacher, a mentor, and a friend. He was open-minded but tough, always ready to listen to new ideas and argue them thoroughly. We think he would have been pleased and intrigued by the many new ideas in this collection. It is with extreme affection, and a profound sense of loss, that we dedicate this book to him.

Figure 1. Everett C. Olson: Mentor, teacher, and friend.

Contributors

Nancy Aguilar
Center for Marine Biotechnology and Biomedicine
Scripps Institute of Oceanography
University of California, San Diego
La Jolla, California 92093
USA

David S Berman
Section of Vertebrate Paleontology
Carnegie Museum of Natural History
Pittsburgh, Pennsylvania 15213
USA

Richard Beerbower
Department of Geological Sciences
Binghamton University
Binghamton, New York
USA

Ramon Diaz-Uriarte
Department of Zoology
University of Wisconsin
430 Lincolin Drive
Madison, Wisconsin 53706
USA

Robert Dudley
Department of Zoology
University of Texas
Austin, Texas 78712
USA

Larry Frolich
Department of Biology
University of St. Thomas
St. Thomas, Minnesota
USA
Instituto Superior
Quito, Ecuador

Carl Gans
Department of Biology
University of Michigan
Ann Arbor, Michigan 48109
USA

Theodore Garland Jr.
Department of Zoology
University of Wisconsin
430 Lincolin Drive
Madison, Wisconsin 53706
USA

Gary B. Gillis
Department of Ecology and Evolution
University of California, Irvine
Irvine, California 92717
USA

Jeffrey B. Graham
Center for Marine Biotechnology and Biomedicine
Scripps Institute of Oceanography
University of California, San Diego
La Jolla, California 92093
USA

Nicholas Hotton III
Department of Paleobiology
National Museum of Natural History
Smithsonian Institution
Washington, D.C.
USA

George V. Lauder
Department of Ecology and Evolution
University of California, Irvine
Irvine, California 92717
USA

Michel Laurin
Museum of Paleontology
University of California
Berkeley, California 94720

Michael Y. S. Lee
School of Biological Science
Zoology A08
University of Sydney
New South Wales 2006
Australia

R. Eric Lombard
Department of Organismal Biology and Anatomy
The University of Chicago
Chicago Illinois 60637
USA

Karen L. M. Martin
Division of Natural Sciences
Pepperdine University
Malibu, California 90263
USA

Kenneth A. Nagy
Department of Biology
University of California Los Angeles
Los Angeles, California 90024
USA

Everett C. Olson
Department of Biology
University of California Los Angeles
Los Angeles, California 90024
USA

Mary J. Packard
Department of Biology
Colorado State University
Fort Collins, Colorado 80523
USA

Robert R. Reisz
Department of Zoology
University of Toronto, Erindale Campus
Mississauga, Ontario L5L 1C6
Canada

Roger S. Seymour
Department of Zoology
University of Adelaide
Adelaide SA 5005
Australia

Patrick S. Spencer
Department of Geology
University of Bristol
Queens Road, Bristol BS8 1RJ
United Kingdom

James Stewart
Faculty of Biological Science
University of Tulsa
Tulsa, Oklahoma 74104
USA

Stuart S. Sumida
Department of Biology
California State University, San Bernardino
San Bernardino, California 92407
USA

CHAPTER 1

AN INTEGRATED APPROACH TO THE ORIGIN OF AMNIOTES: COMPLETING THE TRANSITION TO LAND

Karen L. M. Martin

Stuart S. Sumida

The origin of amniotes was a critical step in vertebrate evolution. It is apparent that fundamental, macroevolutionary changes in the natural histories and physiological features of tetrapods occurred at or near this transition, which set the stage for the Age of Reptiles and, ultimately, for the radiation of mammals and birds. Despite its tremendous evolutionary importance, the origin of amniotes has been addressed directly by relatively few studies. Other biologically important transitions, for example the transition from water to air breathing or the evolution from ectothermy to endothermy, have been examined independently by many disciplines. However, the origin of amniotes has been, for the most part, studied by paleontologists and thus limited to osteological data. Our goal has been to bring together paleontologists and neontologists specializing on both extant and extinct organisms, to examine this evolutionary transition that is rich in importance, but historically poor in testable hypotheses.

Amniotes were the first vertebrates to complete the transition to land, and this release from life history stages requiring bodies of water provided a profound freedom for new evolutionary radiations. Their descendants accomplished innovations in terrestrial herbivory, behavioral thermoregulatory strategies, respiration in air, water balance and conservation, and new means of locomotion, including flight. In this volume we provide a holistic approach to understanding primitive tetrapods that should increase our understanding of this macroevolutionary change. Our aim is to provide a broad overview of the evolution of the vertebrate body with both paleontologists and neontologists addressing the origin of amniotes and making inferences from a variety of different perspectives. Vertebrate paleontologists have been slow to adopt experimental approaches, and reproducible studies addressing function by vertebrate paleontologists are still disappointingly few. Conversely, functional morphologists and physiologists rarely address the influences that their work on extant organisms might have on understanding the extinct animals that made the transition from anamniote to amniote. We hope that this volume will prove to be heuristic, generating dialogue and forging new connections to stimulate novel ways of thinking about the evolution of the tetrapods.

SCIENTIFIC AND HISTORICAL PERSPECTIVE

Early focus on the origin of amniotes inevitably turned to a search for progressively more primitive, and in some cases hypothetical, ancestors (Romer, 1966; Carroll, 1969a,b, 1970). Over the course of the past thirty years, the context in which primitive tetrapods have been examined has changed radically. New concepts in biogeography and the acceptance of plate tectonics forced a reexamination of the taxa presumed to be close to the origin of amniotes (Hotton, 1992). The advent of cladistic methods of phylogenetic analysis has suggested profoundly different hypotheses regarding the primitive radiations of amniotes (e.g., Heaton, 1980; Holmes, 1984; Heaton and Reisz, 1986; Gauthier *et al.*, 1988a, Berman *et al.*, 1992; Lee, 1993, 1995; Carroll, 1995; Laurin and Reisz, 1995). This volume has grown out of our desire to address new advances in our understanding of the phylogenetic relationships of early amniotes and to add the critically important experimental approaches of functional biology.

The authors gathered here address the integrated influences of morphology, physiology, evolutionary biology, behavior, developmental biology, histology, ecology, and phylogeny on the origin of amniotes. However, caveats are necessary. With fossil animals, only the morphology of hard tissues can typically be studied; the other aspects of their biology must almost always be inferred. Yet morphology may be a poor predictor of physiology; animals with similar morphologies may have very divergent physiologies (Burggren and Bemis, 1990). Moreover, changes in physiology or function may leave few morphological clues in the fossil record (Lauder, 1981). Clearly, many inferences are not testable by experimentation or observation when studying extinct organisms. Nevertheless, we believe that the present can be gainfully considered as a key to the past, provided caution is used (Garland and Carter, 1994). With appropriate care, we believe that useful analogies can be made between extant and extinct animals.

OVERVIEW OF TEXT

A major concern in approaches to the study of the origin of amniotes is the controversy surrounding nomenclature and phylogeny of the groups formerly known as the Amphibia and the Reptilia. In a rigorous use of cladistic terminology, these two groups are radically re-defined (Gauthier *et al.*, 1988b). The word "amphibian" is not used in a taxonomic sense, although it is often used by authors included here as a colloquial means of communicating the concepts of anamniote tetrapods. Groups of animals previously placed under the term "Reptilia" are considered to be paraphyletic. Therefore, we have attempted to focus this work on the earliest amniotes and their immediate sister groups. In order to provide context, however, neontological examples are taken from extant animals that may of necessity be far removed phylogenetically from this transition. Where inferences are drawn, chapter authors have stated their assumptions explicitly. Laurin and Reisz examine current hypotheses about the phylogeny of tetrapods in "A New Perspective on Tetrapod Phylogeny," as do Lee and Spencer in "Crown-Clades, Key

Characters and Taxonomic Stability: When Is an Amniote Not an Amniote?" Significantly, the first chapter suggests that extant lissamphibians may be more closely related to early amniotes than was previously supposed, thus making them potentially much more useful surrogates for testing physiological hypotheses. In "Biogeography of Primitive Amniotes," Berman, Sumida, and Lombard update the biogeographic context for the evolution of early amniotes and point out the importance of Pangea and plate migration in the spread of tetrapod groups.

The physical and physiological transition to land began long before the origin of amniotes. Some of the most primitive fishes may have had lungs for air breathing (Randall *et al.*, 1981). Many primitive tetrapods were semi-terrestrial and may have spent part of their lives on land (Smithson, 1980), perhaps even breeding terrestrially (Carroll, 1988). Their eyes, ears, and lateral lines must have adjusted to the medium of air instead of water (Duellman and Trueb, 1986). These changes may be considered exaptations (Gould and Vrba, 1982). Insects and plants completed the transition to land before the vertebrates came fully ashore, providing habitat, cover, and food for the explorers of this brave new world, but a diet of high-fiber plants probably required changes in digestive physiology (Diamond and Buddington, 1987). Habitat and climate are discussed in the chapters "Biogeography of Primitive Amniotes" and "The Late Paleozoic Atmosphere and the Ecological and Evolutionary Physiology of Tetrapods" by Graham, Aguilar, Dudley and Gans. Lauder and Gillis explore ways that these animals exploited food resources in a chapter on "Origin of the Amniote Feeding Mechanism: Experimental Analysis of Outgroup Clades." Hotton, Olson, and Beerbower take a parallel approach with extinct taxa in the chapter "Amniote Origins and the Discovery of Herbivory." Movement on land required adjustments to the axial and appendicular skeleton (Sumida and Lombard, 1991), discussed in the chapter "Locomotor Features of Taxa Spanning the Origin of Amniotes" by Sumida.

The evolution of an amniotic egg was a key innovation (Thomson, 1992) that defines this clade and provided one of the means of invasion of a new habitat. Aspects of this process are included in the chapters by Packard and Seymour in "Evolution of the Amniote

Egg" and by Stewart in "Morphology and Evolution of the Egg of Oviparous Amniotes." However, the differences between the adults of the earliest amniote and its immediate anamniote "predecessor" may have been minor and apparently unremarkable at the time of the transition, and indeed, the exact identity of the earliest amniotes may still be in doubt (Romer, 1966; Carroll, 1969a,b; Lombard and Sumida, 1992; Smithson *et al.*, 1994).

The differences between present-day lissamphibians and extant "reptiles" go far beyond the evolution of the amniotic egg. It has been suggested that the group Reptilia should not be defined solely on the basis of morphology but also on shared life history and physiological traits (Gans and Pough, 1982). Such traits may permit wide variation in structure. We suggest that this may also be true for the "Amphibia," and that the origin of amniotes may have been a step that left few tangible morphological clues but engendered enormous physiological changes that profoundly altered the course of evolutionary history. These differences are examined as clues to past physiological evolution in the chapters on "The Role of the Skin in the Origin of Amniotes: Permeability Barrier, Protective Covering, and Mechanical Support" by Frolich and "Water Balance and the Physiology of the Amniote Transition" by Martin and Nagy. In the final chapter, "Reconstructing Ancestral Trait Values Using Squared-Change Parsimony: Plasma Osmolarity at the Amniote Transition," Garland, Martin, and Diaz-Uriarte use physiological data to test a historical hypothesis with a phylogenetic analysis.

We are attempting to foster a closer communication between experimental physiologists and morphologists who work on testable models and the paleontologists who study the fossils of animals that lived at the time of the origin of amniotes. The work of each group has implications for the other, and we provide here a framework for that cooperation. We intend this gathering of paleontological and neontological specialists to provide an integrated benchmark collection of studies that will be of use to students of vertebrate paleontology, physiology, herpetology, functional morphology, evolution, and vertebrate biology in general. Furthermore we hope that it will stimulate new work in this long-neglected area. This collaborative effort is an optimistic beginning. As the biological sciences advance with new techniques and integrated approaches, we are hopeful that this volume will invigorate the cooperative analysis, by students of the past and present, of an event that took place at least 350 million years ago.

ACKNOWLEDGMENTS

The authors thank Drs. David Polcyn, Elizabeth Rega, and Thomas Vandergon for reading the text and providing useful comments on its form and construction.

LITERATURE CITED

Berman, D. S, S. S. Sumida, and R. E. Lombard. 1992. Reinterpretation of the temporal and occipital regions in *Diadectes* and the relationships of diadectomorphs. *Journal of Paleontology*, 66:481-499.

Burggren, W. W., and W. Bemis. 1990. Studying physiological evolution: paradigms and pitfalls. Pages 191-228 in: *Evolutionary Innovations* (M. H. Nitecki, ed.). Chicago: The University of Chicago Press.

Carroll, R. L. 1969a. Origin of reptiles. Pages 1-44 in: *Biology of the Reptilia, Volume 1, Morphology* (C. Gans, A. d'A. Bellairs, and T. S. Parsons, eds.). New York: Academic Press.

Carroll, R. L. 1969b. Problems of the origin of reptiles. *Biological Reviews*, 44:393-432.

Carroll, R. L. 1970. The ancestry of reptiles. *Philosophical Transactions of the Royal Society of London* B, 257:267-308.

Carroll, R. L. 1988. *Vertebrate Paleontology and Evolution.* New York: William H. Freeman and Company.

Carroll, R. L. 1995. Problems of the phylogenetic analysis of paleozoic choanates. *Bulletin of the Museum of Natural History, Paris*, 17:389-445.

Diamond, J. M., and R. K. Buddington. 1987. Intestinal nutrient absorption in herbivores and carnivores. Pages 193- 203 in: *Comparative Physiology: Life in Water and on Land* (P. Dejours, L. Bolis, C. R. Taylor, and E. R. Weibel, eds.). Berlin: Springer-Verlag, Liviana Press.

Duellman, W. E., and L. Trueb. 1986. *Biology of Amphibians*. New York: McGraw Hill.

Gans, C., and F. H. Pough. 1982. Physiological ecology: Its debt to reptilian studies, its value to students of reptiles. Pages 1-13 in: *Biology of the Reptilia, Volume 12: Physiological Ecology*. New York: Academic Press.

Garland, T. Jr., and P. A. Carter. 1994. Evolutionary Physiology. *Annual Review of Physiology*, 56:579-621.

Gauthier, J. A., A. G. Kluge, and T. Rowe. 1988a. The early evolution of the Amniota. Pages 103-155 in: *The Phylogeny and Classification of the*

Tetrapods, Volume 1, Amphibians, Reptiles, and Birds, Systematics Association Special Volume No. 35A (M. J. Benton, ed.). Oxford: Clarendon Press.

Gauthier, J., A. J. Kluge, and T. Rowe. 1988b. Amniote phylogeny and the importance of fossils. *Cladistics*, 4:105-209.

Gould, S. J., and E. S. Vrba. 1982. Exaptation -- a missing term in the science of form. *Paleobiology*, 8:4-15.

Heaton, M. J. 1980. The Cotylosauria: a reconsideration of a group of archaic tetrapods. Pages 497-551 in: *The Terrestrial Environment and the Origin of Land Vertebrates*, Systematics Association Special Volume No. 15 (A. L. Panchen, ed.). London: Academic Press.

Heaton, M. J., and R. R. Reisz. 1986. Phylogenetic relationships of captorhinomorph reptiles. *Canadian Journal of Earth Sciences*, 23:402-418.

Holmes, R. 1984. The Carboniferous amphibian *Proterogyrinus scheelei* Romer, and the early evolution of tetrapods. *Philosophical Transactions of the Royal Society of London* B, 306:431-527.

Hotton, N. 1992. Global distribution of terrestrial and aquatic tetrapods, and its relevance to the position of the continental masses. Pages 267-285 in: *New Concepts in Global Tectonics* (Chatterjee, S. and N. Hotton III, eds.). Lubock: Texas Tech University Press:.

Lauder, G. V. 1981. Form and function: structural analysis in evolutionary morphology. *Paleobiology*, 7:430-442.

Laurin, M., and R. R. Reisz. 1995. A reevaluation of early amniote phylogeny. *Zoological Journal of the Linnean Society*, 113:165-223.

Lee, M. S. Y. 1993. The origin of the turtle body plan: bridging a famous morphological gap. *Science*, 261:1716-1720.

Lee, M. S. Y. 1995. Historical burden in systematics and the interrelationships of 'parareptiles'. *Biological Reviews*, 70:459-547.

Lombard, R. E., and S. S. Sumida. 1992. Recent progress in understanding early tetrapods. *American Zoologist*, 32:609-622.

Randall, D. J., W. W. Burggren, A. P. Farrell, and M. S. Haswell. 1981. *The evolution of air breathing in vertebrates*. Cambridge University Press: Cambridge, United Kingdom.

Romer, A. S. 1966. *Vertebrate Paleontology*, Third Editon. Chicago: The University of Chicago Press.

Smithson, T. R. 1980. An early tetrapod fauna from the Namurian of Scotland. Pages 407-438 in: *The Terrestrial Environment and the Origin of Land Vertebrates* (A. L. Panchen, ed.). London: Academic Press.

Smithson, T. R., R. L. Carroll, R. L. Panchen, and S. M. Andrews. 1994. *Westlothiana lizziae* from the Visean of East Kirkton, West Lothian, Scotland, and the amniote stem. *Transactions of the Royal Society of Edinburgh: Earth Sciences*, 84: 383-412.

Sumida, S. S., and R. E. Lombard. 1991. The atlas-axis complex in the late Paleozoic genus *Diadectes* and the characteristics of the atlas-axis complex across the amphibian to amniote transition. *Journal of Paleontology*, 65: 973-983.

Thomson, K. E. 1992. Macroevolution: The morphological problem. *American Zoologist*, 32: 106-112.

CHAPTER 2

A NEW PERSPECTIVE ON TETRAPOD PHYLOGENY

Michel Laurin

Robert R. Reisz

INTRODUCTION

The origin of amniotes has always attracted much attention, and early studies proposed several potential close relatives of amniotes. Most notable among these are the articles by Carroll, who, following Romer (1950), proposed that anthracosaurs, such as gephyrostegids and *Solenodonsaurus*, were the most plausable ancestors of amniotes (Carroll, 1969, 1970a). Recent discussions largely agreed with Carroll and focused almost exclusively on diadectomorphs and seymouriamorphs (Gauthier *et al.*, 1988; Laurin and Reisz, 1995). Similarly, several potential relatives of modern amphibians have been suggested (Parsons and Williams, 1963; Carroll and Currie, 1975; Carroll and Holmes, 1980) but most studies have only dealt with temnospondyls (Bolt, 1979; Trueb and Cloutier, 1991). Milner (1993) was one of the first to compare hypotheses on the origin of lissamphibians in which temnospondyls and lepospondyls were considered.

Relatively few large-scale phylogenetic studies on early tetrapods have been performed, and the origin of amniotes has generally been studied separately from the origin of lissamphibians. Studies on the origin of amniotes usually only dealt with anthracosaurs (Carroll, 1970a; Gauthier *et al.*, 1988) because these have been presumed to be the closest relatives of amniotes. A consensus emerged that Diadectomorpha is the sister-group of Amniota whereas seymouriamorphs are among the next closest relatives of amniotes (Panchen and Smithson, 1988; Laurin and Reisz, 1995). Research on the origins of lissamphibians concluded that "dissorophoids" were the closest relatives of lissamphibians (Trueb and Cloutier, 1991).

The latest and most comprehensive study on the origin of amniotes was performed by Carroll (1995) and was consulted extensively for the present study. Most of the characters used in this study (Appendix 1) were taken from or modified from Carroll's insightful manuscript and from other phylogenetic studies (Gauthier *et al.*, 1988; Milner, 1988; Panchen and Smithson, 1988; Trueb and Cloutier, 1991), but all were recoded from specimens or the descriptive literature. In addition to including early amniotes and anthracosaurs, Carroll (1995) considered lepospondyls and temnospondyls. This inclusive and judicious choice of taxa yielded new insights into the origin of amniotes, such as the hypothesis that lepospondyls are closely related to diadectomorphs and amniotes. If this is accurate, seymouriamorphs are not as closely related to amniotes as previously believed. This conclusion is relevant to studies on the origin and early evolution of amniotes because using lepospondyls instead of seymouriamorphs as a second outgroup could affect the polarity of several characters.

Few phylogenetic studies considered the origin of amniotes and lissamphibians in the context of a global analysis of tetrapod relationships. However, this approach is necessary in order to

---

Figure 1. Most parsimonious tree. The nodes are numbered in alphabetical order from the in-group node up. When a name is given, the letter designating the node is omitted but it is still reserved for this node. Several clades are named in the text but could not be labeled on the figure. These include the following clades: K (Embolomeri), N (Seymouriamorpha), Y (Apoda), Z (Gymnophiona), AA (Salientia), AB (Anura) and AD (Diadectomorpha). ➔

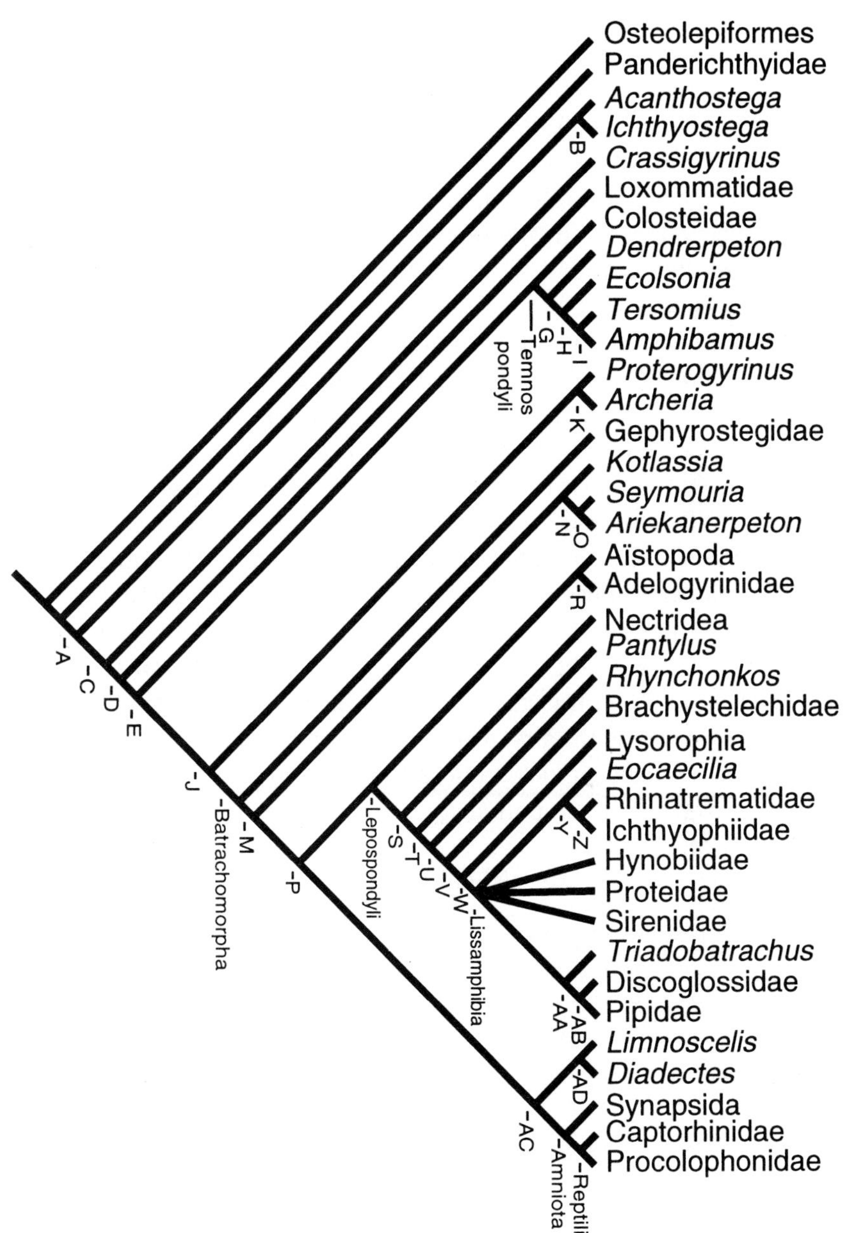
Osteolepiformes
Panderichthyidae
Acanthostega
Ichthyostega
Crassigyrinus
Loxommatidae
Colosteidae
Dendrerpeton
Ecolsonia
Tersomius
Amphibamus
Proterogyrinus
Archeria
Gephyrostegidae
Kotlassia
Seymouria
Ariekanerpeton
Aïstopoda
Adelogyrinidae
Nectridea
Pantylus
Rhynchonkos
Brachystelechidae
Lysorophia
Eocaecilia
Rhinatrematidae
Ichthyophiidae
Hynobiidae
Proteidae
Sirenidae
Triadobatrachus
Discoglossidae
Pipidae
Limnoscelis
Diadectes
Synapsida
Captorhinidae
Procolophonidae
-A
-B
-C
-D
-E
-G
-H
-I
-Temnospondyli
-J
-K
-M
-N
-O
-P
-R
-S
-T
-U
-V
-W
-Y
-Z
-AA
-AB
-AC
-AD
-Batrachomorpha
-Lepospondyli
-Lissamphibia
-Amniota
-Reptilia

determine which early terrestrial choanates are relatives of amniotes,which are relatives of lissamphibians, and which are relatives of the crown group that has been called Tetrapoda (Gauthier *et al.*, 1989). The first large-scale phylogenetic study of tetrapods based on a data matrix (Fracasso, 1983) did not gain general acceptance and only a condensed tree without a supporting matrix was published (Fracasso, 1987). Panchen and Smithson (1988) published the first discussion of tetrapod phylogeny that was extensively supported by polarized characters. Panchen and Smithson concluded that lissamphibians are part of Temnospondyli. According to them, successively more remote relatives of this clade include Microsauria, Colosteidae, Nectridea, and Ichthyostegidae. They considered Diadectomorpha to be the sister group of Amniota, and that Seymouriamorpha, Anthracosauroideae, *Crassigyrinus*, and Loxommatoidea were related to this clade. Therefore, all known terrestrial Paleozoic choanates except ichthyostegids were part of the crown group defined by lissamphibians and amniotes, according to Panchen and Smithson. This implied that the evolutionary lineage leading to amniotes had diverged from the amphibian lineage in the Devonian. Panchen and Smithson's phylogeny was probably the best supported phylogeny of tetrapods, but it was not based on a data matrix (or the matrix was not published). The present study is the first discussion of tetrapod evolution based on a computer-assisted phylogenetic analysis of a data matrix including both amniotes and lissamphibians.

METHODS

Representative taxa of all the major groups of Paleozoic choanates and modern amphibians were selected (Appendix 2). The terminal taxa are restricted to small, well-defined groups whose monophyly is well established. This procedure allows the monophyly of large taxa to be tested and avoids the problematic reconstruction of the primitive condition of relatively diverse taxa. Therefore, taxa such as Seymouriamorpha and Microsauria are not included as terminal taxa but are represented by a sample of genera (Fig. 1). Thirty-eight taxa (including two outgroups) and 157 characters (Appendix 1) were used in this study. Polymorphism within terminal taxa was coded when the primitive condition for a taxon could not be reconstructed with confidence. Partial uncertainty (when a taxon can possess a subset of the states only) was also coded. PAUP 3.1.1 and MacClade 3.0 were used to analyze the data. Because MacClade allows partial uncertainty and polymorphism in a matrix but PAUP does not, the tree lengths reconstructed by these two programs do not coincide and the numbers given below were obtained from MacClade (Maddison and Maddison, 1992). In PAUP, multistate taxa were interpreted as polymorphism (rather than uncertainty). Interpreting the multi state taxa as uncertainty yields the same phylogeny. All characters were equally weighted and unordered. Characters were optimized by the DELTRAN algorithm of PAUP 3.1.1 (Swofford, 1993). Alternate reconstructions (when the optimization is ambiguous) were examined using MacClade 3.0. All the polytomies present in the consensus trees discussed are interpreted as soft. Therefore, character optimizations close to Lissamphibia were examined in MacClade 3.0 and differ somewhat from the optimizations given by PAUP 3.1.1 because the latter supports only hard polytomies. MacClade was also used to examine character optimization at the base of the tree. Therefore, the list of autapomorphies of nodes A-C (Appendix 3) do not follow exactly the PAUP output. The heuristic search algorithm was used with the following parameters: seven trees were kept in the stepwise addition sequence, 20 iterations were done using the random addition sequence, the tree bisection-reconnection swapping algorithm was chosen, zero-length branches were collapsed, and all the most parsimonious trees were saved.

Because of time limitations and the extensive amount of time required to analyze large data matrices, some taxa are only represented by a few members when many more would have been required to illustrate the full range of morphological variations present in a taxon. This is especially true of temnospondyls, amniotes, microsaurs, and lissamphibians.

Temnospondyls are extremely speciose and morphologically diverse (Milner, 1990). Two early, basal temnospondyl taxa (Colosteidae and *Dendrerpeton*) are included to allow determination of the affinities of temnospondyls to other groups of Paleozoic choanates (Fig. 2), whereas three dissorophoids are included to test the hypothesis that temnospondyls gave rise to lissamphibians (Figs. 3 and 4). Three dissorophoids were required because *Amphibamus* and *Tersomius* have been extensively compared to lissamphibians but are relatively poorly known (Carroll, 1964), whereas *Ecolsonia* has not been compared as frequently but is a much better known genus (Berman *et al.*, 1985). No branchiosaur was included in this study because they are probably based on immature specimens (Schoch, 1992).

Amniotes are also very diverse and speciose, and their phylogeny and anatomy have been studied in detail (Laurin and Reisz, 1995). Therefore, this analysis will only consider three taxa (synapsids, procolophonids, and captorhinids) representing the main groups of early amniotes (synapsids, anapsids, and romeriids in the terminology of Gauthier *et al.*, 1988).

Eleven families and two suborders of microsaurs were recognized by Carroll and Gaskill (1978). Three of the best known families and both suborders are represented in this analysis. *Pantylus* and *Rhynchonkos* (the sole representatives of Pantylidae and Rhynchonkidae, respectively) represent the Tuditanomorpha (which includes 7 families) whereas Brachystelechidae represent the Microbrachomorpha (known by 4 families). These taxa were chosen

Figure 2. Skulls of terrestrial choanates in dorsal view. A, *Ichthyostega* sp.; B, *Megalocephalus pachycephalus* (Loxommatidae); C, *Greererpeton burkemorani* (Colosteidae); D, *Tersomius texensis*; E, *Proterogyrinus scheelei*; F, *Seymouria baylorensis*; G, *Limnoscelis paludis*; H, *Captorhinus laticeps* (Captorhinidae). Most of the illustrations were redrawn from the following sources: A, Carroll (1988); B, Beaumont (1977); C, Smithson (1982); D, Carroll (1964); E, Holmes (1984); G, Fracasso (1983); H Heaton (1979). In this figure and all the following, the skulls were drawn at various scales to make their lengths comparable. Scale bar equals 1 cm. Abbreviations used in Figs. 2-4: apf, anterior palatal fenestra; at, anterior tectal; bc, braincase; bo, basioccipital; ch, choana; ep, epipterygoid; ex, exoccipital; f, frontal; fo, fenestra ovalis; fp, frontoparietal; in, internasal; it, intertemporal; j, jugal; l, lacrimal; m, maxilla; mp, maxillopalatine; n, nasal; nc, nasal capsule; o, orbit; ob, os basale; oc, otic capsule; op, opisthotic; os, orbitosphenoid, p, parietal; pal, palatine; pf, postfrontal; pl, pleurosphenoid; pm, premaxilla; po, postorbital; pop, preopercular; pp, postparietal; prf, prefrontal; pro, pro-otic; ps, parasphenoid; pt, pterygoid; q, quadrate; qj, quadratojugal; s, stapes; sm, septomaxilla; so, supraoccipital; sph, sphenethmoid; sq, squamosal; st, supratemporal; t, tabular; tf, tentacular fossa; v, vomer. ➔

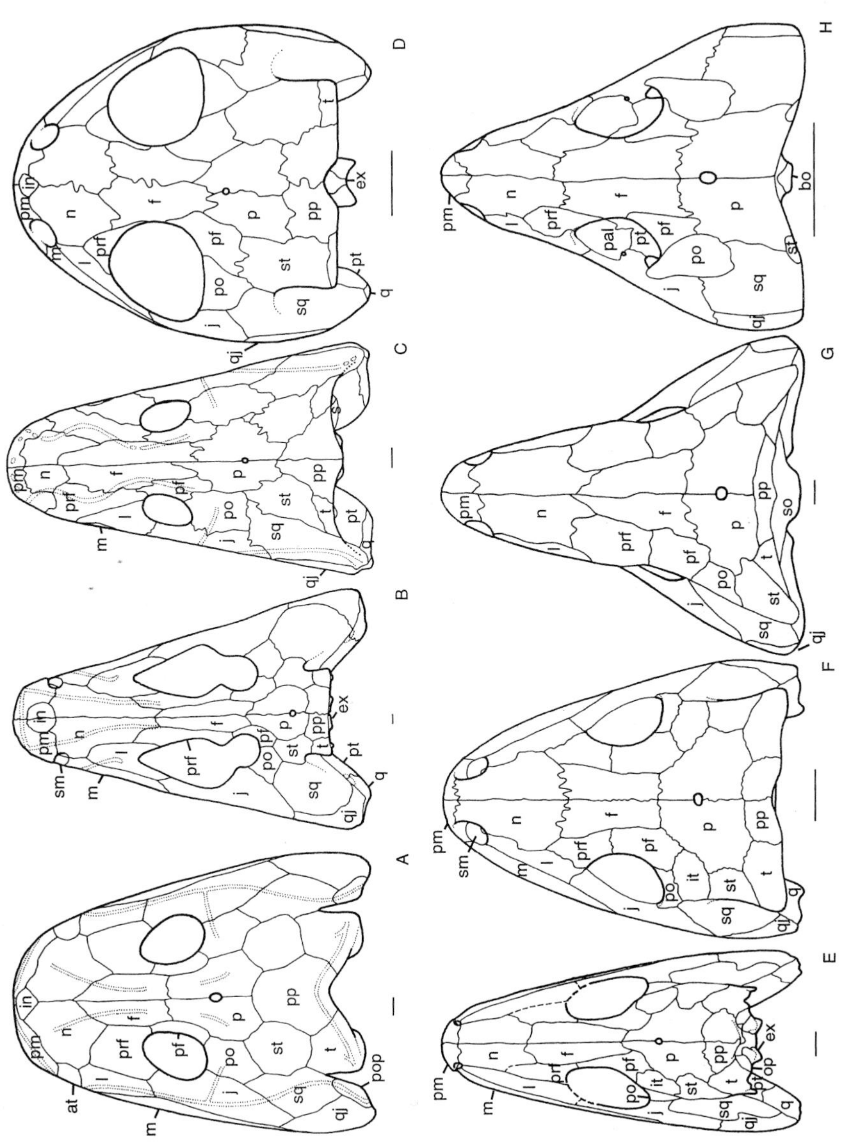
A
B
C
D
E
F
G
H

because they are relatively well known and because *Rhynchonkos* has been closely allied to apodans (Carroll and Currie, 1975).

The three orders of lissamphibians are represented by three terminal taxa each. *Eocaecilia*, the newly described Jurassic apodan, was included because it retains several primitive characters lost in all other known apodans (Jenkins and Walsh, 1993). Rhinatrematidae and Ichthyophiidae represent gymnophiones because these two families are believed to have diverged early from other caecilians (Duellman and Trueb, 1986). Urodeles have a poor fossil record and are represented by three extant families: Hynobiidae, Sirenidae, and Proteidae. The first two families were included because they are believed to have diverged early from other urodeles, whereas proteids are well known (Duellman and Trueb, 1986). Salientia is represented by *Triadobatrachus* (the oldest known salientian) and two basal, extant families (Discoglossidae and Pipidae).

Much of the data come from descriptive literature, but specimens of many taxa were examined, including *Amphibamus*, Lysorophia, *Diplocaulus* (Nectridea), *Seymouria*, *Ariekanerpeton*, *Alytes* (Discoglossidae), *Pipa*, *Xenopus* (Pipidae), *Hynobius* (Hynobiidae), *Necturus* (Proteidae), *Siren*, *Pseudobranchus* (Sirenidae), *Epicrionops* (Rhinatrematidae), *Ichthyophis* (Ichthyophiidae), *Diadectes*, Synapsida, Captorhinidae, and Procolophonidae.

Space limitations prevent us from listing all the primary literature used to gather the anatomical data used in the analysis, but many of these references can be found in Carroll and Gaskill (1978), Duellman and Trueb (1986), Reisz (1986), Carroll (1988), Gauthier *et al.* (1988), Trueb and Cloutier (1991), Clack (1994), and Laurin and Reisz (1995).

Some characters were problematic because of the wide morphological diversity of the taxa. For instance, all lepospondyls except for some aïstopods and some nectrideans have lost a temporal bone (Fig. 3). We have considered the remaining element to be a tabular because this is how it has usually been interpreted in microsaurs and in lysorophians (Carroll and Gaskill, 1978; Wellstead, 1991). Therefore, we assumed the loss of a temporal element in lepospondyls to be homologous. We have also assumed that the tabular-squamosal element of adelogyrinids was a squamosal because the tabular has been lost more frequently in various tetrapods than the squamosal. The composition of the centrum (character 102 in the appendices) of lepospondyls and lissamphibians has been debated (Wake and Lawson, 1973) and partial uncertainty was coded when a single element was present throughout the column. Therefore, two conditions were entered for such taxa: centrum composed of a large, circular intercentrum (1), and centrum composed of a large pleurocentrum (5). This was done in an attempt to determine the composition of the centrum in these taxa by looking at the optimization of this character on the most parsimonious tree.

The robustness of the clades was established by determining how many extra steps were required to collapse the nodes. A similar procedure has occasionally been used in paleontological studies (Laurin and Reisz, 1995), but it has been first and most often employed by botanists, who called it "decay analysis", "decay index", or "Bremer support" (Mishler, 1994).

RESULTS

Only three most parsimonious tree were found. They require 654 steps and have a consistency index of 0.52. The character optimizations and discussions below are all based on the strict consensus tree (Fig. 1). The consensus tree differs substantially from the commonly accepted phylogeny of tetrapods and has important implications for hypotheses of amniote origins. For instance, Lepospondyli includes Lissamphibia and is closely related to diadectomorphs and amniotes. Previously, seymouriamorphs were considered to be among the closest known relatives of amniotes (Gauthier *et al.*, 1988). In the new phylogeny, Diadectomorpha, Amniota and Lepospondyli form the crown group of terrestrial choanates.

Gauthier *et al.* (1989) defined Tetrapoda as the crown-group of terrestrial choanates. According to Gauthier *et al.* (1989), Tetrapoda included temnospondyls, embolomeres, seymouriamorphs, diadectomorphs, and amniotes because temnospondyls were thought to have given rise to modern amphibians, whereas embolomeres and seymouriamorphs were thought to be closely related to amniotes.

The new phylogeny changes drastically the composition of this crown group, but the taxonomic implications of this fact will be dealt with elsewhere. In this study, the crown group of terrestrial choanates will simply be called "crown batrachomorphs". Batrachomorpha Säve-Söderbergh, 1934, is here defined as all choanates sharing a more recent ancestry with amniotes than with embolomeres. This taxon resembles superficially Batrachosauria as defined by Gauthier *et al.* (1988), but it is distinct because Batrachomorpha is not part of Anthracosauria as defined by Gauthier *et al.* (1988), whereas Batrachosauria is. Under the phylogeny presented by Gauthier *et al.* (1988), Batrachomorpha would be a junior synonym of Batrachosauria, but under the new phylogeny Batrachosauria is a junior synonym of Anthracosauria and it cannot be used. Seymouriamorphs, gephyrostegids, Embolomeri, Temnospondyli, loxommatids, *Crassigyrinus*, and the Devonian choanates *Ichthyostega* and *Acanthostega* are successive sister groups of crown batrachomorphs.

The phylogenetic analysis allows a new list of autapomorphies of the included taxa to be given (Appendix 3). The new phylogeny also warrants a new taxonomy of tetrapods, but this will be done elsewhere. Most nodes in the tree are designated by one or two letters (Fig. 1), but previously defined taxa are labeled whenever possible.

DISCUSSION

Comparisons with Previous Phylogenies

The close relationships between amniotes and lissamphibians is probably the most unorthodox result of this analysis. However, close relationships between some lepospondyls and amniotes have been postulated in the past. Vaughn (1962) suggested that "microsaurs" (probably a paraphyletic group) were closely related to "captorhinomorphs" (another paraphyletic group including early relatives of diapsids). Carroll (1995) produced a phylogeny in which lepospondyls were closely related to amniotes. Lepospondyls have repeatedly been argued to be a paraphyletic group (Carroll and Gaskill, 1978; Milner, 1993; Panchen and Smithson, 1988). However, these arguments were based more on suspected convergence between various lepospondyls than on similarities between some lepospondyls and other terrestrial vertebrates.

The only study to which the present phylogeny can be easily compared is Carroll's (1995) phylogenetic analysis of Paleozoic choanates. Most of the eighteen taxa studied by Carroll are also included in the present study, and thus the results are readily compared. Carroll found four equally most parsimonious trees but did not give a consensus tree of these. We produced a strict consensus of Carroll's trees (Fig. 5). Its topology is very similar to the shortest tree found in this study. Indeed, most of the clades found by Carroll have also been found in this study. The only exception consists in the clade including lysorophids, adelogyrinids, and aïstopods found by Carroll (1995). In this study, nectrideans and various microsaurs seem to share more synapomorphies with lysorophids than with adelogyrinids and aïstopods. Of course, these results are difficult to compare because the present study includes three clades of microsaurs instead of Microsauria as a terminal taxon, and Carroll omitted lissamphibians from his study. Otherwise, the agreement between these two studies is remarkable and may result partly from a similar choice of characters.

The inclusion of lissamphibians in the matrix was essential to this analysis. Without modern amphibians, the phylogeny could have been interpreted as suggesting that lepospondyls were early "reptiliomorphs." Such a conclusion would have been erroneous, even if it had been based on an accurate phylogeny. Therefore, studies of the origin of amniotes should not ignore the position of modern amphibians.

Decay Index and Robustness of the Clades

In most previous phylogenetic studies of tetrapods, the strength of various clades was assessed by counting the number of synapomorphies linking its members (Gardiner, 1982; Gauthier *et al.*, 1988; Panchen and Smithson, 1988). However, such an approach is fundamentally incomplete in that it does not consider how many characters support alternative clades. Such homoplasy can seriously weaken the strength of a clade and only a few extra steps may be required to collapse a clade supported by several characters. This has been recently recognized and some studies have discussed characters supporting alternative clades to provide a more realistic approach is insufficient because it may not give a global assessment of the parsimony of the alternatives. For instance, Milner (1993) listed the characters thought to support close affinities between dissorophoids and lissamphibians and the characters supporting close affinities between some lepospondyls and l assessment of the support for various taxa (Milner, 1993). However, even such an lissamphibians. In the present study, an equivalent procedure would have consisted in listing the characters supporting the Lepospondyli or one of its subgroups (as redefined here) and comparing this to a list of synapomorphies linking dissorophoids to lissamphibians. Such an approach only considers the autapomorphies of one node without considering the optimization of all the other characters on the rest of the tree. An even more rigorous approach is to determine exactly how many extra steps are required to collapse the nodes in a phylogeny (Laurin and Reisz, 1995) This number has often been called the decay index (Mishler, 1994). The higher the decay index, the more robust is the node. This method takes into consideration homoplasy and even characters whose optimization is ambiguous. This approach was applied here for selected clades.

Several clades in the new phylogeny are only weakly supported and require only one extra step to collapse (nodes B, I, S, V, Y, and Diadectomorpha). The small amount of support for some of these nodes probably results from the lack of available anatomical data. For instance, nectrideans are a fairly diverse group, but no genus is well known. *Diploceraspis* is represented by well-preserved cranial remains (Beerbower, 1963) but few postcranial remains. Urocordylids are known from several skeletons, but their skulls are poorly preserved and the braincase is often cartilaginous (Bossy, 1976).

Figure 3. Skulls of lepospondyls in dorsal view. A, *Sauropleura pectinata* (Nectridea); B, *Pantylus cordatus*; C, *Rhynchonkos stovalli*; D, *Brachydectes elongatus* (Lysorophia); E, *Epicrionops petersi* (Rhinatrematidae); F, *Siren lacertina* (Sirenidae); and G, *Pipa pipa* (Pipidae). Cartilage is stippled. Some of the illustrations were redrawn from the following sources: A, Bossy (1976); B, Romer (1969); C, Carroll and Currie (1975); D, Wellstead (1991); E, Nussbaum (1977). Scale bar = 1 cm unless otherwise indicated. See the legend to figure 2 for abbreviations. ➔

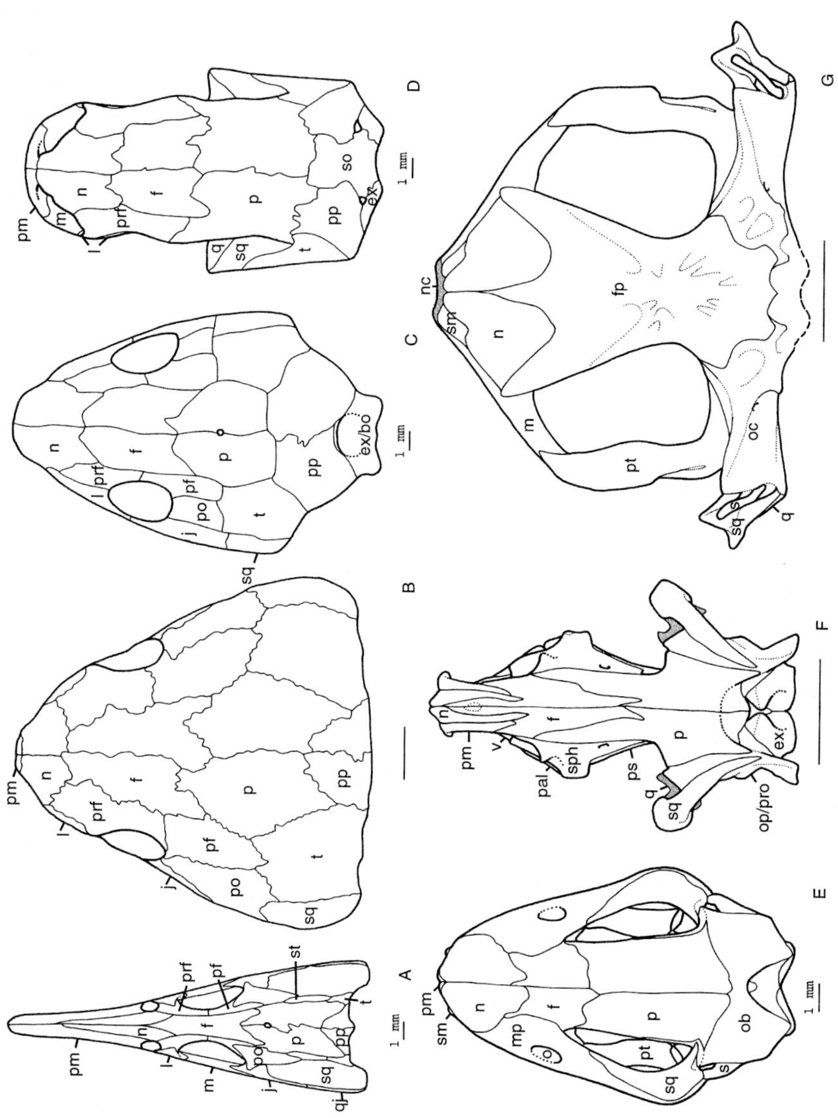
A
pm
n
l
m
j
qj
prf
pf
f
p
po
sq
pp
st
t
1 mm
B
pm
n
l
prf
f
j
pf
p
po
t
sq
pp
C
n
l
prf
f
pf
p
j
po
t
sq
pp
ex/bo
1 mm
D
pm
n
l
m
prf
f
p
q
sq
t
pp
so
ex
1 mm
E
pm
sm
n
mp
f
p
pt
sq
s
ob
1 mm
F
pm
n
v
f
pal
sph
ps
p
q
sq
op/pro
ex
G
nc
sm
n
fp
m
pt
oc
sq
s
q

The apparent paraphyly (or polyphyly) of urodeles is surprising. Urodeles are perhaps not usually perceived as poorly known, but relatively few descriptions of the osteology of the basal families are available, and specimens of these families are not widely available (Duellman and Trueb, 1986). Furthermore, the affinities of urodeles and their phylogeny have often been discussed, but their monophyly has seldom been rigorously tested. Some soft anatomical, biochemical, or behavioral autapomorphies of urodeles may be known, but such characters could not be incorporated into the present study because most of the included taxa are extinct. Therefore, the apparent paraphyly of urodeles is probably an artifact and it is hoped that the present study will stimulate the search for more osteological diagnostic characters of urodeles and publications of osteological descriptions. We do not believe urodeles to be paraphyletic, although this is a remote possibility. Furthermore, the main reason for including lissamphibians in this analysis was to evaluate their relationships to amniotes rather than to study the phylogeny within Lissamphibia. In order to deal adequately with the last topic, many more representatives of extant amphibians should be included.

Several clades are strongly supported. Among the most robust are Seymouriamorpha (node N, seven extra steps), node P (crown batrachomorphs, eight extra steps), Lepospondyli (seven extra steps), node U (five extra steps), node W (five extra steps), and Lissamphibia (five extra steps). Fortunately, these nodes are also among the most important clades in this phylogeny. The strength of these nodes is not surprising considering that the origin of all these clades appears to be associated with a radical shift in morphology.

Figure 4. Skulls of terrestrial choanates in occipital view. A, *Megalocephalus pachycephalus* (Loxommatidae); B, *Greererpeton burkemorani* (Colosteidae); C, *Tersomius texensis*; D, *Proterogyrinus scheelei*; E, *Seymouria baylorensis*; F, *Limnoscelis paludis*; G, *Captorhinus laticeps* (Captorhinidae); H, *Pantylus cordatus*; I, *Rhynchonkos stovalli*; J, *Brachydectes elongatus* (Lysorophia); K, *Siren lacertina* (Sirenidae); L, *Pipa pipa* (Pipidae). For credits, see the legends to Figures 2 and 3, except for F, redrawn from Fracasso (1983). Scale bar = 1 cm unless otherwise indicated. See the legend to Figure 2 for abbreviations. ➔

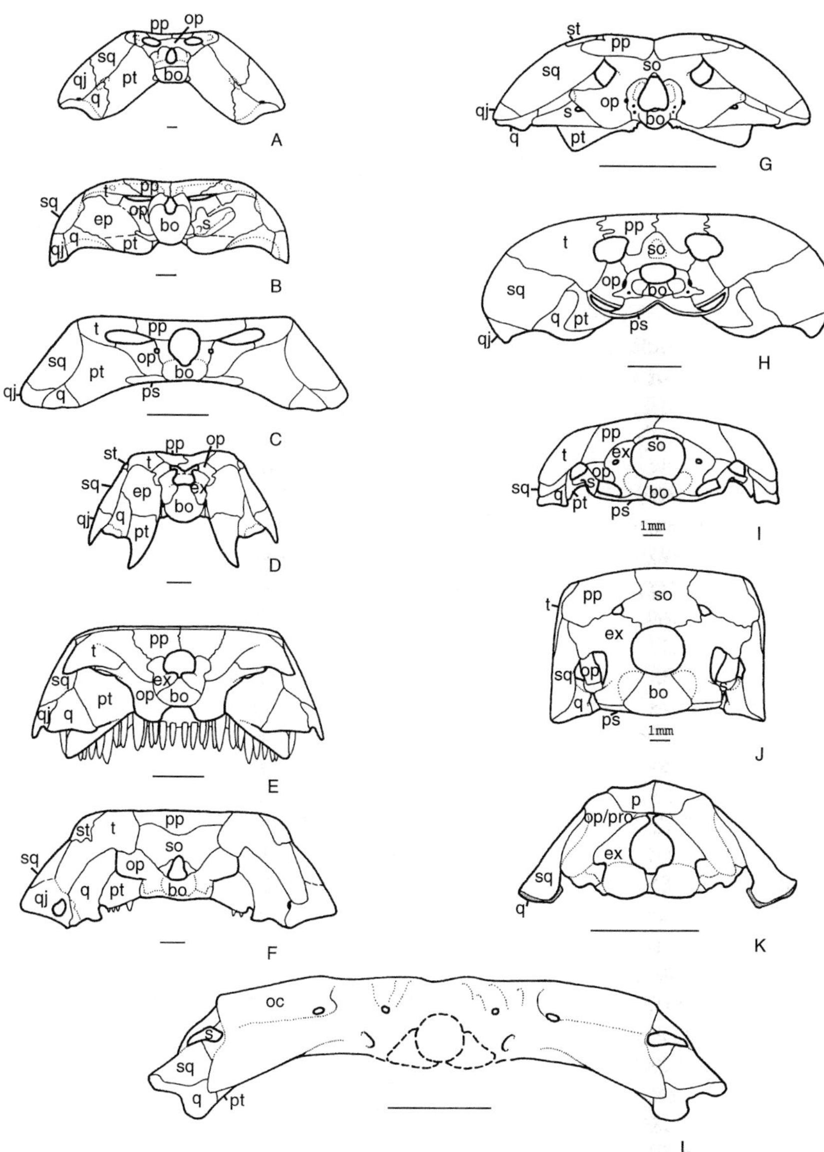
pp
op
t
sq
qj
pt
q
bo
A
st
pp
so
sq
qj
s
op
bo
q
pt
G
sq
t
pp
op
ep
bo
s
qj
q
pt
B
pp
t
so
sq
op
bo
q
pt
ps
qj
H
t
pp
sq
pt
op
bo
qj
q
ps
C
st
pp
op
t
sq
ep
ex
bo
qj
q
pt
D
pp
ex
so
t
op
sq
q
s
bo
pt
ps
1mm
I
pp
t
sq
ex
pt
op
bo
qj
q
E
t
pp
so
ex
sq
op
bo
q
ps
1mm
J
st
t
pp
so
sq
op
qj
q
pt
bo
F
p
op/pro
ex
sq
q
K
oc
s
sq
q
pt
L

Origin of Lissamphibians

One of the most surprising results of this analysis is that Lissamphibia seems to be only remotely related to dissorophoids. Most of the discussions during the past quarter century on the origins of modern amphibians have focused on temnospondyls, especially dissorophoids (Bolt, 1969, 1977, 1979; Bolt and Lombard, 1985; Milner, 1988; Trueb and Cloutier, 1991). The results of our analysis suggest that lissamphibians are lepospondyls. Indeed, Lepospondyli, as redefined here, is one of the most robust nodes and requires seven extra steps to collapse. Lepospondyl affinities for lissamphibians have been suggested before (Gregory, 1965), but such views never gained much popularity. Carroll and Currie (1975) suggested that *Rhynchonkos* was closely related to apodans, whereas Carroll and Holmes (1980) argued that urodeles evolved from microsaurs. Milner (1993) compared support for the hypothesis that all lissamphibians were derived from temnospondyls with the evidence favoring an origin of gymnophiones from microsaurs and an origin of batrachians from temnospondyls. He concluded that only slightly more evidence supported a monophyletic origin of all lissamphibians from temnospondyls than separate origins of gymnophiones from microsaurs and batrachians from temnospondyls. None of these earlier studies were based on a data matrix, and the number of taxa considered was low. Consequently, only a small subset of all the possible hypotheses were considered. The results of the present study differ from these earlier hypotheses in that modern amphibians seem to form a monophyletic group excluding all the known groups of Paleozoic terrestrial vertebrates. Furthermore, *Rhynchonkos* does not appear to be the closest relative of lissamphibians. Lysorophians share several synapomorphies with lissamphibians, but they have never been considered seriously as close relatives of extant amphibians. Indeed, Wellstead (1991) only discussed affinities between lysorophians and apodans (rather than lissamphibians) and concluded that "there is no

Figure 5. Strict consensus of the four equally most parsimonious trees found by Carroll (1995). This tree is identical to the majority-rule consensus of the four most parsimonious trees. The nodes that are also found in Figure 1 have been labeled on this tree; they do not follow the terminology used in Carroll (1995). ➔

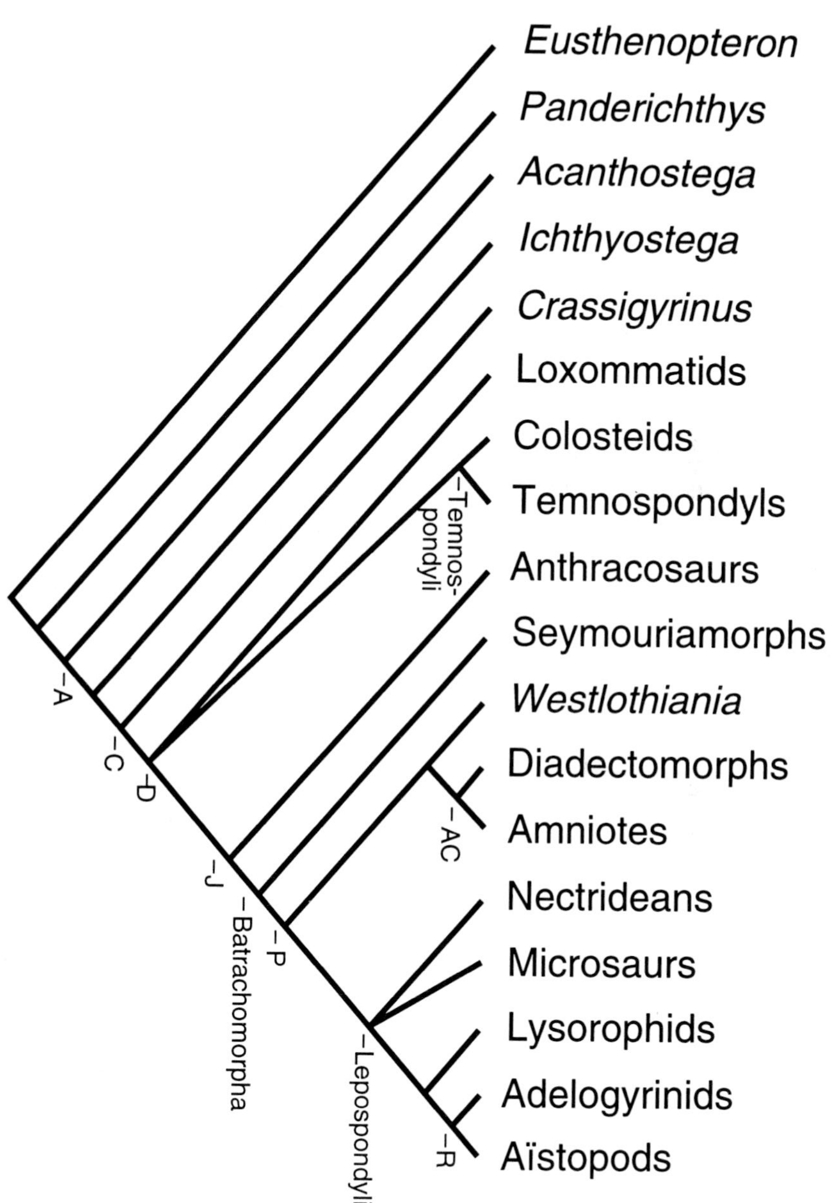
Eusthenopteron
Panderichthys
Acanthostega
Ichthyostega
Crassigyrinus
Loxommatids
Colosteids
Temnospondyls
Anthracosaurs
Seymouriamorphs
Westlothiania
Diadectomorphs
Amniotes
Nectrideans
Microsaurs
Lysorophids
Adelogyrinids
Aïstopods
–Temnos-pondyli
–A
–C
–D
–J
–Batrachomorpha
–P
–AC
–Lepospondyli
–R

autapomorphy linking apodans and lysorophoids." Nevertheless, five extra steps are required to collapse the clade that includes lysorophoids and lissamphibians (node W).

The shortest tree in which lissamphibians are closely related to dissorophoid temnospondyls was searched to determine how many extra steps this hypothesis requires. Surprisingly, the shortest tree in which dissorophoids and lissamphibians form a clade requires 33 extra steps (687 steps). This high number of extra steps is unexpected, considering that few studies have argued for lepospondyl origin of lissamphibians, but is readily explained by the sum of autapomorphies of the taxa that include Lissamphibia but not Temnospondyli (nodes W, V, U, T, S, Lepospondyli, nodes P, M, Batrachomorpha, and node J). Most of the characters supporting these nested clades are convergent in the trees in which lissamphibians are closely related to temnospondyls. Therefore, the hypothesis that lissamphibians are derived from temnospondyls can be rejected with confidence. Some readers might object that some of the temnospondyls believed by Trueb and Cloutier (1991) to be the closest relatives of lissamphibians were not included in this analysis. The authors excluded these taxa because they are too poorly known, because some of them may be represented by juvenile postmetamorphic specimens, and because their morphology did not seem to be distinct enough from the dissorophoids already included.

Status of Anthracosaurs

The paraphyly of anthracosaurs as they were previously understood is unexpected. Our ideas about the phylogeny and relationships of Anthracosauria have varied, but this taxon typically included embolomeres, gephyrostegids, seymouriamorphs, and sometimes diadectomorphs and amniotes (Carroll, 1988; Lombard and Sumida, 1992). Some may be surprised that gephyrostegids do not form a clade with embolomeres, but it takes three extra steps to obtain this topology and only one unambiguous synapomorphy links gephyrostegids to embolomeres in such a tree (the potentially mobile joint between the skull table and the cheek).

Seymouriamorphs have been compared to diadectomorphs and amniotes because they were believed to be closely related to them (Heaton, 1980; White, 1939). Indeed, seymouriamorphs were once believed to be an early group of reptiles (White, 1939). Instead, seymouriamorphs now seem relatively remote from amniotes in that they are as closely related to lissamphibians as they are to amniotes.

Status of the Crown Group of Terrestrial Choanates

The crown group of terrestrial choanates was previously believed to include most known terrestrial choanates as temnospondyls were assumed to be closely related to lissamphibians and anthracosaurs were thought to have given rise to amniotes (Gaffney, 1980; Panchen and Smithson, 1988). Only Devonian taxa, such as *Ichthyostega*, *Acanthostega*, *Tulerpeton*, and the Mississippian choanate *Crassigyrinus*, were thought to be excluded from this crown-group (Panchen and Smithson, 1988; Gauthier *et al.*, 1989). The position of lepospondyls was problematic and there was no consensus on whether or not they were monophyletic or not (Carroll and Gaskill, 1978; Carroll, 1988; Panchen and Smithson, 1988). Therefore, most groups of early terrestrial choanates were compared either with modern amphibians (Bolt, 1977) or with amniotes (Carroll, 1970a) to document the origin of these groups of tetrapods. The present phylogeny suggests that these efforts were largely misdirected because only diadectomorphs and lepospondyls are more closely related to one of the two extant clades of tetrapods (Amniota and Lissamphibia) than to the other. A crown group of a similar composition has been suggested by Gardiner (1982). Gardiner's (1982) phylogeny is similar to the new phylogeny, but Gardiner only considered a few taxa. The terminal taxa included in his crown group were Anura, Urodela, Apoda, Aïstopoda, Nectridea, and Amniota. Therefore, he did not consider diadectomorphs, "microsaurs", adelogyrinids, or lysorophians. Gardiner believed that aïstopods and nectrideans formed a clade that was the sister-group of Lissamphibia. Therefore, his monophyletic Lepospondyli excluded Lissamphibia. Nevertheless, his phylogeny agrees with ours in most other details and he believed temnospondyls and loxommatids to be successive sister groups of the crown group of terrestrial choanates.

The new phylogeny suggests that amniotes and lissamphibians are more closely related and that they share a longer evolutionary history than previously believed. Previous phylogenies (Panchen and Smithson, 1988) suggested that the evolutionary lineage leading to amniotes had diverged from the lineage leading to lissamphibians soon after the first terrestrial choanate appeared, in the Devonian. However, the present phylogeny suggests that a diverse terrestrial vertebrate fauna (represented by *Ichthyostega*, *Acanthostega*, *Crassigyrinus*, loxommatids, temnospondyls, embolomeres, gephyrostegids, and seymouriamorphs) existed for several million years prior to the divergence of the amniote and lissamphibian lineages. Therefore, the evolutionary lineage that led to amniotes diverged from the lineage leading to lissamphibians in the Mississippian at the latest. Previous phylogenies (Panchen and Smithson, 1988) suggested that this divergence had occurred in the Upper Devonian, if not earlier.

Evolution of Selected Characters

The new phylogeny has interesting implications for the evolution of four characters frequently used in considerations of amniote relationships.

Tabular-Parietal Contact

The tabular-parietal contact has usually been considered to unite embolomeres and seymouriamorphs to diadectomorphs and amniotes (Gauthier *et al.*, 1988), but this character is also present in all lepospondyl groups that have a tabular (Figs. 2 and 3). Therefore, this character is diagnostic of node J (embolomeres and batrachomorphs). Of course, the identity of the temporal element in most lepospondyls is difficult to determine and it has been assumed to be a tabular in this analysis, but even if this element were a supratemporal the presence of a tabular-parietal contact would be an autapomorphy of node J. Some nectrideans also have a supratemporal and the tabular-parietal contact is definitely present in this group (Fig. 3).

Transverse Flange of the Pterygoid

The transverse flange of the pterygoid has been usually considered to be diagnostic of amniotes and diadectomorphs (Gauthier *et al.*, 1988; Panchen and Smithson, 1988), but it is a synapomorphy of seymouriamorphs, gephyrostegids, diadectomorphs, and amniotes and also appeared convergently in dissorophoids. The transverse flange was lost in lepospondyls (see Appendix 3).

Supraoccipital

The supraoccipital has often been considered to be a synapomorphy of amniotes and diadectomorphs (Gauthier *et al.*, 1988) and its presence in some lepospondyls (Fig. 4) was often assumed to result from convergence, but it seems to be a synapomorphy of all these taxa. Among the lepospondyls included in this analysis, a supraoccipital is present in *Pantylus*, *Rhynchonkos*, brachystelechids, and lysorophians. Nectrideans and lissamphibians appear to have lost this element (this character could not be coded in aïstopods and in adelogyrinids).

Parasternal Process of the Interclavicle

The parasternal process (long posterior stem) of the interclavicle has been thought to be a synapomorphy of amniotes, diadectomorphs, and seymouriamorphs (Carroll, 1968), but this character seems to be an autapomorphy of the clade that includes all these taxa and lepospondyls. A parasternal process may have been present primitively in lepospondyls because it is present in *Pantylus*, brachystelechids, and some lysorophians. However, the optimization of this character at the base of Lepospondyli is ambiguous because the interclavicle of adelogyrinids and nectrideans lacks a parasternal process (this character could not be coded in aïstopods). Lissamphibians have lost the interclavicle.

Origin of Amniotes

The significance of this new phylogeny to the origin of amniotes lies in the recognition of an entirely new set of outgroups beyond diadectomorphs. The new phylogeny indicates that future phylogenetic studies of amniote phylogeny will have to include lepospondyls as outgroups. This has not been a common practice, but it will be necessary if the phylogeny presented here is accepted. Using lepospondyls could certainly help to polarize several characters used in phylogenetic studies of amniotes because they are very different from the seymouriamorphs that were previously used as outgroups.

Carroll (1970b, 1991) suggested that early amniotes showed evidence of having passed through a phase in which their skull was reorganized to be adapted to small size (snout-vent length 100 mm).

Carroll's is the most explicit functional scenario on the origin of amniotes and needs to be summarized before it is discussed. Carroll suggested that amniotes evolved from anamniote tetrapods that had direct development and laid their eggs on land, like plethodontid salamanders. According to Carroll, the size of an anamniotic egg is limited by the absence of extraembryonic membranes and by the absence of a mineralized shell because support and gas exchange would be problematic beyond a certain size (10 mm in diameter). Caecilians have large eggs, but this may be possible because of the elastic fibers in the egg capsule and the large filamentous gills of the embryos. Carroll suggests that external gills would not have been present in the embryos of the ancestors of amniotes because the gas exchange is performed by the allantois. These constraints on the egg size of the ancestors of amniotes, together with the correlation between egg size and adult size found in modern amniotes and plethodontids, suggests that our anamniotic ancestors would have been small and that their size could have increased only after the acquisition of the amniotic egg (according to Carroll, extant amphibians can only reach large adult sizes through a long growth period in a larval stage). This apparently explained why the groups suggested by Carroll as the earliest amniotes and their sister groups were generally small. Furthermore, the extensive ossification found in the vertebrae, carpus, and tarsus of early amniotes suggests that the adults were small, like plethodontids in which precocious ossification appears to result in small adult size (growth stops when all the cartilage is replaced by bone). Therefore, there were theoretical reasons to expect the earliest amniotes to be small (Carroll 1970b, 1991), and there was some support in favor of this scenario (such as the small size and extensive ossification of the earliest amniotes).

One problem with this scenario is that factors other than egg size may constrain plethodontids. The most obvious of these is the absence of lungs. Without lungs, plethodontids are restricted to small adult size because the surface/volume ratio of the animal must remain high enough to allow the animal's needs in oxygen to be accommodated by cutaneous respiration. Therefore, the small size of terrestrial plethodontids does not necessarily result from the small size of their eggs and from their direct development. Indeed, the polarity of this argument may be reversed. Direct development and terrestrial eggs are also found in some gymnophiones (Duellman and Trueb, 1986). The small size of many gymnophiones may yield further support for Carroll's scenario, but again, other factors may constrain the size of many gymnophiones. The fossorial existence of most of these tetrapods may select for a small size. Furthermore, the correlation between egg size and adult size claimed by Carroll (1991) for most tetrapods is not strong among lissamphibians (Duellman and Trueb, 1986).

The suggestion that lepospondyls are close relatives of amniotes may at first appear to lend support to Carroll's scenario because most lepospondyls are small. However, optimization of size on the tetrapod phylogeny indicates that lepospondyls are the only small relatives of amniotes and that small size (snout-vent length 100 mm) does not appear to be primitive for Amniota or larger clades that also include diadectomorphs and lepospondyls. Diadectomorphs, embolomeres, and some, if not all, seymouriamorphs are large. *Utegenia*, *Ariekanerpeton*, and *Discosauriscus*, the only small seymouriamorphs, seem to be only represented by incomplete growth series in which the adults are missing. It is possible that the relatively small size of the earliest amniotes may simply be coincidental or result from taphonomic biases. For instance, most amniotes found at Joggins and Florence, Nova Scotia (two of the localities in which the earliest amniotes were found) were preserved in hollow tree stumps. As the largest stumps had a diameter of only about 30-60 cm, the vertebrates preserved in them were necessarily relatively small. Even here, some of the amniote remains found at Joggins and Florence (Reisz, 1972) were only moderately small (centrum length 5-9 mm), with an estimated snout-vent length of up to 350 mm. In fact, the smallest presumed amniote from Joggins, *Archerpeton* with an estimated snout-vent length of 60 mm (Carroll, 1964, 1970b), is a tiny microsaur (R. Reisz, personal observation). A survey of Middle Pennsylvanian amniote bearing localities indicates that the size of amniotes found in these sites ranges from 130 to 300 mm in snout-vent length. Furthermore, Carroll's scenario was predicated on the assumption that the earliest known amniotes were basal, an assumption that is not supported by the most recent phylogeny (Laurin and Reisz, 1995).

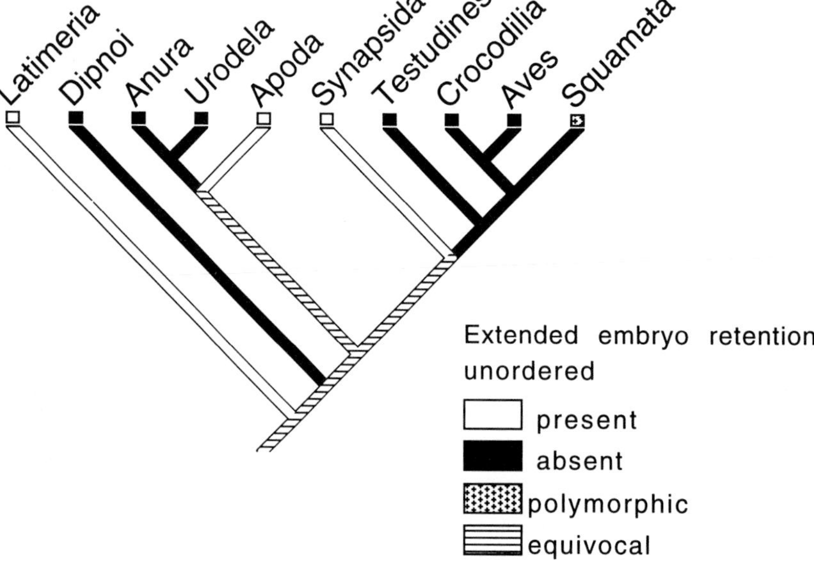

Figure 6. Evolutionary history of egg retention in extant choanates. This cladogram indicates that the primitive state for egg retention cannot be determined for amniotes or their extant sister taxon on the basis of the available evidence. Therefore, the hypothesis that the origin of amniotes included an intermediate stage in which anamniotic eggs were laid on land is not more parsimonious than the suggestion that the extraembryonic membranes evolved to facilitate extended egg retention.

Therefore, very small size is probably not a primitive character for amniotes and this aspect of the scenario proposed by Carroll (1970b, 1991) is not supported by the fossil record.

Carroll's hypothesis of transformation from the presumed basal amphibian to the amniotes includes a number of evolutionary modifications of the reproductive process. These involve internalfertilization, abbreviation and elimination of larval stages, laying of eggs in terrestrial locations; and development of extraembryonic membranes and shell. According to Carroll, all of these changes are more or less interdependent and must have occurred concurrently. However, he assumed that the evolution of the amniote reproductive pattern proceeded via an intermediate stage in which anamniotic eggs were laid in damp places on land. We are convinced that this assumption cannot be made because the evolutionary innovation of internal fertilization immediately raises the possibility of embryo retention and embryo-mother interaction (Lombardi, 1994). This is supported by the development of placentation in extant synapsids and frequent placentation in squamates, regardless of whether they lay eggs or not. It is quite reasonable to suggest (Lombardi, 1994) that some of the extraembryonic membranes evolved in an embryo-retaining form, as a pathway for fetal-maternal interaction, rather than a protection from the harsh external environment. Carroll's hypothesis of amniote origins also assumes that the primitive amniote condition involves the laying of eggs on land. Unfortunately, we cannot deduce the reproductive pattern of diadectomorphs, the nearest sister taxon of amniotes. We also have no evidence on the reproductive strategies of the numerous fossil members of the amniote clade because neither embryo-retaining specimens nor nests of Paleozoic or Early Mesozoic amniotes have been found. If we use a cladogram of extant tetrapods, with *Latimeria* and Dipnoi as outgroups, and investigate the evolutionary history of embryo retention among amniotes, we see the following ambiguous distribution pattern of the two character states (extensive embryo retention absent or present): The nearest relatives of tetrapods have both conditions, with *Latimeria* showing extensive embryo retention, whereas extant Dipnoi lay eggs at an early developmental stage. Lissamphibia have both conditions. Among extant squamates we see a bewildering array of reproductive strategies, with frequent placentation as well as extended embryo retention (Blackburn, 1992), but egg laying without extended embryo retention appears to be primitive for sauropsids because all other extant members of this clade crocodiles, birds, and turtles lack extended embryo retention. All extant synapsids (including monotremes and therians) show extended embryo retention. The resultant pattern does not allow us to determine the primitive condition for amniotes (Fig. 6). Therefore, the scenario that the evolution of the amniotic condition involved the intermediate stage of anamniotic eggs being laid on land is not more parsimonious than the alternative suggested here (that extraembryonic membranes appeared to facilitate extensive embryo retention). It is obvious that this aspect of the origin of amniotes needs further study.

The fossil record does not support the thesis that the earliest amniotes and their ancestors were small. If the evolutionary innovation of the amniotic egg permitted an increase in the size of the adult, as suggested by Carroll, the fossil record should provide a test for this hypothesis. However, the fossil record indicates that the adult size of amniotes does not show a significant increase in size from the earliest appearance of amniotes in the lower part of the Middle Pennsylvanian until the upper part of the Late Pennsylvanian. Therefore, the main adaptive advantage of the amniotic egg was probably not to allow increase in the size of the adults in the absence of extended larval stages but rather increased chances of survival of the hatchling by reducing the probability of starvation or predation. It is generally assumed that tetrapods primitively practiced a reproductive strategy involving large numbers of eggs laid in water, with little parental investment per egg. This is the presumed reproductive strategy of lepospondyls and seymouriamorphs. In contrast, the assumed primitive condition for amniotes involves a reproductive strategy of producing fewer eggs than their anamniotic relatives with greater parental investment in each egg, including nourishment and possibly parental care. Perhaps the origin of amniotes includes a shift between these two reproductive strategies. Therefore, though the origin of amniotes is probably associated with terrestrial reproduction, it may also be associated with a shift toward increased parental investment in each egg. Early amniotes may have laid larger eggs than their distant ancestors, and they may have laid them later if they had extensive embryo retention. Their hatchlings may have been larger and vulnerable to predators to a lesser degree or for a shorter period than the eggs and hatchlings of groups that laid smaller eggs and that lacked embryo retention. By investing energy into parental care, early amniotes would have ensured high survival chances to their offspring.

The new phylogeny and the scenario of the origin of amniotes associated with a shift in reproductive strategies do not specify when the amniotic egg appeared. However, according to the new phylogeny, the amniotic egg can only have appeared in the lineage that led to Amniota, no sooner than the divergence of this lineage from the lineage that led to lissamphibians, and at least slightly before the divergence between synapsids and sauropsids. Determining more accurately the timing of the appearance of the amniotic egg requires knowledge of the type of egg laid by diadectomorphs (or *Westlothiania*, if it is a close relative of amniotes). Unfortunately, fossil eggs are rare in the Paleozoic (and amniote eggs have not been discovered) and only indirect arguments may allow us to solve this question. Aquatic larvae are known in temnospondyls and seymouriamorphs, and lateral-line canal grooves are known in several groups of early terrestrial choanates. Therefore, the status of diadectomorphs may one day be established using such evidence. No diadectomorph larva is known, and as far as we know diadectomorphs lacked lateral-line canal grooves. These facts, as well as the large size of diadectomorphs, do not preclude the possibility that diadectomorphs may have laid amniotic eggs, but the fossil record is currently insufficient to safely settle this question.

The new phylogeny presented here is of course incomplete and more detailed phylogenies will no doubt appear in the future, but it is hoped that the present study will stimulate research on the anatomy and phylogeny of early choanates. The large amount of missing and partially uncertain data present in the matrix (Appendix 2) clearly illustrates that progress in this field will require more anatomical data.

SUMMARY

A phylogenetic analysis based on a data matrix of 38 taxa and 157 osteological characters has produced a dramatically new hypothesis of tetrapod phylogeny. Amniota and diadectomorphs are closely related, as previously reported, but this clade is closely related to a monophyletic Lepospondyli that includes Lissamphibia. In contrast to previous hypotheses, diadectomorphs, amniotes, and Lepospondyli include all extant tetrapods and constitute the crown group of terrestrial choanates. This crown group is much smaller that previously suggested. Seymouriamorpha, long considered closely related to diadectomorphs and amniotes, is the sister group of tetrapods according to this analysis. Gephyrostegids, embolomeres, temnospondyls, loxommatids, and *Crassigyrinus* are more distant relatives of the crown group.

ACKNOWLEDGMENTS

We thank Ms. Diane Scott for drawing the reconstructions of the skulls. We sincerely thank all the curators and other personnel who lent us specimens for this study. The people include (in alphabetical order) Dr. David S Berman (Carnegie Museum), Dr. John R. Bolt (Field Museum of Natural History), Dr. Richard S. Laub (Buffalo Museum of Science), Dr. Robert Murphy (Royal Ontario Museum), Dr. Timothy B. Rowe (Texas Memorial Museum), Dr. B. Sanchiz, (National Museum of Natural Sciences, Madrid) Mr. C. R. Schaff (Museum of Comparative Zoology), and Mr. William F. Simpson (Field Museum of Natural History). We are grateful to Dr. Robert L. Carroll (Redpath Museum, McGill University) for sending us a manuscript of his unpublished phylogenetic study of choanates and some articles that were unavailable at Toronto. We thank Dr. D. Wake and Dr. M. Wake for reading the manuscript, for stimulating conversations on lissamphibian anatomy, and for letting one of us (M. L.) study the extensive collection of lissamphibian dried skeletons and cleared and stained specimens in the Museum of Vertebrate Zoology and in the Marvalee Wake Collection. Personal communications were obtained from David S Berman on *Ecolsonia* and from Mr. Michael deBraga on Procolophonidae. We thank Dr. J. Gauthier for stimulating conversations on systematics. We thank Dr. K. Padian for interesting conversations on systematics and for his hospitality. We also thank the staff of the document delivery and interlibrary loan services of the Erindale Campus (especially Mrs. Pamela King, Mrs. Andrea Pillo, and Mrs. Patricia Lai) for their help. We are indebted to Mr. Xu Jianping for translating parts of a Chinese article on hynobiid salamanders into English. This research was funded by Natural Sciences and Engineering Research Council of Canada (NSERC), Fonds pour la Formation de Chercheurs et l'Aide à la Recherche (FCAR), the Fondation Desjardins, the University of Toronto, and a grant to Robert R. Reisz. This is UCMP Contribution 1648.

APPENDIX 1

List of Characters Used in this Study

1 Lateral-line canals location in adults: surrounded by bone (0); in grooves at the surface (1); in soft tissues or absent (2).
2 Lateral-line canal grooves in ontogeny: always present (0); in larvae only, lost in adults (1); never present (2).
3 Dermal sculpturing: high ridges and pits, "temnospondyl pattern" (0); cosmine (1); shallow, widely spaced pits, "anthracosaur pattern" (2); low rugosities, pustules (3); none, smooth bone (4).
4 Rostrum: absent (0); short, high (1); long (2).
5 Orbit: elliptical, surrounded by bone (0); with large anterior fenestra (1); expanded posteriorly (2).
6 Jaw joint position relative to occiput (i.e., the occipital condyle): behind occiput (0); close to level of occiput (1); in front of occiput (2).
7 Frontal and parietal: distinct, with at least three large ossifications (0); both present as paired ossifications (1); fused into a frontoparietal (2); paired parietal and median frontal (3).
8 Frontal: excluded from orbit (0); contacts orbit between prefrontal and postfrontal (1); contacts orbit, postfrontal absent (2).
9 External naris: marginal, reaches ventral skull margin (0); well above skull margin (1).
10 Anterior tectal: present (0); absent (1).
11 Skull table-cheek joint: sutured (0); potentially mobile, smooth joint present (1); potentially mobile, slit or fenestra present (2).
12 Parietal: restricted to skull table (0); with superficial occipital flange (1).
13 Parietal-squamosal contact: absent (0); present (1).
14 Postorbital-supratemporal contact: present (0); absent (1).
15 Pineal foramen: present (0); absent (1).
16 Intertemporal: present (0); absent (1).

17 Supratemporal: present (0); absent (1).
18 Tabular: present (0); absent (1).
19 Tabular occipital flange: absent (0); single, extends ventromedially from posterolateral corner of skull table (1); two broad, low laminae or a lamina and a tabular horn (2); long, curved flange with ventromedial expansion (3); single, broadly attached to posterior edge of dorsal flange (4); single, overlapped by postparietal posteriorly (5).
20 Subdermal tabular horn: absent (0); present (1).
21 Ventrally bent, sculptured tabular horn: absent (0); present (1).
22 Postparietal number: two (0); one (1); absent (2).
23 Postparietal position: on skull table or on skull table and on occiput (0); only on occiput (1).
24 Tabular-parietal contact: absent (0); present (1).
25 Prefrontal: present (0); absent (1).
26 Prefrontal anterior contacts: anterior tectal and lacrimal, minimally (0); nasal and lacrimal only (1); external naris and/or septomaxilla, in addition to other elements (2).
27 Postfrontal: reaches orbit (0); excluded from orbit (1); absent (2).
28 Postorbital: reaches orbit (0); excluded from orbit (1); absent (2).
29 Lacrimal: reaches orbit (0); excluded from orbit (1); absent (2).
30 Jugal: reaches orbit (0); excluded from orbit (1); absent (2).
31 Temporal emargination: absent, temporal area covered by postspiracular and opercular series (0); present between squamosal, tabular, and supratemporal (1); absent, squamosal, tabular and supratemporal occupy temporal area (2); present behind quadrate (3); present behind squamosal (4).
32 Maxilla-quadratojugal contact: present (0); absent (1).
33 Palatal shelf of maxilla : absent (0); present (1).
34 Tentacular foramen in maxilla: absent (0); present (1).
35 Quadratojugal: present (0); absent (1).
36 Pterygoid-squamosal contact: absent (0); present (1).
37 Occipital flange of squamosal: absent (0); present, convex (1); present, concave, in notch (2).
38 Quadrate dorsal process: long, high (0); short, low (1).
39 Lateral palatal tooth row: present, uninterrupted (0); interrupted (1); absent (2).
40 Posteromedial transverse vomerine tooth row: absent (0); present (1).
41 Palatal recess: present (0); divided medially (1); covered by bone (2).
42 Vomerine fangs: present (0); absent (1).
43 Vomerine shagreen of denticles: absent (0); present (1).
44 Palatine: discrete (0); fused or absent (1).
45 Lateral exposure of palatine: absent (0); present (1).
46 Palatine fangs: present (0); absent (1).
47 Palatine shagreen of denticles: absent (0); present (1).
48 Ectopterygoid: discrete (0); fused or absent (1).
49 Ectopterygoid fangs: present (0); absent (1).
50 Ectopterygoid shagreen of denticles: absent (0); present (1).

51 Transverse flange of pterygoid: absent (0); present, extends into adductor chamber (1); present, extends transversely to ectopterygoid (2).
52 Transverse vertical flange of pterygoid: absent (0); present (1).
53 Pterygoid shagreen of denticles: present (0); absent (1).
54 Pterygoid-maxilla contact: absent (0); present (1).
55 Parasphenoid: contacts vomer (contact visible on palate) (0); does not reach vomer (1).
56 Interpterygoid vacuity: narrow (0); broad (1); absent (2).
57 Palatine position: excluded from interpterygoid vacuity (0); contacts vacuity (1).
58 Maxilla: excluded from interpterygoid vacuity (0); contacts vacuity (1).
59 Parasphenoid morphology: narrow cultriform process, body moderately broad (0); broad cultriform process merging into broad body (1); narrow cultriform process, body expanded into lateral wings (2).
60 Parasphenoid denticles: present (0); absent (1).
61 Parasphenoid posterior edge: transverse or indented (0); forms a posteromedial process (1).
62 Ventral cranial fissure: exposed on the palate (0); covered by the parasphenoid (1).
63 Posttemporal fenestra: large (0); reduced to a foramen or absent (1).
64 Paroccipital process: absent (0); present (1).
65 Otic tube: absent (0); present (1).
66 Exoccipital contact with dermatocranium: absent (0); with postparietal (1); with parietal (2).
67 Braincase endochondral roof: ossified as a unit (0); unossified (no supraoccipital is present) (1); discrete supraoccipital present (2).
68 Basioccipital condyle: deeply concave, with notochord extension into braincase (0); concave condyle with shallow notochordal pit (1); paired (2); median, broad, convex (3); narrow, convex (4).
69 Basioccipital and exoccipital: indistinguishably fused in the adult (0); suturally distinct throughout ontogeny (1); basioccipital never distinct (2).
70 Basioccipital position: reaches edge of foramen magnum (0); excluded from edge of foramen magnum (1).
71 Parasphenoid and endochondral braincase elements (other than basisphenoid): suturally distinct in adults (0); indistinguishably fused in adults (1).
72 Gap between sphenethmoid and oticooccipital portions of braincase: absent (0); present (1).
73 Preopercular: present (0); absent (1).
74 Subopercular: present (0); absent (1).
75 Epipterygoid ossification: present (0); absent (1).
76 Epipterygoid and pterygoid: fused (0); suturally distinct or epipterygoid absent (1).
77 Basicranial articulation: not fused (0); sutured (1).
78 Stapes position: links braincase to quadrate (0); without bony contact distally (1).

79 Stapes: without a dorsal process (0); with a dorsal process confluent with the footplate (1); with a dorsal process distinctly set off from the footplate by the notch (2).
80 Proximal stapedial footplate: absent (0); present, discrete (1); represented by a gentle proximal swelling (2).
81 Stapedial foramen: present (0); absent (1).
82 Mandibular fenestra: absent (0); some small fenestrae in splenial, postsplenial, and angular (1); between angular, postsplenial, splenial, and prearticular (2).
83 Anterior coronoid: present (0); fused or absent (1).
84 Central coronoid: present (0); fused or absent (1).
85 Posterior coronoid: present (0); fused or absent (1).
86 Anterior splenial: present (0); fused or absent (1).
87 Postsplenial: present (0); fused or absent (1).
88 Prearticular: present (0); fused or absent (1).
89 Angular: present (0); fused or absent (1).
90 Surangular: present (0); fused or absent (1).
91 Parasymphysial fangs: present (0); absent (1).
92 Coronoid fangs: present (0); absent (1).
93 Coronoid denticles: absent (0); present (1).
94 Dentary: dentigerous (0); edentulous (1).
95 Retroarticular process: small to mid-sized, extends posteriorly (0); very large (1); medium sized, extends posteroventrally (2).
96 Medial mandibular tooth row: present (0); absent (1).
97 Crowns of marginal teeth: monocuspid (0); bicuspid (1); tricuspid (2); bulbous with a labially convex cutting edge and a lingual cavity (3).
98 Marginal teeth: nonpedicellate (0); pedicellate (1).
99 Maxillary tooth size: at least as large as dentary teeth (0); much smaller than dentary teeth (1).
100 Labyrinthine infolding: present (0); absent (1).
101 Zygapophyses: absent from most of the tail (0); present almost to the tip of the tail (1).
102 Presacral centra: large, crescentic intercentrum, small, paired pleurocentrum (0); large, circular intercentra, no pleurocentra (1); large, crescentic intercentrum, no pleurocentrum (2); circular pleurocentrum, crescentic intercentrum (3); circular intercentrum and pleurocentrum (4); large pleurocentrum, no intercentrum (5).
103 Neural arches and spines: all paired or fused only at tip of spine (0); fused in at least some segments (1); separated by a thin layer of cartilage (2).
104 Neural arches and centra: discrete elements (0); fused in adults, traces of a neurocentral suture may remain (1); indistinguishably fused early in ontogeny (2).
105 Canal for supraneural ligament: present in at least some arches (0); absent (1).
106 Large foramina in centrum or pedicle: absent (0); present in centrum (1); in pedicle (2).
107 Anterior dorsal neural arches: flat or concave (0); swollen (1).

108 Atlantal arch: paired (0); fused (median) (1).
109 Distinct odontoid process: absent (0); present as a convex median process medial to two concave articular surfaces (1); formed by the apex of two angled articular surfaces (2).
110 Anterior articular surface of atlantal centrum: no broader than posterior surface of centrum (0); broader than posterior surface (1).
111 Anterior process of axial intercentrum-atlantal pleurocentrum complex: absent (0); present (1).
112 Atlantal pleurocentrum and axial intercentrum: discrete elements (0); fused (1).
113 Atlantal intercentrum: present (0); absent (1).
114 Atlantal pleurocentrum: paired (0); median (1).
115 Transverse processes in mid-presacral vertebrae: absent (0); short, robust (1); long (2).
116 Parapophyses: short (0); almost as long as the diapophyses (1); absent or not distinct from the diapophyses (2).
117 Capitulum articulation in mid-presacral vertebrae: with the intercentrum of its segment (0); intercentrum and pleurocentrum of its segment (1); pleurocentrum of its segment (2); pleurocentrum of segment in front (3); transverse process (4).
118 Number of presacral vertebrae: 23-32 (0); fewer than 23 (1); 33-60 (2); more than 60 (3).
119 Number of sacral vertebrae: none (0); one (1); two or more (2).
120 Ribs: short, straight (0); long, curved (1).
121 Discrete dorsal fin: present (0); absent (1).
122 Radials for caudal fin: present (0); absent (1).
123 Uncinate processes: absent (0); present (1).
124 Interclavicle: without a parasternal process (0); with a parasternal process (1); absent (2).
125 Clavicle: present (0); absent (1).
126 Clavicle shape: with a distinctly expanded ventral plate (0); without a ventral plate, or with narrow ventral plate (1).
127 Cleithrum: broad dorsally (0); slender (1); broad ventrally (2); absent (3).
128 Anocleithrum: present (0); absent (1).
129 Lateral extrascapular: present (0); absent (1).
130 Median extrascapular: present (0); absent (1).
131 Number of scapulocoracoid ossifications: one (0); two (1); three (2).
132 Humerus: present (0); absent (1).
133 Supinator process: absent, shaft with rounded anterior edge (0); absent, anterior ridge present on shaft (1); present (2).
134 Ectepicondyle: low, indistinct, or absent (0); high, distinct (1).
135 Deltopectoral crest: distinct from proximal articular surface (0); confluent with proximal articular surface (1).
136 Entepicondyle: not greatly expanded (0); greatly expanded (1).
137 Ectepicondylar foramen: present (0); absent (1).
138 Entepicondylar foramen: present (0); absent (1).
139 Radius and ulna: discrete elements (0); fused to each other (1); absent (2).

140 Olecranon process in adult: absent or cartilaginous (0); ossified (1).
141 Number of manual digits: none (0); eight (1); five (2); four (3).
142 Number of pelvic ossifications: one (0); three (1); two (2).
143 Iliac blade: very short (0); consists of dorsal and posterior flanges (1); posterodorsal flange only (2); dorsal flange (3); long anterodorsal flange (4).
144 Internal trochanter: absent (0); present and separated from articular surface by periosteal bone (1); confluent with proximal articular surface (2).
145 Intertrochanteric fossa: present, proximal head of femur concave ventrally (0); absent, proximal head convex ventrally (1).
146 Adductor crest: absent (0); present (1).
147 Tibia and fibula: discrete elements (0); fused into tibiofibula (1).
148 Astragalus: absent, discrete tibiale, intermedium, and proximal centrale are present (0); present (1).
149 Tibiale and fibulare: short (0); long (1).
150 Number of pedal digits: none (0); seven (1); five (2); four (3).
151 Number of phalanges in second toe: two (0); three (1).
152 Number of phalanges in third toe: three (0); four (1).
153 Number of phalanges in fourth toe: four (0); five (1); three (2).
154 Number of phalanges in fifth tow: four (0); five (1); three (2); two (3).
155 Dorsal scales: rhombic or ovoid (0); circular (1); absent (2); sculptured osteoderms (3); long and slender (4); quadrangular with rounded corners (5).
156 Ventral scales: quadrangular, long, and slender (0); sculptured osteoderms (1); mosaic of polygonal plates (2); elliptical (3); absent (4).
157 Lepidotrichia: present (0); absent (1).

APPENDIX 2

Data matrix. On pages 43-53 taxa, characters, and character-states utilized in discussion of results and hypotheses of relationships in this chapter. Polymorphism and partially uncertain data are indicated by two or more states separated by the symbols "&" and "/," respectively. Unknown condition is indicated by "?".

Taxa\Characters	1	2	3	4	5	6	7	8	9	10	11	12	13	14
Osteolepiformes	0	0	0&1	0	0	1	0	0	1	0	1	0	0	0
Panderichthyidae	0	0	0	2	0	?	1	0	0	0	0	0	0	0
Acanthostega	0	0	0	0	0	0	1	0	0	1	0	0	0	0
Ichthyostega	0	0	0	0	0	0	1	0	0	0	0	0	0	0
Crassigyrinus	1	0	0	0	0	0	1	?	1	0	1	0	0	?
Loxommatidae	1	0	0	0	1	0	1	0	1	1	0	0	0	0
Colosteidae	1	0	0	0	0	0	1	0	0	1	0	0	0	0
Dendrerpeton	2	1/2	0	0	0	0	1	0	1	1	0	0	0	0
Tersomius	2	1/2	0	0	0	1	1	1	1	1	0	0	0	0
Amphibamus	?	?	0	0	0	0	1	0	?	?	0	0	0	0
Ecolsonia	2	1/2	0	0	0	0	1	1	1	1	0	0	0	0
Proterogyrinus	1	0&1	2	0	0	0	1	0	0	1	1	0	0	?
Archeria	1	0	2	0	0	0	1	0	0	1	1	0	0	?
Gephyrostegidae	1	1/2	3	0	0	1	1	1	1	1	1	0	0	0
Kotlassia	2	1/2	0	0	0	1	1	0	1	1	0	0	0	1
Seymouria	2	1/2	0	0	0	1	1	0	1	1	0	0	0	1
Ariekanerpeton	2	1	0	0	0	1	1	0	1	1	0	0	0	1
Aistopoda	?	?	?	0	0	1&2	1&2	0&1	1	1	0	0	0	-
Adelogyrinidae	1	0	0	0	0	1	1	0	1	1	0	0	1	-
Nectridea	1&2	0&1	0	0	0	0&1 &2	1&3	0&1	1	1	0&1	0	0&1	0
Pantylus	2	1/2	0	1	0	1	1	0	1	1	0	0	0	-
Rhynchonkos	2	1/2	4	1/2	0	2	1	0	1	1	0	0	0	-
Brachystelechida	2	1/2	4	1/2	0	2	1	1	1	1	0	1	1	-
Lysorophia	2	1/2	4	0	2	2	1	0	0	1	0	0	1	-
Triadobatrachus	2	1/2	3	0	2	1	2	2	?	1	0	0	1	-
Discoglossidae	2	2	0&4	0	2	1	2	2	1	1	0	0	0	-
Pipidae	2	2	4	0	2	1	2	2	1	1	0	0	0	-
Hynobiidae	2	2	4	0	2	2	1	2	1	1	0	1	1	-
Sirenidae	2	2	4	0	2	2	1	2	-	1	0	1	0&1	-
Proteidae	2	2	4	?	2	2	1	2	-	1	0	1	0	-
Eocaecilia	2	2	4	0	0	2	1	0	1	1	0	1	1	-
Rhinatrematidae	2	2	4	0	0	2	1	0	1	1	2	1	0	-
Ichthyophiidae	2	2	4	0	0	2	1	0	1	1	2	1	1	-
Limnoscelis	2	1/2	3	1	0	1	1	0	1	1	0	0	0	0
Diadectes	2	1/2	3	0	0	1	1	0	1	1	0	0	0	0
Synapsida	2	2	3&4	0&1	0	1	1	1	1	1	0	0	0	0
Captorhinidae	2	2	0	0&1	0	1	1	1	1	1	0	0	1	1
Procolophonidae	2	2	4	1	0	2	1	1	1	1	0	1	0	0

Taxa\Characters	15	16	17	18	19	20	21	22	23	24	25	26	27	28
Osteolepiformes	0	0	0	0	0	0	0	0	0	0	0	0	0	0
Panderichthyidae	0	0&1	0	0	0	0	0	0	0	0	0	0	0&1	0
Acanthostega	0	1	0	0	0	0	0	0	0	0	0	1	0	0
Ichthyostega	0	1	0	0	1/2	0	0	1	0	0	0	1	0	0
Crassigyrinus	0	0	0	0	1	1	0	0	0	0	0	0	0	0
Loxommatidae	0	0&1	0	0	1	0	0	0	0	0	0	1	0	0
Colosteidae	0	0&1	0	0	1	0	0	0	0	0	0	2	0	0&1
Dendrerpeton	0	0	0	0	?	0	1	0	0	0	0	1	0	0
Tersomius	0	1	0	0	1	?	?	0	0	0	0	1	0	0
Amphibamus	0	1	0	0	?	?	?	0	0	0	0	1	0	0
Ecolsonia	0	1	0	0	2	0	1	0	0	0	0	2	0	0
Proterogyrinus	0	0	0	0	2	1	0	0	0	1	0	1	0	0
Archeria	0	0	0	0	2	1	0	0	0	1	0	1	0	0
Gephyrostegidae	0	0	0	0	?	?	?	0	0	1	0	1	0	0
Kotlassia	0	0	0	0	2	0	0	0	0	1	0	1	0	0
Seymouria	0	0	0	0	2	0	1	0	0	1	0	1	0	0
Ariekanerpeton	0	0	0	0	2	0	0	0	0	1	0	1	0	0
Aistopoda	0	1	0&1	0&1	?	0	0	0&1&2	0	?	0	1	0&2	0&1
Adelogyrinidae	0	1	1	1	-	-	-	0	0	-	0	?	0	1
Nectridea	0	1	0&1	0	0&3&4	0	0	0	0	1	0	2	0	0&1&2
Pantylus	1	1	1	0	3	0	0	0	0	1	0	1	0	0
Rhynchonkos	0	1	1	0	3	0	0	0	0	1	0	1	0	0
Brachystelechida	0	1	1	0&1	0	0	0	2	-	1	0	1&2	0	0
Lysorophia	1	1	1	0	0	0	0	0	1	1	0	2	2	2
Triadobatrachus	0	1	1	1	-	-	-	2	-	-	0	?	2	2
Discoglossidae	1	1	1	1	-	-	-	2	-	-	1	-	2	2
Pipidae	0&1	1	1	1	-	-	-	2	-	-	1	-	2	2
Hynobiidae	1	1	1	1	-	-	-	2	-	-	0	1	2	2
Sirenidae	1	1	1	1	-	-	-	2	-	-	1	-	2	2
Proteidae	1	1	1	1	-	-	-	2	-	-	1	-	2	2
Eocaecilia	1	1	1	1	-	-	-	2	-	-	0	2	0	0
Rhinatrematidae	1	1	1	1	-	-	-	2	-	-	1	-	2	2
Ichthyophiidae	1	1	1	1	-	-	-	2	-	-	1	-	2	2
Limnoscelis	0	1	0	0	5	0	0	1	1	1	0	1	0	0
Diadectes	0	1	0	0	5	0	0	1	1	0	0	1	0	0
Synapsida	0	1	0	0	5	0	0	0	1	1	0	1	0	0
Captorhinidae	0	1	0	1	-	-	-	0	1	-	0	1	0	0
Procolophonidae	0	1	0	1	-	-	-	2	-	-	0	1	0	0

Taxa\Characters	**29**	**30**	**31**	**32**	**33**	**34**	**35**	**36**	**37**	**38**	**39**	**40**	**41**	**42**
Osteolepiformes	0	0	0	0	0	0	0	0	0	-	0	0	0	0
Panderichthyidae	0	0&1	0	0	0	0	0	0	0	?	0	0	0	0
Acanthostega	1	0	1	0	0	0	0	1	0	0	0	0	1	0
Ichthyostega	1	0	1	0	0	0	0	1	0	?	0	0	0	0
Crassigyrinus	1	0	1	0	0	0	0	1	0	0	0	0	1	?
Loxommatidae	0	0	1	0&1	0	0	0	1	0	0	2	0	0&2	0
Colosteidae	0	0	2	0	0	0	0	1	1	1	0	0	1	0
Dendrerpeton	0	0	1	0	0	0	0	1	0	?	2	0	2	0
Tersomius	0	0	1	0	0	0	0	1	0	0	2	0	2	0
Amphibamus	0	0	1	?	0	0	0	?	0	?	1	0	2	0
Ecolsonia	0	0	1	0	0	0	0	?	0	0	2	0	2	0
Proterogyrinus	0	0	1	1	0	?	0	1	0	0	1	?	?	?
Archeria	1	0	1	1	0	0	0	1	0	0	1	?	?	?
Gephyrostegidae	0	0	1	0&1	0	0	0	?	0	0	2	0	2	0
Kotlassia	0	0	1	?	0	0	0	1	0	0	0	0	2	?
Seymouria	0	0	1	1	0	0	0	1	0	0	2	0	2	0
Ariekanerpeton	0	0	1	?	0	0	0	1	0	?	2	0	2	0
Aistopoda	0	0&2	2	0&1	0	0	0	0	?	?	1	?	2	1
Adelogyrinidae	?	0	4	0&1	0	0	0	0	2	?	2	?	?	1
Nectridea	0	0	2	0&1	0	0	0	?	0&1	0	0	0	2	1
Pantylus	0	0	2	1	0	0	0	0	1	0	2	0	2	1
Rhynchonkos	0	0	2	?	0	0	?	0	?	1	0	0	2	1
Brachystelechida	0	0	2	?	0	0	1	0	1	0	?	0	2	?
Lysorophia	0	2	2	-	0	0	1	0	1	0	1	0	2	1
Triadobatrachus	?	2	4	?	0	?	?	?	0	?	1/2	?	?	?
Discoglossidae	2	2	4	0	0	0	0	0	0	1	2	1	0	1
Pipidae	2	2	4	-	0	0	1	0	0	1	2	0	0	1
Hynobiidae	0	2	2	-	0	0	1	0	0	0	2	1	0	1
Sirenidae	2	2	2	-	0	0	1	-	0	0	1	0	0&1	1
Proteidae	2	2	2	-	-	-	1	0	0	0	0	0	0&2	1
Eocaecilia	0	0	2	1	0	1	0	?	0	0	0	0	2	1
Rhinatrematidae	2	2	2	-	1	1	1	0	0	0	0	0	2	1
Ichthyophiidae	2	2	2	-	1	1	1	0	0	0	0	0	2	1
Limnoscelis	0	0	2	1	0	0	0	0	0	0	2	0	2	1
Diadectes	0	0	3	1	0	0	0	0	0	0	2	0	2	1
Synapsida	0	0	2	0	0	0	0	0	1	0	2	0	2	1
Captorhinidae	0	0	2	1	0	0	0	0	1	0	2	0	2	1
Procolophonidae	0	0	3	1	0	0	0	1	2	0	2	0	2	1

Taxa\Characters	43	44	45	46	47	48	49	50	51	52	53	54	55	56
Osteolepiformes	0	0	0	0	0	0	0	0	0	0	0	0	0	0
Panderichthyidae	?	0	0	0	?	0	0	0	?	0	?	0	0	0/2
Acanthostega	1	0	0	0	1	0	1	0	2	0	0	0	1	0
Ichthyostega	?	0	0	1	?	0	1	?	2	0	?	0	1	0
Crassigyrinus	?	0	0	0	0	0	0	?	0	0	0	0	?	0
Loxommatidae	?	0	0	0	?	0	0	?	0	0	0	0	1	0
Colosteidae	0	0	0	0	0	0	0	0	0	0	0	0	1	1
Dendrerpeton	1	0	0	0	1	0	0	1	?	?	0	0	0	1
Tersomius	0	0	1	0	0	0	0	0	1	0	1	0	0	1
Amphibamus	1	0	?	0	1	0	0	?	1	0	0	?	0	1
Ecolsonia	1	0	1	0	1	0	0	0	1	?	0	0	0	1
Proterogyrinus	?	0	0	0	1	0	0	1	?	0	0	0	1	0
Archeria	?	?	0	?	?	0	?	0	0	0	0	0	1	0
Gephyrostegidae	1	0	0	0	1	0	0	1	1	0	0	0	?	0
Kotlassia	?	0	0	?	1	0	?	?	1	1	0	0	1	0
Seymouria	1	0	0	0	1	0	1	1	1	1	0	0	1	2
Ariekanerpeton	1	0	0	0	1	0	1	1	1	1	0	0	1	2
Aistopoda	?	0	0	1	?	?	?	?	0	0	?	0	?	0
Adelogyrinidae	1	0	0	1	1	0	1	0	0	0	0	0	?	0
Nectridea	0&1	0	0	1	0&1	0&1	1	0	0&2	0	0&1	0	0&1	0&1
Pantylus	1	0	0	1	1	1	-	-	0	0	0	0	1	0
Rhynchonkos	0	0	0	1	0	0	1	-	0	0	1	0	?	1
Brachystelechida	0	0	0	?	0	0	1	0	0	0	1	0	1	1
Lysorophia	0	0	-	1	0	1	-	-	0	0	1	0	0	0
Triadobatrachus	?	0	0	1	0	1	-	?	0	0	1	1	?	1
Discoglossidae	0	1	-	-	-	1	-	-	0	0	1	1	1	1
Pipidae	0	1	-	1	-	1	-	-	0	0	1	1	?	1
Hynobiidae	0	1	-	-	-	1	-	-	0	0	1	0	0	1
Sirenidae	0&1	0	-	1	1	1	-	-	-	0	-	-	0	-
Proteidae	0	1	-	-	-	1	-	-	0	0	1	-	0	0
Eocaecilia	0	0	0	1	0	1	-	-	0	0	0	0	0	1
Rhinatrematidae	0	1	-	1	-	1	-	-	1	0	1	1	0	1
Ichthyophiidae	0	1	-	1	-	1	-	-	0	0	1	1	0	1
Limnoscelis	1	0	0	1	1	0	1	0	1	0	0	0	1	0
Diadectes	1	0	0	1	1	0	1	0	1	0	1	0	1	0
Synapsida	1	0	0	1	1	0	1	0	1	0	0	0	1	0
Captorhinidae	0	0	0	1	1	1	-	-	1	0	0	0	1	0
Procolophonidae	1	0	0	1	1	0	1	0	1	0	0	0	1	0

Taxa\Characters	57	58	59	60	61	62	63	64	65	66	67	68	69	70
Osteolepiformes	0	0	0	0	0	0	0	0	0	0	0	0	0	?
Panderichthyidae	0	0	?	0	?	?	?	?	?	?	?	?	?	?
Acanthostega	0	0	0	0&1	0	1	?	1	0	?	?	0	1	?
Ichthyostega	0	0	0	1	0	0	?	1	0	?	?	0	?	?
Crassigyrinus	0	0	0	?	0	1	?	?	?	?	?	1	?	?
Loxommatidae	0	0	0	?	0	1	0	1	0	0	0	1	1	?
Colosteidae	0	0	0	0	0	1	0	1	0	1	1	1	0	0
Dendrerpeton	0	0	0	0	0	1	?	?	?	1	?	?	?	?
Tersomius	1	0	0	0	0	1	0	1	0	1	1	2	0	0
Amphibamus	1	?	0	0	0	1	?	?	?	?	?	?	?	?
Ecolsonia	0	0	0	1	0	1	1	1	0	1	1	2	?	0
Proterogyrinus	0	0	0	0	0	1	-	0	0	0	1	1	1	0
Archeria	0	0	0	1	0	1	-	0	0	0	1	1	1	0
Gephyrostegidae	0	0	0	0	?	1	?	?	0	?	?	?	?	0
Kotlassia	0	0	0	?	1	1	1	1	1	1	1	1	?	?
Seymouria	-	-	2	1	1	1	1	1	1	1	1	1	1	0
Ariekanerpeton	-	-	2	?	1	1	1	1	1	?	?	1	1	?
Aistopoda	0	0	0	?	?	1	-	0	0	-	0/2	1	0/2	?
Adelogyrinidae	0	0	0	1	0	1	?	?	0	1	?	1	?	?
Nectridea	0&1	0	0	1	1	1	0	1	0	1	1	2	1	0
Pantylus	0	0	2	1	1	1	0	1	0	0	2	2	1	0
Rhynchonkos	0	0	0	0	0&1	1	1	0	0	1	2	2	1	0
Brachystelechida	0	0	0	1	0&1	1	1	0	0	-	2	2	0	0
Lysorophia	0	0	1	1	0	1	1	0	0	1	2	2	1	0
Triadobatrachus	1	1	2	1	0	1	?	0	1	-	?	?	2	?
Discoglossidae	-	1	2	1	1	1	1	0	1	-	1	2	2	-
Pipidae	-	1	1&2	1	1	1	1	0	1	-	1	2	2	-
Hynobiidae	-	1	1	1	1	1	1	0	0	-	1	2	2	-
Sirenidae	-	-	1	1	0&1	1	1	0	0	-	1	2	2	-
Proteidae	-	-	1	1	0	1	1	0	0	2	1	2	2	-
Eocaecilia	1	0	1	0	0	1	1	0	0	-	?	2	0/2	-
Rhinatrematidae	1	0	1	1	1	1	1	0	0	-	1	2	2	-
Ichthyophiidae	1	0	1	1	-	1	1	0	0	-	1	2	2	-
Limnoscelis	0	0	0	0	0	1	?	0	0	0	2	3	1	?
Diadectes	0	0	0	1	0	1	1	1	0	0	0/2	3	1	1
Synapsida	0	0	0	0&1	0	1	0&	1	0	0	0/2	4	0&1	1
Captorhinidae	0	0	0	1	1	1	0	1	0	0	2	4	0	?
Procolophonidae	0	0	0	1	0	1	0	1	0	0	2	4	1	?

Taxa\Characters	**71**	**72**	**73**	**74**	**75**	**76**	**77**	**78**	**79**	**80**	**81**	**82**	**83**	**84**	**85**
Osteolepiformes	0	0	0	0	0	0	0	0	0	0	0	?	0	0	0
Panderichthyidae	?	?	0	0	0	0	0	0	0	0	1	0	0	0	0
Acanthostega	0	0	0	?	0	0	0	0	1	1	0	?	?	?	?
Ichthyostega	0	?	0	0	0	0	?	?	?	?	?	0	0	0	0
Crassigyrinus	0	?	0	1	0	0	0	?	?	?	?	0	0	0	0
Loxommatidae	0	0	1	1	0	0	0	?	?	?	?	1	0	0	0
Colosteidae	0	1	1	1	0	0	0	0	0	1	0	2	0	0	?
Dendrerpeton	?	?	1	1	?	?	0	?	?	?	0	2	?	?	?
Tersomius	0	1	1	1	0	1	1	?	?	?	?	2	0	0	0
Amphibamus	0	?	1	1	0	?	?	?	?	1	0	2	?	?	0
Ecolsonia	0	?	1	1	?	0	0	1	?	?	?	2	0	0	0
Proterogyrinus	0	1	1	1	0	1	0	?	?	?	?	2	0	0	0
Archeria	0	0	1	1	0	0	0	?	?	?	?	2	0	0	0
Gephyrostegidae	0	?	1	1	?	?	0	?	?	?	?	2	?	?	0
Kotlassia	0	1	1	1	0	?	0	1	0	2	0	2	0	0	0
Seymouria	0	1	1	1	0	1	0	1	0	2	1	0	0	0	0
Ariekanerpeton	0	?	1	1	0	1	0	?	?	?	?	2	0	0	0
Aistopoda	?	0	1	1	0	0	0	1	0	1	1	0	1	1	1
Adelogyrinidae	?	?	1	1	?	?	?	?	?	?	?	?	?	?	?
Nectridea	0	?	1	1	?	1	0&1	?	?	?	?	2	1	1	0&1
Pantylus	0	0	1	1	0	1	0	?	?	?	1	2	?	0	?
Rhynchonkos	0	0	1	1	0	1	0	0	0	1	0&1	2	1	0	0
Brachystelechida	0&1	0	1	1	0	1	0	0	0	1	0&1	2	1	1	1
Lysorophia	0	1	1	1	1	1	0	0	0	1	1	0	1	1	1
Triadobatrachus	0	1	1	1	1	1	?	1	0	1	1	0	1	1	1
Discoglossidae	0	1	1	1	1	1	1	1	0	1	1	0	1	1	1
Pipidae	0	1	1	1	1	1	1	1	0	1	1	0	1	1	1
Hynobiidae	0	1	1	1	1	1	0	0	0	1	1	0	1	1	1
Sirenidae	0	1	1	1	1	1	-	0	0	1	?	0	1	1	1
Proteidae	0	?	1	1	1	1	0	0	0	1	1	0	1	0	1
Eocaecilia	1	?	1	1	?	?	0	0	0	1	0	?	1	1	1
Rhinatrematidae	0	0	1	1	1	1	0	0	0	1	0	0	1	1	1
Ichthyophiidae	1	0	1	1	1	1	0	0	0	1	1	0	1	1	1
Limnoscelis	0	1	1	1	0	1	0	?	?	?	?	2	?	0	0
Diadectes	0	1	1	1	?	?	1	?	2	1	0	2	?	0	?
Synapsida	0	1	1	1	0	1	0	0	2	1	0	0&2	1	0	0
Captorhinidae	0	1	1	1	0	1	?	0	2	1	0	2	1	1	0
Procolophonidae	0	1	1	1	0	1	0	1	0	1	1	2	1	1	0

Taxa\Characters	**86**	**87**	**88**	**89**	**90**	**91**	**92**	**93**	**94**	**95**	**96**	**97**	**98**	**99**	**100**
Osteolepiformes	0	0	0	0	0	0	0	0	0	0	0	0	0	0	0
Panderichthyidae	0	0	0	0	0	0	0	0	0	0	0	0	0	0	0
Acanthostega	?	?	0	?	0	0	?	?	0	0	?	0	0	0	0
Ichthyostega	0	0	0	0	0	0	?	?	0	0	?	0	0	0	0
Crassigyrinus	0	0	0	0	0	0	0	1	0	0	1	0	0	0	0
Loxommatidae	0	0	0	0	0	0	1	1	0	0	1	0	0	0	0
Colosteidae	0	0	0	0	0	0	0	1	0	0	1	0	0	1	0
Dendrerpeton	0	0	0	0	0	0	1	1	0	0	1	0	0	0	0
Tersomius	0	0	0	0	0	0	1	0	0	0	1	0&1	1	0	?
Amphibamus	0	0	0	0	0	?	1	?	0	0	1	1	1	?	?
Ecolsonia	0	0	0	0	0	?	1	1	0	0	1	0	0	0	0
Proterogyrinus	0	0	0	0	0	0	0	1	0	0	1	0	0	0	0
Archeria	0	0	0	0	0	0&1	1	1	0	0	1	0	0	0	0
Gephyrostegidae	0	0	0	0	0	0	1	1	0	0	1	0	0	0	0
Kotlassia	0	0	0	0	0	1	1	1	0	0	1	0	0	?	0
Seymouria	0	0	0	0	0	1	1	1	0	0	1	0	0	0	0
Ariekanerpeton	0	0	0	0	0	1	1	1	0	0	1	0	0	0	?
Aistopoda	1	1	1	0	1	1	-	-	0	0	1	0	0	0	1
Adelogyrinidae	0	0	0	0	0	1	?	?	0	0	1	0	0	0	1
Nectridea	0	1	0	0	0	1	1	0&1	0	0	0&1	0	0	0	1
Pantylus	0	0	0	0	0	1	1	1	0	0	1	0	0	0	1
Rhynchonkos	0	0	0	0	0	1	1	0	0	0	0	0	0	0	1
Brachystelechida	0	?	0	0	?	1	-	-	0	0	1	0&1 &2	0	0	1
Lysorophia	0	1	0	0	0	1	-	-	0	0	1	0	0	0	1
Triadobatrachus	1	1	0	1	1	?	-	-	1	?	-	?	?	?	?
Discoglossidae	1	1	0	1	1	1	-	-	1	0	1	1	1	0	1
Pipidae	1	1	0	1	1	1	-	-	1	0	1	0	0	0	1
Hynobiidae	1	1	0	0	1	1	-	-	0	0	1	0	1	0	1
Sirenidae	1	1	0	1	1	1	1	1	1	2	1	3	0&1	0	1
Proteidae	1	1	0	1	1	1	-	-	0	0	1	1	1	-	1
Eocaecilia	1	1	0	1	1	1	-	-	0	1	0	1	1	?	1
Rhinatrematidae	1	1	0	1	1	1	-	-	0	1	0	0	1	0	1
Ichthyophiidae	1	1	0	1	1	1	-	-	0	1	0&1	0	1	0	1
Limnoscelis	0	1	0	0	0	0	1	1	0	0	1	0	0	0	0
Diadectes	0	1	0	0	0	1	?	?	0	0	?	0&2	0	0	1
Synapsida	0	1	0	0	0	1	1	0	0	0	1	0	0	0	1
Captorhinidae	0	1	0	0	0	1	1	0	0	0	1	0	0	0	1
Procolophonidae	0	1	0	0	0	1	1	0	0	0	1	0&1	0	0	1

Taxa\Characters	**99**	**100**	**101**	**102**	**103**	**104**	**105**	**106**	**107**	**108**	**109**	**110**	**111**	**112**	**113**
Osteolepiformes	0	0	0	0&1	0	0	0	0	0	?	0	0	0	0	0
Panderichthyidae	0	0	0	1	0	?	?	0	?	0	?	?	?	-	?
Acanthostega	0	0	?	?	?	?	?	?	?	?	?	?	?	?	?
Ichthyostega	0	0	0	0	?	0	0	0	?	?	?	?	?	?	?
Crassigyrinus	0	0	0	2	0	0	-	0	0	0	0	0	0	-	0
Loxommatidae	0	0	?	?	?	?	?	?	?	?	?	?	?	?	?
Colosteidae	1	0	1	0	1	0	0	0	0	0	?	?	?	0	0
Dendrerpeton	0	0	?	0	1	0	?	0	0	?	?	?	?	?	?
Tersomius	0	?	?	0&3	1	0	?	0	?	?	?	?	?	?	?
Amphibamus	?	?	1	0&3	1	0	?	0	0	?	?	?	?	?	?
Ecolsonia	0	0	1	0	1	0	1	0	0	?	?	?	?	?	?
Proterogyrinus	0	0	1	3	1	0	0	0	0	0	0	0	0	0	0
Archeria	0	0	1	4	1	0	0	0	0	0	0	0	0	0	0
Gephyrostegidae	0	0	?	3	1	0	1	0	0	0	0	0	0	0	0
Kotlassia	?	0	1	3	1	1/2	1	0	1	0	?	?	?	0	0
Seymouria	0	0	1	3	1	1	1	0	1	0	0	0	0	0	0
Ariekanerpeton	0	?	1	3	1	0/1	1	0	1	0	?	?	?	?	?
Aistopoda	0	1	1	1/5	1	2	1	2	0	0	0	0	0	-	1
Adelogyrinidae	0	1	?	5	1	0	1	1	0	0	?	0	0	-	1
Nectridea	0	1	1	1/5	1	2	1	0	0	1	1	1	0	-	1
Pantylus	0	1	1	5	1	2	1	0	?	1	1	1	0	-	1
Rhynchonkos	0	1	1	3	1	2	1	0	0	1	1	1	0	0	1
Brachystelechida	0	1	1	5	1	2	1	0	?	?	1	1	0	-	1
Lysorophia	0	1	1	1/5	2	0	1	1	0	0	1	1	0	-	1
Triadobatrachus	?	?	?	1/5	1	2	1	0	0	1	0	1	0	-	1
Discoglossidae	0	1	-	1/5	1	2	1	0	0	1	0/2	1	0	-	1
Pipidae	0	1	-	1/5	1	2	1	0	0	1	0/2	1	0	-	1
Hynobiidae	0	1	1	1/5	1	2	1	2	0	1	1	1	0	-	1
Sirenidae	0	1	1	1/5	1	2	1	2	0	1	1	1	0	-	1
Proteidae	-	1	1	1/5	1	2	1	2	0	1	1	1	0	-	1
Eocaecilia	?	1	1	3	1	?	?	2	0	1	1	1	0	0	1
Rhinatrematidae	0	1	1	1/5	1	2	1	?	0	1	2	1	0	-	1
Ichthyophiidae	0	1	1	1/5	1	2	1	?	0	1	2	1	0	-	1
Limnoscelis	0	0	1	3	1	1/2	1	0	1	0	?	?	1	1	0
Diadectes	0	1	1	3	1	1/2	1	0	1	0	?	0	1	1	0
Synapsida	0	1	1	3	1	1	1	0	0	0	0	0	0	0&	0
Captorhinidae	0	1	1	3	1	1/2	1	0	1	0	0	0	0	1	0
Procolophonidae	0	1	1	3	1	1/2	1	?	1	0	0	0	0	1	0

Taxa\Characters	114	115	116	117	118	119	120	121	122	123	124	125	126	127
Osteolepiformes	0	0	0	0	0	0	0	0	0	0	0	0	0	0
Panderichthyidae	?	0	0	0	?	?	0	1	0	0	0	0	?	0
Acanthostega	?	?	?	?	?	?	?	?	?	?	0	0	0	0
Ichthyostega	?	1	0	0	0	1	1	1	0	1	1	0	0	0
Crassigyrinus	-	1	0	0	?	?	1	1	1	0	0	0	0	0
Loxommatidae	?	?	?	?	?	?	?	?	?	?	?	?	?	?
Colosteidae	0	2	0	0	2	1	0	1	1	1	0	0	0	0
Dendrerpeton	?	?	0	0	0	?	0	1	1	?	0	0	0	0
Tersomius	?	?	?	?	?	?	0	1	1	0	?	?	?	?
Amphibamus	?	?	0	0	1	1	0	1	1	0	0	0	?	1
Ecolsonia	?	2	0	0	?	?	0	1	1	1	0	0	0	0
Proterogyrinus	0	1	0	1	0	1	1	1	1	0	0	0	0	0
Archeria	1	1	0	1	2	1	1	1	1	0	0	0	0	0
Gephyrostegidae	?	1	0	?	0	?	1	1	1	1	1	0	0	1
Kotlassia	?	1	0	0/1	0	1/2	1	1	1	1	1	0	0	0
Seymouria	0	1	0	1	0	1&2	1	1	1	0	1	0	0	1
Ariekanerpeton	?	1	0	0/1	0	1	1	1	1	0	1	0	0	1
Aistopoda	1	1	0	2	3	?	0	1	1	1	0/1	1	-	3
Adelogyrinidae	1	1	0	3	3	-	0	1	1	1	0	0	0	0
Nectridea	1	1&2	0&1	2	0&1	1	0	1	1	0	0	0	0	1
Pantylus	1	1	0	3	0	1	1	1	1	0	1	0	1	1
Rhynchonkos	1	1	?	?	2	2	0	1	1	0	?	0	0	?
Brachystelechida	1	1	0	3	1	1	1	1	1	0	1	0	1	1
Lysorophia	1	1	0	3	3	1	0	1	1	0&1	0&1	0	1	1
Triadobatrachus	1	2	2	2	1	1	0	1	1	0	?	0	?	1
Discoglossidae	1	2	2	4	1	1	0	1	1	0	2	0	1	2
Pipidae	1	2	2	-	1	1	-	1	1	-	2	0	1	2
Hynobiidae	1	2	1	2	1	1	0	1	1	0	2	1	-	3
Sirenidae	1	2	1	2	2	-	0	1	1	0	2	1	-	3
Proteidae	1	2	1	2	1	1	0	1	1	0	2	1	-	3
Eocaecilia	1	1	0	?	2	?	0	1	1	0	?	?	?	?
Rhinatrematidae	1	1	0	2	3	-	0	1	1	0	2	1	-	3
Ichthyophiidae	1	1	0	2	3	-	0	1	1	0	2	1	-	3
Limnoscelis	?	2	0	1/2	0	2	1	1	1	0	?	0	0	0/1
Diadectes	1	1	0	2	0	2	1	1	1	1	1	0	0	0
Synapsida	1	1&2	0	2	0	2	1	1	1	0	1	0	0	0
Captorhinidae	1	1&2	0	2	0	2	1	1	1	0	1	0	0	1&3
Procolophonidae	1	1	0	2	0	2	1	1	1	0	1	0	1	3

Taxa\Characters	128	129	130	131	132	133	134	135	136	137	138	139	140	141	142
Osteolepiformes	0	0	0	0	0	0	0	0	0	0	0	0	0	0	0
Panderichthyidae	0	0	0	?	0	0	0	?	0	?	?	0	0	0	?
Acanthostega	0	1	1	0	0	1	0	?	1	0	0	0	0	1	?
Ichthyostega	1	1	1	0	0	1	0	0	1	0	0	0	1	?	0
Crassigyrinus	1	1	1	?	0	1	1	1	0	0	0	0	0	?	1/2
Loxommatidae	?	1	1	?	0	?	?	?	?	?	?	0	?	?	?
Colosteidae	1	1	1	0	0	1	0&1	1	1	1	0	0	1	3	1
Dendrerpeton	1	1	1	0	0	0	1	?	0	1	0	0	1	?	1
Tersomius	?	1	1	?	0	?	?	?	?	?	?	0	?	?	?
Amphibamus	1	1	1	0	0	?	?	?	0	1	1	0	1	3	1
Ecolsonia	1	1	1	0	0	2	1	0	0	1	1	0	1	?	1
Proterogyrinus	1	1	1	0	0	1	1	0	1	1	0	0	1	2	1
Archeria	1	1	1	0	0	1	1	0	1	1	0	0	1	2	1
Gephyrostegidae	1	1	1	0	0	1	1	?	1	1	0	0	1	2	1
Kotlassia	1	1	1	?	0	2	1	?	1	1	1	0	1	?	1
Seymouria	1	1	1	1	0	2	1	0	1	1	0	0	0	2	1
Ariekanerpeton	1	1	1	1	0	0/2	0	1	?	1	0	0	?	2	1
Aistopoda	1	1	1	-	1	-	-	-	-	-	-	2	-	0	-
Adelogyrinidae	1	1	1	-	1	-	-	-	-	-	-	2	-	0	-
Nectridea	1	1	1	0	0	2	1	?	1	1	0&1	0	?	3	1
Pantylus	1	1	1	0	0	0	0	0	0	1	0	0	1	3	1
Rhynchonkos	?	1	1	?	0	0	0	0	0	1	1	0	1	?	1
Brachystelechida	1	1	1	0	0	0	?	0	0	1	0	0	1	?	1
Lysorophia	1	1	1	0	0	0	0	1	0	1	1	0	0	3	1/2
Triadobatrachus	1	1	1	?	0	?	?	?	0	1	1	0	1	?	2
Discoglossidae	1	1	1	1	0	0	0	?	0	1	1	1	1	3	1
Pipidae	1	1	1	1	0	0	0	1	0	1	1	1	1	3	1
Hynobiidae	1	1	1	0	0	0	0	1	0	1	1	0	0	3	1
Sirenidae	1	1	1	0&1	0	0	0	1	0	1	1	0	0	3	-
Proteidae	1	1	1	0	0	0	0	1	0	1	1	0	0	3	2
Eocaecilia	1	1	1	?	0	?	?	?	0	1	1	0	?	?	?
Rhinatrematidae	1	1	1	-	1	-	-	-	-	-	-	2	-	-	-
Ichthyophiidae	1	1	1	-	1	-	-	-	-	-	-	2	-	-	-
Limnoscelis	1	1	1	1	0	2	1	0	1	1	0	0	1	2	1
Diadectes	1	1	1	1	0	2	1	?	1	1	0	0	1	2	1
Synapsida	1	1	1	2	0	2	1	0	1	1	0	0	1	2	1
Captorhinidae	1	1	1	2	0	0	1	0	1	1	0	0	1	2	1
Procolophonidae	1	1	1	2	0	2	0	0	1	1	0	0	0	2	1

Taxa\Characters	**143**	**144**	**145**	**146**	**147**	**148**	**149**	**150**	**151**	**152**	**153**	**154**	**155**	**156**	**157**
Osteolepiformes	0	0	0	0	0	0	?	0	-	-	-	-	0&1	0	0
Panderichthyidae	?	?	?	?	?	0	?	0	-	-	-	-	0	0	0
Acanthostega	?	?	?	?	?	?	?	?	?	?	?	?	0	0	0
Ichthyostega	1	0	0	1	0	0	0	1	?	?	?	?	1	0	0
Crassigyrinus	1	?	?	?	0	?	?	?	?	?	?	?	?	0	1
Loxommatidae	?	?	?	?	?	?	?	?	?	?	?	?	?	?	1
Colosteidae	2	1	0	1	0	0	0	2	0	0	0	?	0	0	1
Dendrerpeton	2	1	0	1	0	0	0	?	?	?	?	?	0	0	1
Tersomius	?	?	?	?	0	?	?	?	?	?	?	?	2	?	1
Amphibamus	3	?	?	1	0	0	0	2	0	0	0	0	?	0	1
Ecolsonia	?	1	0	1	0	?	0	?	?	?	?	?	3	1	1
Proterogyrinus	1	2	0	1	0	0	0	2	1	?	?	1	2	0	1
Archeria	1	2	0	1	0	0	0	2	?	?	?	?	?	0	1
Gephyrostegidae	1	?	?	1	0	0&1	-	2	1	1	?	1	1	0	1
Kotlassia	1	1	0	1	0	0	0	?	?	?	?	?	?	?	1
Seymouria	1	2	0	1	0	0	0	2	?	?	?	?	?	?	1
Ariekanerpeton	3	?	?	?	0	?	?	2	1	1	1	2	1	3	1
Aistopoda	-	-	-	-	-	-	-	?	-	-	-	-	1	0	1
Adelogyrinidae	-	-	-	-	-	-	-	-	-	-	-	-	0	0	1
Nectridea	2	?	?	1	0	1	?	2	1	1	0	2	2	0	1
Pantylus	2	1	0	1	0	0	0	2	?	?	?	?	2	2	1
Rhynchonkos	2/3	1	0	1	0	0	0	2/3	1	0	?	?	2	?	1
Brachystelechida	3	1	0	1	0	1	-	2	1	0&1	0&1	0&2	2	0	1
Lysorophia	3	2	0	1	0	?	?	2	1	0	2	2	2	4	1
Triadobatrachus	4	0	1	0	0	0	1	?	?	?	?	?	2	4	1
Discoglossidae	4	0	1	0	1	0	1	2	0	0	0	3	2	4	1
Pipidae	4	0	1	0	1	0	1	2	0	0	0	3	2	4	1
Hynobiidae	3	1	1	0	0	0	0	2&3	0	0&1	2	2	2	4	1
Sirenidae	-	-	-	-	-	-	-	0	-	-	-	-	2	4	1
Proteidae	3	1	1	0	0	0	0	3	0	0	2	-	2	4	1
Eocaecilia	?	1	?	?	0	?	?	?	?	?	?	?	?	?	1
Rhinatrematidae	-	-	-	-	-	-	-	0	-	-	-	?	5	3	1
Ichthyophiidae	-	-	-	-	-	-	-	0	-	-	-	?	5	3	1
Limnoscelis	1	1	0	1	0	0	0	2	1	1	1	0	?	?	1
Diadectes	2	1	0	1	0	?	?	?	?	?	?	?	?	?	1
Synapsida	2&3	1	0	1	0	1	-	2	1	1	1	0	2	0	1
Captorhinidae	2	1	0	1	0	1	-	2	1	1	1	0	4	0	1
Procolophonidae	3	1	0	0	0	1	-	2	1	1	1	0	2&3	0	1

Appendix 3

The following is a list of autapomorphies of the clades and terminal taxa. In this list, ambiguous characters are indicated by an asterisk and the state of the character is indicated in parentheses when it is not "1." Reversals (any transition to a state of lower numerical value than the state present at the next more inclusive node) are indicated by a "minus" and by a number in parentheses when the reversal is not to the state "0".

Node A: -6*, 31, 36, 43*, 55, 64*, 69*, 80, 115, 119*, 120, 129, 130, 133, 143*, 146*.
Node B: 16, 26*, 29*, 49, 51(2), 136*.
Acanthostega: 10, 41*, 47*, 141*.
Ichthyostega: 19*(1/2), 22, 46, 60*, 123*, 124, 128*, 140, 150*, 155.
Node C: 1, 9*, 19*, 62*, 68, 74, 93*, 96*, 122, 128*, 134, 142, 157.
Crassigyrinus: 11, 20, 29*, 41*, 102(2), 135.
Node D: 10, 26*, 39(2), 73, 92.
Loxommatidae: 5, 82*.
Node E: 41*(2), 67, 72, 82*(2), 101*, 103*, 137*, 140*, 144*, 150*(2).
Temnospondyli: 56, 66*, -69, 115(2), -120, 141*(3), 143(2).
Colosteidae: -9, 26(2), 31(2), 37, 38*, -39, -41(1), -43, 79*, -92, 99, 118(2), 123*, 135, 136*.
Node G: 1(2), 2(1/2), 21, 47*, -55.
Dendrerpeton: 50, -133*.
Node H: 16, 45, 51*, 68*(2), 138.
Ecolsonia: 8*, 19(2), 26(2), 60, 63, 78*, 105*, 123*, 133*(2), 155*(3), 156.
Node I: 57, 98.
Tersomius: 6, 8*, -43, -47, 53, 76*, 77*, -93*, 155*(2).
Amphibamus: -39(1), 97*, 118*, 127*, 143*(3).
Node J: 19(2), 24, 32, 47*, 102(3), 117, 136*, 141*(2), 151*, 154*.
Embolomeri: 3(2), -9, 11*, 20, -39(1), -64, 144(2).
Proterogyrinus: 50*, 76*, -92, 155*(2).
Archeria: 29, 60, -72, 102(4), 114, 118(2).
Batrachomorpha: 6, 51, 105, 124, 152*, 155*.
Gephyrostegidae: 2*(1/2), 3(3), 8, 11*, 50*, 123, 127.
Node M: 1(2), 49, 60, 76*, 91, 104, 133(2), 154(2).
Seymouriamorpha: 2*, 14, 52, 61, 63, 65, 66*, 78, 80(2), 107*.
Kotlassia: -39, 123, 138.
Node O: 50*, 56(2), 59(2), 127, 131*.
Seymouria: 21, 81*, -82, -140*, 144*(2).
Ariekanerpeton: -134, 135, 143(2), 156*(3).
Node P: 16, 31(2), -36, 42, 46, 67(2), 83, 100, 114, 117(2), 143(2).

Lepospondyli: 17, -51, 66*, -72, 81, 102(5), 104(2), 113, -120.
Node R: 118(3), 123, 132, 139(2), -141*.
Aistopoda: -39(1), -64*, 69*(0/2), 78*, -82*, 84*, 85*, 86, 87*, 88, 90, 106*(2), 125, 127(3).
Adelogyrinidae: -1*(1), 13, 18*, 28*, 31(4), 37*(2), -104, 106*, 117(3), -124*, -155*.
Node S: 19*(3), 68(2), 108, 110, 127, 141*(3), 155*(2).
Nectridea: 26(2), -39*, 61*, -67(1), 84*, 87*, 109*(2), -124*, 148.
Node T: 2*(2), 4, 37*, 109*, 117(3), -133, -134, -136.
Pantylus: 15, 48, 59(2), 61*, -66, 120, 126*, 156(2).
Node U: 3(4), 6(2), -43, -47, 53, 56*, 63, -64, -152*.
Rhynchonkos: 38, -39*, -60, -93, -96, -102(3), 118*(2), 119(2), 138*.
Node V: 13, -19, 35*, 84*, 85*, 118*, 126*, 143*(3).
Brachystelechidae: 8, 12, 22*(2), -69, 120, 148.
Node W: -4, 5*(2), 15, 27(2), 28(2), 30(2), 48, -55*, 59, 72*, 75, -82, 87*, 135, 138*, -140*, 153*(2), 156(4).
Lysorophia: -9, 23*, 26*(2), -39(1), -56*, 103(2), -104, 106, -108, 118*(3), 144(2).
Lissamphibia: 8*(2), 12*, 18, 22*(2), 29*(2), -37, 57*, -67(1), 69(2), 86, 89*, 90, 98*, 106*(2), 115*(2), -117(2), 124(2), 125*, 127*(3), 145*, 146*, -151*.
Hynobiidae: -29*, 40*, -41*, 44*, 58*, 61*, -89*, 116*.
Sirenidae: 25*, 39*, 47, 94, 95(2), 97*(3), 116*, 118*(2), -150*.
Proteidae: -13*, 25*, -39*, 44*, -56*, 66*(2), -84, 97*, 116*, 142*(2), 150*(3).
Apoda: -5*, -8*, 34, -39*, 95, -96, -115*(1), 118*(2).
Eocaecilia: 26*(2), -27, -28, -29*, -30, -35, -53, -60, 71*, -81*, 97*, -102*(3).
Gymnophiona: 11(2), 25*, 33, 44*, 54*, -72*, 109(2), 118*(3), 132, 139(2), -150*, 155*(5), 156*(3).
Rhinatrematidae: -13, 51, 61*, -81*.
Ichthyophiidae: 71*.
Salientia: -6(1), 7(2), -12*, 31(4), 54*, 58*, 59(2), 65, 78, 94*, -109, 116(2), -125*, 140*, 143*(4), -144*, 149*.
Triadobatrachus: -3(3), -15, -127*(1), 142*(2).
Anura: -13*, 25*, 38*, -41*, 44*, 61*, 77*, 127(2), 131*, 139, 147, -153*, 154*(3).
Discoglossidae: -32*, -35, 40*, 55*, 97*, 117*(4).
Pipidae: -98*.
Node AC: 2*(2), 19*(5), 23, 70, 79(2), 87*, 112, 119(2), 153*, -154.
Diadectomorpha: 3*(3), 22, 68*(3), 107*, 111, 131*.
Limnoscelis: 4*, -60, -64, -91, -100, 115(2), -143(1).
Diadectes: -24, 31(3), 53, 63*, 77, 123.
Amniota: 8, 37*, 68*(4), -93, 131*(2), 148, 155*(2).
Synapsida: 3*(3/4), -32.
Reptilia: 18, 84, 107*, 127(3).
Captorhinidae: 13, 14, -43, 48, 61, -69, -133, 155(4).
Procolophonidae: 3*(4), 4*, 6(2), 12, 22(2), 31(3), 36, 37(2), 78, -79, 81, 126, -134, -140, 143(3), -146.

LITERATURE CITED

Beaumont, E. H. 1977. Cranial morphology of the Loxommatidae (Amphibia: Labyrinthodontia). *Philosophical Transactions of the Royal Society of London,* B280: 29-101.

Beerbower, J. R. 1963. Morphology, paleoecology, and phylogeny of the Permo-Pennsylvanian amphibian *Diploceraspis*. *Bulletin of the Museum of Comparative Zoology,* 130:33-108.

Berman, D. S, R. R. Reisz, and D. A. Eberth. 1985. *Ecolsonia cutlerensis*, an Early Permian dissorophid amphibian from the Cutler Formation of north-central New Mexico. *Circular of the New Mexico Bureau of Mines and Mineral Resources,* 191:1-31.

Blackburn, D. G. 1992. Convergent evolution of viviparity, matrotrophy, and specializations for fetal nutrition in reptiles and other vertebrates. *American Zoologist*, 32:313-321.

Bolt, J. R. 1969. Lissamphibian origins: possible protolissamphibian from the Lower Permian of Oklahoma. *Science,* 166:888-891.

Bolt, J. R. 1977. Dissorophoid relationships and ontogeny, and the origin of the Lissamphibia. *Journal of Paleontology,* 51:235-249.

Bolt, J. R. 1979. *Amphibamus grandiceps* as a juvenile dissorophid: evidence and implications. Pages 529-563 in: *Mazon Creek Fossils* (M. H. Nitecki, ed.). London: Academic Press.

Bolt, J. R. and R. E. Lombard. 1985. Evolution of the amphibian tympanic ear and the origin of frogs. *Zoological Journal of the Linnean Society,* 24:83-99.

Bossy, K. V. H. 1976. Morphology, paleoecology, and evolutionary relationships of the Pennsylvanian urocordylid nectrideans (subclass Lepospondyli, class Amphibia). pp. 370, Unpublished Ph.D. dissertation, Yale University, New Haven, Connecticut.

Carroll, R. L. 1964. Early evolution of the dissorophid amphibians. *Bulletin of the Museum of Comparative Zoology,* 131:161-250.

Carroll, R. L. 1968. The postcranial skeleton of the Permian microsaur *Pantylus*. *Canadian Journal of Zoology,* 46:1175-1192.

Carroll, R. L. 1969. Problems of the origin of reptiles. *Biological Reviews of the Cambridge Philosophical Society,* 44:393-432.

Carroll, R. L. 1970a. The ancestry of reptiles. *Philosophical Transactions of the Royal Society,* B257:267-308.

Carroll, R. L. 1970b. Quantitative aspects of the amphibian-reptilian transition. *Forma et Functio,* 3:165-178.

Carroll, R. L. 1988. *Vertebrate Paleontology and Evolution.* New York: W. H. Freeman and Company.

Carroll, R. L. 1991. The origin of reptiles. Pages 331-353 in: *Origins of the Higher Groups of Tetrapods—Controversy and Consensus*, (H.-P. Schultze and L. Trueb eds.). Ithaca: Comstock Publishing Associates.

Carroll, R. L. 1995. Phylogenetic analysis of Paleozoic choanates. *Bulletin du Muséum National d'Histoire naturelle de Paris, 4ème série*, 17:389-445.

Carroll, R. L. and P. J. Currie. 1975. Microsaurs as possible apodan ancestors. *Zoological Journal of the Linnean Society,* 57:229-247.

Carroll, R. L. and P. Gaskill. 1978. *The Order Microsauria.* Memoirs of the American Philosophical Society, 210 pages, Philadelphia: American Philosophical Society.

Carroll, R. L. and R. Holmes. 1980. The skull and jaw musculature as guides to the ancestry of salamanders. *Zoological Journal of the Linnean Society,* 68:1-40.

Clack, J. A. 1994. *Acanthostega gunnari,* a Devonian tetrapod from Greenland; the snout, palate and ventral parts of the braincase, with a discussion of their significance. *Meddelelser om GrØnland, Geoscience,* 31:1-24.

Duellman, W. E. and L. Trueb. 1986. *Biology of Amphibians.* New York: McGraw-Hill.

Fracasso, M. A. 1983. Cranial osteology, functional morphology, systematics and paleoenvironment of *Limnoscelis paludis* Williston. pp. 624 Unpublished Ph.D. dissertation, Yale University, New Haven, Connecticut.

Fracasso, M. A. 1987. Braincase of *Limnoscelis paludis* Williston. *Postilla,* 201:1-22.

Gaffney, E. S. 1980. Phylogenetic relationships of the major groups of amniotes. Pages 593-610 in: *The Terrestrial Environment and the Origin of Land Vertebrates* (A. L. Panchen, ed.). London: Academic Press.

Gardiner, B. G. 1982. Tetrapod classification. *Zoological Journal of the Linnean Society,* 74:207-232.

Gauthier, J., D. C. Cannatella, K. De Queiroz, A. G. Kluge, and T. Rowe. 1989. Tetrapod phylogeny. Pages 337-353 in: *The Hierarchy of Life* (B. Fernholm, K. Bremer, and H. Jornvall, eds.). New York: Elsevier Science Publishers B. V. (Biomedical Division).

Gauthier, J., A. G. Kluge, and T. Rowe. 1988. The early evolution of the Amniota. Pages 103-155 in: *The Phylogeny and Classification of the Tetrapods, Volume 1: Amphibians, Reptiles, Birds* (M. J. Benton, ed.). Oxford: Clarendon Press.

Gregory, J. T. 1965. Microsaurs and the origin of captorhinomorph reptiles. *American Zoologist,* 5:277-286.

Heaton, M. J. 1979. Cranial anatomy of primitive captorhinid reptiles from the Late Pennsylvanian and Early Permian Oklahoma and Texas. *Bulletin of the Oklahoma Geological Survey,* 127:1-84.

Heaton, M. J. 1980. The Cotylosauria: A Reconsideration of a Group of Archaic Tetrapods. Pages 497-551 *The Phylogeny and Classification of the Tetrapods, Volume 1: Amphibians, Reptiles, Birds* (M. J. Benton, ed.). Oxford: Clarendon Press.

Holmes, R. 1984. The Carboniferous amphibian *Proterogyrinus scheelei* Romer, and the early evolution of tetrapods. *Philosophical Transactions of the Royal Society,* B306:431-527.

Jenkins, F. A., Jr. and D. M. Walsh. 1993. An Early Jurassic caecilian with limbs. *Nature,* 365:246-249.

Laurin, M. and R. R. Reisz. 1995. A reevaluation of early amniote phylogeny. *Zoological Journal of the Linnean Society,* 113:165-223.

Lombard, R. E. and S. S. Sumida. 1992. Recent progress in understanding early tetrapods. *American Zoologist,* 32:609-622.

Lombardi, J. 1994. Embryo retention and evolution of the amniote condition. *Journal of Morphology*, 220 (3):368.

Maddison, W. P. and D. R. Maddison. 1992. *MacClade: Analysis of phylogeny and character evolution.* Sunderland, Massachusetts: Sinauer Associates.

Milner, A. R. 1988. The relationships and origin of living amphibians. Pages 59-102 in: *The Phylogeny and Classification of the Tetrapods, Volume 1: Amphibians, Reptiles, Birds* (M. J. Benton, ed.). Oxford: Clarendon Press.

Milner, A. R. 1990. The radiations of temnospondyl amphibians. Pages 321-349 in: *Major Evolutionary Radiations* (P. D. Taylor and G. P. Larwood, eds.). Oxford: Clarendon Press.

Milner, A. R. 1993. The Paleozoic relatives of lissamphibians. *Herpetological Monographs,* 7:8-27.

Mishler, B. D. 1994. Cladistic analysis of molecular and morphological data. *American Journal of Physical Anthropology,* 94:143-156.

Nussbaum, R. A. 1977. Rhinatrematidae: A new family of caecilians (Amphibia: Gymnophiona). *Occasional Papers of the Museum of Zoology, University of Michigan,* 682:1-30.

Panchen, A. L. and T. R. Smithson. 1988. The relationships of the earliest tetrapods. Pages 1-32 in: *The Phylogeny and Classification of the Tetrapods, Volume 1: Amphibians, Reptiles, Birds* (M. J. Benton, ed.). Oxford: Clarendon Press.

Parsons, T. S. and E. E. Williams. 1963. The relationships of the modern Amphibia: A re-examination. *Quarterly Review of Biology,* 38:26-53.

Reisz, R. R. 1972. Pelycosaurian reptiles from the Middle Pennsylvanian of North America. *Bulletin of the Museum of Comparative Zoology,* 144:27-62.

Reisz, R. R. 1986. Pelycosauria. *Encyclopedia of Paleoherpetology,* 17A:1-102.

Romer, A. S. 1950. The nature and relationships of the Paleozoic microsaurs. *American Journal of Science,* 248:628-654.

Romer, A. S. 1969. The cranial anatomy of the Permian amphibian *Pantylus. Breviora,* 314:1-37.

Schoch, R. R. 1992. Comparative ontogeny of Early Permian branchiosaurid amphibians from Southwestern Germany. *Palaeontographica. Abteilung A. Palaeozoologie-Stratigraphie,* 222:43-83.

Smithson, T. R. 1982. The cranial morphology of *Greererpeton burkemorani* Romer (Amphibia: Temnospondyli). *Zoological Journal of the Linnean Society,* 76:29-90.

Swofford, D. L. 1993. *PAUP: Phylogenetic Analysis Using Parsimony.* Champaign, Illinois: Illinois Natural History Survey.

Trueb, L. and R. Cloutier. 1991. A phylogenetic investigation of the inter- and intrarelationships of the Lissamphibia (Amphibia: Temnospondyli). Pages 223-313 in:, *Origins of the Higher Groups of Tetrapods—Controversy and*

Consensus (H.-P. Schultze and L. Trueb, eds.). Ithaca: Cornell University Press.

Vaughn, P. P. 1962. The Paleozoic microsaurs as close relatives of reptiles, again. *American Midland Naturalist,* 67:79-84.

Wake, D. B. and R. Lawson. 1973. Developmental and adult morphology of the vertebral column in the plethodontid salamander *Eurycea bislineata*, with comments on vertebral evolution in the Amphibia. *Journal of Morphology,* 139:251-300.

Wellstead, C. F. 1991. Taxonomic revision of the Lysorophia, Permo-Carboniferous lepospondyl amphibians. *Bulletin of the American Museum of Natural History,* 209:1-90.

White, T. E. 1939. Osteology of *Seymouria baylorensis* Broili. *Bulletin of the Museum of Comparative Zoology,* 85:325-409.

CHAPTER 3

CROWN-CLADES, KEY CHARACTERS AND TAXONOMIC STABILITY: WHEN IS AN AMNIOTE NOT AN AMNIOTE?

Michael S. Y. Lee

Patrick S. Spencer

INTRODUCTION

The traditional definition of the taxon Amniota is based on a key character: anything with an amniote egg is an amniote (e.g., Kingley, 1917; Parker and Haswell, 1940). In more recent times, however, there has been a significant shift in systematics towards defining taxa based on monophyly and ancestry (Hennig, 1966; de Queiroz and Gauthier, 1992). Phylogenetic taxonomy–the application of taxon names to clades stemming from a particular ancestor–would refine the traditional definition of Amniota to the following: the clade consisting of the first animal to possess the amniote egg and all its descendants. De Queiroz and Gauthier (1990, 1992) term this an apomorphy-based definition. The content of Amniota, so defined, is straightforward when only extant taxa are being considered. The phylogenetic relationships of extant tetrapods are well established, and the amniote egg is morphologically distinctive and readily observable. Thus, turtles, lepidosaurs, archosaurs (including birds), and mammals are unequivocally amniotes, whereas lissamphibians and all other organisms are excluded from the Amniota. Fossils pose a problem, however, as eggs are rarely fossilised, and are very difficult to assign to taxa even when they are well preserved (e.g., see Norell *et al.*, 1994). When dealing with fossil forms, therefore, it becomes more difficult to make the distinction between amniotes and non-amniotes, and the content of Amniota therefore becomes uncertain. For this reason (among others), many workers have advocated restricting Amniota to the clade bounded by extant amniotes (e.g. Gauthier *et al.*, 1988a, b; Sumida and Lombard, 1991; Berman *et al.,* 1992; Laurin and Reisz, 1995). Amniota, in this sense, denotes the clade composed of the most recent common ancestor of extant amniotes and all its descendants. This type of approach has been termed a node-based, crown-clade approach (de Queiroz and Gauthier, 1992).

It has been suggested (e.g., Gauthier *et al.*, 1988a) that restriction of widely known names to crown-clades increases both precision and stability, compared to traditional (apomorphy-based) definitions. "Precision," as used here, means the following: Given a particular phylogeny, and a particular phylogenetic definition of a taxon name, how accurately can we delineate its boundaries? Because phylogenetic definitions of taxon names attach names to clades stemming from a particular ancestor, the question of "precision" ultimately boils down to how accurately can we pinpoint the ancestor implicated in the definition of the taxon name. "Stability," on the other hand, refers to the contents of the clade: In the face of changing ideas of phylogenetic relationships, how constant is the composition of this clade? Even if we can identify the ancestor, and thus the boundary of the clade, we still have to determine which organisms fall within the clade and which organisms fall outside it. There is a third related concept, nomeclatural priority: What name should be applied to this clade? These three rather different concepts have been implicated in previous discussions, but they have not been clearly defined, and have often been confused with each other. For instance, Bryant (1994) uses the same term, "stability," to refer all three concepts. He states that a problem with apomorphy-based definitions is that "references to biological significance necessarily allow for disagreement as to what constitutes a 'significant' character...and therefore perpetuate instability" (p. 127: "imprecision" according to our terminology), crown-clades are highly corroborated by hard and soft anatomical, molecular, and behavioural characters and thus "promote...stability" (p.124: "stability" according to our terminology), and "stability regarding the use of names is probably best achieved through the codes of nomenclature" (p.129: "priority" according to our terminology). Failure to specify the different senses in which the word stability is used in different parts of the paper creates confusion and is often misleading. Bryant concludes that "stability regarding use of names is probably best achieved through codes of nomenclature" (p.129). Based on his other usages of stability in the paper, this statement implies that introduction of some rigid code of nomenclature (specially concerned with the naming of clades) will also make the named clades more precisely definable and their contents more stable despite changing views of phylogeny. However, determining by some new decree what are the rightful names for the taxa such as "the first organism with a dentary-squamosal jaw joint and all its descendants" and "crown-group birds" will not help us delineate the boundaries of those clades more precisely or help us resolve whether enigmatic (often fossil) taxa fall within or outside the boundaries of those named clades. Precision, stability, and priority are therefore three very different problems.

Here, we use the problem of the definition of Amniota to investigate the first two problems. We provide a revised phylogeny of "crown-clade" amniotes and their nearest relatives and evaluate the two widely used definitions of Amniota, the traditional apomorphy-based definition and a more recent, node-based, crown-clade definition. We present evidence that some taxa lying outside the extant crown-clade possessed the amniote egg: These taxa are amniotes under the traditional (apomorphy-based) definition of Amniota but are excluded from Amniota under the crown-clade definition. Using our phylogeny as a case study, we then examine critically the suggestion that crown-clade definitions are more precise (with respect to boundaries) and stable (with respect to content) than more traditional definitions. We agree that, in almost all cases, crown-clade definitions are more precise, but we suggest that both types of definitions are equally unstable.

It should be emphasised that we do not question here the significant arguments in favor of phylogenetic, rather than typological, definitions of taxon names (see deQueiroz and Gauthier, 1990, 1992). Rather, we are concerned with the following question: Given that one wishes to adopt phylogenetic definitions, are crown-clade definitions superior to other types of phylogenetic definitions?

PHYLOGENETIC FRAMEWORK

Our revised phylogeny of Anthracosauria (*sensu* Gauthier *et al.*, 1988b), the clade consisting of amniotes and their nearest amphibian-grade relatives, is depicted in Figure 1. This is based on a cladistic analysis of anthracosaurians summarised in the appendix. This study considered 45 characters for seven terminal taxa: Embolomeri, Gephyrostegidae, Seymouriamorpha, Solenodonsauridae, Diadectomorpha, Synapsida, and Sauropsida. It was extracted from a much larger cladistic analysis that considered both the external relationships of amniotes and the systematics of basal amniotes (Spencer, 1994; P. Spencer and M. Wilkinson, manuscript in preparation).

The topology of our most parsimonious cladogram is similar to those proposed in other recent studies, such as Gauthier *et al.* (1988b), Panchen and Smithson (1988), and Berman *et al.* (1992). However, the characters diagnosing the nodes are slightly different. Some characters are new; of the characters proposed by previous workers, some are used unchanged, and others are reinterpreted as applying to different nodes, others have been rejected because they are rather ill-defined or highly variable within basic taxa. Previous cladistic studies that have used the characters are identified: where an asterisk precedes the citation, the character was used to diagnose a different grouping than it does in the current study. A full listing of primitive and

Figure 1. A revised phylogeny of reptiliomorph tetrapods, based on the cladistic analysis discussed in Appendix 1. The Bremer index and bootstrapping frequency for each clade is shown. The synapomorphies diagnosing each clade are discussed in the text. ➔

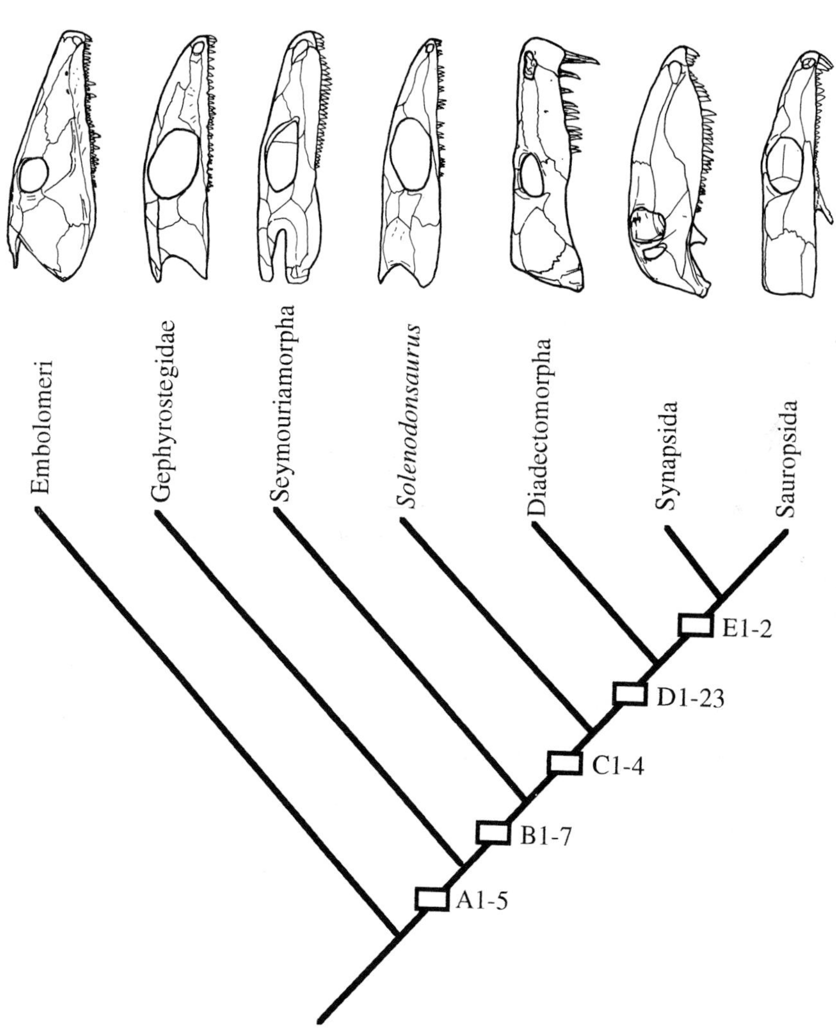
Embolomeri
Gephyrostegidae
Seymouriamorpha
Solenodonsaurus
Diadectomorpha
Synapsida
Sauropsida
E1-2
D1-23
C1-4
B1-7
A1-5

derived states for all characters, and the distribution of these states, and the details of the cladistic analysis, are presented in the appendix.

Groupings are not formally named, as the application of names to clades is the subject of this paper. They are instead denoted by letters. Also, we do not consider the enigmatic *Westlothiana*, because this animal is currently undergoing further study by other workers.

Node A.

A1: A transverse flange of the pterygoid is present (character 17: 0➔1, ci = 1) (Panchen and Smithson, 1988).

A2: The posterior coronoid bears a prominent, discrete posterodorsal process–the "coronoid process" (20: 0➔1, ci = 1) (Gauthier *et al.*, 1988b).

A3: There are 25 or fewer presacral vertebrae (31: 0➔1, ci = 1).

A4: The atlas pleurocentrum is a single ossification (33: 0➔1, ci = 1) (Sumida and Lombard, 1991).

A5: The forelimbs and hindlimbs are both well-developed, being at least 20% longer than the skull (44: 0➔1, ci = 1).

Node B.

B1: The skull table and cheek are united in a firm sutural contact (7: 0➔1, ci = 1).

B2: The tabular is partly exposed on the occiput (10: 0->1, ci = 1) (Gauthier *et al.,* 1988b).

B3: A cranioquadrate passage (space) is present between the paroccipital process of the opisthotic and the quadrate ramus of the pterygoid (16: 0➔1, ci = 1).

B4: A caniniform region ("canine peak") is present, consisting of a series (4-6) of enlarged teeth near the anterior end of the maxilla (25: 0➔1, ci = 1) (Gauthier *et al.*, 1988b).

B5: The large anterior tusks on the dentary, medial to the marginal teeth, are lost (30: 0➔1, ci = 1) (Gauthier *et al.*, 1988b).

B6: The intercentra are reduced, being 40% as long as the (pleuro)centra, or shorter (38: 0->1, ci = 1).

B7: The ilium is an extensive plate, the bifurcation of the ilium being reduced (43: 0➔1, ci = 1) (Gauthier *et al.*, 1988b; see also Sumida, this volume).

Node C.
C1: The posterolateral (maxillary) ramus of the premaxilla is a well-developed process, and forms much of the ventral border of the external naris and much of the anterior margin of the internal naris (2: 0➔1, ci = 1).
C2: The intertemporal is lost (6: 0➔1, ci = 1) (*Gauthier *et al.*, 1988b; *Berman *et al.*, 1992; *Smithson *et al.*, 1994).
C3: The supratemporal forms the posterolateral corner of the skull table (8: 0➔1, ci = 1) (*Berman *et al.*, 1992; *Smithson *et al.*, 1994).
C4: The small palatal denticles are arranged in a discrete anterolateral row across the pterygoid and palatine (27: 0➔1, ci = 1).
C5: The large tusks on the palatine are lost (29: 0➔1, ci = 1) (Gauthier *et al.*, 1988b).

Node D.
D1: The lateral line grooves on the skull table are completely absent (1: 0➔1, ci = 1) (*Gauthier *et al.*, 1988b).
D2: The septomaxilla lies entirely within the external naris (3: 0➔1, ci = 1) (*Gauthier *et al.*, 1988b).
D3: The dorsal ("amphibian") temporal notch is reduced or absent (5: 0➔1, ci = 1) (Gauthier *et al.*, 1988b).
D4: The tabular is located entirely on the occiput (10: 1➔2, ci = 1) (Gauthier *et al.*, 1988b).
D5: The postparietal is located on the occiput (11: 0➔1, ci = 1) (Gauthier *et al.*, 1988b).
D6: A discrete supraoccipital ossification is present (12: 0➔1, ci = 1) (Gauthier et al. 1988b; *Panchen and Smithson, 1988; Berman *et al.*, 1992).
D7: The opisthotic does not contact the postparietal or parietal (13: 0➔1, ci = 1).
D8: The groove or foramen on the occiput for the vena capitis dorsalis is lost (15: 0➔1, ci = 1).
D9: A row of teeth is present on the margin of the transverse flange of the pterygoid (18: 0➔1, ci = 1).
D10: The ectopterygoid is short (anteroposterior dimension) and broad (lateromedial dimension); its length:breadth ratio is 2 or less (19: 0➔1, ci = 1).

D11: The palate is "open," with wide interpterygoid vacuities and long, narrow pterygoids (20: 0➔1, ci = 1) (Clack, 1988).
D12: The first (anteriormorst) coronoid is lost (21: 0➔1, ci = 1) (Gauthier *et al.*, 1988b; Berman *et al.*, 1992).
D13: The presplenial is lost (22: 0➔1, ci = 1). (Gauthier et al. 1988b).
D14: There is a well-developed caniniform tooth on the maxilla (26: 0➔1, ci = 1) (*Gauthier *et al.*, 1988b).
D15: A longitudinal tooth row is present on the ectopterygoid (28: 0➔1, ci = 1).
D16: There are two or more pairs of robust sacral ribs (42: 0➔1, ci = 1). Primitively, only one pair of enlarged sacral ribs is present, although, in some seymouriamorphs, a second, very slender pair of ribs contacts the ilium (Sumida, 1990) (Gauthier *et al.*, 1988b).
D17: The atlas pleurocentrum is positioned directly above the axis intercentrum (32: 0➔1, ci = 1) (Sumida and Lombard, 1991; Sumida *et al.*, 1992).
D18: The atlas neural spine is modified into small, posterodorsally directed epipophysis (36: 0➔1, ci = 1) (Sumida and Lombard, 1991).
D19: The axis neural arch and axis pleurocentrum are sutured or fused to each other (37: 0➔1, ci = 1) (Sumida and Lombard, 1991).
D20: The rib cage is ventrally complete (39: 0➔1, ci = 1). We have not confirmed this character for ourselves: It was mentioned in a textbook (Zug, 1993) that did not cite a primary source.
D21: There are one or two discrete coracoid ossifications in shoulder girdle (40: 0➔1, ci = 1) (*Gauthier *et al.*, 1988b; Smithson *et al.*, 1994).
D22: The interclavicle possesses a narrow, rod-like posterior median stem - the parasternal process (41: 0➔1, ci = 1) (Clack, 1988).
D23: The bifurcation of the ilium is completely lost (43: 1➔2, ci = 1) (*Gauthier *et al.*, 1988b; see also Sumida, this volume).

Node E.
E1: The squamosal enters the margin of the posttemporal opening (9: 0➔1, ci = 1) (Gauthier *et al.*, 1988b).
E2: The sulcus cartilaginous meckelii is exposed as an anterior groove on the medial surface of the mandible (23: 0➔1, ci = 1).

E3: The atlas pleurocentrum and axis intercentrum are sutured or fused to each other (34: 0➔1. ci = 1). This trait is polymorphic in diadectomorphs, and thus might instead diagnose node D (Gauthier *et al.*, 1988b; Sumida and Lombard, 1991).
E4: An astragalus is present, ossifying from a single center (45: 0➔1, ci = 1). Among diadectomorphs, large individuals of *Diadectes* occasionally possess an astragalus-like structure, but this is a compound ossification and not equivalent to the structure in synapsids and sauropsids (Rieppel, 1993; Gauthier *et al.*, 1988b; *Panchen and Smithson, 1988; Smithson *et al.*, 1994; Sumida, this volume).

Our conclusions differ from other analyses (e.g. Gauthier *et al.*, 1988b; Panchen and Smithson, 1988; Sumida *et al.*, 1992) in that we do not recognise “Anthracosauroideae”, a group consisting of Embolomeri and Gephyrostegidae. Gephyrostegids are more closely related to higher reptiliomorphs than are embolomeres. Also, nycteroleterids, nyctiphruretids, and lanthanosuchids, often considered to be amphibian-grade relatives of amniotes (e.g., Milner, 1993), are not specifically shown on our cladogram because they have been demonstrated by us to be unequivocally amniotes: They lie on the chelonian stem within Sauropsida (Spencer, 1994; Lee, 1995, 1996).

Gauthier *et al.* (1988b) proposed many other characters to diagnose node E (crown-clade amniotes), but these are invalid. Loss of the intertemporal also characterizes diadectomorphs (Berman *et al.*, 1992) and *Solenodonsaurus,* and thus diagnoses node C. The convex occipital condyle is not found in many basal amniotes: For instance, it is concave in turtles, lanthanosuchids, pareiasaurs, procolophonids, and nyctiphruretids (Lee, 1995). The caniniform maxillary tooth and the two coracoid ossifications are both present in diadectomorphs and diagnose node D. The absence of dorsal osteoderms also occurs in diadectomorphs and most seymouriamorphs; moreover, dermal armor occurs in some basal amniotes (pareiasaurs, *Sclerosaurus,* turtles, and some captorhinids; Lee, 1995).

It is clear that the groupings vary in the amount of corroboration. Clade D is diagnosed by a very large number of synapomorphies, and the Bremer index (Bremer, 1988) and bootstrapping (Sanderson, 1995) show that it is very robust (Fig. 1) -

more robust than previously proposed (e.g., Panchen and Smithson, 1988; Gauthier *et al.*, 1988b). Clade E, on the other hand, is, contrary to Gauthier *et al.* (1988b), very weakly corroborated. The significance of this will be discussed in the following section.

DEFINING AMNIOTA: KEY-CHARACTER AND CROWN CONCEPTS

Given the above described phylogeny, applying the crown-clade definition of Amniota is very straightforward. Amniota would consist of the most recent common ancestor of clade E and all its descendants. Thus, the boundaries of this clade are very clear, and this definition is therefore very precise (Fig. 2).

Drawing the boundary of Amniota under the traditional definition is more problematical, because it requires knowledge of when the amniote egg arose. With fossil taxa, this requires making complex and slightly dubious paleobiological inferences. Clearly, parsimony dictates that the latest time the amniote egg could have arisen is in the most recent common ancestor of living amniotes (i.e., clade E). The earliest possible time can also be determined. Seymouriamorphs had free-swimming larvae with external gills (Špinar, 1952) and thus clearly lacked the amniote egg, which must be deposited terrestrially. Thus, the amniote egg must have arose after the latest common ancestor of node B. This leaves diadectomorphs and *Solenodonsaurus* in purgatory. However, a few inferences can be made about diadectomorphs. Diadectomorphs include forms such as *Diadectes,* which have complex dental adaptations indicative of a diet of high fiber terrestrial vegetation, an interpretation supported by their capacious, barrel-shaped chests which were presumably used for fermentation of ingested material (Hotton *et al.*, this volume). No

Figure 2. Alternative definitions of Amniota. The circle denotes the ancestor specified by the crown-clade, node-based definition of Amniota. The thick line denotes where the amniote egg could have arisen, and thus, the possible position of the ancestor specified by an apomorphy-based definition. The amniote egg is most likely to have arose along the portion of the phylogeny denoted by the solid shading, and is less likely to have arose along the portion denoted by the cross-hatched shading. The alternative position of Diadectomorpha, proposed by Berman *et al.* (1992), is indicated. ➔

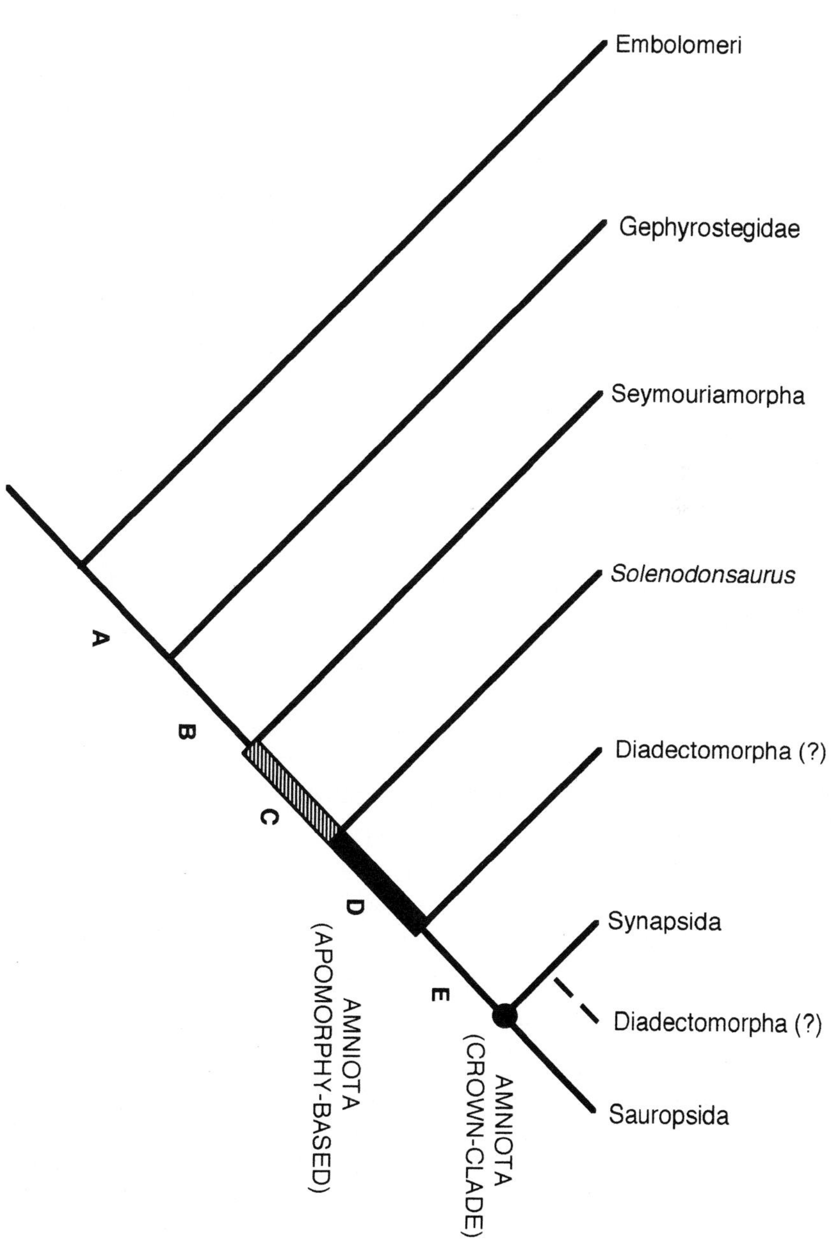
Embolomeri
Gephyrostegidae
Seymouriamorpha
Solenodonsaurus
Diadectomorpha (?)
Synapsida
Diadectomorpha (?)
Sauropsida
A
B
C
D
E
AMNIOTA
(APOMORPHY-BASED)
AMNIOTA
(CROWN-CLADE)

amphibian (anamniote)-grade taxon has ever managed to invade the niche of terrestrial herbivore, and a causal reason has been suggested for this. The endosymbionts required for digestion of high fiber vegetation must be obtained immediately after hatching fromterrestrial microbial decomposers (Modesto, 1992; Zug, 1993; Hotton and Beerbower, 1994; Hotton *et al.*, this volume). This constraint therefore prevents anamniote (aquatically reproducing) taxa from feeding on high-fiber vegetation. The fact that advanced diadectomorphs invaded this niche suggests that diadectomorphs had evolved the amniote egg.

Further circumstantial evidence exists, although this evidence is more dubious. Diadectomorphs, like crown-clade amniotes and unlike more basal anthracosaurians, have a complete rib cage and thus may have been capable of pulmonary respiration. Their large size and consequent low surface area to volume ratios might also have hindered the use of cutaneous respiration (S. Sumida, personal communication). This hypothesised switch in respiratory mechanisms suggests that the skin in diadectomorphs might have been, as in crown-clade amniotes, heavily keratinized and less permeable (Zug, 1993). Therefore, diadectomorphs probably had obligate terrestrial habits and thus terrestrial reproduction. Similarly, the loss of the lateral line system at node D suggests a switch to a more terrestrial existence.

The evidence that diadectomorphs possessed the amniote egg is strong, although not overwhelming. This leaves only *Solenodonsaurus* unaccounted for. Unfortunately, little can be said about the likely reproductive mode of this animal. Larvae are not known (which might reveal amphibian reproductive habits) nor does it have any known adaptations suggestive of amniote reproduction. However, a large number of character changes separate diadectomorphs and higher reptiliomorphs (node D) from more basal forms such as *Solendonsaurus*, suggesting that some radical ecological and/or developmental breakthrough occurred at the base of node D. Based on the morphological gap separating diadectomorphs from *Solenodonsaurus,* the amniote egg probably arose at some point along the amniote stem after *Solenodonsaurus* but before diadectomorphs. This, of course, is little more than an educated guess, but it is the best we can do.

Thus, the boundaries of Amniota under the traditional definition are imprecise. The amniote egg could have evolved anywhere between the last common ancestor of node B and the last ancestor of node D (Figure 2), although it probably evolved toward the later end of this period. Amniota therefore might or (more likely) might not include *Solenodonsaurus,* and any other taxa discovered in the future which lie in the area of uncertainty. This uncertainty is a result solely of our inability to determine exactly when the amniote egg arose–and thus, where the boundary separating Amniotes and nonamniotes lies. Even if we could directly observe the evolution of this structure, however, imprecision would remain. It is unlikely that all the complex adaptations of the amniote egg (chorion, amnion, allantois, porous calcified shell, air space, and large yolk sac) arose instantaneously: There would still be a problem of having to arbitrarily decide when the egg had evolved enough of these structures to be called "amniote". Thus, the traditional definition of Amniota is much more imprecise than the crown-clade definition.

As mentioned previously, crown-clade definitions have also been suggested to be more stable in content than more inclusive, apomorphy-based definitions in the face of changing ideas about phylogenetic relationships.

In our phylogeny, the most robust clade is clade D: the diadectomorph-synapsid-sauropsid clade. This clade constitutes Amniota under the most reasonable interpretation of the traditional definition (see above) and is corroborated by over twenty synapomorphies. The Bremer index (Bremer, 1988) and bootstrapping frequency (Sanderson, 1995) indicate that this clade is very robust (Fig. 1). The most weakly corroborated clade is clade E, Amniota under the crown-clade definition. Only three unambiguous synapomorphies unite synapsids and sauropsids, to the exclusion of diadectomorphs, and there are two characters that conflict with this arrangement. The Bremer index shows that only an extra step is needed to break up this clade, and the bootstrapping frequency of this clade is also very low (Fig. 1). As expected, an alternative arrangement has recently been proposed (Fig. 2): Berman *et al.*, (1992) unite diadectomorphs with synapsids on the basis of one character, the "otic trough," a flange of the opisthotic that projects ventrolaterally from the posterior margin of the fenestra ovalis. The alternative phylogenetic arrangement of Berman *et al.* does not affect the content of Amniota as traditionally defined: No matter which phylogeny is accepted, apomorphy-based Amniota contains diadectomorphs, synapsids, and sauropsids. However, the content of the crown-clade definition of Amniota is affected. Under our phylogeny, crown-clade Amniota excludes diadectomorphs, whereas under Berman *et al.*'s phylogeny, diadectomorphs are part of crown-clade Amniota. The reason for this is clear–the apomorphy-based definition of Amniota attaches the taxon name to a highly corroborated clade that is unlikely to be altered, wheareas the crown-clade definition of Amniota attaches the name to a weakly corroborated clade which is susceptible to being dismantled.

It seems, therefore, that previous suggestions that crown-clade definitions are more stable in content are not true, at least in our particular example. The structure of these arguments must now be examined to see if they are valid in general.

Gauthier (1984, 1986) and Gauthier *et al.* (1988a, 1989) have suggested that crown-groups are more highly corroborated and stable in content because "there should be a larger number of synapomorphies diagnostic of extant sister groups, because some kinds of evidence will not be preserved in fossils" (Gauthier *et al.*, 1988a: p. 182). Similarly, Laurin and Reisz (1995) apply well-known names such as Amniota to crown-clades "to ensure maximum stability of their content and diagnosis" (p. 172). An early discussion of this argument was presented in Gauthier (1986). He suggested that restricting Aves to crown-clade members increases stability in content, as the relationships of crown-clade birds to one another, and to other extant diapsids, are stable. Also, compared to other extant diapsids, extant birds are diagnosed by a large suite of synapomorphies, including soft anatomical structures observable in all these forms. Thus, crown-clades are diagnosed by a large number of synapomorphies and are stable in content.

However, as noted by Gauthier (1986), fossils bridge this large morphological gap separating extant birds from other extant diapsids. Many of the synapomorphies separating modern birds from other modern diapsids actually diagnose larger groups, such as "Avialae" (crown-group birds, *Archaeopteryx,* and some other fossil forms), Theropoda, and Dinosauria. Also, the phylogenetic position of many fossils is controversial, often because of incomplete or poor preservation and paucity of nonosteological data. For instance, some workers (e.g., Thulborn, 1984) have claimed that tyrannosaurs are actually closer to modern birds than is *Archaeopteryx.* A traditional definition of Aves, based on the origin of feathers, is therefore undesirable, as fewer features separate this clade ("Avialae" of Gauthier, 1986) from derived theropods, the phylogenetic position of many fossil taxa (e.g., tyrannosaurs) with respect to this clade is uncertain, and the content of this clade is thus prone to change.

Gauthier (1986) therefore argues that the *extant* content of crown-clades is highly corroborated and stable, whereas the *total* (extant plus fossil) content of more inclusive, apomorphy-based clades is less corroborated and more stable. It is true that the interrelationships of extant taxa with each other will usually be well established, and that the phylogenetic position of many fossil taxa will be uncertain. However, this applies whether we adopt crown-clade or more inclusive, "traditional" definitions. Regardless of which definition of Aves we adopt, the total (extant plus fossil) content of Aves will be uncertain, because of arguments over the correct phylogenetic position of controversial fossil taxa. Similarly, when fossils are considered, the number of features separating Aves (however defined) from non-Aves will be small. For instance, there are few synapomorphies separating crown-clade Aves from taxa such as *Sinornis* and *Ichthyornis,* and the position of many fragmentary fossils such as *Alexornis* and Enantiornithes within or outside this crown-clade is uncertain (Cracraft, 1986).

However, when only extant taxa are considered, the synapomorphies diagnosing Aves (whether crown-clade or more inclusive) will be numerous and the content correspondingly stable. There will be fewer problems with missing data, soft anatomical features will be known for all taxa, and a large morphological gap separates modern birds from all other amniotes. For instance, even if we adopt a more inclusive (traditional, apomorphy-based) definition of Aves, as the clade composed of the first feathered tetrapod and all its descendants, the extant content of this clade is stable, and numerous apomorphies–many of them soft anatomicalñdistinguish the extant members of this clade from other extant amniotes.

When both extant and fossil forms are considered, the content of both crown and more inclusive (often apomorphy-based) clades will usually be unstable and the number of diagnostic synapomorphies small. When only extant taxa are considered, the content of both crown and more inclusive clades will usually be more stable, and the number of diagnostic synapomorphies large. Gauthier and colleages (Gauthier, 1984, 1986; Gauthier *et al.*, 1988a, 1989) conflate the two issues by arguing that, when only extant taxa are considered, the content of crown-clades is stable and corroborated by numerous characters, but when extant and fossil forms are considered, the content of more inclusive clades is unstable and poorly corroborated. For instance, Gauthier (1986 p. 36) states that although the membership of "of Avialae [a more inclusive clade]... is expected to vary *according to the discovery of new fossils,* the contents and diagnosis of the taxon Aves [a crown clade] *in the context of extant amniotes* will remain unchanged" [italics added]. The two alternative approaches are therefore being evaluated on different criteria. There is no reason why a crown-clade should be more highly corroborated than a "traditional," more inclusive, apomorphy-based clade. As our amniote example shows, this is often not the case. Thus, adopting a crown-clade approach does not increase taxonomic stability. Indeed, because the boundaries of traditional, more inclusive clades are usually defined on anatomical features or morphological gaps perceived (rightly or wrongly) to be significant, such clades would probably tend to be more highly corroborated than crown-clades.

CONCLUSIONS

As expected, when attempting to ascertain the boundaries of Amniota, the crown-clade approach is much more precise than the apomorphy-based approach. This is not surprising, because the former approach is more theory neutral, depending merely on the topology of the cladogram, whereas the latter approach requires much paleobiological inference and consequent uncertainty. However, the crown-clade definition of Amniota might be objectionable to some biologists: As we have shown in the previous discussion, this definition excludes from the Amniota some taxa that appear to have possessed the amniote egg. Similar issues have recently been raised regarding the taxonomy of other groups: Crown-clade definitions would exclude *Archaeopteryx* from Aves (Gauthier, 1986), *Ichthyostega* and *Acanthostega* from Tetrapoda (Lebedev and Coates, 1995), and *Morganucodon* from Mammalia (Rowe 1988). The desirability or otherwise of such redefinitions have been discussed elsewhere (de Queiroz and Gauthier, 1992; Lucas, 1992; Bryant, 1994) and will not be repeated here. However, contrary to previous suggestions, we argue that crown-clade definitions are neither more stable in content nor highly corroborated, than traditional, apomorphy-based definitions: Both are subject to changes induced by changing views of phylogeny. The contents of clades, and thus taxa, will always be subject to revision. However, this instability might be viewed as an asset, rather than a liability. If systematics is to be informative, then changes in our knowledge of life's diversity should be reflected by corresponding changes in our taxonomies (Gaffney, 1979). Indeed, the most stable taxa are notoriously uninformative, paraphyletic wastebaskets such as "invertebrates" and "fish".

ACKNOWLEDGMENTS

We thank Mike Coates, Jenny Clack, Stuart Sumida, and Mark Wilkinson for helpful comments on the manuscript, and Jacques Gauthier for discussion. We thank Stuart Sumida and Karen Martin for the invitation to contribute to this volume. During preparation of this manuscript, MSYL was supported by the Association of Commonwealth Universities, the Cambridge Philosophical Society, and the Australian Research Council. PSS was supported by the Natural Environment Research Council (United Kingdom) and hte Nuffield Foundation.

APPENDIX 1 : A PRELIMINARY ANALYSIS OF THE EXTERNAL RELATIONSHIPS OF AMNIOTES.

The following are the details of the cladistic analysis on which the arguments in the present chapter are based. The in-group taxa considered in this analysis are Embolomeri, Gephyrostegidae, Seymouriamorpha, Solenodonsauridae,

Diadectomorpha, Synapsida, and Sauropsida (*sensu* Laurin and Reisz, 1995). The in-group taxa together form a monophyletic grouping, and the nearest outgroups are the Baphetidae ("Loxommatoidea"), Crassigyrinidae, and Colosteoidea (Spencer, 1994). The characters are listed below, grouped according to anatomical regions. The distribution of the various character states across the ingroup taxa are listed in Table 1. Character states were polarised by outgroup comparison, and the primitive state for the ingroup is listed as "0".

The data were analyzed using the exhaustive search procedure in PAUP 3.1.1 (Swofford, 1993). The tree was rooted by the out-group method with an all-zero ancestor (a hypothetical taxon possessing all the states inferred to be primitive for the in-group). Characters polymorphic in terminal taxa were treated as uncertainty (i.e., the primitive character state in the terminal taxon was uncertain). Whether multistate characters are ordered according to morphoclines or unordered (see Wilkinson, 1992) does not affect the topology and length of the most parsimonious tree. Only one tree was found [51 steps, consistency index (ci) 0.922, retention index 0.949], and characters were optimized onto this tree using delayed transformation, which assumes that apomorphies diagnose the least inclusive clades consistent with the phylogeny. The characters diagnosing each clade are discussed in the main text. Bootstrapping and the Bremer (or support) index (Bremer, 1988; Eernisse and Kluge, 1993) were ascertained for each clade (Fig. 1): Multistate characters were ordered during these calculations.

Skull Roof

1. Lateral line grooves on skull roof: Ddeply etched (0); weak or absent (1).
2. Posterolateral (maxillary) ramus of premaxilla: short, largely or entirely excluded from margins of internal and external nares (0); long, forming the ventral margin of the external naris and anterior margin of the internal naris (1).
3. Septomaxilla: exposed on skull roof immediately behind external naris (0); lies entirely within external naris (1).
4. Frontal: excluded from orbital margin (0); enters orbital margin (1).
5. Dorsal ("amphibian") temporal emargination: deeply incised (0); weak or absent (1).
6. Intertemporal: present (0); absent (1).
7. Contact between skull table and cheek: squamosal (part of cheek) bears a flange that abuts (but does not suture with) the supratemporal (part of skull table) (0); sutural contact (1).
8. Posterolateral corner of skull table: formed by tabular (0); formed by supratemporal (1).
9. Squamosal: excluded from margin of posttemporal fenestra (0); enters posterior margin of posttemporal fenestra (1).
10. Tabular: confined to skull table (0); partly occipital (1); entirely occipital (2).
11. Postparietal: on skull table (0); on occiput (1).

Braincase

12. Supraoccipital ossification: absent (0); present (1).
13. Opisthotic: contacts parietal and postparietal dorsally (0); does not contact parietal or postparietal (1).
14. Otic trough: absent (0); present (1).
15. Groove or foramen on occiput for vena capitis dorsalis: present (0); absent (1).
16. Cranioquadrate passage: absent, paroccipital process and quadrate ramus of pterygoid closely apposed (0); present, paroccipital process and quadrate ramus of pterygoid separated by large space (cranioquadrate passage) (1).

Palate

17. Transverse flange on posterolateral margin of pterygoid: absent (0); present (1).
18. Posterolateral margin of pterygoid: edentulous (0); with single row of denticles (1).
19. Ectopterygoid: long and narrow, anteroposterior dimension more than twice mediolateral dimension (0); short and broad, anteroposterior dimension less than twice mediolateral dimension (1).
20. Palate: "Closed" interpterygoid vacuities short and narrow, pterygoids wide (0); "open" interptergyoid vacuities long and wide, pterygoids narrow (1).

Lower Jaw

21. Coronoid process formed by posteriormost coronoid bone: absent or weak (0); prominent (1).
22. Coronoid elements: three coronoids present (0); two or fewer coronoids present (anteriormost coronoid lost) (1).
23. Presplenial: present (0); absent (1).
24. Sulcus cartilaginous meckelii: not exposed on medial surface of mandible (covered by splenial) (0); exposed anterior as a groove on medial surface of mandible (1).

Dentition

25. Marginal teeth: without lingual foramina for replacement teeth (0); with lingual foramina for replacement teeth (1).
26. Maxillary teeth: anterior 4-6 teeth the same size as, or only slightly larger than, posterior teeth (0); anterior 4-6 teeth greatly enlarged, forming a caniniform region or "canine peak" on the anterior end of the maxilla (1).
27. Single, distinct, enlarged caniniform tooth on maxilla: absent (0); present (1).
28. Palatal denticles: scattered evenly over entire palatal surface (0); arranged in a discrete row that extends anterolaterally across the pterygoid and palatine (1).
29. Longitudinal tooth row on ectopterygoid: absent (0); present (1).
30. Large tusks on palatine: present (0); absent (1).
31. Dentary tusks: large tusks present on anterior portion of dentary, medial to marginal teeth (0); tusks absent (1).

Axial Skeleton

32. Presacral vertebrae: 28 or more (0); 25 or fewer (1).
33. Atlas pleurocentrum: positioned between the atlas and axis intercentra (0); positioned directly above the axis intercentrum (1).
34. Atlas pleurocentrum: paired ossification (0); single ossification (1).
35. Atlas pleurocentrum and axis intercentrum: separate (0); fused with one another (1).
36. Atlas intercentrum: without posterior median expansion (0); with posterior median expansion projecting beneath atlas pleurocentrum (1).
37. Atlas neural spine: large vertical blade (0); small, posterodorsally inclined epipophysis (1).
38. Axis pleurocentrum and axis neural arch: separate (0); fused with one another (1).
39. Intercentra: well-developed, 50% as long as pleurocentra, or longer (0); reduced, 40% as long as pleurocentra, or shorter (1).
40. Rib cage: incomplete, ribs do not extend ventrally (0); complete, ribs extend ventrally (1).

Appendicular Skeleton

41. Discrete coracoid ossifications: absent, scapulocoracoid is a single ossification (0); present, scapulocoracoid consists of a scapula and one or two coracoids (1).
42. Interclavicle: no parasternal process (0); with distinct, rod-like posterior median stem ("parasternal process") (1).
43. Sacral ribs: single pair (0); two or more pairs (1).
44. Ilium: deeply bifurcated, consisting of a posterior, horizontally oriented ramus and an anterior, posterodorsally oriented ramus (0); bifurcation reduced (1); bifurcation absent (2).
45. Forelimbs and hindlimbs: short, not longer than skull (0); long, at least 20% longer than skull (1).
46. Astragalus: absent (0); present (1).

Table 1. The data matrix used in the cladistic analysis. ? denotes character indeterminate because of incomplete preservation. "A" denotes that states 0 and 1 both occur within a terminal taxon. ➔

Taxon\Characters	**1**	**2**	**3**	**4**	**5**	**6**	**7**	**8**	**9**	**10**	**11**	**12**	**13**	**14**	**15**	**16**	**17**
Ancestor	0	0	0	0	0	0	0	0	0	0	0	0	0	0	0	0	0
Embolomeri	0	0	?	0	0	0	0	0	0	0	0	0	0	0	0	0	0
Gephyrostegidae	0	?	0	1	0	0	0	0	?	0	0	?	?	0	?	?	1
Seymouriamorpha	0	?	0	0	0	0	1	0	0	1	0	0	A	0	0	1	1
Solenodonsauridae	0	1	?	0	?	1	1	1	?	?	0	?	?	0	?	?	1
Diadectomorpha	1	1	1	0	1	1	1	1	0	2	1	1	1	1	1	1	1
Synapsida	1	1	1	A	1	1	1	1	1	2	1	1	1	1	1	1	1
Sauropsida	1	1	1	1	1	1	1	1	1	2	1	1	1	0	1	1	1

Taxon\Characters	**18**	**19**	**20**	**21**	**22**	**23**	**24**	**25**	**26**	**27**	**28**	**29**	**30**	**31**	**32**	**33**	**34**
Ancestor	0	0	0	0	0	0	0	0	0	0	0	0	0	0	0	0	0
Embolomeri	0	0	0	0	0	0	0	0	0	0	0	0	A	0	0	0	0
Gephyrostegidae	0	0	1	0	0	0	0	0	0	0	0	0	0	1	0	1	0
Seymouriamorpha	0	0	1	0	0	0	A	1	0	0	0	0	1	1	0	?	0
Solenodonsauridae	?	?	?	?	?	?	0	1	0	1	?	1	1	1	?	?	?
Diadectomorpha	1	1	1	1	1	0	1	1	1	1	1	1	1	1	1	1	A
Synapsida	1	1	1	1	1	1	1	1	1	1	1	1	1	1	1	1	1
Sauropsida	1	?	1	1	1	1	0	1	A	1	1	1	1	1	1	1	1

Taxon\Characters	**35**	**36**	**37**	**38**	**39**	**40**	**41**	**42**	**43**	**44**	**45**
Ancestor	0	0	0	0	0	0	0	0	0	0	0
Embolomeri	1	0	0	0	0	0	0	0	0	0	0
Gephyrostegidae	0	0	0	0	0	0	0	0	0	1	0
Seymouriamorpha	1	0	0	1	0	0	0	0	1	1	0
Solenodonsauridae	?	?	?	1	0	?	0	0	?	?	?
Diadectomorpha	0	1	1	1	1	1	1	1	2	1	0
Synapsida	0	1	1	1	1	1	1	1	2	1	1
Sauropsida	0	1	1	1	1	1	1	1	2	1	1

LITERATURE CITED

Berman, D. S, S. S. Sumida, and R. E. Lombard. 1992. Reinterpretation of the temporal and occipital regions in *Diadectes* and the relationships of the diadectomorphs. *Journal of Paleontology,* 66:481-499.

Bremer, K. 1988. The limits of amino acid sequence data in angiosperm phylogenetic reconstruction. *Evolution*, 42:795-803.

Bryant, H. N. 1994. Comments on the phylogenetic definition of taxon names and conventions regarding the naming of crown clades. *Systematic Biology,* 43: 124-130.

Clack, J. A. 1988. New material of the early tetrapod *Acanthostega* from the Upper Devonian of East Greenland. *Paleontology,* 31:699-724.

Cracraft, J. 1986. The origin and early diversification of birds. *Paleobiology,* 12:383-399.

de Queiroz, K., and J. A. Gauthier. 1990. Phylogeny as a central principle in taxonomy: phylogenetic defintions of taxon names. *Systematic Zoology,* 39:307-322.

de Queiroz, K., and J. A. Gauthier. 1992. Phylogenetic taxonomy. *Annual Review of Ecology and Systematics,* 23:449-480.

Eernisse, D. J., and A. G. Kluge. 1993. Taxonomic congruence versus total evidence, and amniote phylogeny inferred from fossils, molecules, and morphology. *Molecular Biology and Evolution,* 10:1170-1195.

Gaffney, E. S. 1979. An introduction to the logic of phylogeny reconstruction. Pages 79-111 in: *Phylogenetic Analysis and Paleontology* (J. Cracraft and N. Eldredge, eds.). New York: Columbia University Press.

Gauthier, J. A. 1984. A cladistic analysis of the higher systematic categories of the Diapsida. Unpublished Ph.D. Dissertation, University of California, Berkeley, California.

Gauthier, J. A. 1986. Saurischian monophyly and the origin of birds. *Memoirs of the California Academy of Science,* 8:1-55.

Gauthier, J., D. Canatella, K. De Queiroz, A. G. Kluge, and T. Rowe. 1989. Tetrapod phylogeny. Pages 337-353 in: *The Hierarchy of Life* (B. Fernholm, K. Bremer and H. Jornwall, eds). London: Elsevier Science Publishers.

Gauthier, J., A. G. Kluge, and T. Rowe. 1988a. Amniote phylogeny and the importance of fossils. *Cladistics,* 4:105-209.

Gauthier, J. A., A. G. Kluge, and T. Rowe. 1988b. The early evolution of the Amniota. Pages 103-155 in: *The Phylogeny and Classification of the Tetrapods* (M. J. Benton, ed.). Oxford: Clarendon Press.

Heaton, M. J. 1980. The Cotylosauria: a reconsideration of a group of archaic tetrapods. Pages 497-551 in: *The Terrestrial Environment and the Origin of Land Vertebrates* (A. L. Panchen, ed). London: Academic Press.

Hennig, W. 1966. *Phylogenetic Systematics.* University of Illinois Press, Urbana.

Hotton, N., III, and J. R. Beerbower. 1994. Tetrapod herbivory: what took them so long? *Journal of Morphology,* 220:355.

Kingley, J. S. 1917. *Outline of Comparative Anatomy of the Vertebrates*. Second Edition. London: John Murray.

Laurin, M., and R. R. Reisz. 1995. A reevaluation of early amniote phylogeny. *Zoological Journal of the Linnean Society*, 113:165-223.

Lebedev, O., and M. I. Coates. 1995. The postcranial skeleton of the Devonian tetrapod *Tulerpeton curtum* Lebedev. *Zoological Journal of the Linnean Society*, 114:307-348.

Lee, M. S. Y. 1995. Historical burden in systematics and the interrelationships of "parareptiles". *Biological Reviews*, 70:459-547.

Lee, M. S. Y. 1996. Correlated progression and chelonian origins. *Nature*, 379:812-815.

Lucas, S. G. 1992. Extinction and the definition of the class Mammalia. *Systematic Biology*, 41:370-371.

Milner, A. R. 1993. Biogeography of Palaeozoic tetrapods. Pages 324-353 in: *Palaeozoic Vertebrate Biostratigraphy and Biogeography* (J. A. Long, ed.). London: Belhaven.

Modesto, S. P. 1992. Did herbivory foster early amniote diversification? *Journal of Vertebrate Paleontology*, 12:44A.

Norell, M. A., J. M. Clark, D. Dashzeveg, R. Barsbold, L. M. Chiappe, A. R. Davidson, M. C. McKenna, A. Perle, and M. J. Novacek. 1994. A theropod dinosaur embryo and the affinities of the flaming cliffs dinosaur eggs. *Science*, 266:779-782.

Panchen, A. L., and T. R. Smithson. 1988. The relationships of the earliest tetrapods. Pages 1-32 in: *The Phylogeny and Classification of the Tetrapods* (M. J. Benton, ed). Oxford: Clarendon Press.

Parker, T. J., and W. A. Haswell. 1940. *A Text-Book of Zoology. Volume 2*. London: Macmillan.

Rieppel, O. 1993. Studies on skeleton formation in reptiles. IV. The homology of the reptilian (amniote) astragalus revisited. *Journal of Vertebrate Paleontology*, 13,:31-47.

Rowe, T. 1988. Definition, diagnosis, and origin of Mammalia. *Journal of Vertebrate Paleontology*, 8:241-264.

Rowe, T., and J. A. Gauthier. 1992. Ancestry, paleonotology, and the definition of the name Mammalia. *Systematic Biology*, 41:372-378.

Sanderson, M. J. 1995. Objections to bootstrapping phylogenies: a critique. *Systematic Biology*, 44:299-320.

Smithson, T. R., R. L. Carroll, A. L. Panchen, and S. M. Andrews. 1994. *Westlothiana lizziae* from the Visean of East Kirkton, West Lothian, Scotland, and the amniote stem. *Transactions of the Royal Society of Edinburgh: Earth Sciences*, 84:383-412.

Spencer, P. S. 1994. The Early Interrelationships and Morphology of Amniota. Unpublished Ph.D. Thesis, University of Bristol, United Kingdom.

Špinar, Z. V. 1952. Revision of some Moravian Discosauriscidae. *Rozpravy ustrededniho Uštavu Geologickeho*, 15:1-160.

Sumida, S. S. 1990. Vertebral morphology, alternation of neural spine height, and structure in Permo-Carboniferous tetrapods, and a reappraisal of primitive modes of terrestrial locomotion. *University of California Publications in Zoology*, 122:1-133.

Sumida, S. S. 1994. Morphological features of the locomotor apparatus of taxa spanning the amphibian to amniote transition. *Journal of Morphology,* 220:398-399.

Sumida, S. S., and R. E. Lombard. 1991. The atlas-axis complex in the late Paleozoic genus *Diadectes* and the characteristics of the atlas-axis complex across the amphibian to amniote transition. *Journal of Paleontology,* 65:973-983.

Sumida, S. S., R. E. Lombard, and D. S Berman. 1992. Morphology of the atlas-axis complex of the late Palaeozoic tetrapod suborders Diadectomorpha and Seymouriamorpha. *Philosophical Transactions of the Royal Society of London, Series B,* 336:259-273.

Swofford, D. L. 1993. *PAUP: Phylogenetic Analysis Using Parsimony.* Version 3.1.1. Champagne, Illinois: Illinois Natural History Survey.

Thulborn, R. A. 1984. The avian relationships of *Archaeopteryx*, and the origin of birds. *Zoological Journal of the Linnean Society,* 82:115-158.

Wilkinson, M. 1 992. Ordered versus unordered characters. *Cladistics*, 8:375-385.

Zug, G. R. 1993. *Herpetology.* San Diego: Academic Press.

CHAPTER 4

BIOGEOGRAPHY OF PRIMITIVE AMNIOTES

David S Berman

Stuart S. Sumida

R. Eric Lombard

INTRODUCTION

The past three decades have seen a revolution in the way terrestrial vertebrate fossils are studied. This is due, primarily, to two significant influences: (1) the advent of cladistic methods of phylogenetic analysis, and (2) the acceptance of plate tectonics and plate movement (Lombard and Sumida, 1992). Marking the beginning of this period of innovative study were Romer's (1966) classic textbooks in vertebrate paleontology and Carroll's (1969a,b,c, 1970) series of landmark studies of the origins of reptiles. Portions of the results of these studies have since been challenged, but they are significant in illustrating the initial cusp of integration of biological, geological, and paleogeographic data bearing on the origins of amniotes. Since then, a deeper understanding of the geography of ancient continents and a series of aggressively cladistic analyses of basal amniotes have provided a dual, overlapping context in which to consider animals associated with the origin of the amniote clade.

The origin of amniotes must have begun by the Middle Pennsylvanian or earlier (Carroll, 1988; Gordon and Olson, 1994). Most of the taxa considered here are known from the Middle Pennsylvanian (Late Carboniferous) to the Early Permian (Table 1). The recent development of more detailed paleogeographic maps (Scotese and McKerrow, 1990; Scotese and Golanka, 1992; Scotese and Langford, 1995) for these periods now facilitates a much clearer survey of the biogeographic distributions of these taxa. Specifically, paleogeographic reconstructions now include not only the outlines of continental masses but major elevational data as well. Any consideration of the biogeography of tetrapods must account not only for the land masses but also the possible physical barriers within them, such as mountain chains, which could have affected dispersals. Furthermore, as a clearer picture of ancient global climates develops, the opportunity to investigate the influence climate had on world-wide distributions of vertebrates becomes available (Olson, 1984; Parrish *et al.*, 1986; Milner, 1993).

Until recently, the vast majority of the major Late Carboniferous-Early Permian groups of terrestrial vertebrates were best known from North America, with only rare occurrences from Europe. However, the discovery of the early stem-amniote *Westlothiana* in Visean deposits of Scotland (Smithson, 1989; Smithson *et al.*, 1994) and the recent discriptions of the first representatives of the seymouriamorphs and diadectomorphs, as well as a variety of amniotes, from Europe (Berman and Martens, 1993; Sumida *et al.*, 1996) demands a more global perspective for the origin of this group.

PHYLOGENETIC CONTEXT

Studies by Carroll (1964, 1969a,b,c, 1970) have proposed a transitional sequence of taxa that included gephyrostegids and soleonodonsaurids and led to a primitive amniote bauplan exemplified by the small, presumably insectivorous members of the family Protorothyrididae ("Protorothyridae" of Carroll). During the past 15 years, a general-consensus, cladistically based hypothesis has been reached, primarily as a result of the work of Heaton (1980), Reisz (1980a, 1986), Gauthier *et al.* (1988), Berman *et al.* (1992), Lee

(1993), Smithson *et al.* (1994), Laurin and Reisz (1995), Lee and Spenser (this volume), and Reisz and Laurin (this volume). Figure 1 illustrates this hypothesis. Though the protorothyridids are still considered among the more primitive amniotes, the most primitive members of the Amniota are now generally accepted to be the pelycosaurian-grade synapsids (Reisz, 1981, 1986; Heaton and Reisz, 1986; Gauthier *et al.*, 1988; Laurin and Reisz, 1995; Reisz and Laurin, this volume). Progressively more recent clades are the Captorhinidae, Protorothyrididae, and Araeoscelidia (Heaton and Reisz, 1986). Gauthier *et al.* (1988) and Laurin and Reisz (1995) interpolate between the Synapsida and Captorhinidae clades an enigmatic group of less well-known amniotes, the Parareptilia, that includes the Procolophonia, Millerosauroidea, Pariesauroidea, and possibly turtles. The parareptilians, however, are currently undergoing intensive restudy and therefore are not considered in this survey, as well as the mesosaurs because of their aquatic nature.

Most of the authors indicated above consider Diadectomorpha and Seymouriamorpha to be successively more distant outgroups to the Amniota. In most recent work the Seymouriamorpha has always been considered to be a stem amniote group and no evidence exists for their inclusion within the Amniota. On the other hand, although most workers consider the Diadectomorpha to be a sister clade of Amniota, a minority of studies (Berman *et al.*, 1992; Lee and Spencer, this volume) have suggested that the diadectomorphs might actually be amniotes, possibly sharing a most recent common ancestor with Synapsida. Perhaps most significantly, the idea that *Diadectes* and its allies could be amniotes is one with a long history (Williston, 1911a; Romer, 1956, 1964). Regardless of assignment within or without the amniote clade, it is clear that the Diadectomorpha is extremely close to the origin of amniotes.

Despite minor differences in recent phylogenetic analyses, the phylogenetic relationships and biogeographic distributions of the following taxa are considered by all to be critical to an understanding of the origin of amniotes: Seymouriamorpha, Diadectomorpha, pelycosaurian-grade Synapsida, Captorhinomorpha, and primitive Diapsida. Although members of the Parareptilia may prove to be important in the origin of amniotes, they are currently not well known and no occurrences earlier than the Upper Permian have yet been reported. Additional knowledge of the Parareptilia could very well alter our collective view of the origin of amniotes. Figure 1 provides a consensus working hypothesis of the phylogenetic relationships of the taxa considered in this study as well as their stratigraphic ranges. Laurin and Reisz (this volume) and Sumida (this volume) provide a summary of the morphological characteristics of the taxa included as well as characters supporting the phylogeny. Finally, it should be pointed out that in this survey only taxa represented by skeletal remains are considered and, thus, ichnites are excluded.

PALEOGEOGRAPHIC-CLIMATIC SETTING

As depicted in the paleogeographic maps of Figures 2-8, global geography during the Late Carboniferous-Early Permian time span remained rather uniform. The region central to this discussion, as dictated by the taxa being considered, encompasses two major mountain ranges. The first and most important is depicted on the paleogeographic maps as forming an unbroken range of Himalayan magnitude (D. Rowley, personal communication), approximating the position of the paleoequator and extending across central Pangaea. This range resulted from three separate historical events of collisional orogenesis between the conjoined North American and Baltic continents, referred to as Euramerica, and the southern supercontinent of Gondwana. These three major orogenic belts are referred to as the Appalachian, Mauretanide, and Variscan (Scotese and Langford, 1995). The mountains of the Appalachian Orogeny extended across much of Central America, the northern margin of South America, and the southern and southeastern margins of North America, and then continued eastward across north Africa as the Mauretanide Orogeny.

Figure 1. Consensus hypothesis of relationships of basal amniote taxa and their successive sister out-groups Diadectomorpha and Seymouriamorpha. Heavy bars on the tree branches are approximations of the temporal distributions of the taxa surveyed. The arrows at the bottom of the figure indicate the approximate time slices for the Late Carboniferous and Early Permian distribution maps. Range is considered to extend through the entire stage in which a representative group occurs. The continuity of each group is inferred, despite the fact that a continuous series of fossils is not available for any of the groups. ➔

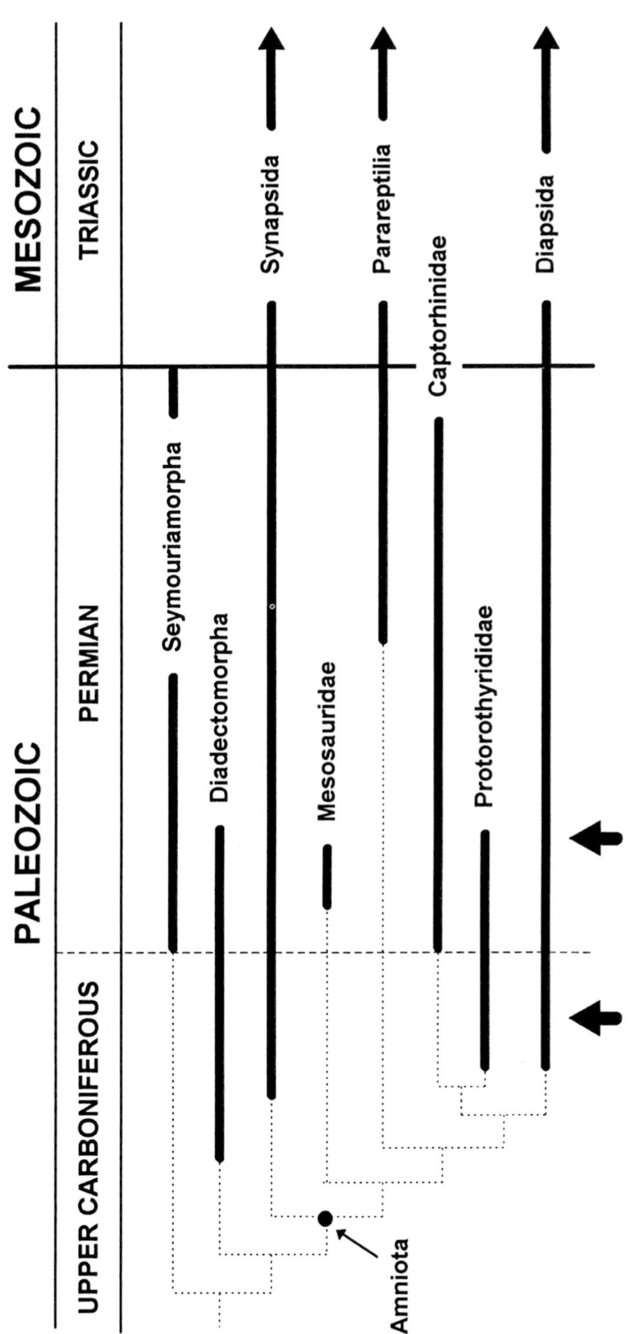
MESOZOIC
PALEOZOIC
TRIASSIC
PERMIAN
UPPER CARBONIFEROUS
Seymouriamorpha
Diadectomorpha
Synapsida
Mesosauridae
Parareptilia
Captorhinidae
Protorothyrididae
Diapsida
Amniota

The Mauretanide mountains were in turn continued north-eastward into the Baltic region of Euramerica by the mountains of the Variscan Orogeny. The Appalachian-Mauretanide-Variscan mountains reached their greatest development in the Early Permian, but by the Late Permian were already in decline. During the Late Carboniferous-Early Permian the sutured Gondwana and Euramerica drifted northward about 10 degrees as a result of a clockwise rotation of Pangaea, sothat the Appalachian-Mauretanide-Variscan mountain chain marking their union now straddled the paleoequator more directly.

In the eastern portion of central Pangaea a second prominent mountain range, consisting of the Altay-Sayan and Uralian orogenies, arose during the Late Carboniferous and Early Permian in western Asia. The previously widely separated continents of Siberia and Kazakhstan collided in the Late Carboniferous to form the Altay-Sayan mountains, and by the Early Permian the cojoined Siberian-Kazakhstan continent collided with the Baltic region of Euramerica to complete the formation of the northeastern region of Pangaea and produce the Uralian mountains. Due to the clockwise rotation of Pangaea during the Late Carboniferous-Early Permian western Asia drifted southward about 10°.

Almost all of the vertebrates whose Late Carboniferous-Early Permian distributions are plotted here were preserved in sediments derived from the paleoequatorial mountain ranges of central Pangaea and deposited along the southern, equatorial margin of Euramerica. During the Late Carboniferous these sediments typically resulted in locally extensive coal deposits that developed under hot, humid, climatic conditions in shallow-water, swamp-like basins adjacent to or on the flanks of the mountains. Northward, at about 30° paleolatitude, this environment was replaced by evaporitic basins, indicating generally warmer conditions marked by strong seasonal aridity (Parrish *et al.*, 1986). Beginning in the latest Pennsylvanian, the equatorial coal deposits that typify this period were replaced, usually gradually, by predominately red-bed and evaporitic deposits of the Early Permian, indicating a trend toward drier climates with marked seasonal aridity (Olson and Vaughn, 1970).

In addition to the Late Carboniferous-Early Permian tetrapods of Euramerica, this survey includes several localities in western Asia that have yielded discosauriscid seymouriamorphs. These localities, generally regarded as Early Permian, occur within the Kazakhstan continent and in deposits derived from the mountains of the Altay-Sayan and Uralian orogenies. Deposition occurred in a wet, temperate climate which resulted in the formation of coal deposits in the Late Carboniferous. The cessation of coal deposition was perhaps due to not only global climatic changes but also the southward drift of this region into a warmer, more seasonably drier climatic belt (Parrish *et al.*, 1986).

CARBONIFEROUS-PERMIAN BOUNDARY

Critical to understanding the plotting of the paleogeographic distribution patterns of the Late Carboniferous and Early Permian tetrapod groups discussed here is the stratigraphic level we have adopted as the Carboniferous-Permian (C-P) boundary in some of the major Late Paleozoic depositional sections in North American and Europe. The placement of the C-P boundary in Late Paleozoic sections is frequently quite controversial, with widely differing opinions. Typically this stems from the difficult problem of exact correlation between marine and continental sections in which there is no mutual interbedding of the strata. Before proceding with any discussion of the late Carboniferous and Early Permian distribution patterns of vertebrates, it is essential to establish what we are adopting as the C-P boundary in those terrestrial sections in which its placement been disputed.

In several papers describing the vertebrates and sediments of the commonly referred to Early Permian red-bed deposits of New Mexico, particularly the lowermost levels of the Cutler Formation, the somewhat ambiguous age of Permo-Carboniferous has been applied (Eberth and Miall, 1991; Berman, 1993; Eberth and Berman, 1993). Although no exact age determination is available, these deposits, as well as those at equivalent or near equivalent levels in the red-bed sections of the Abo and Sangre de Cristo formations (Berman, 1993), will be regarded as lowermost Permian. The only major exception to this age assignment is the well-known, fossil-bearing beds of the Cutler Formation of El Cobre Canyon in north-central New Mexico. Here the exposed stratigraphic section of the Cutler Formation spans the Permo-Pennsylvanian boundary from at least the Upper Pennsylvanian Virgilian to the Lower Permian Wolfcampian, with the majority of the fossils coming from the Late Pennsylvanian levels on the canyon floor (Eberth and Miall, 1991; Eberth and Berman, 1993).

There has also been considerable debate regarding the precise position of the C-P boundary in the late Paleozoic section of the Dunkard Basin of the tri-state region of Ohio, Pennsylvania, and West Virginia. Assessments of the position of the C-P boundary, based on a great variety of paleontological evidence, have ranged from somewhere in the Monongahela Group to various levels in the overlying Washington and Greene formations of the Dunkard Group (see Barlow, 1975). In the present review, the C-P boundary is considered to coincide approximately with the contact between the Dunkard and Monongahela groups. This has not only been the most widely accepted placement, but is supported by correlations using terrestrial vertebrates (Berman and Berman, 1975; Olson, 1975).

Most critical to the purpose of this review, however, is the placement of the C-P boundary in the Upper Carboniferous-Lower Permian continental sediments derived from the Variscan Orogeny and deposited in the well-known, numerous, small and discontinuous intramontane basins in central and western Europe. In this region, the C-P boundary has been traditionally recognized at the Stephanian-Rotliegend boundary. Some very challenging evidence has been presented, however, to suggest a much higher placement that would include most of the Lower Rotliegend in the Late Carboniferous.

The continental Permian of central and western Europe, the Rotliegend, is in a strict sense a lithostratigraphic term that refers to sediments underlain by the uppermost part of the Carboniferous (i.e., Stephanian C) and overlain by the marine beds of the Zechstein (i.e., Late Permian). The Rotliegend, therefore, cannot be considered either a biostratigraphic or chronostratigraphic unit. The Rotliegend is divided into two stages (also lithostratigraphic units), a Lower, also called the Autunian (derived from the Permian basin in Autun, France), and an Upper Rotliegend, also called the Saxonian (derived from the Sachsen region in central Germany). The C-P boundary has traditionally been established as the lowest stratigraphic occurrence of a macroflora that includes most significantly *Callipteris conferta* and

C. naumanni. Their appearance in different basins, or even within the same basin, is irregular, making the tracing of the C-P boundary impossible. In such instances the C-P boundary, as well as the boundary between the Lower and Upper Rotliegend, is recognized by lithologic marker beds, in most cases conglomerates, which indicate the onset of a rejuvenation of the sediment source relief, the Variscan Orogeny. Precise establishment of the C-P boundary in the type region of the continental section in Europe is also made difficult by the lack of interbedded, easily dated marine sediments. In several reviews of these problems Kozur (1984, 1988, 1989) has strongly rejected the widely accepted notion that the Rotliegend marks the beginning of the Early Permian in central and western Europe and that this event can be recognized by the first occurrence of the plant fossil *Callipteris*. He points out that *Callipteris* occurs not only in the stratotype of the Stephanian C but also in continental sediments that can be correlated directly with interbedded, unambiguously Carboniferous marine sediments, such as the upper Missourian Stage in Kansas and the middle and upper Gzhelian Stage of the Donets basin in Russia. Alternatively, Kozur redefined the C-P boundary in central and western Europe on the basis of published accounts of abrupt, coincidental changes in the flora and fauna that occur at a high level in the Lower Rotliegend (within the Lower Oberhof Formation in the Saale Basin in the Thurigian Forest and at the base of the Soetern Formation in the Saar-Nahe Basin, Germany).

If Kozur's reassignment of the C-P boundary is adopted, it profoundly affects the widely accepted age of many of the European vertebrates discussed here. For example, most, if not all, of the discosauriscid seymouriamorphs, the diadectomorphs, many of the pelycosaurian-grade synapsids, such as *Haptodus* plus its closely related genera, and the edaphosaurids are from levels in the Lower Rotliegend below his proposed C-P boundary and would have to be considered Pennsylvanian in age. As corroborative evidence for his proposed boundary, Kozur (1984) notes that the European occurrences of *Haptodus* would now agree in age with those from upper Missourian sediments in Kansas that are dated by interbedded marine deposits.

Support for Kozur's proposed reassignment of the C-P boundary has come from a recently described assemblage of Early Permian terrestrial or semi-terrestrial vertebrates from the well-known Bromacker quarry in the Tambach Formation, lowermost unit of the Upper Rotliegend of the Saale Basin, central Germany (Berman and Martens, 1993; Sumida *et al.*, 1996). The Bromacker assemblage is not only of great interest because it contains elements found heretofore only in the Lower Permian of North America, but also in suggesting an Early Permian Wolfcampian age for the base of the Upper Rotliegend and, therefore, a Late Pennsylvanian age for most, if not all, of the underlying Lower Rotliegend. However, until there is universal agreement for revising the European C-P boundary, we will accept the base of the Rotliegend as approximating the boundary.

DISTRIBUTION OF SEYMOURIAMORPHA

No Late Carboniferous seymouriamorphs are recognized, but members of two families, Discosauriscidae and Seymouriidae, were widely distributed during the Early Permian, although by the lower Upper Permian their record is reduced to two occurrences of the former from China (Zhang *et al.*, 1984) and Russia (Ivakhnenko, 1981) and, one of the latter from Oklahoma (Olson, 1980). The taxa and their ages and localities, as well as the bibliographic sources of this data, used to plot the Early Permian distributions of the Discosauriscidae and Seymouriidae in Figure 2 are as follows:

- ***Discosauriscidae***

  Ariekanerpeton (Tadghikistan), *Discosauriscus* (Czech Republic, France, and Germany), *Letoverpeton* (Czech Republic and Germany), and *Utegenia* (Kazakhstan); Ivakhnenko (1981), Kuznetsov and Ivakhnenko (1981), and Werneburg (1989, 1990).

- ***Seymouriidae***

  Seymouria (Germany, New Mexico, Oklahoma, Prince Edward Island, Texas, and Utah); Langston (1963), Vaughn (1966), Olson (1979a, 1980), Berman *et al.*(1987), and Berman and Martens, (1993).

Although the discosauriscids listed above are assigned an Early Permian age, this is not certain for some of their occurrences. For example, *Utegenia* is from beds that range in age from the end of the Carboniferous to the beginning of the Permian, and no precise age is known for *Ariekanerpeton*. These Asian discosauriscids have been given an Early Permian age apparently on the basis on correlations with European forms. In addition, as already noted above in the discussion of the Carboniferous-Permian Boundary, many of the European discosauriscid localities may actually be latest Carboniferous rather than Early Permian. Regarding the age of this group, it is also impotant to note that *Discosauriscus* has been reported from the lower Upper Permian in the southern cis-Uralian forelands of Russia (Ivakhnenko, 1981) and *Urumqia* from China is Upper Permian (Zhang *et al.*, 1984). With the exception of one suspicious description, *Seymouria* is restricted to the Early Permian. On the basis of a single partial skeleton that includes only a small portion of the skull, Olson (1980) described a new species of *Seymouria* from the lower Upper Permian of Oklahoma. Generic identification was based almost entirely on the shape of the presacral neural arches, which it could conceivably share with more distantly related groups, particularly the diadectomorphs. In addition, there are aspects of the appendicular skeleton which are unlike those of any known seymouriid.

Despite the described age and taxomonic resolution difficulties, there exists an interesting stratigraphic replacement in central Europe of the discosauriscids (*Discosauriscus* and *Letoverpeton*) by *Seymouria*. The smaller and morphologically more primitive, aquatic discosauriscids (Laurin, 1995), preserved typically in lacustrine grey sediments and black shales (as are the Russian and Asian forms), have been known for well over a century and reported, often in large numbers, from numerous localities in the Lower and possibly the basalmost level of the Upper Rotliegend (Werneburg, 1989, 1990; Ivakhnenko, 1990). On the other hand, until very recently *Seymouria* was restricted to the Lower and possibly the lowermost Upper Permian of North America (Olson, 1980; Berman *et al.*, 1987). From the well-known Bromacker locality in central Germany two specimens of *Seymouria* were described from the Lower Permian Tambach Formation, which lies near the base of the Upper Rotliegend (Berman and Martens, 1993). Not only does the Tambach Formation occupy a stratigraphic level distinctly above those yielding the

European discosauriscids, but it also consists of typical fluvial red-bed sediments like those in which the Permian *Seymouria* specimens of North America are preserved. The replacement of the basically upper Lower Rotliegend discosauriscids by the later, basal Upper Rotliegend *Seymouria* in Europe may reflect either an evolutionary event or response to an ecological succession, or both.

The absence of discosauriscids in North America seems unexpected in view of that region's widespread occurrences of extensive, Late Pennsylvanian-Early Permian deposits similar to those of the Lower Rotliegend of central and western Europe. One explanation can be offered here to explain this distribution: the origin of the seymouriamorphs took place in Europe with the appearance first of the strictly aquatic, Lower Rotliegend discosauriscids, which subsequently gave rise to the semiterrestrial Early Permian seymouriids. With the exception of the few Asian and Russian discosauriscids, specimens of this family are preserved in sediments that accumulated during the the Late Carboniferous and Early Permian in elongate, fault-bounded, northeast-trending, intramontane basins distributed in a broad band across central and western Europe from western Poland and the Czech Republic southwest through central Germany and south-central France. The basins formed on the northern margin of the Variscan mountains of the Baltic region and apparently represented a physical environmental setting unique from those of other regions in Euramerica in which fossil-bearing sediments of the same age accumulated. This has prompted the suggestion that the Late Paleozoic intramontane basins of Europe may have been a site of endemism for several families of amphibians and the stem amniote discosauriscids as well (Milner, 1993). On the basis of this premise, it can be further speculated that with the gradual replacement of the coal swamps by the terrestrial, fluvial red-bed environments in the intramontane basins of the Baltic region of Europe the discosauriscids gave rise to and were replaced by *Seymouria*, the earliest occurring and morphologically most primitive member of the Seymouriidae.

Figure 2. Early Permian distributions of the seymouriamorph families Discosauriscidae (D) and Seymouriidae (S). Localities, latitidue and longitude may be found in Appendix 1. ➔

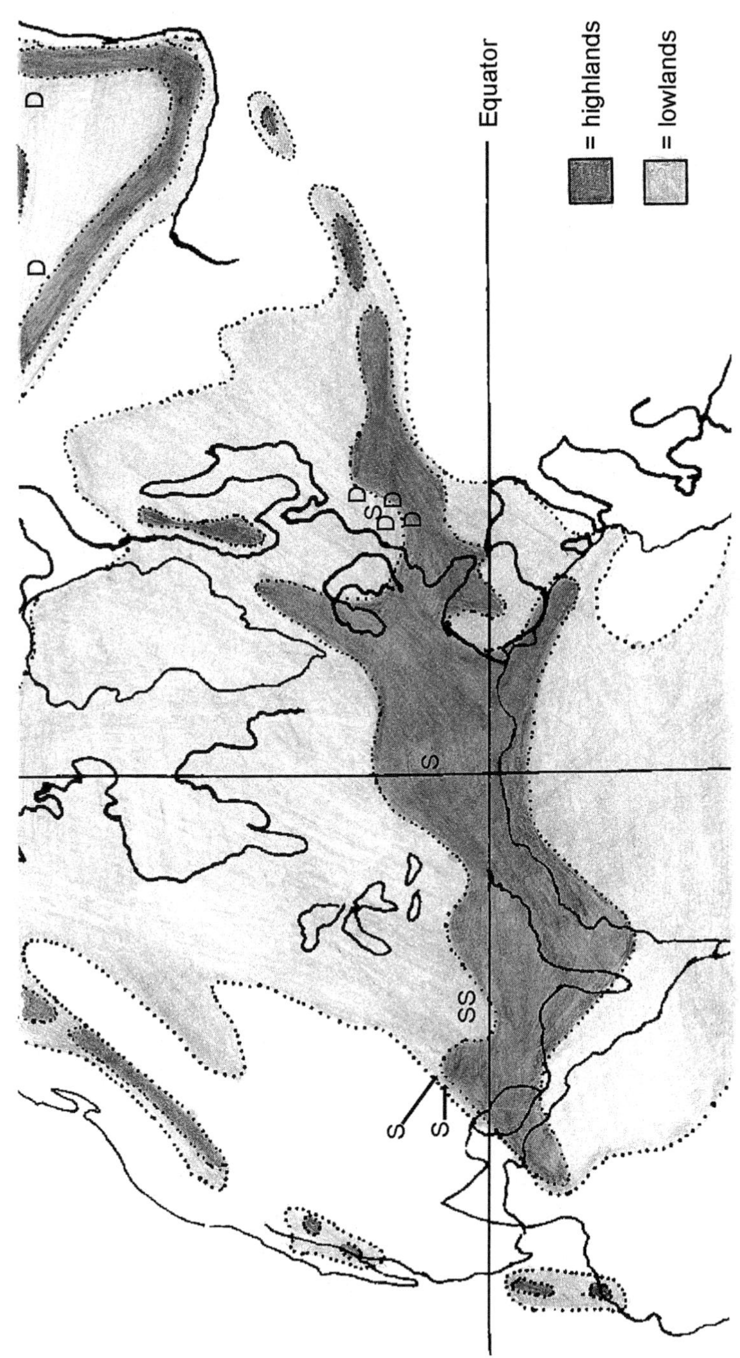
Equator
= highlands
= lowlands
D
D
D
S
D
D
D
S
SS
S
S

This event, coupled with the widespread development of red-bed environments, may have allowed the semiterrestrial *Seymouria* the opportunity to expand its range to North America.

If the discosauriscids originated in the Baltic region of Europe, then, as indicated by recent paleogeographic reconstructions (Scotese and Langford, 1995; Ziegler, 1996), they could not have reached the Kazakhstan continent and the cis-Uralian forelands of Russia until at least the Early Permian. Although Kazakhstan and Euramerica were widely separated during the Late Carboniferous, even after their Early Permian union a persistant shallow sea extended across the eastern margin of Euramerica to intervene between the terrestrial areas of Euramerica to the west and the Uralian mountains and their forelands and Kazakhstan to the East. However, a narrow corridor between these two regions was available for faunal interchange. During the Permian a broadening isthmus of emergent land surrounding a string of high mountains that was an eastward extension the Variscan Orogeny, spanned the shallow sea to join the Baltic region of Euramerica with the southern margin of the Kazakhstan. The hypothesis that discosauriscids were endemic to the Baltic region of Euramerica and expanded their range to Kazakhstan and Russia during the Permian becomes more plausible if, as discussed earlier, the Lower Rotliegend deposits of Europe containing the discosauriscids are considered to be at least in part latest Carboniferous.

Of important consideration to this discussion is an explanation for why *Seymouria* remained undetected in Europe until recently (Berman and Martens, 1993), despite a long history of intensive prospecting of the productive deposits of the Rotliegend. The same question can also be asked about the recent discoveries at the Bromacker locality of the diadectomorph *Diadectes*, discussed below, and a trematopid temnospondyl amphibian (Sumida *et al.*, 1994, 1996). Both are terrestrially adapted forms known almost exclusively as common elements in the Early Permian red-bed assemblages of North America (a few Late Pennsylvanian, North American trematopids are known). Martens (1988, 1989) and Berman and Martens (1993) have suggested that the rare occurrences of representatives of North American Early Permian terrestrial tetrapods in the Rotliegend deposits of Europe are due to a bias in exploration that has greatly ignored the red-bed sediments where such forms are most likely to be discovered. Poor exposures of the Rotliegend red-bed deposits and the long-standing, widely accepted misconception that they represent an inhospitable dry climate in which preservation of vertebrate skeletal remains would have been unlikely has fostered a history of little interest in their exploration. This has had the expected results of there being only a paucity of vertebrates collected from the red-beds of the Rotliegend, because most investigators have concentrated on the lacustrine gray sediments and black shales that are noted for their highly productive sites yielding mainly obligatory aquatic amphibians and stem amniote discosauriscids. It seems very apparent that the explanation for the similarity between the widely separated Early Permian assemblages of the Bromacker locality and those of the United States is that similar environments are being sampled, and that the fluvial red-bed sediments are the most likely source of terrestrial tetrapods in the Rotliegend.

DISTRIBUTION OF DIADECTOMORPHA

The suborder Diadectomorpha consists of three families. The Late Pennsylvanian-Early Permian Limnoscelidae and Early Permian Tseajaiidae have a strictly North American record, whereas members of the Late Pennsylvanian-Early Permian Diadectidae are found in both North America and central Europe. It is important to note here two taxonomic reassessments of reported limnoscelids. Initially, Martens (1989) reported the occurrence of a limnoscelid-like form from the Bromacker locality in the Lower Permian, Tambach Formation at the base of the Upper Rotliegend of central Germany. However, during the course of study of recently discovered, undescribed *Diadectes* specimens from the same locality, it became apparent that the limnoscelid-like specimen should instead be referred to that genus. Second, *Romeriscus periallus*, described as a limnoscelid from the Pennsylvanian of Nova Scotia by Baird and Carroll (1967), was restudied by Laurin and Reisz (1992), who reassessed the genus as a *nomen dubium*. It should also be noted here that our distribution survey does not include the North American diadectids *Empedocles*, *Enpedias*, *Bolbodon*, *Chilonyx*, *Diadectoides*, *Nothodon*, *Helodectes*, and *Animasaurus*, as they very likely represent junior synonyms of *Diadectes* (Olson, 1947; Berman and Sumida, 1995).

The taxa and their ages and localities, as well as the bibliographic sources of this data, used to plot the Late Carboniferous and Early Permian distributions of the diadectomorph families Limnoscelidae, Tseajaiidae, and Diadectidae in Figures 3 and 4 are as follows:

Late Carboniferous (Fig. 3)

- ***Limnoscelidae***

Limnoscelis (New Mexico and Colorado) and *Limnostegis* (Nova Scotia); Williston (1911b), Carroll (1967), Fracasso (1980), Berman and Sumida (1990), and Berman (1993).

- ***Diadectidae***

Desmatodon (Colorado, New Mexico, and Pennsylvania) and *Diasparactus* (New Mexico); Case (1908), Case and Williston (1913), Romer (1952), Vaughn (1969, 1972), Fracasso (1980), Berman (1993), Eberth and Berman (1993), and Berman and Sumida (1995).

Early Permian (Fig. 4)

- ***Limnoscelidae***

Limnosceloides (New Mexico and West Virginia) *Limnoscelops* (Colorado), and limnoscelid (Utah); Romer (1952), Vaughn (1962), Lewis and Vaughn (1965), Langston (1966), and Berman (1993).

- ***Tseajaiidae***

Tseajaia (New Mexico and Utah); Vaughn (1964), Moss (1972), and Berman (1993).

- ***Diadectidae***

Diadectes (Colorado, Germany, New Mexico, Ohio, Oklahoma, Prince Edward Island, Texas, Utah, and West Virginia), *Phanerosaurus* (Germany), and *Stephanospondylus* (Germany); Meyer

Figure 3. Late Carboniferous distributions of the diadectomorph families Limnoscelidae (L) and Diadectidae (D). Localities, latitidue and longitude may be found in Appendix 1. ➔

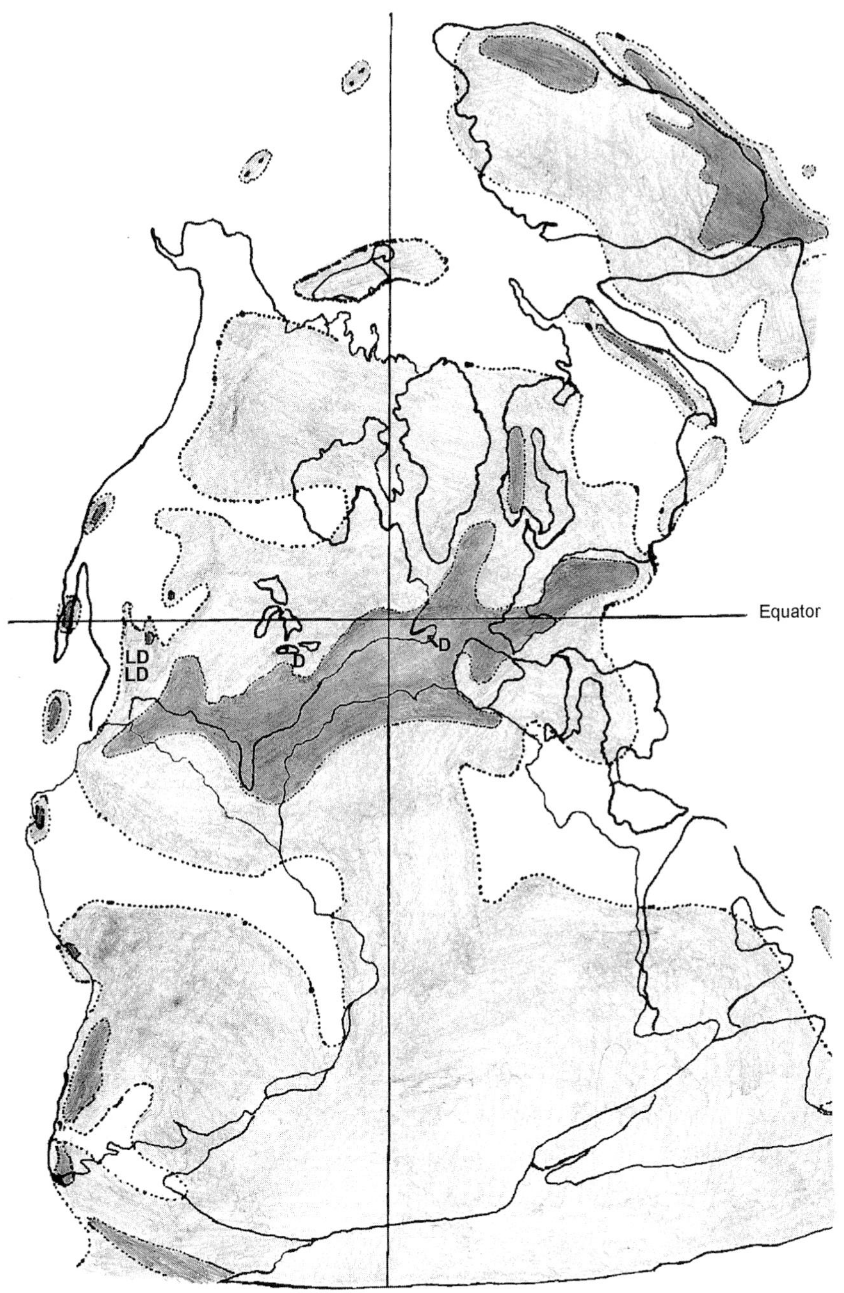
Equator
LD
LD
D
D

(1860), Geinitz and Deichmüller (1882), Olson (1947, 1967, 1975), Langston (1963), Lewis and Vaughn (1965), Berman (1971, 1993), and Berman and Martens (1993).

The diadectomorphs exhibit a distribution that is nearly restricted to North America. Of the three families only the diadectids are represented in Europe and then by only three genera from Germany. Two of these, *Phanerosaurus* (Meyer, 1860) and *Stephanospondylus* (Geinitz and Deichmüller, 1882), were described over a century ago on the basis of very fragmentary specimens from the Lower Rotliegend. No other materials have been found that can be referred to them, and *Stephanosaurus* may be a junior synonym of *Phanerosaurus* (Romer, 1925). A total of four specimens of *Diadectes*, a complete skeleton, an isolated skull, and the greater portions of two postcrania, have been recently recovered from the Lower Permian Tambach Formation of the Upper Rotliegend, Bromacker locality in central Germany. However, only the isolated skull has been briefly reported and was referred to the Diadectidae without description or generic assignment (Martens, 1993; Berman and Martens, 1993). A thorough examination of the three Bromacker specimens reveals, however, that they differs from the widely distributed Early Permian *Diadectes* species of North America in only very minor ways and, therefore, should be considered a new species of that genus. Our explanation of why *Diadectes* has remained undedected for so long in the Early Permian deposits of Europe is the same as that offered above to account for the similar, single European occurrence of *Seymouria* at the Bromacker locality.

DISTRIBUTION OF PELYCOSAURIAN-GRADE SYNAPSIDA

The paraphyletic pelycosaurian-grade synapsids, which can be referred to as simply primitive synapsids, have a predominately North America record, with rare occurrences in Europe. Most recent classifications of the order Synapsida divide it into two sister

Figure 4. Early Permian distributions of the diadectomorph families Limnoscelidae (L), Tseajaiidae (T), and Diadectidae (D). Localities, latitidue and longitude may be found in Appendix 1. ➔

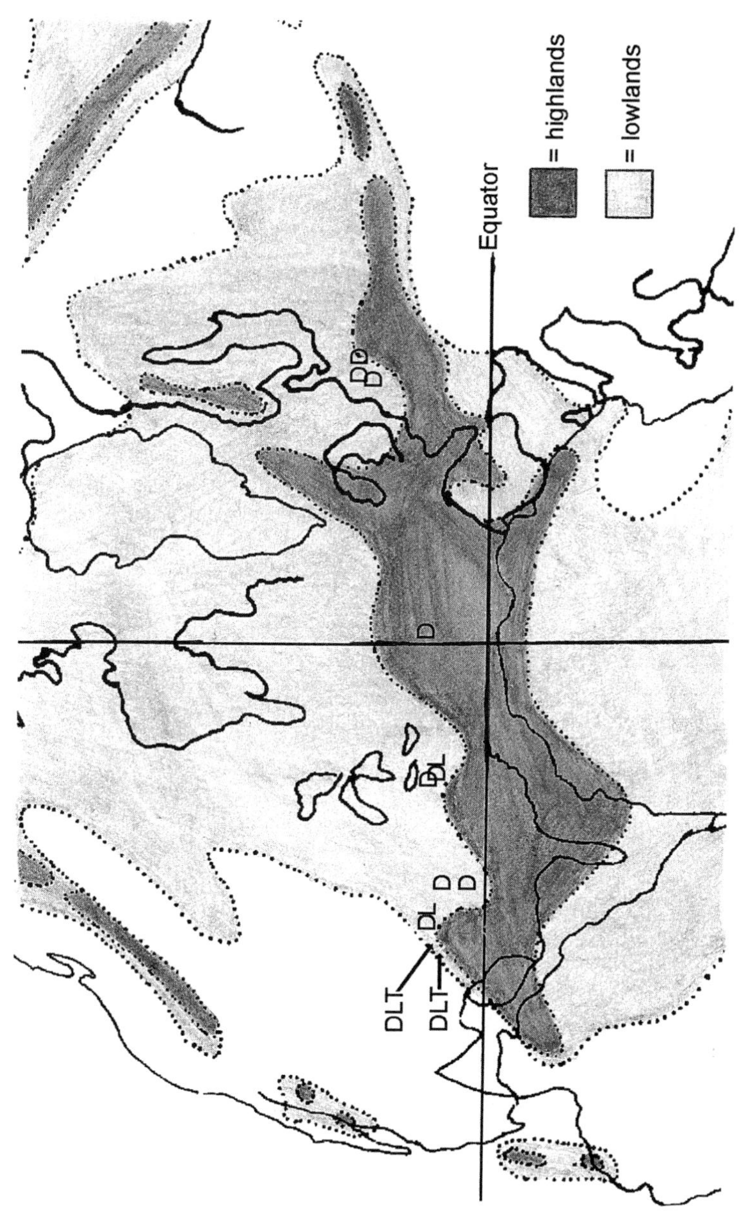
Equator
= highlands
= lowlands
D
DL
DLT

taxa, the suborders Caseasauria and Eupelycosauria (Kemp, 1982; Reisz, 1986). The Caseasauria, with the more primitive members,includes the families Caseidae and Eothyridae, whereas the more advanced Eupelycosauria includes all other primitive synapsid families, Varanopseidae, Ophiacodontidae, Edaphosauridae, and Sphenacodontidae, as well as the advanced synapsids, the Therapsida and their descendants, the mammals. One modification of this taxonomic scheme, however, is made here to more accurately express the relationships of *Haptodus* and closely related genera. These taxa were formerly considered members of the Sphenacodontidae, but recent cladistic analyses (Gauthier *et al.*, 1988; Reisz *et al.*, 1992) have demonstrated that they form a clade that falls outside this family as its nearest sister taxon. For this reason *Haptodus* and closely allied genera will be referred unofficially to the family "haptodontids." Several other taxonomic reassessments of the primitive synapsids have been adopted here and require a brief explanation. With the description of the small, primitive edaphosaurid *Ianthosaurus*, based on excellent specimens from the Late Pennsylvanian of Kansas (Reisz and Berman, 1986; Modesto and Reisz, 1989), it became apparent that the taxonomic validity of the small edaphosaurids *Edaphosaurus raymondi*, *E. mirabilis*, and *E. credneri* was in serious doubt (Modesto and Reisz, 1990). The first two forms are known only by neural spine fragments and the third by a small portion of the postcranial skeleton. As these materials do not allow distinction from *Ianthosaurus*, they therefore cannot be considered diagnostic below the family level. Currie (1977) synonomized all the haptodontids into a single genus, *Haptodus*. Subsequent studies by Laurin (1993, 1994), however, have demonstrated the validity of the *Haptodus*-like genera *Cutleria*, *Palaeohatteria*, and *Pantelosaurus*. *Scoliomus* Williston and Case, 1913, was shown to be a junior synonym of *Sphenacodon* by Reisz and Berman (1985). Finally, following Reisz's (1986) review of the primitive synapsids, several genera are not included in this survey

Figure 5. Late Carboniferous distributions of the primitive eupelycosasurian (e) families Varanopseidae, Ophiacodontidae, and Edaphosauridae; and the advanced eupelycosaurian (E) families "Haptodontidae" and Sphenacodontidae. Localities, latitidue and longitude may be found in Appendix 1. ➔

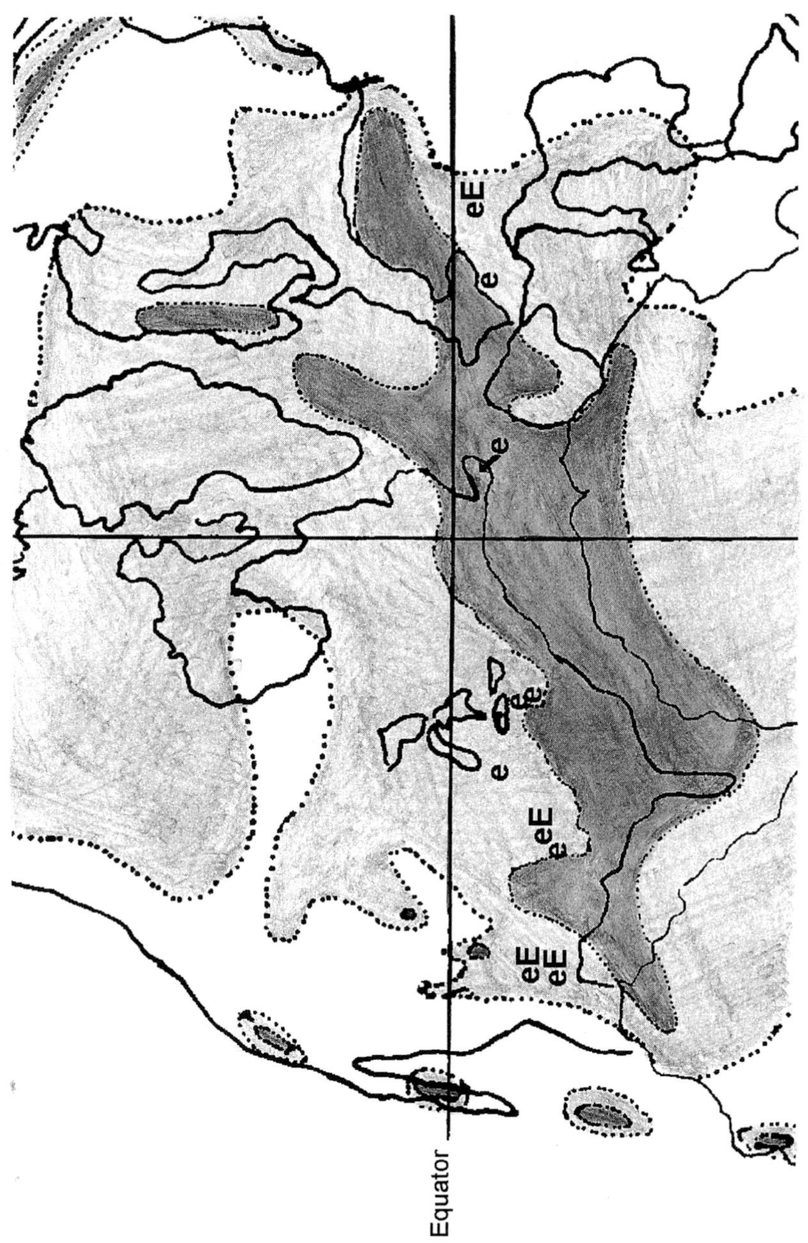
eE
e
e
e
e
e
eE
e
eE
eE
Equator

either because of their uncertain familial assignment (*Delorhynchus*, *Echinerpeton*, *Milosaurus*, *Nitosaurus*, *Protoclepsydrops*, *Thrausmosaurus*, *Trichasaurus*, and *Xyrospondylus*), or because they are not synapsids (*Bayloria* and *Colobomycter*; see also Laurin and Reisz, 1988).

In plotting the paleogeographic distributions of the primitive synapsids, three groups are distinguished: (1) the suborder Caseasauria, (2) the most primitive eupelycosaurian families Varanopsidae, Ophiacodontidae, and Edaphosauridae, and (3) the more derived eupelycosaurian families "haptodontids" and Sphenacodontidae. These taxa form, respectively, progressively more recent sister-group relationships with the Therapsida and their descendants, the mammals. The pelycosaurian-grade synapsid taxa and their ages and localities as well as the bibliographic sources of these data, used to plot the Late Carboniferous and Early Permian distributions of the three groupings proposed above in Figures 5 and 6 are as follows:

Late Carboniferous - More Primitive Eupelycosauria (Fig. 5)

- ***Varanopseidae***

  Aerosaurus and *Ruthiromia* (New Mexico); Reisz (1986).

- ***Ophiacodontidae***

  Archaeothyris (Nova Scotia), *Baldwinonus* (New Mexico), *Clepsydrops* (Illinois and Pennsylvania), *Ophiacodon* (Colorado, Kansas, New Mexico, and Texas), *Stereophallodon* (Texas), and *Stereorhachis* (France); Reisz (1986), and Sumida and Berman (1993).

- ***Edaphosauridae***

  Edaphosaurus (New Mexico and West Virginia), edaphosaurid (?*Edaphosaurus raymondi* and *E. mirabilis*; Czeck Republic and Pennsylvania), and *Ianthosaurus* (Colorado and Kansas); Vaughn

---

Figure 6. Early Permian distributions of the caseosaurian (C) families Eothyrididae and Caseidae; the primitive eupelycosasurian (e) families Varanopseidae, Ophiacodontidae, and Edaphosauridae; and the advanced eupelycosaurian (E) families "Haptodontidae" and Sphenacodontidae. Localities, latitidue and longitude may be found in Appendix 1. ➔

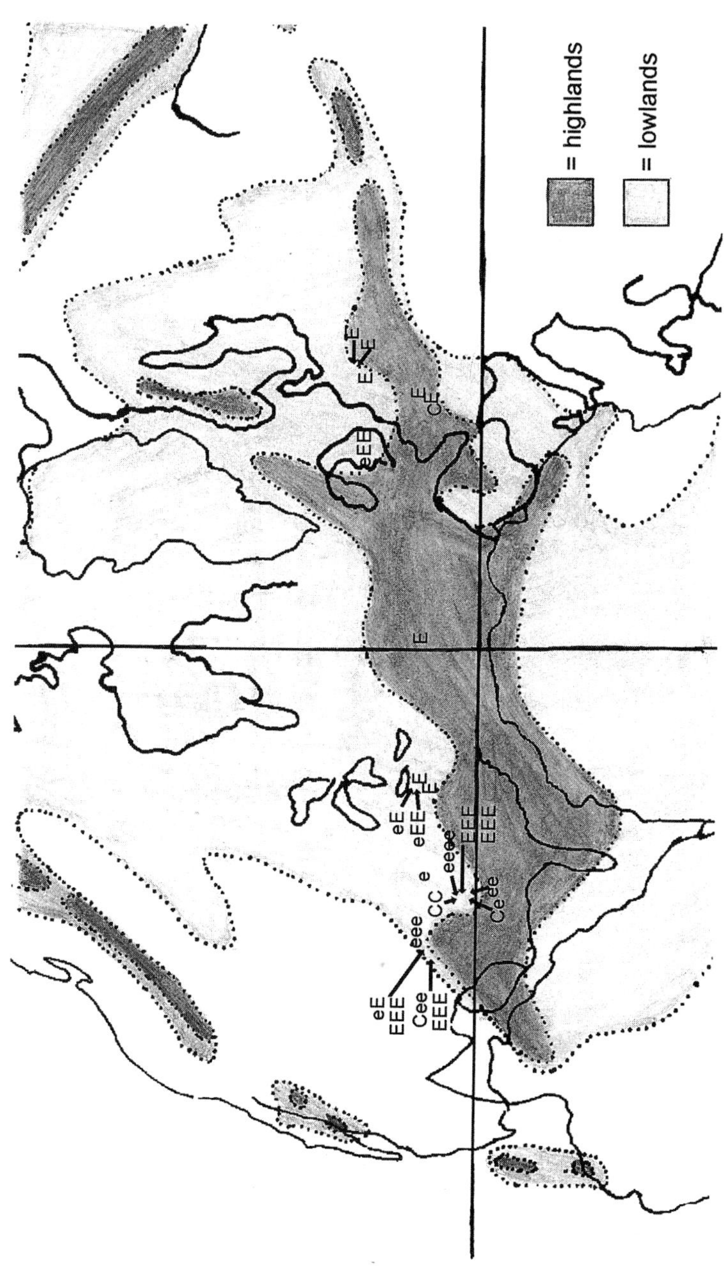
= highlands
= lowlands

(1964), Reisz (1986), Reisz and Berman (1986), Modesto and Reisz (1990), and Sumida and Berman (1993).

Late Carboniferous - More Advanced Eupelycosauria (Fig. 5)

- ***"Haptodontids"***

  Haptodus (Kansas) and a haptodontid (Colorado); Laurin (1993) and Sumida and Berman (1993).

- ***Sphenacodontidae***

  Macromerion (Czech Republic) and *Sphenacodon* (New Mexico); Reisz (1986).

Early Permian - Caseasauria (Fig. 6)

- ***Eothyridae***

  Eothyris (Texas) and *Oedaleops* (New Mexico); Reisz (1986).

- ***Caseidae***

  Casea (Texas and France) and *Cotylorhynchus* (Oklahoma); Reisz (1986).

Early Permian - More Primitive Eupelycosauria (Fig. 6)

- ***Varanopseidae***

  Aerosaurus (New Mexico), *Basicranodon* (Oklahoma), *Mycterosaurus* (Colorado and Texas), and *Varanops* (Texas); Reisz (1986).

- ***Ophiacodontidae***

  Baldwinonus (Ohio), *Ophiacodon* (Colorado, England, Kansas, New Mexico, Oklahoma, Ohio, Texas, and Utah), and *Varanosaurus* (Oklahoma and Texas); Vaughn (1962, 1964), Lewis and Vaughn (1965), Olson (1976, 1975), Riesz (1986), and Berman *et al.* (1995).

- ***Edaphosauridae***

  Edaphosaurus (New Mexico, Ohio, Texas, Utah, and West Virginia), an edaphosaurid (?*Edaphosaurus credneri*, Germany), *Glaucosaurus* (Texas), and *Lupeosaurus* (Texas); Olson and Vaughn (1970), Reisz (1986), Sumida (1989), and Modesto (1994).

Early Permian - More Advanced Eupelycosauria (Fig. 6)

- ***"Haptodontids"***

Cutleria (Colorado), *Haptodus* (France and England)), *Palaeohatteria* (Germany), and *Pantelosaurus* (Germany); Currie (1979), Reisz (1986), and Laurin (1993).

- ***Sphenacodontidae***

Bathygnathus (Prince Edward Island), *Ctenorhachis* (Texas), *Ctenospondylus* (Ohio, Texas, and Utah), *Dimetrodon* (New Mexico, Ohio, Oklahoma, Texas, and Utah), *Neosaurus* (France), *Secodontosaurus* (Texas), and *Sphenacodon* (England, New Mexico, and Utah); Olson (1975), Vaughn (1964), Reisz (1986), and Hook and Hotton (1991).

It is noteworthy that only one therapsid has been reported from the Early Permian, the enigmatic *Tetraceratops* from north-central Texas (Laurin and Reisz, 1990), which was previously considered a primitive synapsid (Reisz, 1986). If this taxonomic reassessment is correct, then serious consideration would have to be given to the hypothesis that the advanced synapsids had their origin no later than earliest Permian and either have not been detected in the classic collecting deposits or those not yet thoroughly prospected or have not left a fossil record. On the other hand, a few primitive synapsids, representing two families, occur in the Late Permian, the eupelycosaurian varanopseids *Elliotsmithia* (South Africa), *Mesenosaurus* (Russia), and *Varanodon* (Oklahoma), and the caseasaurian caseids *Angelosaurus* (Texas), *Caseopsis* (Texas), *Cotylorhynchus* (Oklahoma and Texas), and *Ennatosaurus* (Russia) (Romer and Price, 1940; Reisz, 1986).

The European fossil record of primitive synapsids is very sparse, with a maximum of only ten genera recognized. Of these, five are known by substantial, articulated portions of the cranial and postcranial skeletons, whereas four of the remaining genera are represented by single specimens of tooth-bearing jaw bones and the fifth by a neural spine fragment and a small portion of the postcranial skeleton. In this survey, approximately 26 Late Carboniferous-Early Permian primitive synapsid genera are recognized from North America, many containing several species represented by numerous specimens. To this list can be added at least eight genera of uncertain familial status, as they are known by very fragmentary or imperfect specimens. This great disparity in global representation is undoubtedly due largely to distinctly different preservational environments being sampled. The overwhelming majority of the North American primitive synapsids are from the extensive, well-exposed, thick sections of Lower Permian red-bed fluvial deposits of the midcontinental and southwestern regions of the United States. Occurrences of European primitive synapsids in similar Lower Permian deposits are, on the other hand, rare and include jaw fragments of *Ophiacodon*, *Haptodus*, and *Sphenacodon* from England (Paton, 1974) and a skull and partial postcranial skeleton of *Casea* from France (Sigogneau-Russell and Russell, 1974). The poor exposures of the red-bed sediments and the apparent bias in exploration of sediments reflecting limnic or aquatic environments undoubtedly account greatly for the rarity of primitive synapsids in Europe. However, the association of at least two of the European primitive synapsids, the edaphosaurids and the ophiacodontid *Stereorhachis*, with limnic or aquatic deposits may not be unusual. Although not aquatic, edaphosaurids are typically viewed as inhabiting near-shore limnic environments and ophiacodontids are thought to tend toward piscivorous habits and amphibious adaptation. It should also be kept in mind that some of the rare occurrences may simply represent erratics and do not indicate environmental preference.

DISTRIBUTION OF CAPTORHINOMORPHA

The suborder Captorhinomorpha contains two, for the most part easily distinguishable, families—the Protorothyrididae and Captorhinidae. Whereas the Protorothyrididae is mainly a Late Carboniferous family, with a few Early Permian members, the Captorhinidae is restricted to the Early and Late Permian, though only the Early Permian distributions are plotted here. The Early Permian captorhinids are known to occur only in the southwestern and

Figure 7. Late Carboniferous distributions of the captorhinomorph family Protorothyrididae (P). Localities, latitidue and longitude may be found in Appendix 1. ➔

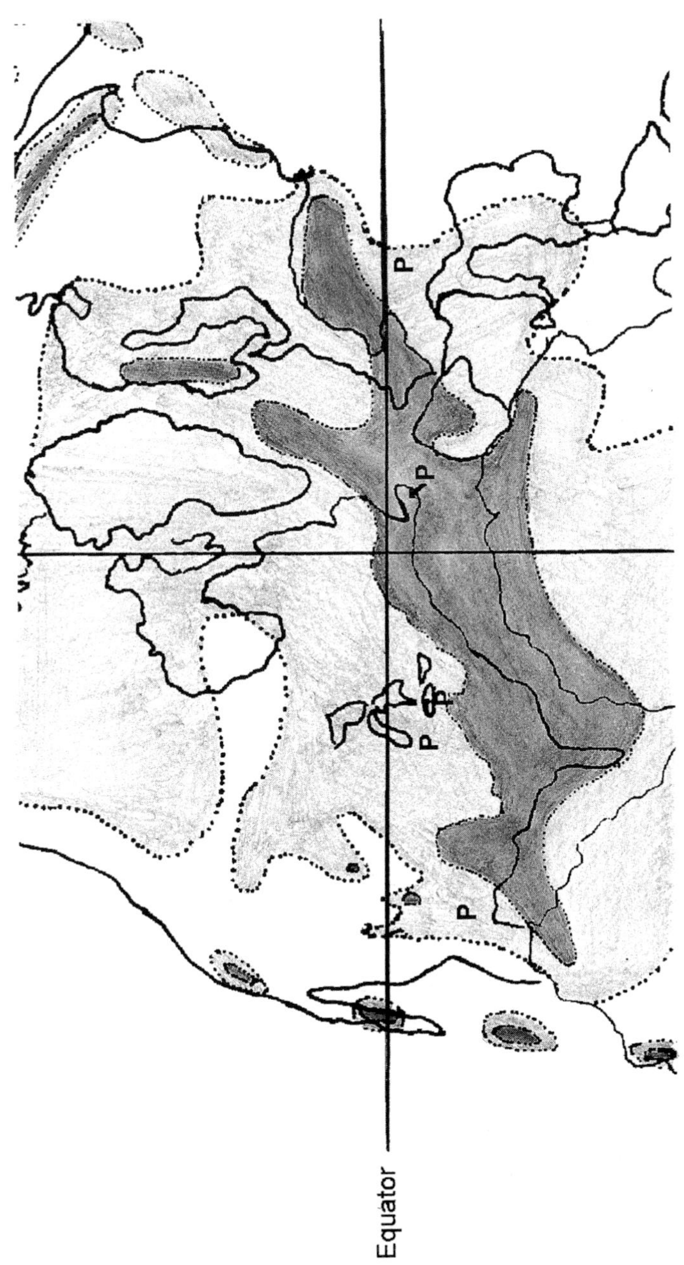
P
P
P
P
P
Equator

midcontinental regions of the United States, but persist into the Late Permian, where they are represented by several, mostly new forms which are widely distributed, with occurrences in Oklahoma and Texas (*Kahneria* and *Rothaniscus*), Russia (*Hecatogomphius*), Nigeria (*Moradisaurus*), Zimbabwe (*Protocaptorhinus*, also in Texas), and India (a captorhinid).

The captorhinomorph taxa and their ages and localities, as well as the bibliographic sources of this data, used to plot the Late Carboniferous and Early Permian distributions of the protorothyridids and captorhinids in Figures 7 and 8 are as follows:

Late Carboniferous (Fig. 7)

- ***Protorothyrididae***

Anthracodromeus (Ohio), *Archerpeton* (Nova Scotia), *Brouffia* (Czech Republic), *Cephalerpeton* (Illinois and Ohio), *Coelostegus* (Czech Republic), *Hylonomus* (Nova Scotia), a protorothyridid (New Mexico), and *Paleothyris* (Nova Scotia); Olson (1958, 1967, 1970), Carroll (1964, 1969c), Carroll and Baird (1972), Reisz and Baird (1983), and Lucas and Hunt (1991).

Lower Permian (Figure 8)

- ***Protorothyrididae***

A protorothyridid (Oklahoma), *Protorothyris* (Texas and West Virginia), and *Thuringothyris* (Germany); Clark and Carroll (1973), Reisz (1980b), and Boy and Martens (1991).

- ***Captorhinidae***

Captorhinikos (Oklahoma and Texas), *Captorhinus* (New Mexico, Oklahoma, and Texas), *Captorhinoides* (Texas), *Eocaptorhinus* (Oklahoma), *Labidosaurikos* (Oklahoma and Texas), *Labidosaurus* (Oklahoma and Texas), *Rhiodenticulatus* (New Mexico), *Romeria* (Texas), and *Protocaptorhinus* (Texas); Seltin (1959), Olson (1962b), Olson (1970), Clark and Carroll (1973), Heaton (1979), and Berman and Reisz (1986).

Figure 8. Early Permian distributions of the captorhinomorph families Protorothyrididae (P) and Captorhinidae (C). Localities, latitidue and longitude may be found in Appendix 1. ➔

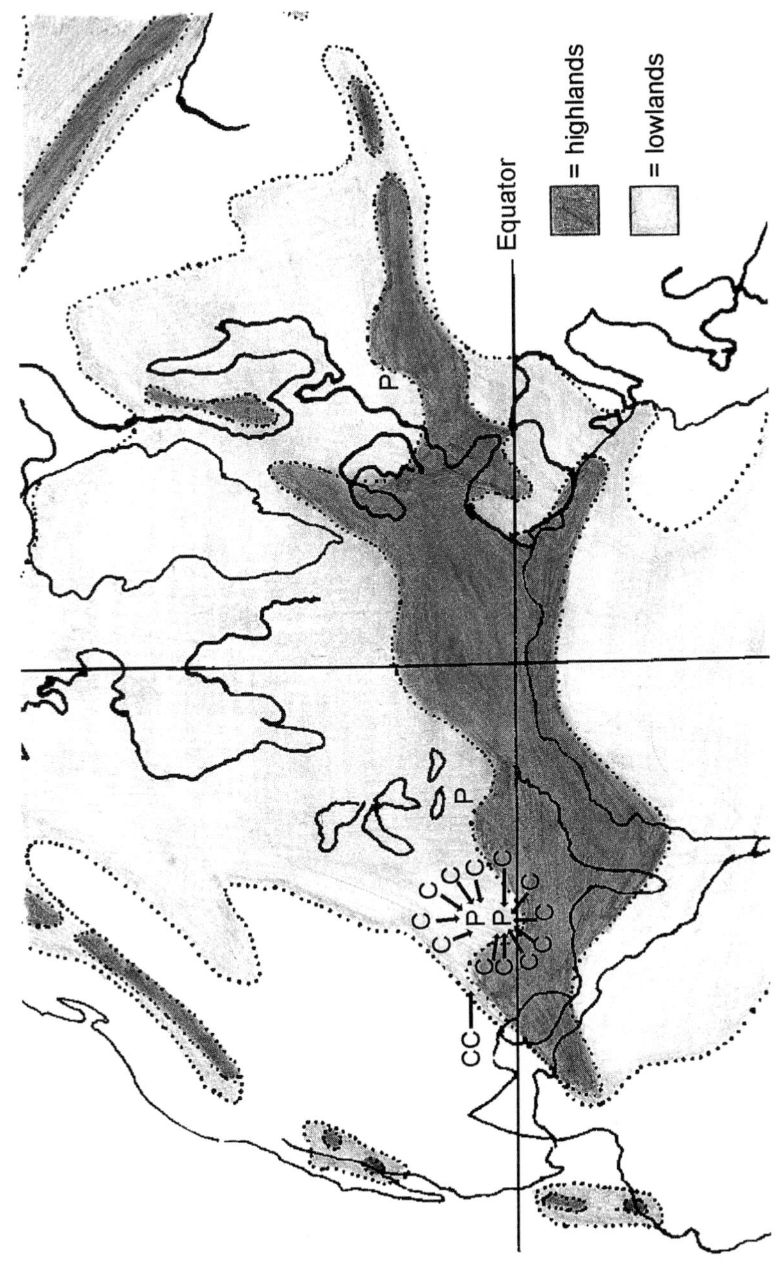
Equator
= highlands
= lowlands
P
P
C
C
C
C
C
C
P
P
C
C
C
C
C
C
C
CC

With the exception of two genera, *Protorothyris* and *Thuringothyris*, the protorothyridids are restricted to the Late Carboniferous, whereas none of the captorhinids occur before the Early Permian. Classically, this near-exclusive, temporal displacement has been taken as evidence that protorothyridids might be ancestral to the captorhinids. The nearly exclusive, distribution patterns of the two families, however, are not as easily explained. Both families are highly terrestrial, yet the protorothyridids are found in sediments representing a variety of environments, although mostly limnic, whereas the captorhinids are found almost exclusively in the fluvial red-bed sediments of the southwestern and midcontinental regions of the United States. Although Boy and Martens (1991) assigned *Thuringothyris* to the Protorothyrididae, they described it as possessing a number of captorhinid features. With this in mind, its occurrence in the red-bed sediments of the Lower Permian Tambach Formation of the Upper Rotliegend Bromacher locality of central Germany is of interest. Perhaps, as was suggested to explain the poor representation of seymouriids, diadectids, and primitive synapsids in Europe, the absence of captorhinids in Europe is more apparent than real. The long history of collecting in Europe has been characterized seemingly by an emphesis on the richly fossiliferous limnic or aquatic facies, whereas the terrestrial red-bed facies have received far less attention due to the false perception that they are not likely to preserve fossils in worthwhile numbers or quality.

DISTRIBUTION OF EARLY DIAPSIDA

Late Pennsylvanian and Early Permian representatives of the order Diapsida are few, and their distributions can be easily described without depiction on paleogeographic maps. The suborder Araeoscelidia, containing the Late Pernnsylvanian Petrolacosauridae and the Early Permian Araeoscilidae, represents the earliest stage in the adaptive radiation of the Diapsida (Laurin, 1991). In addition to the excellent specimens of the Late Pennsylvanian *Petrolacosaurus* of Kansas (Reisz, 1981) and the Early Permian *Araeoscelis* of Texas (Reisz *et al.*, 1984), the Araeoscelidia is otherwise represented by only two other, poorly known, Early Permian genera: *Kadaliosaurus* of Germany (Credner, 1889) and *Zarcasaurus* of New Mexico (Brinkman *et al.*, 1984). Olson (1970) has described a possible araeoscelidian, *Dictybolus*, from the Lower Permian of Oklahoma based on isolated bones of numerous individuals, however, familial assignment was not possible. More advanced, Late Paleozoic primitive diapsids, intermediate between the Araeoscelidia and the Neodiapsida (Laurin, 1991), are restricted to a single genus, *Apsisaurus*, from the Early Permian of Texas (Laurin, 1991). Two nearly complete skeletons of an unnamed, nonaraeoscelidian from the Late Pennsylvanian of Kansas were briefly described by Reisz (1988); one of these specimens, however, is now considered by him (R. Reisz, personal communication) to be a varanopseid synapsid.

Except for *Kadaliosaurus*, the early, primitive diapsids are, as far as known, restricted to the southwestern and midcontinental regions of the United States. It is highly suspicious that the distribution of the highly agile, terrestrial, lizard-like early, primitive diapsids includes only one European representative. Again, the obvious explanation for this pattern is that it is an artifact in exploration in Europe.

DISCUSSION

With the possible exception of some Asian discosauriscids, all the Late Carboniferous and Early Permian tetrapod groups in this survey are distributed within a narrow latitudinal belt of approximately 10^0 north and south of the paleoequator, where presumably a warm, humid, tropical climate existed. This distribution extended across the southern region of Euramerica, with the southern extent being bounded by the prominent Appalachian-Mauretinide-Variscan mountain chain of central Pangaea that marks the continental union with Gondwana. Terrestrial tetrapods have not been recorded during this time in the several regions of terrestrial deposits found in Russian, India, South America, and South Africa that were distributed between approximately 10-30^0 north and 10-60^0 south of the paleoequator. Yet, these regions have yielded rich and diverse assemblages of Late Permian tetrapods. Early Permian climates outside of the tropical, equatorial belt are believed to have ranged from dry subtropical belts in the lower latitudes (generally 10-30^0), characterized by evaporite deposits, to fairly wet, temperate belts at higher latitudes (generally

50-60^{0}), characterized by coal deposits (Parrish *et al.*, 1986). Undoubtedly the aquatic habit of the discosauriscids permited them to occupy also the wet, temperate climates of the higher latitudes of Asia.

Within the groups surveyed only a few exceptions can be noted that do not exhibit a cosmopolitan, Euramerican distribution during the Late Carboniferous and Early Permian. The ubiquitous presence of discosauriscids in the Early Permian of Europeas well as their sparse presence in Asia, and their apparent absence in North America is the most striking example and is suspected of reflecting a true biogeographical pattern. The explanation offered here for this distributional pattern is that the aquatic discosauriscids were endemic to central and western Europe, evolving during the latest Pennsylvanian or earliest Permian in the unique physical setting of the intramontane basins of that region. Occurrences in the wet, temperate climates of the higher paleolatitudes of the lower Upper Permian of the cis-Uralian forelands of Russia and the Upper Permian of Asia suggest subsequent dispersals. Furthermore, we speculate that the stratigraphic replacement of the discosauriscids by the semi-terrestrial *Seymouria* in the Rotliegend deposits of Germany may indicate a response to the gradual shift from limnic to predominately terrestrial environmental conditions, or an evolutionary event, or both. If *Seymouria* first evolved from discosauriscids in Europe, then it is plausible that the widespread development of terrestrial, red-bed environments during the Early Permian permitted an extention of its range across Euramerica.

Evidence of cosmopolitan distributions of Late Carboniferous-Early Permian Euramerican tetrapods has been bolstered greatly by the recent discovery of the diadectomorph *Diadectes* in Germany, as this highly terrestrial genus is widely distributed throughout North America. On the other hand, the absence of representatives of the limnoscelid and tseajaiid diadectomorphs in Europe cannot be considered strongly opposing evidence to cosmopolitism in view of their rarity even in North America, where they occur mainly in the southwestern United States. Certainly, some regional uniquenesses in faunal assemblages is to be expected over as wide an area as Euramerica. With regard to the Diadectomorpha, it is also important to note that the fragmentary specimens on which the rare German forms *Phanerosaurus* and *Stephanospondylus* are based do not allow confident familial assignment, and, although widely accepted as diadectids, they could just as likely pertain to either Limnoscelidae or Tseajaiidae. Although the primitive synapsids exhibit a very one-sided distribution pattern in which the greatest number of taxa and specimens are known from North America, there is sufficient evidence to indicate that the majority of families were Euramerican in distribution. Whereas the protorothyridids appear to have been Euramerican in distribution, there is currently no convincing evidence to contradict the empirical data that the restriction of the Early Permian captorhinids to North America and their absence in Europe does not reflect a true biogeographical pattern. However, *Thuringothyris* from the Lower Permian of Germany, though assigned to the Protorothyrididae (Boy and Martens, 1991), does exhibit several captorhinid features and may represent an intermediate form. Finally, the poorly documented early, primitive diapsids, though represented in the Late Pennsylvanian only in North America, are suspected of having had an Euramerican distribution during the Early Permian.

It has been argued throughout this discussion that the paucity of specimens and the relative low degree of taxonomic diversity of the seymouriids, diadectomorphs, primitive synapsids, possibly the protorothyridid captorhinomorphs, and probably the early, primitive diapsids in Europe, as compared to North America, is due largely to marked differences in the environments of deposition that have been traditionally explored. This pattern, therefore, does not necessarily infer limited biogeographical distributions.

The Late Permian documents not only pronounced changes in the terrestrial tetrapod assemblages, but also in the global climates and distribution patterns of terrestrial vertebrates (Olson, 1979b, 1984; Parrish *et al.*, 1986; Hotton, 1992; Milner, 1993). Coal deposits, which dominated the paleoequatorial zone and the northern, high paleolatitudes between 30 and 60° during the Late Carboniferous and Early Permian, occurred only at very high latitudes in the northern and southern hemispheres in the Late Permian. Whereas existing evaporite deposits on the western margin of equatorial Pangaea became more prominent, extensive evaporite deposits appear for the first time along the eastern margin of Pangaea of central and northern Europe between approximately 15 and 40° paleolatitude. Tilites, which were widely distributed in the high latitudes of the southern supercontinent of Gondwana during the Late Carboniferous and Early Permian, are essentially gone (except for a small area in Australia) at the beginning of the Late Permian. These sedimentary changes indicate a dramatic expansion of the dry subtropical belts and a marked narrowing or loss of the tropical, paleoequatorial belt (Parrish *et al.*, 1986).

Among the groups, or their subgroups, surveyed here, most were waning or had disappeared by the beginning of the Late Permian. Of the Seymouriamorpha families Discosauriscidae and Seymouriidae, there are only three instances of persistence into the lower Upper Permian: a specimen of *Seymouria* from Oklahoma (Olson, 1980), although, as noted above, its identity is suspect; *Discosauriscus* specimens from a site in the cis-Uralian forelands of Russia (Ivakhnenko, 1981); and the discosauriscid *Urumqia* from China (Zhang *et al.*, 1984). However, several members of the seymouriamorph families Seymouriidae and Kotlassiidae have been reported from the Late Permian of Russia. There are no Late Permian representatives of the suborder Diadectomorpha. On the other hand, of the primitive synapsids, only two very closely related genera (*Cotylorhynchus* and *Ennatosaurus*) of the caseasaurian family Caseidae and three very closely related genera (*Elliotsmithia*, *Mesenosaurus*, and *Varanodon*) of the eupelycosaurian family Varanopseidae have been reported from Late Permian deposits of the widely distant areas of the cis-Uralian forelands of Russia, South Africa, and Oklahoma (Reisz, 1986). One of these five genera (*Cotylorhynchus*) also occurs in the Lower Permian of Oklahoma.

A similar biogeographical history is seen among the members of the Captorhinidae of the suborder Captorhinomorpha. Seven Late Permian forms have been described from the widely distant areas of India, Russia, United States, and Zimbabwe, with one of these (*Protocaptorhinus*) also known from the Early Permian of the United States. Also of interest, certain members of this group exhibit closer relationships than is otherwise expressed by their confamilial assignment, yet they occur in widely distant areas: for example *Protocaptorhinus* from Zimbabwe and Oklahoma (Gaffney and Mckenna, 1979), a captorhinid from India (Kutty, 1972), and

Riabininus from Russia (Ivakhnenko, 1990); *Moradisaurus* from Nigeria (Ricqlés and Taquet, 1982) and *Rothianiscus* from Oklahoma and Texas (Olson and Beerbower, 1953; Olson, 1962a, 1965; Olson and Barghusen, 1962); and *Kahneria* from Oklahoma (Olson, 1962a) and *Hecatogomphius* from Russia (Vyushkov and Chudinov, 1957; Olson, 1962a). These distributions suggest that during the latest Early or earliest Late Permian surviving lineages of the seymouriamorphs, primitive synapsids, and captorhinids dispersed long distances both north and south of the paleoequator to occupy climatic regimes ranging from warm, dry, subtropical to wet temperate. In addition, representatives of the continuations of these groups into the upper Lower and lower Upper Permian of the midcontinental region of the United States (Texas and Oklahoma) were then also living under similar climatic conditions. According to Olson and Vaughn (1970), in this region the Lower through Upper Permian sections and their vertebrates document a history in which generally warm, humid conditions were gradually replaced by increasingly seasonal and drier climates.

The emergence of the advanced synapsids of the order Therapsida, unarguably the descendants of the advanced eupelycosaurian primitive synapsids, at the beginning of the Late Permian undoubtedly represents the most dramatic change in the Permian, global faunal assemblages. All the undoubted, early Late Permian members of the Therapida are from two regions–the Russian cis-Uralian forelands and South Africa. Several very fragmentary specimens from deposits of approximately the same age in the midcontinent of the United States have been described as therapsids (Olson and Beerbower, 1953; Olson, 1962a, 1974). These assignments, however, are considered very uncertain, as the specimens could conceivably be primitive synapsids. An indepth study of paleoclimatological evidence by Parrish *et al.* (1986: p. 125) concludes that the Russian and South African therapsid-bearing deposits "were in temperate and fairly moist climates that resulted from the joint effects of fairly high paleolatitudes and locations near the warmer eastern side of Pangaea (in the Russian deposits)." The therapsids of these two regions, though separated by almost 90° of latitude in which there are almost no terrestrial deposits of the same age, are, for the most part, very similar. Therapsid distributions at this time, therefore, may actually reflect the distribution of deposits suitable for their preservation rather than a true biogeographical pattern.

During the Late Permian to the Early Triassic, Pangaea continued to drift northward for approximately another 10°, as well as rotate somewhat clockwise, and undergo considerable accretion topography accumulation in western Laurasia. In addition, the once prominent Permo-Carboniferous mountain chains that extended across central Pangaea had become reduced to small remnants. Parrish *et al.* (1986) and Hotton (1992) have given excellent accounts of the therapsid global distributions during this period. Very similar therapsid assemblages in the uppermost Permian through the Triassic are depicted as expanding their ranges longitudinally but remaining restricted mainly to high northern and southern paleolatitudes except in the Upper Triassic. The Upper Triassic therapsids exhibit distributions that noticeably more closely approach the paleoequator, reaching paleolatitudes as low as 15° north and 30° south. It cannot be resolved confidently, however, whether the absence of equatorial and subtropical therapsids during the latest Permian (as in the early Late Permian) through the Middle Triassic reflects a true biogeographical event. The few areas within the equatorial and subtropical climatic belts containing terrestrial sediments during this time, mostly Lower and Middle Triassic and lying between the northern paleolatitudes of about 5-30°, have not yielded any therapsids. It is, therefore, uncertain whether the wide latitudinal separation of the therapsids to northern and southern regions of warm to temperate climates was due to an intolerance to the hot, seasonally arid conditions of the intervening equatorial or near-equatorial belts, or the absence or near absence of terrestrial, tetrapod-bearing beds (Parrish *et al.*, 1986). Most current opinions (Olson, 1984; Parrish *et al.*, 1986; Hotton, 1992), however, interpret the global similarities of the therapsids, despite their wide separations, as indicating the absence of any major barriers to the probable occurrence of Pangaean-wide migrations and a cosmopolitan distribution. Olson (1979b) has ruled out endothermy as a prerequisite to therapsid dispersal but does not discount that incipient endothermy may well have enhanced its rapidity and extent. As noted above, several undoubtedly ectothermic groups exhibit identical distribution patterns during the Late Permian.

ACKNOWLEDGMENTS

The authors acknowledge Drs. F. Zeigler and D. Rowley for helpful insights into the Permian of China. Sincere thanks are expressed to Mr. Rick Everett for software assistance and aid in production of the paleogeographic maps when all else seemed lost. Dr. Elizabeth Rega helped to translate important material from German and British into American English. Support for field work that produced some of the German specimens reported herein was provided by Grant 5182-94 from the National Geographic Society (to SSS and DSB), Grant CRG.940779 from the North Atlantic Treaty Organization (NATO) (to SSS), and a California State University San Bernardino Minigrant (to SSS).

APPENDIX 1: PALEOGEOGRAPHIC DATA

Appendix 1 on pages 122-130 includes taxonomic, locality, latititudinal, and longitudinal data for the maps in Figures 2-8.

Genus		Locality	Lat.	Long.	sy	map #	References
Late Carboniferous Maps							
Limnoscelidae							
Limnoscelis	New Mexico	El Cobre Canyon	+36.2	-106.3	L	3	Williston (1911), Carroll (1967),
Limnoscelis	Colorado	Badger Creek	+38.2	-105.8	L	3	Fracasso (1980), Berman and
Limnostegis	Nova Scotia	Florence	+46.2	-60.4	L	3	Sumida (1990), and Berman (1993).
							Case (1908), Case and Williston
Diadectidae							
Desmatodon	Colorado	Badger Creek	+38.2	-105.8	D	3	(1913), Romer (1952), Vaughn
Desmatodon	New Mexico	El Cobre Canyon	+36.2	-106.3	D	3	(1969, 1972), Fracasso (1980),
Desmatodon	Pennsylvania	Pittsburgh	+40.4	-80.0	D	3	Berman (1993), and Berman and
Diasparactus	New Mexico	El Cobre Canyon	+36.2	-106.3	D	3	Sumida (1995).
Eupelycosauria I							
Varanopseidae							
Aerosaurus	New Mexico	El Cobre Canyon	+36.2	-106.3	e	5	Reisz (1986)
Ruthiromia	New Mexico	El Cobre Canyon	+36.2	-106.3	e	5	
Ophiacodontidae							
Archaeothyris	Nova Scotia	Florence	+46.2	-60.4	e	5	Reisz (1986), Reisz and Berman

Genus	Locality		Lat.	Long.	sy	map #	References
Baldwinonus	New Mexico	El Cobre Canyon	+36.2	-106.3	e	5	(1986), and Sumida and Berman,
Clepsydrops	Illinois	Danville	+40.1	-87.7	e	5	1993).
Clepsydrops	Pennsylvania	Pittsburgh	+40.4	-80.0	e	5	
Ophiacodon	Colorado	Badger Creek	+38.2	-105.8	e	5	
Ophiacodon	Kansas	Garnett	+38.2	-95.2	e	5	
Ophiacodon	New Mexico	El Cobre Canyon	+36.2	-106.3	e	5	
Ophiacodon	Texas	Archer City	+33.6	-98.6	e	5	
Stereophallodon	Texas	Windthorst	+33.6	-98.5	e	5	
Stereorhachis	France	Autun	+46.9	+4.3	e	5	

Edaphosauridae

Genus	Locality		Lat.	Long.	sy	map #	References
Edaphosaurus	New Mexico	El Cobre Canyon	+36.2	-106.3	e	5	Vaughn (1964) and Reisz (1986),
Edaphosaurus	West Virginia	Cameron	39.8	-80.6	e	5	Reisz and Berman (1986), Modesto
edaphosaurid	Czeck Republic	Nyrany	+49.4	-13.1	e	5	and Reisz (1990), and Sumida and
edaphosaurid	Pennsylvania	Pittsburgh	+40.4	-80.0	e	5	Berman (1993).
Ianthosaurus	Colorado	Badger Creek	+38.2	-105.8	e	5	
Ianthosaurus	Kansas	Garnett	+38.2	-95.2	e	5	

Eupelycosauria II

"Haptodontids"

Genus	Locality		Lat.	Long.	sy	map #	References
Haptodus	Kansas	Garnett	+38.2	-95.2	E	5	Laurin (1992) and Sumida and
haptodontid	Colorado	Badger Creek	+38.2	-105.8	E	5	Berman (1993).

Genus	Locality		Lat.	Long.	sy	map #	References
Sphenacodontidae							
Macromerion	Czech Republic	Nyrany	+49.4	-13.1	E	5	Reisz (1986).
Sphenacodon	New Mexico	El Cobre Canyon	+36.2	-106.3	E	5	
Protorothyrididae							
Anthracodromeus	Ohio	Linton	40.2	-80.7	P	7	Olson (1958, 1967, 1970), Carroll (1964; 1969c), Carroll and Baird (1972), Reisz and Baird (1983), and Lucas and Hunt (1991).
Archerpeton	Nova Scotia	Florence	+46.2	-60.4	P	7	
Brouffia	Czech Republic	Nyrany	+49.4	-13.1	P	7	
Cephalerpeton	Illinois	Danville	+40.1	-87.7	P	7	
Cephalerpeton	Ohio	Linton	40.2	-80.7	P	7	
Coelostegus	Czech Republic	Nyrany	+49.4	-13.1	P	7	
Hylonomus	Nova Scotia	Florence	+46.2	-60.4	P	7	
protorothyridid	New Mexico	El Cobre Canyon	+36.2	-106.3	P	7	
Paleothyris	Nova Scotia	Florence	+46.2	-60.4	P	7	

Early Permian Maps

Genus	Locality		Lat.	Long.	sy	map #	References
Discosauriscidae							
Ariekanerpeton	Tadghikistan	Sarytaypan	+40.0	+70.0	D	2	Ivakhnenko (1981), Kuznetsov and Ivakhnenko (1981), and Werneburg (1989, 1990).
Discosauriscus	Czech Republic	Brno	+49.2	+16.6	D	2	
Discosauriscus	France	Decize	+46.8	+3.4	D	2	
Discosauriscus	Germany	Dresden	+51.1	+13.7	D	2	

Genus	Locality		Lat.	Long.	sy	map #	References
Irumqia	China	Urumqi	+43.0	+88.0	D	2	
Letoverpeton	Czech Republic	Brno	+49.2	+16.6	D	2	
Letoverpeton	Germany	Dresden	+51.1	+13.7	D	2	
Utegenia	Kazakhstan	Alma-Ata	+43.0	+76.0	D	2	
Seymouriidae							
Seymouria	Germany	Tambach-Dietharz	+50.9	+10.7	S	2	Langston (1963), Vaughn (1966),
Seymouria	New Mexico	Coyote	+36.1	-106.6	S	2	Olson (1979, 1980), Berman et al.
Seymouria	Oklahoma	Oklahoma City	+35.4	-97.5	S	2	(1987), and Berman and Marterns,
Seymouria	Prince Edward Is	Charlottetown	+46.2	-63.0	S	2	(1992).
Seymouria	Texas	Archer City	+33.6	-98.6	S	2	
Seymouria	Utah	Mexican Hat	+37.1	-109.8	S	2	
Limnoscelidae							
Limnosceloides	New Mexico	Coyote	+36.1	-106.6	L	4	Romer (1952), Vaughn (1962),
Limnosceloides	West Virgina	Ripley	38.8	-81.8	L	4	Lewis and Vaughn (1965),
Limnoscelops	Colorado	Placerville	+38.0	-108.0	L	4	Langston (1969), and Berman
limnoscelid	Utah	Mexican Hat	+37.1	-109.8	L	4	(1993).
Tseajaiidae							
Tseajaia	New Mexico	Coyote	+36.1	-106.6	T	4	Vaughn (1964), Moss (1972), and
Tseajaia	Utah	Mexican Hat	+37.1	-109.8	T	4	Berman (1993).

Genus	Locality		Lat.	Long.	sy	map #	References
Diadectidae							
Diadectes	Colorado	Placerville	+38.0	-108.0	D	4	Meyer (1860), Geinitz and Deichmueller (1882), Olson (1947, 1967, 1975), Langston (1963), Lewis and Vaughn (1965), Berman (1971, 1993), and Berman and Martens (1992).
Diadectes	Germany	Tambach-Dietharz	+50.9	+10.7	D	4	
Diadectes	New Mexico	Coyote	+36.1	-106.6	D	4	
Diadectes	Ohio	Belpre	+39.3	-81.6	D	4	
Diadectes	Oklahoma	Oklahoma City	+35.4	-97.5	D	4	
Diadectes	Prince Edward Is	Charlottetown	+46.2	-63.0	D	4	
Diadectes	Texas	Archer City	+33.6	-98.6	D	4	
Diadectes	Utah	Mexican Hat	+37.1	-109.8	D	4	
Diadectes	West Virginia	Speed	38.7	-81.4	D	4	
Phanerosaurus	Germany	Dresden	+51.1	+13.7	D	4	
Stephanospondyl	Germany	Dresden	+51.1	+13.7	D	4	
Caseasauria							
Eothyridae							
Eothyris	Texas	Archer City	+33.6	-98.6	C	6	Reisz (1986).
Oedaleops	New Mexico	Coyote	+36.1	-106.6	C	6	
Caseidae							
Casea	Texas	Archer City	+33.6	-98.6	C	6	Reisz (1986).
Casea	France	Rodez	+44.3	+2.5	C	6	

Genus		Locality	Lat.	Long.	sy	map #	References
Cotylorhynchus	Oklahoma	Oklahoma City	+35.4	-97.5	C	6	
Eupelycosauria I							
Varanopseidae							
Aerosaurus	New Mexico	Coyote	+36.1	-106.6	e	6	Reisz (1986).
Basicranodon	Oklahoma	Oklahoma City	+35.4	-97.5	e	6	
Mycterosaurus	Colorado	Placerville	+38.0	-108.0	e	6	
Mycterosaurus	Texas	Archer City	+33.6	-98.6	e	6	
Varanops	Texas	Archer City	+33.6	-98.6	e	6	
Ophiacodontidae							
Baldwinonus	Ohio	Woodsfield	+39.8	-81.1	e	6	Vaughn (1962, 1964), Lewis and Vaughn (1965), Olson (1976, 1975),
Ophiacodon	England	(use Coventry)	+52.4	-1.5	e	6	
Ophiacodon	Colorado	Placerville	+38.0	-108.0	e	6	Riesz (1986), and Berman *et al.* (1995).
Ophiacodon	Kansas	Winfield	+37.3	-97.0	e	6	
Ophiacodon	New Mexico	Coyote	+36.1	-106.6	e	6	
Ophiacodon	Oklahoma	Oklahoma City	+35.4	-97.5	e	6	
Ophiacodon	Ohio	Belpre	+39.3	-81.6	e	6	
Ophiacodon	Texas	Archer City	+33.6	-98.6	e	6	
Ophiacodon	Utah	Mexican Hat	+37.1	-109.8	e	6	
Varanosaurus	Oklahoma	Oklahoma City	+35.4	-97.5	e	6	

Genus	Locality		Lat.	Long.	sy	map #	References
Varanosaurus	Texas	Archer City	+33.6	-98.6	e	6	
Edaphosauridae							
Edaphosaurus	New Mexico	Coyote	+36.1	-106.6	E	6	Olson and Vaughn (1970), Reisz (1986), Sumida (1989), and Modesto (1994).
Edaphosaurus	Ohio	Belpre	+39.3	-81.6	E	6	
Edaphosaurus	Texas	Archer City	+33.6	-98.6	E	6	
Edaphosaurus	Utah	Mexican Hat	+37.1	-109.8	E	6	
Edaphosaurus	West Virginia	Burton	39.7	-80.4	E	6	
Edaphosaurus	West Virginia	Ripley	38.8	-81.6	E	6	
edaphosaurid	Germany	(Gotha?)	+50.9	+10.7	E	6	
Glaucosaurus	Texas	Archer City	+33.6	-98.6	E	6	
Lupeosaurus	Texas	Archer City	+33.6	-98.6	E	6	
Eupelycosauria II							
"Haptodontids"							
Cutleria	Colorado	Placerville	+38.0	-108.0	E	6	Currie (1979), Reisz (1986), and Laurin (1993).
Haptodus	France	Autun	+46.9	+4.3	E	6	
Haptodus	England	Kenilworth	+52.4	-1.5	E	6	
Palaeohatteria	Germany	Dresden	+51.1	+13.7	E	6	
Pantelosaurus	Germany	Dresden	+51.1	+13.7	E	6	

Genus	Locality		Lat.	Long.	sy	map #	References
Sphenacodontidae							
Bathygnathus	Prince Edward Island		+46.2	-63.0	E	6	Olson (1975), Vaughn (1964), Reisz (1986), and Hook and Hotton (1991).
Ctenorhachis	Texas	Archer City	+33.6	-98.6	E	6	
Ctenospondylus	Ohio	Woodsfield	+39.8	-81.1	E	6	
Ctenospondylus	Texas	Archer City	+33.6	-98.6	E	6	
Ctenospondylus	Utah	Mexican Hat	+37.1	-109.8	E	6	
Dimetrodon	New Mexico	Coyote	+36.1	-106.6	E	6	
Dimetrodon	Ohio	Belpre	+39.3	-81.6	E	6	
Dimetrodon	Texas	Archer City	+33.6	-98.6	E	6	
Dimetrodon	Utah	Mexican Hat	+37.1	-109.8	E	6	
Neosaurus	France	Besancon	+47.3	+6.0	E	6	
Secodontosaurus	Texas	Archer City	+33.6	-98.6	E	6	
Sphenacodon	England	Kenilworth?	+52.4	-1.5	E	6	
Sphenacodon	New Mexico	Coyote	+36.1	-106.6	E	6	
Sphenacodon	Utah	Mexican Hat	+37.1	-109.8	E	6	
Protorothyrididae							
protorothyridid	Oklahoma	Oklahoma City	+35.4	-97.5	P	8	Clark and Carroll (1973), Reisz (1980a), and Boy and Martens (1991).
Protorothyris	Texas	Archer City	+33.6	-98.6	P	8	
Protorothyris	West Virginia	Burton	39.7	-80.4	P	8	
Thuringothyris	Germany	Tambach-Dietharz	+50.9	+10.7	P	8	

Genus	Locality		Lat.	Long.	sy	map #	References
Captorhinidae							
Captorhinikos	Oklahoma	Oklahoma City	+35.4	-97.5	C	8	Seltin (1959), Olson and Barghusen (1962), Olson (1970), Clark and Carroll (1973), Heaton (1979), and Berman and Reisz (1986).
Captorhinikos	Texas	Archer City	+33.6	-98.6	C	8	
Captorhinus	New Mexico	Coyote	+36.1	-106.6	C	8	
Captorhinus	Oklahoma	Oklahoma City	+35.4	-97.5	C	8	
Captorhinus	Texas	Archer City	+33.6	-98.6	C	8	
Captorhinoides	Texas	Archer City	+33.6	-98.6	C	8	
Eocaptorhinus	Oklahoma	Oklahoma City	+35.4	-97.5	C	8	
Labidosaurikos	Oklahoma	Oklahoma City	+35.4	-97.5	C	8	
Labidosaurikos	Texas	Archer City	+33.6	-98.6	C	8	
Labidosaurus	Oklahoma	Oklahoma City	+35.4	-97.5	C	8	
Labidosaurus	Texas	Archer City	+33.6	-98.6	C	8	
Rhidenticulatus	New Mexico	Coyote	+36.1	-106.6	C	8	
Romeria	Texas	Archer City	+33.6	-98.6	C	8	
Protocaptorhinus	Texas	Archer City	+33.6	-98.6	C	8	

LITERATURE CITED

Baird, D., and R. L. Carroll. 1967. *Romeriscus*, the oldest known reptile. *Science*, 157:56-59.

Barlow, J. A. 1975. *The Age of the Dunkard. Proceedings of the First I. C. White Memorial Symposium.* West Virginia Geologic and Economic Survey, Morgantown, West Virginia.

Berman, D. S. 1971. A small skull of the Lower Permian reptile *Diadectes* from the Washington Formation, Dunkard Group, West Virginia. *Annals of Carnegie Museum*, 43: 33-46.

Berman, D. S. 1993. Lower Permian vertebrate localities of New Mexico and their assemblages. Pages 11-21 in: *Vertebrate Paleontology in New Mexico* (S. G. Lucas and J. Zidek, eds.). New Mexico Museum of Natural History and Science, Bulletin 2.

Berman, D. S, and S. L. Berman. 1975. *Broiliellus hektopos* sp. nov. (Temnospondyli: Amphibia) Washington Formation, Dunkard Group, Ohio. Pages 69-78 in: *The Age of the Dunkard*; *Proceedings of the First I. C. White Memorial Symposium* (J. A. Barlow, ed.). West Virginia. Geological and Economic Survey, Morgantown, West Virginia.

Berman, D. S, and T. Martens. 1993. First occurrence of *Seymouria* (Amphibia: Batrachosauria) in the Lower Permian Rotliegend of central Germany. *Annals of Carnegie Museum*, 62:63-79.

Berman, D. S, and R. R. Reisz. 1986. Captorhinid reptiles from the early Permian of New Mexico, with description of a new genus and species. *Annals of Carnegie Museum*, 55:1-28.

Berman, D. S, R. R. Reisz and D. A. Eberth. 1987. *Seymouria sanjuanensis* (Amphibia, Batrachosauria) from the Lower Permian Cutler Formation of north-central New Mexico and the occurrence of sexual dimorphism in that genus questioned. *Canadian Journal of Earth Sciences*, 24:1769-1784.

Berman, D. S, R. R. Reisz, J. R. Bolt, and D. Scott. 1995. The cranial anatomy and relationships of the synapsid *Varanosaurus* (Eupelycosauria: Ophiacodontidae) from the Early Permian of Texas and Oklahoma. *Annals of Carnegie Museum*, 64:99-133.

Berman, D. S, S. S. Sumida and R. E. Lombard. 1992. Reinterpretation of the temporal and occipital regions in *Diadectes* and the relationships of the diadectomorphs. *Journal of Paleontology*, 66:481-499.

Berman, D. S, and S. S. Sumida. 1990. A new species of *Limnoscelis* (Amphibia, Diadectomorpha) from the Late Pennsylvanian Sangre de Cristo Formation of central Colorado. *Annals of Carnegie Museum*, 59:303-341.

Berman, D. S, and S. S. Sumida. 1995. New cranial material of the rare diadectid *Desmatodon hesperis* (Diadectomorpha) from the Late Pennsylvanian of central Colorado. *Annals of Carnegie Museum*, 64:315-336.

Boy, J. A., and T. Martens. 1991. Ein neues captorhinomorphes Reptil aus dem thuringischen Rotliegend (Unter-Perm; Ost-Deutschland). *Paleontologische Zeitschrift*, 65:363-389.

Brinkman, D. B., D. S Berman, and D. A. Eberth. 1984. A new araeoscelid reptile, *Zarcasaurus tanyderus*, from the Cutler Formation (Lower Permian) of north-central New Mexico. *New Mexico Geology*, May 1984:34-40.

Carroll, R. L. 1964. The earliest reptiles. *Journal of the Linnean Society* (Zoology), 45:61-83.

Carroll, R. L. 1967. A limnoscelid reptile from the Middle Pennsylvanian. *Journal of Paleontology*, 41:156-1261.

Carroll, R. L. 1969a. The origin of reptiles. Pages 1-44 in: *The Biology of the Reptilia 1 (Morphology A)* (Gans, C, A. d'A. Bellairs, and T. S. Parsons, eds.), London: Academic Press.

Carroll, R. L. 1969b. Problems of the origin of reptiles. *Biological Reviews*, 44:393-432.

Carroll, R. L. 1969c. A Middle Pennsylvanian captorhinomorph and the interrelationships of primitive reptiles. *Journal of Paleontology*, 43:151-170.

Carroll, R. L. 1970. The ancestry of reptiles. *Philosophical Transactions of the Royal Society of London* (B), 257:267-308.

Carroll, R. L. 1988. The primary radiation of terrestrial vertebrates. *Annual Review of Earth and Planetary Science*, 20:45-84.

Carroll, R. L., and D. Baird. 1972. Carboniferous stem-reptiles of the Family Romeriidae. *Bulletin of the Museum of Comparative Zoology*, 143:321-364.

Case, E. C. 1908. Description of vertebrate fossils from the vicinity of Pittsburgh, Pennsylvania. *Annals of Carnegie Museum*, 4:234-241.

Case, E. C., and S. W. Williston. 1913. A description of *Aspidosaurus novomexicanus*. *Publications of the Carnegie Institute of Washington*, 181:7-9.

Clark, J., and R. L. Carroll. 1973. Romeriid reptiles from the Lower Permian. *Bulletin of the Museum of Comparative Zoology*, 144:353-407.

Credner, H. 1889. Die Stegocephalen und Saurier aus dem Rothliegenden des Plauen'schen Grundes bei Dresden. VIII. Theil. *Kadaliosaurus priscus* Cred. *Zeitschrift deutsche geologische Gesellschraft*, 41:319-342.

Currie, P. J. 1977. A new haptodontine sphenacodont (Reptilia; Pelycosauria) from the Upper Pennsylvanian of North America. *Journal of Paleontology*, 51:927-942.

Currie, P. J. 1979. The osteology of haptodontine sphenacodonts (Reptilia: Pelycosauria). *Palaeontographica* Abt. A, 163:130-168.

Eberth. D. A., and D. S Berman. 1993. Stratigraphy, sedimentology and vertebrate paleoecology of the Cutler Formation redbeds (Pennsylvanian-Permian) of north-central New Mexico. Pages 33-48 in: *Vertebrate Paleontology in New Mexico*. (S. G. Lucas and J. Zidek, eds.), New Mexico Museum of Natural History and Science Bulletin 2.

Eberth. D. A., and A. D. Miall. 1991. Stratigraphy, sedimentology and evolution of a vertebrate-bearing, braided to anastomosed fluvial system, Cutler Formation (Permo-Pennsylvanian), north-central New Mexico. *Sedimentary Geology*, 72:225-252.

Fracasso, M. 1980. Age of the Permo-Carboniferous Cutler Formation vertebrate fauna from El Cobre Canyon, New Mexico. *Journal of Paleontology*, 54:1237-1244.

Gaffney, E. S., and M. C. McKenna. 1979. A Late Permian captorhinid from Rhodesia. *American Museum Novitates*, 2688:1-15.

Gauthier, J. A., A. G. Kluge, and T. Rowe. 1988. The early evolution of the Amniota. Pages 103-155 in: *The Phylogeny and Classification of the Tetrapods, Volume 1, Amphibians, Reptiles, Birds*, Systematics Association Special Volume No. 35A (M. J. Benton, ed.), Oxford: Clarendon Press.

Geinitz, H. B., and J. V. Deichmüller. 1882. Die Saurier der unteren Dyas von Sachsen. *Palaeontographica*, 29:1-49.

Gordon, M. S., and E. C. Olson. 1994. *Invasions of the Land*. New York: Columbia University Press.

Heaton, M. J. 1979. Cranial anatomy of primitive captorhinid reptiles from the late Pennsylvanian and early Permian, Oklahoma and Texas. *Bulletin of the Oklahoma Geological Survey*, 127:1-84.

Heaton, M. J. 1980. The Cotylosauria: a reconsideration of a group of archaic tetrapods. Pages 497-551 in: *The Terrestrial Environment and the Origin of Land Vertebrates*, Systematics Association Special Volume No. 15 (A. L. Panchen, ed.). London : Academic Press.

Heaton, M. J., and R. R. Reisz. 1986. Phylogenetic relationships of captorhinomorph reptiles. *Canadian Journal of Earth Sciences*, 23:402-418.

Hook, R. W., and N. Hotton, III. 1991. A new sphenacodontid pelycosaur (Synapsida) from the Wichita Group, Lower Permian of North-Central Texas. *Journal of Vertebrate Paleontology*, 11:37-44.

Hotton, N. III. 1992. Global distribution of terrestrial and aquatic tetrapods, and its relevance to the position of the continental masses. Pages 267-285 in: *New Concepts in Global Tectonics* (S. Chatterjee and N. Hotton III, eds.). Lubbock : Texas Tech University Press.

Ivakhnenko, M. F. 1981. Discosauriscidae from the Permian of Tadzhikstan. *Paleontological Journal* 1981, 1:90-102.

Ivakhnenko, M. F. 1990. Elements of the Early Permian tetrapod faunal assemblages of Eastern Europe. *Paleontological Journal 1990*, 2:102-111

Kemp, T. 1982. *Mammal-like Reptiles and the Origin of Mammals*. London: Academic Press.

Kozur, H. 1984. Carboniferous-Permian boundary in marine and continental sediments. *Comptes Rendues 9e Congres International Stratigraphie et Geologie Carbonifere*, Washington and Champaign-Urbana, 1979, 2:577-586.

Kozur, H. 1988. The age of the Central European Rotliegendes. *Zeitschrift für geologische Wissenschaften*, 16:907-915.

Kozur, H. 1989. Biostratigraphic zonations in the Rotliegendes and their correlations. *Acta Musei Reginaehradecensis*, Ser. A, 22:15-20, Hradec Kralove.

Kutty, T. S. 1972. Permian reptilian fauna from India. *Nature*, 237:462-463.

Kuznetsov, V. V., and M. F. Ivakhnenko. 1981. Discosauriscids from the Upper Paleozoic in southern Kazakhstan. *Paleontological Journal*, 15:101-108.

Langston, W., Jr. 1963. Fossil vertebrates and the late Palaeozoic red beds of Prince Edward Island. *National Museum of Canada Bulletin*, 187:1-36.

Langston, W., Jr. 1966. *Limnosceloides brachycoles* (Reptilia: Captorhinomorpha), a new species from the Lower Permian of New Mexico. *Journal of Paleontology*, 40:690-695.

Laurin, M. 1991. The osteology of a Lower Permian eosuchian from Texas and a review of diapsid phylogeny. *Zoological Journal of the Linnean Society*, 101:59-95.

Laurin, M. 1993. Anatomy and relationships of *Haptodus garnettensis*, a Pennsylvanian synapsid from Kansas. *Journal of Vertebrate Paleontology*, 13:200-229.

Laurin, M. 1994. Re-evaluation of *Cutleria wilmarthi*, an Early Permian synapsid from Colorado. *Journal of Vertebrate Paleontology*, 14:134-138.

Laurin, M. 1995. Comparative cranial anatomy of *Seymouria sanjuanensis* (Tetrapoda: Batrachosauria) from the Lower Permian of Utah and New Mexico. *PaleoBios*, 16:1-8.

Laurin, M., and R. R. Reisz. 1988. Taxonomic position and phylogenetic relationships of *Colobomyter pholeter*, a small reptile from the Lower Permian of Oklahoma. *Canadian Journal of Earth Sciences*, 26:544-560.

Laurin, M., and Reisz, R. R. 1990. *Tetraceratops* is the oldest known therapsid. *Nature*, 345:249-250.

Laurin, M., and R. R. Reisz. 1992. A reassessment of the Pennsylvanian tetrapod *Romeriscus*. *Journal of Vertebrate Paleontology*, 12:524-527.

Laurin, M., and R. R. Reisz. 1995. A reevaluation of early amniote phylogeny. *Zoological Journal of the Linnean Society*, 113:165-223.

Lee, M. S. Y. 1993. The origin of the turtle body plan: bridging a famous morphological gap. *Science*, 26:1716-1720.

Lewis, G. E., and P. P. Vaughn. 1965. Early Permian vertebrates from the Cutler Formation of the Placerville area Colorado. *U. S. Geological Survey Professional Paper*, 503-C:1-50.

Lombard, R. E., and S. S. Sumida. 1992. Recent progress in understanding early tetrapods. *American Zoologist*, 32: 609-622.

Lucas, S. G., and A. P. Hunt. 1991. A Middle Pennsylvanian reptile from New Mexico. *New Mexico Journal of Science*, 31:21-25.

Martens, T. 1988. Die Bedeutung der Rotsedimente für die Analyse der Lebewelt des Rotliegenden *Zeitschrift für geologische Wissenschaften*, 16: 933-938.

Martens, T. 1989. First evidence of terrestrial tetrapods with North-American faunal elements in the red beds of Upper Rotliegendes (Lower Permian, Tambach beds) of the Thuringian Forest (G.D.R.) - first results. *Acta Musei Reginaehradecensis S. A.: Scientiae Naturales*, 22:99-104.

Martens, T. 1993. Ein besonderes Fossil. *Palaeontolgische Zeitschrift*, 66:197-198.

Meyer, H. Von. 1860. *Phanerosaurus naumanni* aus dem Sandstein des Rotliegendes in Deutschland. *Palaeontolgraphica*, VII, 248-252.

Milner, A. R. 1993. Biogeography of Paleozoic tetrapods. Pages 324-353 in: *Palaeozoic Vertebrate Biostratigraphy and Biogeography* (J. A. Long, ed.). London: Belhaven Press.

Modesto, S. P. 1994. The Lower Permian synapsid *Glaucosaurus* from Texas. *Palaeontology*, 37:51-60.

Modesto, S. P., and R. R. Reisz. 1989. A new skeleton of *Ianthasaurus hardestii*, a primitive edaphosaur (Synapsida: Pelycosauria) from the Upper Pennsylvanian of Kansas. *Canadian Journal of Earth Sciences*, 27:834-844.

Modesto, S. P., and R. R. Reisz. 1990. Taxonomic status of *Edaphosaurus raymondi* Case. *Journal of Paleontology*, 64:1049-1051.

Moss, J. L. 1972. The morphology and phylogenetic relationships of the Lower Permian tetrapod *Tseajaia campi* Vaughn (Amphibia: Seymouriamorpha). *University of California Publications in Geological Science*, 98:1-63.

Olson, E. C. 1947. The family Diadectidae and its bearing on the classification of reptiles. *Fieldiana: Geology*, 11:1-53.

Olson, E. C. 1958. Fauna of the Vale and Choza: 14 summary, review, and integration of the geology and faunas. *Fieldiana: Geology*, 10:397-448.

Olson, E. C. 1962a. Late Permian terrestrial vertebrates, U.S.A. and U.S.S.R. *Transactions of the American Philosophical Society* (n. s.), 52:1-224.

Olson, E. C. 1962b. Permian vertebrates from Oklahoma and Texas: Part I. Vertebrates from the Flowerpot Formation. Permian of Oklahoma (Olson and H. Barghusen). Part II. The osteology of *Captorhinikos chozaensis* (Olson). *Oklahoma Geological Survey* No. 59, 68 pages

Olson, E. C. 1965. Chickasha vertebrates. *Oklahoma Geological Survey Circular*, 70: 1-70.

Olson, E. C. 1967. Early Permian vertebrates. *Oklahoma Geological Survey Circular*, 74:1-111.

Olson, E. C. 1970. New and little known genera and species from the Lower Permian of Oklahoma. *Fieldiana: Geology*, 18:357-434.

Olson, E. C. 1974 On the source of therapsids. *Annals of the South African Museum*, 64:27-46.

Olson, E. C. 1975. Vertebrates and the biostratigraphic position of the Dunkard. Pages 155-171 in: *The Age of the Dunkard*; *Proceedings of the First I. C. White Memorial Symposium* (J. A. Barlow, ed.). West Virginia Geological and Economic Survey, Morgantown, West Virginia.

Olson, E. C. 1976. The exploitation of land by early tetrapods. Pages 1-30 in: *Morphology and Biology of Reptiles* (A. d'A. Bellairs and C. B. Cox, eds.). Linnean Sociedty Symposium Series No. 3.

Olson, E. C. 1979a. *Seymouria grandis* n. sp. (Batrachosauria: Amphibia) from the middle Clear Fork (Permian) of Oklahoma and Texas. *Journal of Paleontology*, 53:720-728.

Olson, E. C. 1979b. Biological and physical factors in the dispersal of Permo-Carboniferous vertebrates. Pages 227-238 in: *Historical Biogeography, Plate Tectonics, and the Changing Environment* (J. Gray and A. J. Boucot, eds.). Corvallis: Oregon State University Press.

Olson, E. C. 1980. The North American Seymouriidae. Pages 137-152 in: *Aspects of Vertebrate History*. (L. L. Jacobs, ed.). Flagstaff: Museum of Northern Arizona Press.

Olson, E. C. 1984. Nonmarine vertebrates and Late Paleozoic climates. *IXth International Carboniferous Congress, Competus Rendus*, 5:403-414. University of Illinois Press, Carbondale.

Olson, E. C., and H. Barghusen. 1962. Permian vertebrates Oklahoma and Texas. *Oklahoma Geological Survey Circular*, 59:1-68.

Olson, E. C., and J. R. Beerbower. 1953. The San Angelo Formation, Permian of Texas, and its vertebrates. *Journal of Geology*, 61:389-423.

Olson, E. C., and P. P. Vaughn. 1970. The changes of terrestrial vertebrates and climates during the Permian of North America. *Forma et Functio*, 3:113-138.

Parrish, J. M., J. T. Parrish and A. M. Ziegler. 1986. Permian-Triassic paleogeography and paleoclimatology and implications for therapsid distribution. Pages 109-131 in: *The Ecology and Biology of Mammal-like Reptiles* (N. Hotton III, P. D. MacLean, J. J. Roth and E. C. Roth, eds.). Washington D. C.: Smithsonian Institution Press.

Paton, R. L. 1974. Lower Permian pelycosaurs from the English Midlands. *Palaeontology*, 17:541-552.

Reisz, R. R. 1980a. The Pelycosauria: a review of the phylogenetic relationships. Pages 553-592 in: *The Terrestrial Environment and the Origin of Land Vertebrates*, Systematics Association Special Volume No. 15 (A. L. Panchen, ed.), London: Academic Press.

Reisz, R. R. 1980b. A protorothyridid captorhinomorph reptile from the Lower Permian of Oklahoma. *Life Sciences Contributions Royal Ontario Museum*, 121:1-16.

Reisz, R. R. 1981. A diapsid reptile from the Pennsylvanian of Kansas. *Special Publication of the Museum of Natural History, University of Kansas*, 7:1-74.

Reisz, R. R. 1986. Pelycosauria. *Handbuch der Paleoherpetologie*, Teil 17A:1-102.

Reisz, R. R. 1988. Two small reptiles from a Late Pennsylvanian quarry near Hamilton, Kansas. *Kansas Geological Survey Guidebook Series*, 6:189-194.

Reisz, R. R., and D. Baird. 1983. Captorhinomorph stem reptiles from the Pennsylvanian coal-swamp deposit of Linton, Ohio. *Annals of Carnegie Museum*, 52:393-411.

Reisz, R. R., and D. S Berman. 1985. *Scoliomus puercensis* Williston and Case, 1913, identified as a junior synonym of *Sphenacodon ferox* Marsh (Reptilia, Pelycosauria). *Canadian Journal of Earth Sciences*, 22:1236-1239.

Reisz, R. R., and D. S Berman. 1986. *Ianthasaurus hardestii* n. sp., a primitive edaphosaur (Reptilia, Pelycosauria) from the Upper Pennsylvanian Rock Lake Shale near Garnett, Kansas. *Canadian Journal of Earth Sciences*, 23:77-91.

Reisz, R. R., D. S Berman, and D. Scott. 1984. The anatomy and relationships of the Lower Permian reptile *Araeoscelis*. *Journal of Vertebrate Paleontology*, 4:57-67.

Reisz, R. R., D. S Berman, and D. Scott. 1992 The cranial anatomy and relationships of *Secodontosaurus*, an unusual mammal-like reptile (Synapsida: Sphenacodontidae) from the early Permian of Texas. *Zoological Journal of the Linnean Society*, 104:127-184.

Ricqlés, A. de, and P. Taquet. 1982. La faune de vertebres du Permien supurieur du Niger. I. Le Captorhinomorphe *Moradisaurus grandis* (Reptilia, Cotylosauria)--Le Crane. *Annales de Paleontologie*, 68:33-106.

Romer, A. S. 1925. Permian amphibian and reptilian remains described as *Stephanospondylus*. *Journal of Geology*, 33:447-463.

Romer, A. S. 1952. Fossil vertebrates of the tri-state area. Article 2. Late Pennsylvanian and Early Permian vertebrates of the Pittsburgh-West Virginia region. *Annals of Carnegie Museum*, 33:47-110.

Romer, A. S. 1956. *The Osteology of the Reptiles*. Chicago: The University of Chicago Press.

Romer, A. S. 1964. *Diadectes* and amphibian? *Copeia*, 4: 718-719.

Romer, A. S. 1966. *Vertebrate Paleontology*. Chicago: The University of Chicago Press.

Romer, A. S., and L. I. Price. 1940. Review of the Pelycosauria. *Special Papers of the Geological Society of America*, 28:1-538.

Scotese, C. R. and J. Golanka. 1992. *PALEOMAP Progress Report #20*. Department of Geology, University of Texas at Arlington.

Scotese, C. R. and W. S. McKerrow. 1990. Revised world maps and introduction. Pages 1-21 in: *Palaeozoic Palaegeography and Biogeography* (W. S. McKerrow and C. R. Scotese, eds.). Geological Society Memoir No. 12.

Scotese, C. R. and R. P. Langford. 1995. Pangea and the paleogeography of the Permian. Pages 3-19 *in: Volume I: Paleogeography, Paleoclimates, Stratigraphy* (P. A. Scholle, T. M. Peryt, and D. S. Ulmer-Scholle, eds.). Berlin and New York: Springer-Verlag.

Seltin, R. J. 1959. A review of the family Captorhinidae. *Fieldiana: Geology*, 10:461-509.

Sigogneau-Russell, D., and D. E. Russell. 1974. Étude du premier Caseide (Reptilia, Pelycosauria) d'Europe occidentale. *Bulletin du Muséum National d'Histoire Naturelle*, 230:145- 216.

Smithson, T. R. 1989. The earliest known reptile. *Nature*, 342:676-678.

Smithson, T. R., R. L. Carroll, R. L. Panchen, and S. M. Andrews. 1994. *Westlothiana lizziae* from the Visean of East Kirkton, West Lothian, Scotland, and the amniote stem. *Transactions of the Royal Society of Edinburgh: Earth Sciences*, 84:383-412.

Sumida, S. S. 1989. New information on the pectoral girdle and vertebral column in *Lupeosaurus* (Reptilia, Pelycosauria). *Canadian Journal of Earth Sciences*, 26:143-1349.

Sumida, S. S., and D. S Berman. 1993. The pelycosaurian (Amniota: Synapsida) assemblage from the Late Pennsylvanian Sangre de Cristo Formation of central Colorado. *Annals of Carnegie Museum*, 62:293-310.

Sumida, S. S., D. S Berman, and T. Martens. 1994. A trematopid amphibian from a terrestrial red-bed deposit of the Lower Permian of central Germany. *Journal of Vertebrate Paleontology*, 1994:48A.

Sumida, S. S., D. S Berman, and T. Martens. 1996. Biostratigraphic correlations between the Lower Permian of North America and central Europe using the first record of an assemblage of terrestrial tetrapods from Germany. *Paleobios*, in press.

Vaughn, P. P. 1962. Vertebrates from the Halgaito tongue of the Cutler Formation, Permian of San Juan County, Utah. *Journal of Paleontology*, 36:529-539.

Vaughn, P. P. 1964. Vertebrates from the Organ Rock Shale of the Cutler Group, Permian of Monument Valley and vicinity, Utah. *Journal of Paleontology*, 38:567-583.

Vaughn, P. P. 1966. *Seymouria* from the Lower Permian of southeastern Utah and possible sexual dimorphism in that genus. *Journal of Paleontology*, 40:603-612.

Vaughn, P. P. 1969. Upper Pennsylvanian vertebrates from the Sangre de Cristo Formation of central Colorado. *Los Angeles County Museum Contributions in Science*, 164:1-27.

Vaughn, P. P. 1972. More vertebrates, including a new microsaur, from the Upper Pennsylvanian of central Colorado. *Los Angeles County Museum Contributions in Science*, 223:1-30.

Vjushkov, B. P., and P. K. Chudinov. 1957. (The discovery of Captorhinidae in the Upper Permian of the USSR). In Russian. *C. R. Acad. Sci.* URSS, 112:523-526.

Werneburg, R. 1988. Labyrinthodontier (Amphibia) aus dem Oberkarbon und Unterperm Mitteleuropas - Systematik, Phylogenie und Biostratigraphie. *Freiburger Forschungshefte*, C 436:7-57

Werneburg, R. 1989. Some notes to systematic, phylogeny and biostratigraphy of labyrinthodont amphibians from the Upper Carboniferous and Lower Permian in central Europe. *Acta Musei Reginaehradecensis series A: Scientiae Naturales*, 27:117-129.

Williston, S. W. 1911a. *American Permian Vertebrates*. Chicago: The University of Chicago Press.

Williston, S. W. 1911b. A new family of reptiles from the Permian of New Mexico. *American Journal of Science*, 31:378-398.

Zhang, F., Y. Li and X. Wang. 1984. A new occurrence of Permian seymouriamorphs in Xinjiang, China. *Vertebrata PalAsiatica*, 22:294-304.

Ziegler, A. M., M. L. Hulver, and D. B. Rowley. 1996. Permian world topography and climate In: *Late Glacial and Postglacial Environmental Changes–Quaternary, Carboniferous-Permian and Proterozoic* (I. P. Martini, ed.). New York: Oxford University Press.

CHAPTER 5

THE LATE PALEOZOIC ATMOSPHERE AND THE ECOLOGICAL AND EVOLUTIONARY PHYSIOLOGY OF TETRAPODS

Jeffrey B. Graham

Nancy Aguilar

Robert Dudley

Carl Gans

INTRODUCTION

A dramatic chapter in vertebrate history took place in the approximately 140 million-year span of the Late Paleozoic era extending from the Middle Devonian (about 370 million years before present, mybp) to the end of the Permian (245 mybp). During this interval amphibians originated from the fishes, vertebrates invaded the land, and there was an explosive radiation of the amniotes. By the end of the Permian, amniotes had completely diversified and all of the major terrestrial-vertebrate lineages had appeared (Fig. 1) (Carroll, 1992, 1995; Lombard and Sumida, 1992; Thomson, 1991, 1993; Ahlberg and Milner, 1994; Laurin and Reisz, 1995).

Although this period of vertebrate evolution was enabled largely by the proliferation of the terrestrial biosphere (Shear, 1991;

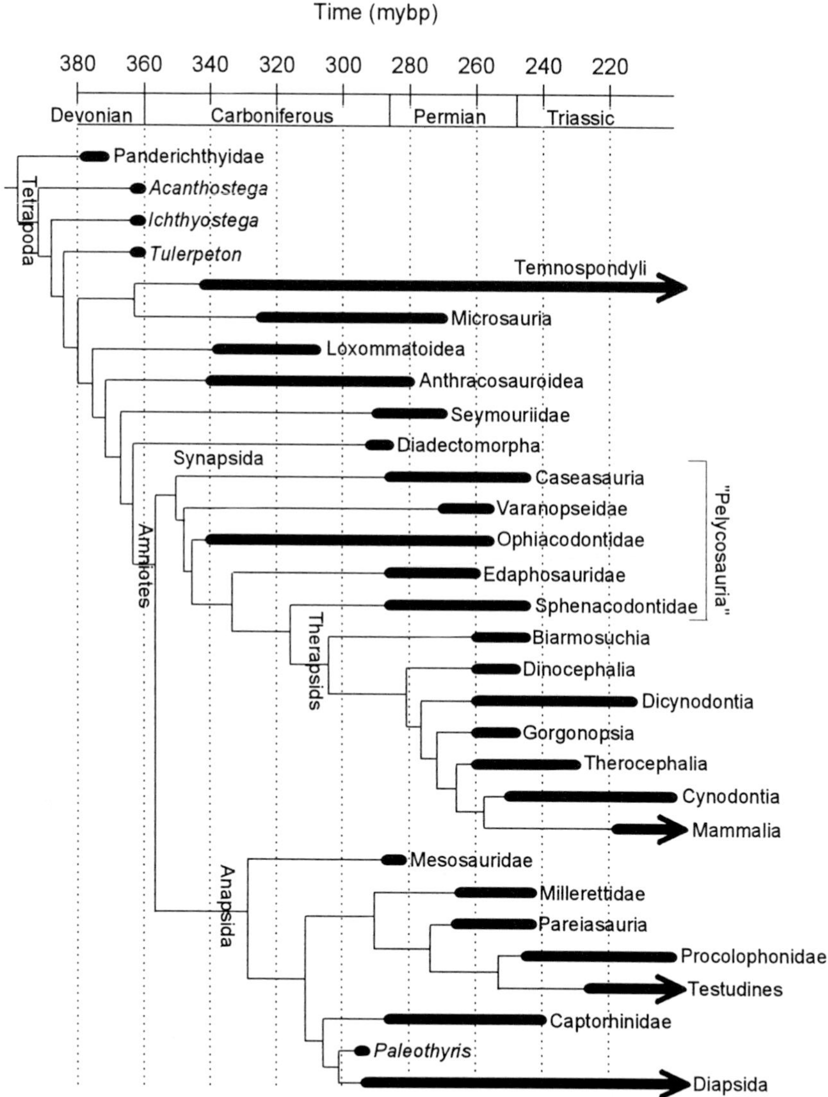

Figure 1. A synoptic view of the tetrapod radiations extending from the panderichthyid fishes of the Late Devonian to the major lineages present at the beginning of the Triassic. This synopsis highlights lineages leading to modern tetrapods and is not all-inclusive. Nodes are not intended to show time of origin. Data compiled from information contained in Benton (1990), Lombard and Sumida (1992), Ahlberg (1995), Laurin and Reisz (1995), and Modesto (1995).

Behrensmeyer *et al.*, 1992; Erwin, 1993, 1994), recent geochemical models of atmosphere evolution show that the late Paleozoic also saw marked changes in atmospheric oxygen and carbon dioxide levels (Berner and Canfield, 1989; Berner, 1993). It has recently been hypothesized that fluctuations in these two molecules, both critically important to the basic biological processes of primary production and respiration, could have significantly influenced biosphere evolution (Graham *et al.*, 1995).

The objectives of this chapter are to consider how changes in atmospheric oxygen may have affected the major evolutionary radiations of the Late Paleozoic tetrapods which included the invasion of land, the transition from amphibians to amniotes, and specialization of the pelycosaurian grade Synapsida and the primitive Therapsida which were ancestral to mammals. We hypothesize that, during the time when the late Paleozoic atmospheric oxygen level rose and subsequently fell, oxygen served as an ecological and energetic resource that facilitated the optimization of terrestrial air breathing by the vertebrates and allowed tetrapods to specialize into modes requiring greater oxidative capacity. We additionally hypothesize that the decline in atmospheric oxygen in the Permian, and the concurrent rise in carbon dioxide and changes in the earth's climate had a further selective influence on several aspects of vertebrate physiology; possibly including the evolution of separate pulmonary and systemic circulations in the synapsid lineage.

MODELS OF THE LATE PALEOZOIC ATMOSPHERE

Most geochemical models of paleoatmospheric oxygen and carbon dioxide have their basis in estimates of fixed carbon exchange and transfer processes with the inorganic, carbonate-silicate cycle (Berner and Canfield, 1989; Robinson, 1991; Berner, 1993; Kasting, 1993; Mora et al., 1996). Figure 2 shows the Phanerozoic atmospheric oxygen model of Berner and Canfield (1989) together with the estimate for atmospheric CO_2 presented by Berner (1993). This shows that there was about a 120 my pulse in atmospheric oxygen. In the late Devonian (360 mybp), atmospheric oxygen was at about 18%

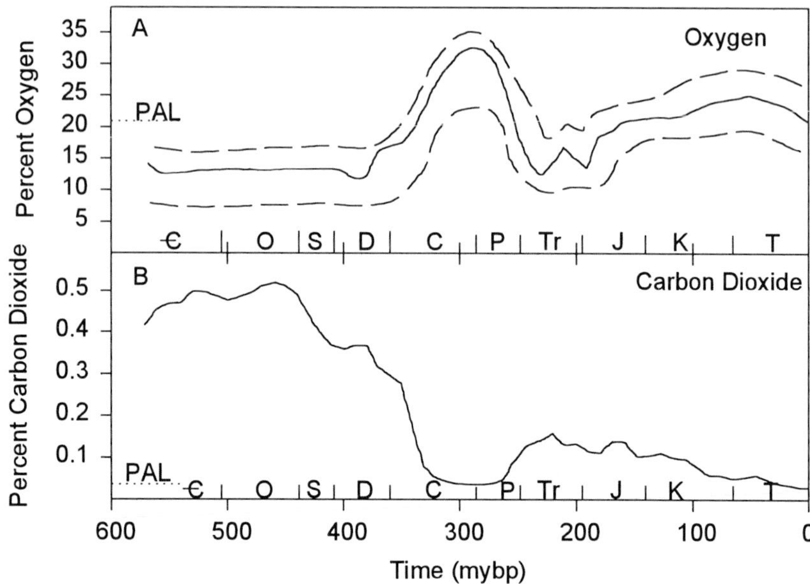

Figure 2. (A) The relative atmospheric concentration of oxygen in relation to geologic time, based on the model of Berner and Canfield (1989). (B) The history of atmospheric carbon dioxide level proposed by Berner (1993). The present atmospheric level (PAL) of each gas is indicated.

[the present atmospheric level (PAL) of oxygen is 21%), but had increased to about 35% by the late Carboniferous (290 mybp), and declined throughout most of the Permian, reaching a value of about 15% in the Early Triassic (240 mybp). Figure 2 also shows that pronounced, but generally opposite changes in atmospheric carbon dioxide occurred at the same time as those of oxygen. From a value of about 0.36% in the Late Devonian, atmospheric carbon dioxide dropped by a factor of 10 to about 0.035 in the Mid-Carboniferous, and then increased in the Late Permian. (The PAL of CO_2 is 0.036%.) As reflected in the error limits surrounding the values for atmospheric oxygen in Figure 2, estimates for this gas differ depending on the sets of underlying assumptions that are made (Berner and Canfield, 1989).

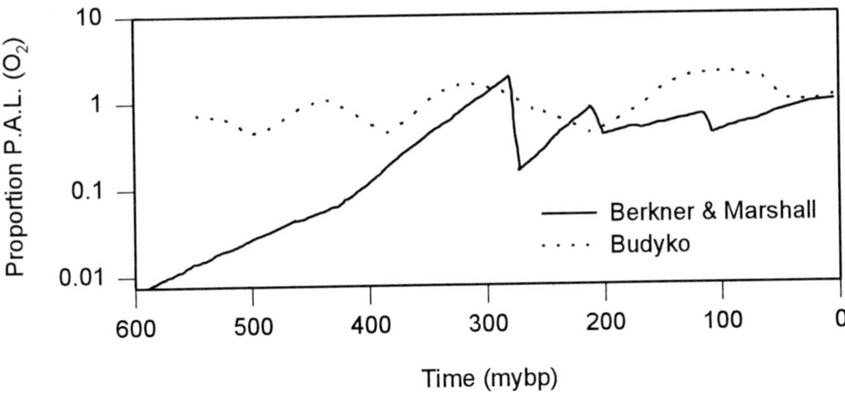

Figure 3. Estimates of paleoatmospheric oxygen based on the models of Berkner and Marshall (1965) and Budyko *et al.* (1987).

This is also the case for the CO_2 model (Berner, 1993; Mora *et al.*, 1996).

Figure 3 shows the atmospheric O_2 reconstructions proposed by Berkner and Marshall (1965) and by Budyko *et al.* (1987). Comparison of these data with Figure 2 reveals a general similarity among the three models with respect to the postulation that a Late Paleozoic oxygen pulse did occur. The greatest similarity is seen forthe contructions of Budyko *et al.* (1987) and Berner and Canfield (1989). However, due to the different data sets, weighting factors, and assumptions, the three models are not in agreement concerning either the estimated temporal occurrence or the magnitude of the atmospheric-oxygen increase. Specifically, Berner and Canfield show a Mid-Carboniferous oxygen peak of between about 1.2 and 1.8 PAL, whereas Budyko *et al*. show an Early Carboniferous oxygen peak of about 1.7 PAL and Berkner and Marshall show a 3.0 PAL peak occurring in the Permian.

Figure 4 shows the paleoatmospheric oxygen constructions of Cloud (1976) and Tappan (1968, 1974). These models are not based

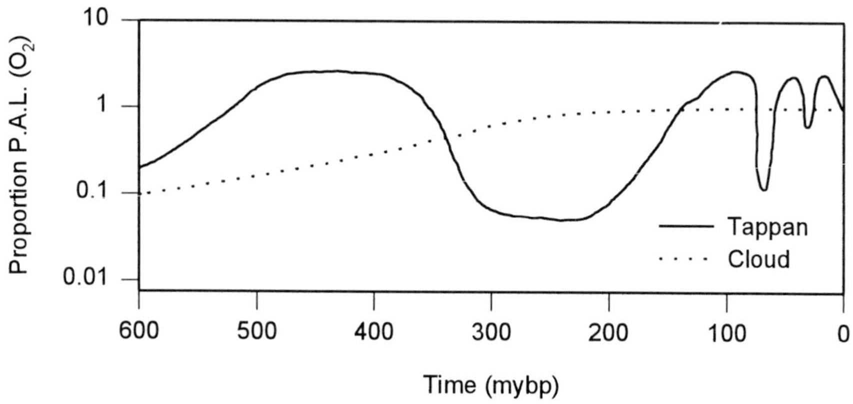

Figure 4. Paleoatmospheric oxygen levels estimated by Tappan (1968, 1974) and Cloud (1976).

on geochemistry and they differ markedly from those in Figures 2 and 3. Cloud assumed a gradual but steady rise in oxygen with time, and set Late Paleozoic values between about 0.5 and 0.9 PAL. Tappan's reconstruction was based on the abundance of phytoplankton fossils in sediments as an index of global primary productivity and thus atmospheric oxygen. Whereas the geochemical models of Berner and Canfield (1989), Berkner and Marshall (1965), and Budyko *et al.*(1987) all indicate a rise in atmospheric O_2 during the late Carbonifernous, Tappan's paleontological model shows nearly the inverse. She postulated a pronounced oxygen decline that began in the Devonian and continued into the Late Jurassic; the lowest oxygen levels estimated by Tappan were <0.1 PAL, during the Permian (Fig. 4). Because of its nearly exclusive dependence on the estimation of global rates of biological oxygen production, it is likely that the Tappan model is not as robust as the geochemical models which consider oxygen flux, organic carbon burial in sediments, and related processes.

POSSIBLE EFFECTS OF HYPEROXIA ON TETRAPOD RADIATIONS

As reviewed by Graham *et al.* (1995), the Late Paleozoic atmospheric oxygen pulse may have broadly affected ecological and evolutionary processes in both terrestrial and aquatic ecosystems. One example of this is the apparent relationship between hyperoxia and arthropod gigantism. A number of insect orders attained large size during the Carboniferous, for example, the dragonfly *Meganeura* had a wing span of over 70 cm and a thoracic diameter of 2.8 cm (May, 1982). In addition to insects, gigantism occurred among diplopods, arthropleurids, and scorpions (Kraus, 1974; Rolfe, 1980; Briggs, 1985; Kukalova-Peck, 1985, 1987; Selden and Jeram, 1989; Shear and Kukalova-Peck, 1990).

Although other factors also influenced body size, it is hypothesized that insect reliance upon a diffusion-dependent tracheal respiratory system enabled certain forms to increase body size in the hyperoxic Carboniferous and that hyperoxia also contributed to the evolution of intensely oxidative processes such as insect flight (Graham *et al.*, 1995). Correspondingly, environmental "permissiveness" toward a large body size disappeared when oxygen fell in the Permian; none of the giant arthropods survived beyond the end of this period (Graham *et al.*, 1995).

In the case of vertebrates, however, the direct delivery of oxygen to respiring cells via the respiratory and circulatory systems lessens the importance of diffusion-dependent influences on body size. Thus, although the Paleozoic oxygen pulse would not be expected to leave a "signature" as clear as insect body size on the vertebrates, we now discuss possible ways in which this pulse may have left indelible marks on tetrapod evolutionary and ecological physiology.

A Synopsis of Late Paleozoic Tetrapod Evolution and Radiation

Figure 1 provides an overview of vertebrate evolution between the Late Devonian and the Mid-Triassic. The diversity of amphibian fossils suggests that several different clades may have independently invaded the land (Gordon and Olson, 1994; Carroll, 1995). However, tetrapods are hypothesized to have evolved from a single ancestral lineage; the osteolepiform panderichthyid fishes appear to be most closely related to the tetrapods. It is speculated that these fishes were bimodal breathers (i.e., both water and air) and had a lung or lung-like air-breathing organ as well as internal gills (Gans, 1970a, 1970b; Romer, 1972; Carroll, 1992; Thomson, 1991; 1993). The first tetrapods, thought to be similar to *Elginerpeton* of the Late Frasnian (368 mybp), were probably capable of bimodal breathing and could have been sympatric with lungfish and the panderichthyids in certain habitats (Thomson, 1993). Accordingly, these early tetrapods resided in water and fed on aquatic invertebrates and vertebrates. Based on the definitive aquatic traits of Late Famennian (360 mybp) tetrapod genera such as *Ventastega, Acanthostega, Ichthyostega*, and others, it is probable that these early tetrapods underwent a 10 to 20 million year period of primarily aquatic existence (Thomson, 1993; Ahlberg and Milner, 1994; Carroll, 1995).

Although the direct fossil links have not been established, the stratigraphic record suggests that amniotes derived monophyletically from an amphibian lineage in the early Carboniferous (345 mybp). By the late Carboniferous (300 mybp), the amniotes had expanded their feeding repertoire to include both insectivory and herbivory, and this group had diversified into the three major lineages ancestral to all of the higher vertebrates. Although the nomenclature for these lineages is currently undergoing change (Laurin and Reisz, 1995), the names by which the three are most widely recognized are the Synapsida, the basal lineage leading to mammals; the "anapsida", from which the turtles would come; and the Diapsida, which would give rise to most of the other reptiles and the birds (Fig. 1 and references therein).

The adaptive radiation of these groups continued throughout the Permian. Within the synapsids, the sphenacodontid pelycosaurs, appeared in the mid-Carboniferous and extended throughout the Permian (Fig. 1). These were the dominant terrestrial predators of the Lower Permian. The best known pelycosaur, *Dimetrodon*, had elongated neural spines that supported a membranous "sail" which may have functioned for heat transfer (Romer, 1948, 1966; Haack, 1986; Tracy *et al.*, 1986). A number of pelycosaurs had dorsal sails. With few exceptions, the terrestrial pelycosaurs had generally large bodies with relatively short limbs. They also had cranial specializations for feeding on large terrestrial prey; however, their dentition suggests that pelycosaurs of the families Edaphosauridae and Caseidae were herbivorous (Modesto, 1995; Hotton et al., this volume). Other pelycosaurs (*Ophiacodon*) were semiaquatic and fed on fishes. The sphenacodontids gave rise to the therapsids or mammal-like reptiles which, as the name implies, were ancestral to mammals.

The early therapsids included both carnivores and herbivores and resembled the sphenacodonts in many skeletal features. Some therapsids could, however, hold their limbs in a somewhat more parasagittal position (i.e., under their bodies) and could thus move them fore and aft. Posture is one of a suite of characters suggested as an indication that therapsid metabolic and thermoregulatory capacity had evolved in the direction of mammals, however, as reviewed by Bennett and Ruben (1986), many of the features that have been interpreted as evidence of metabolic specialization are highly equivocal.

By the late Permian, therapsids had radiated into at least six lineages. The end-Permian mass extinction eliminated about 75% of the terrestrial vertebrates (i.e., 6 of 9 amphibian families and 21 of 27 reptile families), including four of the six therapsid groups (Shear, 1991; Lombard and Sumida, 1992; Erwin, 1993). One surviving therapsid group, the cynodonts, gave rise to the Mammalia in the Late Triassic.

POSSIBLE EFFECTS OF VARIABLE OXYGEN CONCENTRATION

The following discussion deals with possible atmospheric oxygen effects on three aspects of the tetrapod radiation: the invasion of land, the amphibian to amniote transition, and the implications of the Permian oxygen reduction for vertebrate evolution and physiology. The hypothetical scenario we envision is portrayed by Figure 5. This shows the late Paleozoic changes in O_2 and CO_2 in relation to the landmark events in vertebrate evolutionary history as indicated by the fossil record and the important morphological and physiological traits affecting vertebrate metabolism.

The Invasion of Land

The evolution of early amphibians toward a greater dependence on terrestrial life imposed at least three basic physiological problems: an increased dependence upon the air-breathing organ for aerial respiration, the need to minimize desiccation and conserve water, and the need to overcome gravity and move efficiently on the land. These are now discussed in the context of increases in Paleozoic atmospheric oxygen.

Aerial Respiration

Early amphibians were bimodal breathers; however, gills are most effective in water and, in extant bimodal breathers, function primarily for CO_2 release (Graham, 1994). Because gills do not function effectively in air, increased terrestriality was restricted to animals capable of air breathing and possessing an air-breathing organ suitable for both for the uptake of O_2 as well as the release of CO_2 (Gans, 1970a). Moreover, because the early tetrapods had a thick body covering, cutaneous respiration may not have offered an effective alternative to a diminished branchial function in air; these forms were thus heavily dependent upon lungs (Gans, 1970a,b; Romer, 1972; Thomson, 1993).

We suggest that Paleozoic changes in atmospheric O_2 and CO_2 could have greatly mitigated the problems caused by aerial exposure. Just as an elevated atmospheric O_2 would heighten its diffusion gradient across the lung surface into the body, the decline in CO_2 would also increase the gradient for its diffusion from the body to the atmosphere. We do not know how hyperoxia might have affected the blood respiratory properties of these animals. Assuming that air breathing resulted in a right-shifted hemoglobin (Hb)-oxygen dissociation curve (Graham, 1997) tissue respiration would have

---

Figure 5. Relationships between the estimates of paleoatmospheric oxygen (Berner and Canfield, 1989) and carbon dioxide (Berner, 1993) between the Devonian and Triassic Periods and the timing of major events in vertebrate evolution, as indicated by the fossil record. The top panels suggest the hypothetical temporal sequence for the evolution of major physiological features related to ventilatory mechanisms, respiratory control, metabolism, and heart structure. ➔

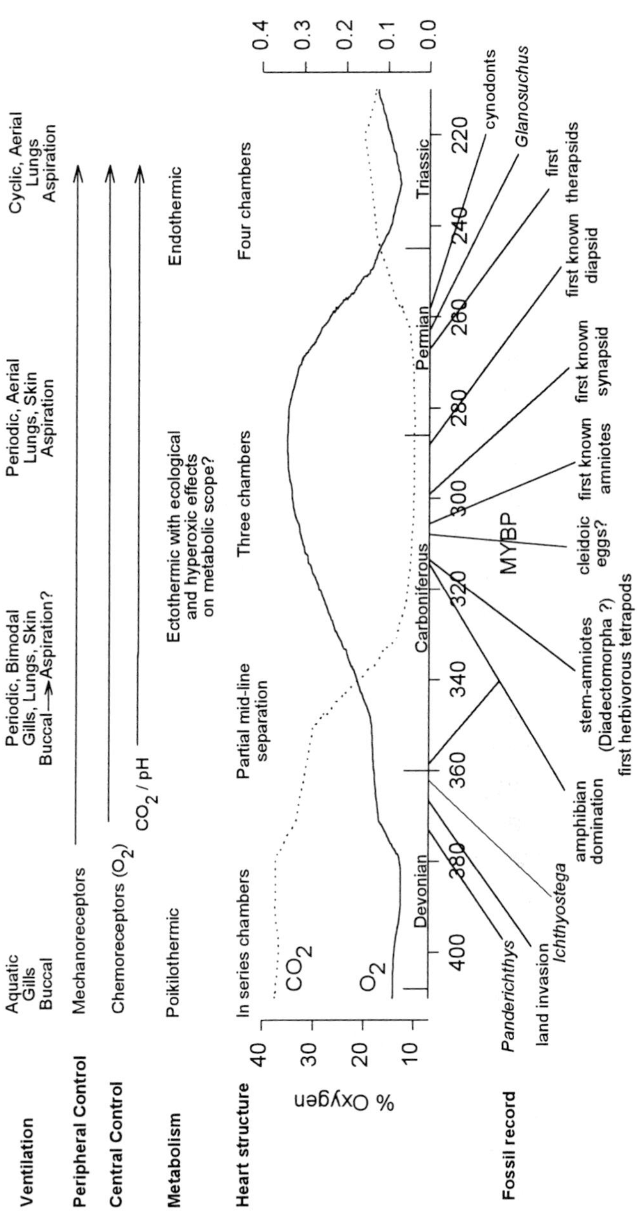

Ventilation
Aquatic
Gills
Buccal
Periodic, Bimodal
Gills, Lungs, Skin
Buccal → Aspiration?
Periodic, Aerial
Lungs, Skin
Aspiration
Cyclic, Aerial
Lungs
Aspiration
Peripheral Control
Mechanoreceptors
Central Control
Chemoreceptors (O_2)
CO_2 / pH
Metabolism
Poikilothermic
Ectothermic with ecological
and hyperoxic effects
on metabolic scope?
Endothermic
Heart structure
In series chambers
Partial mid-line
separation
Three chambers
Four chambers
% Oxygen
40
30
20
10
% Carbon Dioxide
0.4
0.3
0.2
0.1
0.0
CO_2
O_2
Devonian
Carboniferous
Permian
Triassic
400
380
360
340
320
300
280
260
240
220
MYBP
Fossil record
Panderichthys
land invasion
Ichthyostega
amphibian
domination
stem-amniotes
(Diadectomorpha ?)
first herbivorous tetrapods
cleidoic
eggs?
first known
amniotes
first known
synapsid
first known
diapsid
first
therapsids
Glanosuchus
cynodonts

been augmented by higher oxygen partial pressures in the capillaries (i.e., the P50 or unloading oxygen tension of Hb would be higher). This would have been further enhanced by elevated atmospheric oxygen and a reduced carbon dioxide; the latter would in turn favor an increased oxygen transport by Hb and the overall effect could have been to reduced red cell number.

Water Conservation

Amphibians radiating to a more terrestrial existence needed to cope with the increased risk of dehydration. Because many of these forms had a thick armor layer (Romer, 1972), their gill and lung surfaces were the most vulnerable sites for desiccation. Critical features of respiratory surface efficacy are the capacity for respiratory gas diffusion and the impedance to water loss. Accordingly, the capacity of a respiratory surface is defined by the ratio of water loss to oxygen uptake (Withers, 1992). The rise in Paleozoic oxygen would have lowered the threat of evaporative water loss by reducing the ratio of water loss to oxygen gain for each breath taken (i.e., more O_2 uptake per unit loss of water) and by also requiring fewer lung ventilations to satisfy oxygen demand.

Overcoming Gravity

The tetrapod land transition significantly impacted locomotion energetics. Because water is 1000 times more dense than air, a tetrapod that was essentially weightless in water became nearly a thousand times heavier on land. Another important consideration for these early tetrapods may have been the need for body mass displacement during lung breathing in air (Radinsky, 1987).

The capacity for sustained terrestrial locomotion may have played a role in the Late Paleozoic radiation of the tetrapods and their exploitation of newly emerging land resources. We envision hyperoxia as having an initially supportive and ultimately sustaining role in the evolution of tetrapod metabolism and energetics. In the early tetrapods, hyperoxia would have contributed to increased terrestriality by providing a greater quantity of oxygen to sustain the power requirements for locomotion (including body support and equilibrium) or to hasten the metabolic recovery from anaerobic bursts of terrestrial activity. Combined with natural selection for more efficient terrestrial locomotor mechanisms, hyperoxia would have also expanded the terrestrial performance of tetrapods by extending aerobic capacity and endurance.

Evolution and Radiation of the Amniotes

The major changes ushered in with the amniotes included the elimination of the need to lay eggs in water, the acquisition of a greater terrestrial mobility through modifications in skeletal structure, the sensory motor control system, skin armor, and an expansion of the diet to include both insectivory and herbivory (Romer, 1966; Carroll, 1970, 1988; Shear, 1991; Packard and Seymour, this volume; Stewart, this volume; Hotton *et al.*, this volume).

Evolution of the Amniote Egg

With specialized membranes to prevent water loss (amnion), enhance respiratory gas exchange (chorion), and collect waste products (allantois), the amniotic egg (also termed the cleidoic egg), which appeared in the Carboniferous, eliminated amphibian reliance upon aquatic egg laying and larval development. However, because of the requirement for a sufficient quantity of yolk for the entire developmental period, direct-developing eggs needed to be larger, which in turn imposed greater problems in terms of water economy, respiratory gas diffusion, and waste discharge (see Stewart, this volume; and Packard and Seymour, this volume).

Accordingly, natural selection leading to the development of the three amniote egg membranes enabled a further increase in egg size. Moreover, the hyperoxic Carboniferous atmosphere would have allowed the development of large amniote eggs by minimizing the ratio of water loss to oxygen uptake, the same mechanism previously discussed in connection with pulmonary gas exchange (see Withers, 1992).

Morphological Changes

As indicated in Figure 1, the more recent phylogenetic reconstructions place the Synapsida as the basal lineage of the amniotes. Both the Protorothyrididae (Carroll, 1969, 1970) and the pelycosaurs (Reisz, 1980, 1986; Laurin and Reisz, 1995, this volume) have been suggested as the first amniotes. Berman *et al.* (1992) suggest that the Diadectomorpha may have shared a more recent common ancestor with the Synapsida than any other group, effectively making them amniotes as well.

A number of lineages of primitive amniotes such as the Protorothyrididae (Carroll, 1969; 1970), the Captorhinidae (Heaton, 1979; Sumida, 1989; 1990) and the Araeoscelidia (including the earliest Diapsid reptiles, Reisz, 1981) fed on insects and other small prey (Radinsky, 1987). Carroll (1988) suggests that the Carboniferous radiation of the amniotes was linked to the increasing diversity of small terrestrial arthropods. The skeletal changes characteristic of the early reptiles reflect an expanding niche in terms of both diet and requirements for agility. A significant component of the latter is the increased musculoskeletal coordination afforded by the evolution of muscle stretch receptors; which, together with the cleidoic egg, distinguishes amniotes from anamniotes (Carroll, 1986, 1988).

Relative to Paleozoic amphibians, the skulls of the early reptiles are smaller, narrower, and deeper; the latter difference presumably reflects alterations in jaw muscle attachment. Also, the bones of the early reptiles are more slender (lighter) than those of the ancestral amphibians, and both the wrists and ankles of early amniotes were strengthened by the reduction or fusion of bony elements and increased ossification (Sumida, this volume). The pelvic girdle of these reptiles also contacts the vertebral column across two sacral vertebrae (only one in the amphibians), reflecting a proportionately greater propulsive force generated by the hind limbs during locomtion (Radinsky, 1987; Carroll, 1988).

Although many ancestral amphibians had a dense covering of bony dermal scales and plates, a trend in early amniotes was for a continuous covering composed of a horny epidermal layer. This layer may have been formed into scales for mechanical and protective reasons. Early reptiles had localized dermal scales and ventral gastralia; however, their skin was much thinner than that of the early amphibians (Romer, 1972; Olson, 1984; Carroll, 1988).

Amniote Metabolism and Atmospheric Oxygen

Locomotion.–The structural modifications previously detailed suggest that early amniotes had a greater locomotor capacity than their amphibian ancestors. However, increased mobility may have raised metabolic costs and, correspondingly, required a greater access to energetic resources. Nothing is known about the energetics of the early amniotes. However, because they are regarded as similar to extant lizards in many respects (Carroll, 1988), inferences may be drawn based on locomotion and energetics data for lower vertebrates. Metabolic comparisons, for example, show that fishes, amphibians, and reptiles (assuming the same body size, body temperature, and thermal acclimation states for each of these ectotherms) have approximately the same (i.e., within an order of magnitude) metabolic rate (Bennett, 1991; Withers, 1992), but that the energetic cost of transport via swimming is much less than that for crawling, walking, running, or jumping (Schmidt-Nielsen, 1972; Bennett, 1991). Insofar as locomotory capacity is concerned, most studies of sustained locomotion (reviewed by Bennett, 1991) show that amphibians and reptiles have much lower endurance capacities than do fishes; the large increases in terrestrial endurance capacity are associated with the acquisition of higher rates of aerobic metabolism such as occur among the endothermic mammals and birds (Bennett, 1991; Ruben, 1995). Thus, although the energetic costs for terrestrial locomotion are significant, the locomotory energetic patterns evident for extant amphibians and reptiles do not suggest mechanisms leading to an increased capacity for sustained terrestrial locomotion. To our knowledge, no studies have examined the effects of hyperoxia on the metabolic performance of amphibians and reptiles. Such investigations would be useful in testing hypotheses about the effects of a hyperoxic Carboniferous atmosphere on the metabolic and locomotory endurance capacity of the early amniotes.

Hyperoxia and circulation.–The hearts of all amphibians and most reptiles have three chambers, a left and a right atrium and a single ventricle. In reptiles the ventricle is partially subdivided by a septal fold and thus may function largely as if the chamber was subdivided into right and left sides. Only the crocodiles, among the reptiles, and the birds and mammals have a four-chambered (i.e., completely separated pulmonary and systemic circulations) heart. The functional inefficiencies of the three-chambered heart structure are the inevitable intra-ventricular admixture of the oxygenated (pulmonary venous) and deoxygenated (systemic venous) blood streams. The latter, termed right-to-left (R-L) shunting, means that the level of

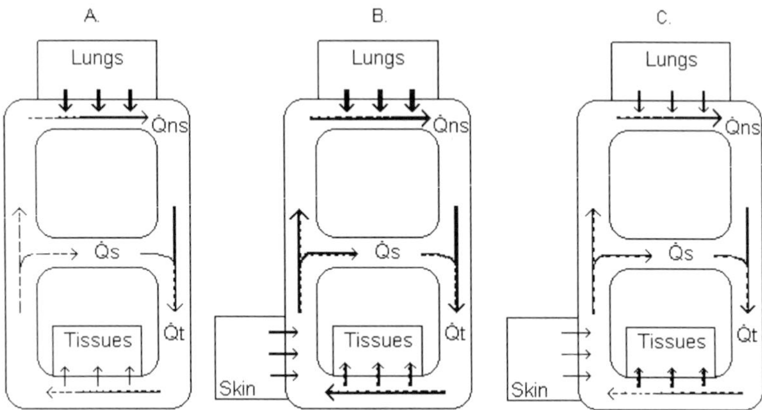

Figure 6. A diagram of the two-compartment oxygen transport model illustrating the effect of intraventricular shunting in a three-chambered heart on aortic oxygen content. The relative thickness of the arrows reflects the quantity of oxygen present in the blood or the amount of oxygen diffusion taking place. (A) The quantity of blood that is not shunted (Qns) passes through the lungs where it is fully oxygenated and is then mixed in the ventricle with the shunted (deoxygenated, venous) blood (Qs), resulting in a less than 100% saturated blood being transported to the tissues (Qt). (B) Hypothetical situation for a Late Carboniferous amniote in the hyperoxic atmosphere and for which the skin now plays a role respiration. Because some oxygen uptake occurs at the skin, neither Qs nor Qt are as low as in A. (C) In a hypoxic atmosphere the same animal would have a lower Qns, Qs, and Qt and would need to compensate by breathing more or reducing shunt flow.

oxygen in the systemic arterial blood will always be less than that in the pulmonary venous blood.

Assuming that the early amniotes had a three-chambered heart with a partially subdivided ventricle, Figure 6 illustrates how shunting would have potentially limited the positive effects of atmospheric hyperoxia on tissue oxygen delivery. Figure 6a shows that the quantity of oxygen that could be delivered to the systemic arteries would be chronically reduced because of the ventricular admixture of the oxygenated pulmonary blood returning from the lungs anddeoxygenated blood returning from the systemic circulation. However, a rise in external oxygen tension would increase lung effectiveness and proportionately elevate the pulmonary venous blood oxygen content. Assuming that these early amniotes also made use of cutanenous respiration (most extant amphibians and certain reptiles use this respiratory mode; Feder and Burggren, 1985), the mechanism by which atmospheric hyperoxia could have influenced aerobic metabolism becomes apparent. If a sizeable fraction of the systemic circulation perfused the skin and took up oxygen from the hyperoxic atmosphere (Fig. 6b), then the systemic venous return would be relatively more oxygenated which, together with the elevated oxygen in the pulmonary circulation, would elevate aortic oxygen content.

Was the Permian Fall in Atmospheric Oxygen a Defining Moment for Amniote Physiology and Evolution?

According to the Berner and Canfield (1989) model, the zenith of atmospheric oxygen occurred at approximately the beginning of the Permian (Figs. 2 and 5). Oxygen fell throughout the Permian, reaching a low of about 15% in the Early Triassic. As described earlier and illustrated in Figure 5, a number of significant changes occurred in the tetrapods during the early Permian, most notable among these was the radiation of the synapsids (pelycosaurs and therapsids) and diapsids (Carroll, 1988) and the end-Permian extinction of about 75% of the entire tetrapod fauna (Radinsky, 1987; Carroll, 1988; Shear, 1991). Erwin (1993, 1994) has documented the diverse factors that contributed to the end-Permian extinction. We have suggested that a global reduction in atmospheric oxygen would have contributed to the extinction of the gigantic arthropods and other groups (Graham *et al*., 1995). We will now discuss some ways that a declining Permian oxygen level could have selected for physiological specializations leading from basal amniotes to mammals.

Figure 5 suggests that many features of tetrapod air-breathing physiology and ventilatory control were probably in place by the beginning of the Permian. Probably included among these were the development of a lung sufficient to sustain aerobic activity (Perry, 1989) and an aspiratory lung ventilatory mechanism [this may have appeared in early amphibians, however, there are differing opinions as to whether or not the thick, overlapping ribs of early tetrapods precluded aspiratory breathing (Romer, 1972; Carroll, 1988; Thomson, 1993; Graham, in press)]. Other respiratory properties likely in place by the Permian were the involvement of central CO_2 or pH chemoreceptors in ventilatory control (this seems to have first appeared in amphibians; Smatresk, 1994), and probably a metabolic rate that, for reasons described previously, was at least equivalent to and perhaps greater than that of extant ectotherms. Physiological features probably not present at the beginning of the Permian included cyclic (continuous) ventilation, endothermy, and a functional four-chambered heart.

We now focus on how the decline in atmospheric oxygen over the 40 million year Permian period could have influenced the evolution of continuous rhythmic ventilation, metabolic rate, and the four-chambered heart in the amniote lineage(s) leading to the mammals.

Ventilation Frequency

The cyclic ventilatory pattern of most ectothermic vertebrate air breathers is intermittent (Milsom, 1990). Non-ventilatory (apneustic) periods become less frequent when metabolic rate is increased by activity or a rise in body temperature. Hypoxia also elevates ventilation frequency. We suggest that the combination of a chronic and progressive atmospheric hypoxia during the Permian and the relatively high metabolic requirements of synapsids could have led to the evolution of continous, rhythmic respiration. Although continuous ventilation would tend to reduce blood CO_2 and thus alter the regulatory set point for this gas in relation to ventilatory control, this rise in atmospheric CO_2 taking place during the Permian could have modulated this effect. It is also likely that, in view of the overriding requirement for oxygen, acclimatory adjustments in acid-base balance would have also mitigated CO_2 effects on ventilatory control.

Metabolic Rate

A significant event in tetrapod ecological physiology was the evolution of endothermy (Bennett, 1991; Ruben, 1995), and biologists have long pondered the how, why, when, and where of this metabolically important specialization. More than any other vertebrate specialization, endothermy has dramatically altered the energetic balance sheet for vertebrates. The standard metabolic rate of ectotherms is about an order of magnitude less than the basal metabolic rate of mammals (Withers, 1992). Endothermy also affects routine activity levels, stamina, and endurance. Bennett and Ruben (1986) have reviewed the numerous hypotheses forwarded regarding the fossil evidence for the acquisition of endothermy in mammals. Ruben (1995) has also reviewed the physiological and metabolic bases for endothermy. The large sails of *Dimetrodon* (from the Lower Permian) and other sphenacodontids suggest the presence of a complex behavioral repertoire revolving around the capacity to regulate heat transfer (Romer, 1948; Haack, 1986; Tracy *et al.,* 1986). The discovery of turbinate bones in the nasal passages of therapsids indicates the presence of a water-conserving mechanism linked to frequent ventilation and endothermy and correspondingly suggests that the evolution of a "mammalian" metabolic rate had occurred by the Late Permian (Hillenius, 1992; 1994).

We suggest a two-part scenario for the evolution of a mammalian-level of metabolism in the hyperoxic Carboniferous-Permian biosphere. First, based on the discussions of Bennett and Ruben (1986), Carroll (1986), Tracy *et al.* (1986), and others, synapsids may have undergone natural selection for a relatively high metabolic rate and also increased their body size (thermal inertia). The sensory and locomotor specializations of these synapsids, as well as their capacities for rapid digestion and assimilation could all have been enhanced. Increased metabolic expenditures such as these, although necessitating a greater rate of energy resource acquisition, would have been favored by an abundance of environmental oxygen. Second, the presence of these metabolically specialized and hyperoxia adapted organisms in a Permian environment characterized by progressive atmospheric hypoxia could have intensified natural selection on certain lineages for an increased ventilation frequency (hence the appearance of turbinal bones in therapsids) and improved cardiac efficiency for oxygen delivery to the tissues (i.e., separation of systemic and pulmonary circulation).

Separation of the Pulmonary and Systemic Circulations

As detailed above, amphibians and most reptiles have a three-chambered heart and an intrinsic rate of intraventricular shunting that causes the admixture of oxygenated and deoxygenated blood (Figure 6a). Even a Paleozoic tetrapod with a systemic cutaneous loop that gained oxygen from the hyperoxic atmosphere would have had its tissue oxygen delivery limited by intraventricular shunting (Figure 6b). Although many species can regulate the quantity of blood that is shunted, this fraction can seldom be reduced to zero (Hicks and Wood, 1989). However, the magnitude of shunting is a function of factors such as the O_2 content of the inspired air. For turtles, Wang and Hicks (manuscript in preparation) report that a slightly more than 50% reduction in inspired oxygen (i.e., from 21% down to below 10%) virtually eliminated the shunt. Assuming that a Permian synapsid experienced a decline in aortic oxygen due to both shunting and a lowered of cutaneous oxygenation (Fig. 6c), this finding suggests a mechanism through which the Permian oxygen decline could have influenced the evolution of a four chambered heart.

DISCUSSION

The Late Paleozoic radiation of tetrapods reflected their successful exploitation of the newly accessible and expanding terrestrial biosphere. The diversity of terrestrial niches accessible to these tetrapods was coupled to a strong selection pressure for an increased performance capacity not only for locomotion, but also for a variety of processes such as feeding, growth, reproduction, and sensory physiology. We have suggested that atmospheric hyperoxia may have played a role in tetrapod radiations by enhancing their metabolic capacity and that this in turn led to increased terrestrial locomotory performance and greater ecological options in the terrestrial environment.

The dramatic shifts in Late Paleozoic atmospheric oxygen and carbon dioxide levels are generally regarded as having been brought about by biospheric changes. However, there has been little consideration of how biosphere evolution may have in turn been affected by alterations in the amounts and ratios of O_2 and CO_2, the key molecules in all life processes. Also, very few writers have considered the possibility that variations in atmospheric oxygen, carbon dioxide, or both have played a role in metazoan evolution; the prevailing view has been one of atmospheric uniformitarianism (Gordon and Olson, 1994). In his comparative physiology text, Schmidt-Nielsen (1993), other than noting present concerns over the greenhouse effect of rising CO_2 levels, stresses the "constancy of atmospheric composition."

That atmospheric change could impact the biosphere is, however, not a novel idea. As early as 1833, Geoffroy Saint-Hilaire wrote "let us suppose that in the course of a slow and gradual advancement of time, the proportions of different components of the atmosphere changed and it was an absolutely indispensible result that the animal world was affected by these changes."

The hypothesis that Late Paleozoic elevations in atmospheric oxygen augmented biosphere evolution and increased natural selection for greater aerobic activity (Graham *et al.*, 1995) elevates the importance of oxygen beyond its fundamental role in heightening the effectiveness of metabolic energy conversion to that of an ecological resource. With regard to the ecological importance of oxygen, G. C. Williams (1992) wrote, "To organisms with aerobic metabolism, oxygen is a resource that allows much higher levels of energy use than is possible for any anaerobe." Budyko *et al.* (1987), also noted the probable importance of an elevated oxygen atmosphere: "However, an increase in the use of energy could also have been achieved by an increase in the oxygen content of the atmosphere. This would have created some advantages in the struggle for existence of more complex organisms, whose vital functions, other things being equal, required more energy." These authors further suggested a role for the hyperoxic atmosphere at critical points in metazoan evolution: "It might be supposed, however, that an increase in metabolism was indispensible for the transition of vertebrates from water to land and the formation of endothermal animals." Regarding oxygen and energetics, they wrote: "animals expending the maximum energy on movement disseminated during the epochs of the greatest increases in oxygen concentration...for example, flying animals such as larged winged insects of the Carboniferous." These authors also suggested an evolutionary role for the hyperoxic atmosphere during the Cretaceous and Paleocene Periods.

Atmospheric changes have clearly impacted the evolution of life (Conway Morris, 1995; Graham *et al.*, 1995). An increase in atmospheric oxygen and the formation of the ultraviolet ozone shield were essential prerequisites to the early radiation of plants (Holland,

1984; Kasting, 1987, 1993). Rising oxygen levels were critical to the evolution of the diffusion limited Ediacaran fauna (Cloud, 1976; Runnegar, 1982a,b), and may have been important in the Cambrian explosion (McMenamin and McMenamin, 1990). Hypoxia has also been suggested as a factor in regional or even mass extinction events (Wignall and Hallam, 1992), and correlations between patterns of extinction and oxidative metabolic requirements have also been suggested for various taxa (McAlester, 1970; Budyko *et al.*, 1987).

As future discoveries permit paleoatmospheric models to resolve more clearly the magnitude and sequence of changes in global oxygen and carbon dioxide over Phanerozoic time, atmospheric variation may take its place alongside numerous other physical factors (including tectonics, sea level regression and transgression, glaciation, and climate) and biotic forces known to have influenced the evolution, radiation, succession, and even the extinction of plants and animals.

SUMMARY

Beginning with the Middle to Late Devonian origin of amphibians and the invasion of land, the Late Paleozoic radiation of the tetrapods was both extensive and relatively rapid. Primitive amniotes had appeared by the Mid-Carboniferous and the origin of major reptilian ancestral lineages took place by the end of that period. By the end of the Permian, a number of specialized stocks had also appeared including the cynodont therapsids from which the mammals evolved. Although the momentous changes that occurred in the late Paleozoic vertebrates can be largely attributed to the growing complexity of the terrestrial biosphere, they may have also been influenced by variations in the atmospheric levels of oxygen. Geochemical models suggest atmospheric oxygen levels increased during the Devonian and Carboniferous, remained elevated until the Mid-Permian, and then decreased. Accompanying these shifts were nearly reciprocal changes in atmospheric carbon dioxide. We hypothesize that, in a manner analagous to the way that nutrient fertilization enhances the growth and luxuriance of a garden, Late Paleozoic elevations in atmospheric oxygen may have enhanced the tempo and mode of terrestrial ecosystem development. Hyperoxia would have augmented vertebrate evolution by permitting natural selection for behavioral patterns requiring an elevated metabolic rate and aerobic scope. In the way that oxygen supplements and enhances the sustained performance level of athletes, an elevated oxygen atmosphere would have favored the initial emergence from water and terrestrial locomotion; it would have influenced many aspects of tetrapod physiology and behavior as well as trophodynamic relationships. In concert with changes in the global climate, reductions in atmospheric oxygen during the Late Permian may have also influenced the evolution of amniote physiology.

ACKNOWLEDGMENTS

This work was supported in part by both the National Science Foundation and the Vetlesen Foundation (JBG), a Ford Foundation Fellowship (NMA), the Leo Lesser Foundation for Tropical Biology (CG), and a Reeder Fellowship (RD). We thank Drs. Karen Martin, Frank Powell, Stuart Sumida, and John West for reading and commenting on manuscript drafts. Dr. Sumida also shared his considerable insight on amniote paleontology and evolutionary relationships and thus contributed significantly to this section of the paper. Dr. J. Hicks also reviewed this paper and shared cardiac shunt data from an "in review" paper and we thank him, T. Wang, and F. Powell for discussions about the implications of Permian hypoxia for vertebrate evolutionary physiology.

LITERATURE CITED

Ahlberg, P. E. 1995. *Elginerpeton pancheni* and the earliest tetrapod clade. *Nature*, 373:420-425.

Ahlberg, P. E. and A. R. Milner. 1994. The origin and early diversification of tetrapods. *Nature*, 368:507-514.

Behrensmeyer, A. K., J. D. Damuth, W. A. DiMichele, H.-D. Sues, and S. L. Wing. 1992. *Terrestrial Ecosystems Through Time: Evolutionary Paleoecology of Terrestrial Plants and Animals*. Chicago: The University of Chicago Press.

Bennett, A. F. 1991. The evolution of activity capacity. *Journal of Experimental Biology*, 160:1-23.

Bennett, A. F. and J. A. Ruben. 1986. The metabolic and thermoregulatory status of therapsids. Pages 207-218 in: *The Ecology and Biology of Mammal-like Reptiles* (N. Hotton, III, P.D. MacLean, J. J. Roth, and E. C. Roth, eds.). Washington, D.C.: Smithsonian Institution Press.

Benton, M. J. 1990. *Vertebrate Paleontology*. London: Harper Collins Academic.

Berkner, L. V., and L. C. Marshall. 1965. On the origin and rise of oxygen concentration in the Earth's atmosphere. *Journal of Atmospheric Sciences*, 22:225-252.

Berman, D. S, S.S. Sumida, and R.E. Lombard. 1992. Reinterpretation of the temporal and occipital regions in *Diadectes* and the relationships of diadectomorphs. *Journal of Paleontology*, 66:481-499.

Berner, R.A. 1993. Paleozoic atmospheric CO_2: Importance of solar radiation and plant evolution. *Science*, 261:68-70.

Berner, R. A. and D. E. Canfield. 1989. A new model for atmospheric oxygen over Phanerozoic time. *American Journal of Science*, 289:333-361.

Briggs, D. E. G. 1985. Gigantism in Palaeozoic arthropods. *Special Papers in Palaeontology*, No.33 p. 157.

Budyko, M. I., A. B. Ronov, and A. L. Yanshin. 1987. *History of the Earth's Atmosphere*. Berlin: Springer-Verlag.

Carroll, R. L. 1969. Origin of reptiles. Pages 1-44 in: *Biology of the Reptilia, Volume 1, Morphology* (C. Gans, A. d'A. Bellairs, and T.S. Parsons, eds.). New York: Academic Press.

Carroll, R. L. 1970. The ancestry of reptiles. *Philosophical Transactions of the Royal Society of London*, Ser. B, 257:267-308.

Carroll, R. L. 1986. The skeletal anatomy and some aspects of the physiology of primitive reptiles. Pages 207-218 in: *The Ecology and Biology of Mammal-like Reptiles* (N. Hotton, III, P.D. MacLean, J. J. Roth, and E. C. Roth, eds.). Washington, D.C.: Smithsonian Institution Press.

Carroll, R. L. 1988. *Vertebrate Paleontology and Evolution.* New York: W. H. Freeman and Company.

Carroll, R. L. 1992. The primary radiation of terrestrial vertebrates. *Annual Review of Earth and Planetary Science*, 20:45-84.

Carroll, R. L. 1995. Between fish and amphibian. *Nature*, 373:389-390.

Cloud, P. 1976. Beginnings of biospheric evolution and their biogeochemical consequences. *Paleobiology*, 2:351-387.

Conway Morris, S. 1995. Ecology in deep time. *Trends in Ecology and Evolution*, 10:263-304.

Erwin, D. H. 1993. *The Great Paleozoic Crisis: Life and Death in the Permian.* New York: Columbia University Press.

Erwin, D. H. 1994. The Permo-Triassic extinction. *Nature*, 267:231-236.

Feder, M. E., and W. W. Burggren. 1985. Cutaneous gas exchange in vertebrates: Design, patterns, control and implications. *Biological Reviews*, 60:1-45.

Gans, C. 1970a. Strategy and sequence in the evolution of the external gas exchangers of ecothermal vertebrates. *Forma et Functio*, 3:61-104.

Gans, C. 1970b. Respiration in early tetrapods–the frog is a red herring. *Evolution*, 24:740-751.

Geoffroy Saint-Hilaire, E. 1833. Memoire sur le degre d'influence du monde ambiant pour modifier les formes animales; question interessant l'origine des especes teleosauriennes et successivement celle des animaux de l'epoque actuelle. *Memoirs de l'Academie des Sciences*, 12:63-92.

Gordon, M. S., and E. C. Olson. 1994. *Invasions of the Land.* New York: Columbia University Press.

Graham, J. B. 1994. An evolutionary perspective for bimodal respiration: A biological synthesis of fish air breathing. *American Zoologist*, 34:229-337.

Graham, J. B. 1997. *Air Breathing Fishes: Evolution, Diversity, and Adaptation.* San Diego: Academic Press.

Graham, J. B., R. Dudley, N. M. Aguilar, and C. Gans. 1995. Implications of the late Palaeozoic oxygen pulse for physiology and evolution. *Nature*, 375:117-120.

Haack, S. C. 1986. A thermal model of the sailback pelycosaur. *Paleobiology*, 12:450-458.

Heaton, M. J. 1979. Cranial anatomy of primitive captorhinid reptiles from the Late Pennsylvanian and Early Permian of Oklahoma and Texas. *Oklahoma Geological Survey Bulletin*, 127:1-84.

Hicks, J. W., and S. C. Wood. 1989. Oxygen homeostasis in lower vertebrates. Pages 311-341 in: *Comparative Pulmonary Physiology, Current Concepts* (S.C. Wood. ed.). New York: Dekker.

Hillenius, W. J. 1992. The evolution of nasal turbinates and mammalian endothermy. *Paleobiology*, 18:17-29.

Hillenius, W. J. 1994. Turbinates in therapsids: evidence for late Permian origins of mammalian endothermy. *Evolution*, 48:207-229.

Holland, H. D. 1984. *The Chemical Evolution of the Atmosphere and Oceans.* Princeton: Princeton University Press.

Kasting, J. F. 1987. Theoretical constraints on oxygen and carbon dioxide concentrations in the Precambrian atmosphere. *Precambrian Research*, 34:205-229.

Kasting, J. F. 1993. Earth's early atmosphere. *Science*, 259:920-926.

Kraus, O. 1974. On the morphology of Paleozoic diplopods. *Symposium of the Zoological Society of London*, 32:13-22.

Kukalova-Peck, J. 1985. Ephemeroid wing venation based on new gigantic Carboniferous mayflies and basic morphology, phylogeny, and metamorphosis of pterygote insects (Insecta, Ephemerida). *Canadian Journal of Zoology*, 63:933-955.

Kukalova-Peck, J. 1987. New Carboniferous Diplura, Monura, and Thysanura, the hexapod ground plan, and the role of thoracic lobes in the origin of wings (Insecta). *Canadian Journal of Zoology*, 65:2327-2345.

Laurin, M. and R. R. Reisz. 1995. A reevaluation of early amniote phylogeny. *Zoological Journal of the Linnean Society*, 113:165-223.

Lombard, R. E. and S. S. Sumida. 1992. Recent progress in understanding early tetrapods. *American Zoologist*, 32:609-622.

May, M. L. 1982. Heat exchange and endothermy in Protodonata. *Evolution*, 36:1051-1058.

McAlester, A. L., 1970. Animal extinctions, oxygen consumption, and atmospheric history. *Journal of Paleontology*, 44:405-409.

McMenamin, M. A. S., and D. L. S. McMenamin. 1990. *The Emergence of Animals - The Cambrian Breakthrough.* New York: Columbia University Press.

Milsom, W. K. 1990. Mechanoreceptor modulation of endogenous respiratory rhythms in vertebrates. *American Journal of Physiology*, 259:R898-R910.

Modesto, S. P. 1995. The skull of the herbivorous synapsid *Edaphosaurus boanerges* from the lower Permian of Texas. *Palaeontology*, 38:213-239.

Mora, C. I., S. G. Driese, and L. A. Colarusso. 1996. Middle to Late Paleozoic atmospheric CO_2 levels from soil carbonate and organic matter. *Science*, 271:1105-1107.

Olson, E. C. 1984. The taxonomic status and morphology of *Pleuristion brachycoelous* Case; referred to *Protocaptorhinus pricei* Clark and Carroll (Reptilia: Captorhinomorpha). *Journal of Paleontology*, 58:1282-1295.

Perry, S. 1989. Structure and function of the reptilian respiratory system. Pages 193-236 in: *Comparative Pulmonary Physiology, Current Concepts* (S.C. Wood. ed.). New York: Dekker.

Radinsky, L. B. 1987. *The Evolution of Vertebrate Design.* Chicago: The University of Chicago Press.

Reisz, R. R. 1980. The Pelycosauria: A review of phylogenetic relationships. Pages 553-592 in: *The Terrestrial Environment and the Origin of Land Vertebrates* (A.L. Panchen, ed.). New York: Academic Press.

Reisz, R. R. 1981. A diapsid reptile from the Pennsylvanian of Kansas. *Special Publication, Museum of Natural History, University of Kansas*, 7:1-74.

Reisz, R. R. 1986. Pelycosauria. *Handbuch der Palaeoherpetologie*, 17:1-102.

Robinson, J.M. 1991. Phanerozoic atmospheric reconstructions: a terrestrial perspective. *Palaeogeography, Palaeoclimatology, and Palaeoecology*, 97:51-62.

Rolfe, W. 1980. Early invertebrate terrestrial fossils. Pages 117-157 in: *The Terrestrial Environment and the Origin of Land Vertebrates* (A. L. Panchen, ed.). New York and London: Academic Press.

Romer, A. S. 1948. Relative growth in pelycosaurian reptiles. Pages 45-55 in: Robert Broom Commemorative Volume (A. L. du Toit, ed.), *Special Publications of the Royal Society of South Africa.*

Romer, A.S. 1966. *Vertebrate Paleontology.* 3rd Edition. Chicago: University of Chicago Press.

Romer, A.S. 1972. Skin breathing - primary or secondary? *Respiration Physiology*, 14:183-192.

Ruben, J. 1995. The evolution of endothermy in mammals and birds: From physiology to fossils. *Annual Reviews of Physiology*, 57:69-95.

Runnegar, B. 1982a. The Cambrian explosion: Animals or fossils? *Journal of the Geological Society of Australia*, 29:395-411.

Runnegar, B. 1982b. Oxygen requirements, biology and phylogenic significance of the late Precambrian worm *Dickinsonia*, and the evolution of the burrowing habit. *Alcheringa*, 6:223-239.

Schmidt-Nielsen, K. 1972. Locomotion: Energy cost of swimming, flying, and running. *Science*, 177:222-228.

Schmidt-Nielsen, K. 1993. *Animal Physiology: Adaptation and Environment.* New York and Cambridge (UK): Cambridge University Press.

Selden, P., and A. Jeram, 1989. Palaeophysiology of terrestrialization in the Chelicerata. *Transactions of the Royal Society of Edinburgh*, 80:303-310.

Shear, W.A. 1991. The early development of terrestrial ecosystems. *Nature*, 351:283-289.

Shear, W. A., and J. Kukalova-Peck, 1990. The ecology of Paleozoic terrestrial arthropods: the fossil evidence. *Canadian Journal of Zoology,* 68:1807-1834.

Smatresk, N. J. 1994. Respiratory control in the transition from water to air breathing in vertebrates. *American Zoologist*, 34:264-279.

Sumida, S. S. 1989. The appendicular skeleton of the Early Permian genus *Labidosaurus* (Reptilia, Captorhinomorpha, Captorhinidae) and the hind limb musculature of captorhinid reptiles. *Journal of Vertebrate Paleontology*, 9:295-313.

Sumida, S. S. 1990. Vertebral morphology, alternation of neural spine height, and structure in Permo-Carboniferous tetrapods, and a reappraisal of primitive modes of terrestrial locomotion. *University of California Publications in Zoology*, 122:1-133.

Tappan, H. 1968. Primary production, isotopes, extinctions, and the atmosphere. *Palaeogeography, Palaeoclimatology, and Palaeoecology*, 4:187-210.

Tappan, H. 1974. Molecular oxygen and evolution. Pages 81-135 in: *Molecular Oxygen in Biology: Topics in Molecular Oxygen Research* (O. Hayaishi, ed.). Amsterdam: North Holland.

Thomson, K. S. 1991. Where did tetrapods come from? *American Scientist*, 79:488-490.

Thomson, K. S. 1993. The origin of the tetrapods. *American Journal of Science*, 293-A:33-62.

Tracy, C. R., J. S. Turner, and R. B. Huey. 1986. A biophysical analysis of possible thermoregulatory adaptations in sailed pelycosaurs. Pages 195-206 in: *The Ecology and Biology of Mammal-like Reptiles* (N. Hotton, III, P.D. MacLean, J. J. Roth, and E. C. Roth, eds.). Washington, D.C.: Smithsonian Institution Press.

Wignall, P. B., and A. Hallam. 1992. Anoxia as a cause of the Permian/Triassic mass extinction: Facies evidence from northern Italy and the western United States. *Palaeogeography, Palaeoclimatology, Palaeoecology*, 93:21-46.

Williams, G. C. 1992. Gaia, nature worship and biocentric fallacies. *The Quarterly Review of Biology,* 67:479-486.

Withers, P. C. 1992. *Comparative Animal Physiology*, New York: Saunders.

CHAPTER 6

ORIGIN OF THE AMNIOTE FEEDING MECHANISM: EXPERIMENTAL ANALYSES OF OUTGROUP CLADES

George V. Lauder

Gary B. Gillis

INTRODUCTION

One of the areas of vertebrate structure and function that has received the most attention during the past 20 years is the study of the feeding system. Due to the relatively good fossil record of bones, the many characters within the jaws used for systematic diagnoses, and interest in the mechanisms used by vertebrates to obtain resources from the environment, functional morphologists and paleontologists have devoted considerable effort to analyzing the vertebrate skull (Bels *et al.*, 1994b; Hanken and Hall, 1993). Investigation of skull design has included characterizing historical transformations of structure and functional patterns within major clades. For example, within the last 15 years, a number of reviews have appeared that deal with aspects of mammalian feeding mechanisms (Novacek, 1993; Russell and Thomason, 1993; Weijs, 1994), as well as jaw function in fishes (Frazetta, 1994; Lauder, 1983a; Liem, 1984), amphibians (Lauder and Reilly, 1994; Roth *et al.*, 1990), and lizards (Bels *et al.*, 1994a; Smith,

1993). However, the study of skull design in relation to several key events in vertebrate evolution, such as the origin of terrestrial feeding systems in tetrapods and the origin of amniote skull structure and function, has been less well analyzed.

This chapter will focus on the origin of the amniote feeding mechanism as a key event in the evolution of the vertebrate skull. However, rather than describe feeding systems within various amniote clades which have been reviewed elsewhere, we will center our analysis around a single general theme. We contend that in order to understand amniote feeding mechanisms and their diversification, it is essential first to understand the structure and function of the feeding mechanism in out-group clades. Thus, we will examine the feeding mechanisms of fishes and amphibians as a means of determining which functional traits are likely to have been primitively present in amniotes. Furthermore, based on this analysis of out-group clades, we believe that many functional attributes of the feeding mechanisms of amniotes are most parsimoniously explained as plesiomorphies retained from anamniote ancestors. Hence, it is important to understand aquatic feeding mechanisms in fishes, as well as aquatic and terrestrial feeding mechanisms in amphibians as a basis for assessing function in amniote taxa that are primitively terrestrial, but in some clades, have secondarily returned to the aquatic environment. Finally, we suggest that further experimental studies of extant amniote and anamniote taxa should provide a better understanding of the evolution of amniote and, more generally, vertebrate feeding mechanisms. For example, understanding general principles of divergence between aquatic and terrestrial feeding systems is an essential step in determining the role that environmental constraints have played in the evolution of vertebrate feeding mechanisms

Figure 1. Lateral and ventral views of cranial movements during prey capture in *Lepomis macrochirus*. The earthworm prey has been dropped through a tube and can be seen emerging from the bottom opening in the first frame. At time = 0 ms, the gape cycle is just beginning. Note that at peak gape (60 msec) lateral expansion of the head (seen in ventral view) has just begun and the upper jaw is maximally protruded. At 80 msec, the jaws have closed on the prey. Further movements of the prey into the mouth are accomplished by transport movements. Modified from Gillis and Lauder (1995). ➔

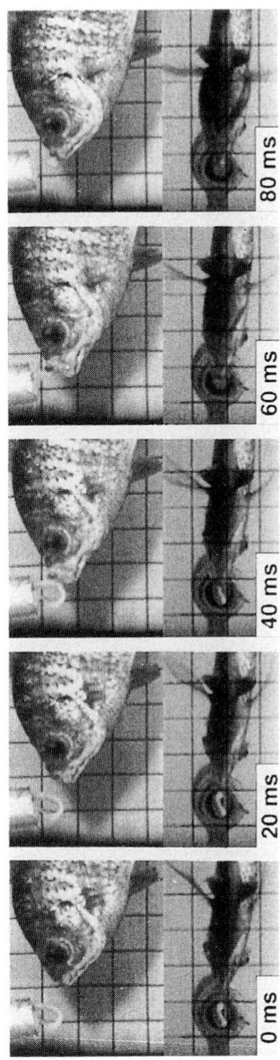
0 ms
20 ms
40 ms
60 ms
80 ms

(Bramble and Wake, 1985; Lauder and Reilly, 1994; Lauder and Schaeffer, 1993).

OUTGROUP PATTERNS: FISHES

The monophyletic clades of extant fishes that form out-group taxa to tetrapods and amniotes are the sharks and relatives (Elasmobranchiomorpha), ray-finned fishes (Actinopterygii), coelacanths (Actinistia), and lungfishes (Dipnoi). The feeding mechanisms of members of all of these taxa have been studied in some form or other during recent years, and a comparative analysis of feeding morphology and function in these clades provides the basis for our subsequent consideration of tetrapod feeding systems.

Initial Prey Capture

Despite the diversity of skull morphology represented by taxa as phylogenetically divergent as sharks, bass, and lungfishes, many common fundamental features of the process of initial prey capture have been observed. Most important is the observation that many taxa capture prey by suction feeding (Grobecker and Pietsch, 1979; Lauder, 1985a; Liem, 1970; Norton and Brainerd, 1993; Nyberg, 1971; Westneat and Wainwright, 1989).

The process of suction feeding involves creating a pressure within the oral cavity that is less than ambient. As shown in figure 1, expansion of oral volume occurs by lateral movement of the suspensoria, elevation of the neurocranium, depression of the lower jaw, and ventral movement of the hyoid region. The result of these movements is a reduction in oral cavity pressure that draws water into the mouth anteriorly carrying the prey toward the gape. The strike may be unsuccessful, in which case the prey escapes; the strike may result in prey being caught between the upper and lower jaws as the mouth closes (as in Fig. 1); or the prey may be completely drawn into the oral cavity. During the time that the mouth is opening, bones covering the gills laterally prevent water influx from the area posterior and lateral to the head and allow an essentially unidirectional flow of water through the mouth from anterior to posterior. Water flows first into the oral cavity, then between and around gill bars and filaments to exit finally in an expanding gap between opercular elements and the side of the head (Fig. 1). In the absence of an appropriate morphological design, the reduction in oral cavity pressure would be expected to draw in water from both posterior and anterior to the head, reducing the effectiveness of suction directed toward the prey.

Direct measurement of pressure changes simultaneously at several sites within the mouth cavity of ray-finned fishes using suction feeding shows that the branchial apparatus may have a significant influence on the function of the feeding mechanism. Figure 2 illustrates the comparative pressures measured at three sites in the oral cavity of a ray-finned fish during suction feeding. Note that, first, negative pressures may be quite large, reaching nearly 600 cm H_2O below ambient. Second, pressures measured anteriorly and posteriorly within the oral cavity are essentially equivalent in magnitude. Third, posterior to the gill bars in the opercular cavity the pressure drop is only about one-fifth that in the oral cavity. Experimental studies have shown that this reduced negative pressure is caused by the gill bars themselves, which are adducted to form a high resistance to flow at the posterior limit of the oral cavity as the mouth opens (Lauder, 1983c). The gill bars are then abducted to allow water to pass posteriorly as the mouth closes.

Although many taxa do not generate large negative pressures during suction feeding (Norton and Brainerd, 1993), fishes as phylogenetically divergent as sharks (Frazetta, 1994; Moss, 1977; Motta *et al.*, 1991), lungfishes (Bemis, 1987; Bemis and Lauder, 1986), and coelacanths (inferred by Lauder; 1980b) are capable of using suction during feeding.

A typical pattern of jaw muscle activity used during suction feeding is illustrated in figure 3. The time from the onset of mouth opening to peak gape is called the expansive phase, and muscles active at the start of this phase include the levator operculi, sternohyoideus (rectus cervicis), and epaxial muscles (Fig. 3). These muscles act to depress the lower jaw and hyoid, and to elevate the neurocranium. Muscles connecting the hyoid to the lower jaw (such as the geniohyoideus) and the adductor mandibulae muscles may also be active during this time. In such cases, there is considerable overlap between the activity of mouth closing and opening muscles. As the mouth closes (the compressive phase), activity continues in the

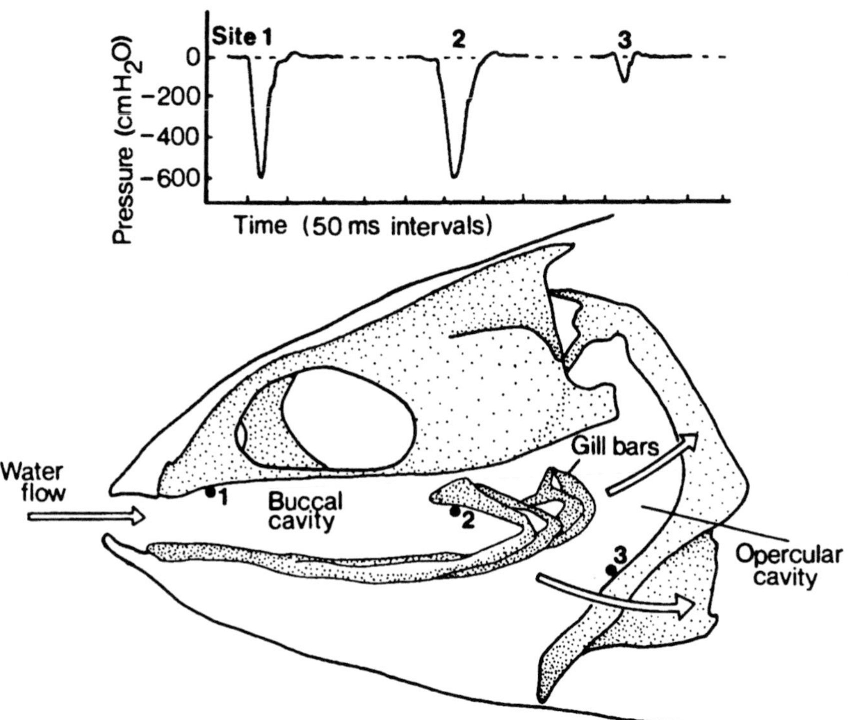

Figure 2. Diagram of the pattern of pressure change in the oral cavity of a percomorph fish during prey capture based on the experimental data from Lauder (1980c; 1983c). Suction feeding is produced by intraoral pressure changes. Note that the negative pressure posterior to the gill bars is greatly reduced compared to both the anterior and posterior sites within the oral cavity (after Lauder 1985c).

adductor mandibulae and geniohyoideus muscles. One consistent kinematic pattern found in almost all teleost fishes studied to date is the peak in hyoid excursion during the compressive phase. This maximal hyoid excursion occurs later than peak gape (Fig. 3) and yet prior to maximal opercular expansion; there is thus an anterior to posterior sequence of peak gape, peak hyoid, and maximum opercular excursion. The recovery phase (defined as the time

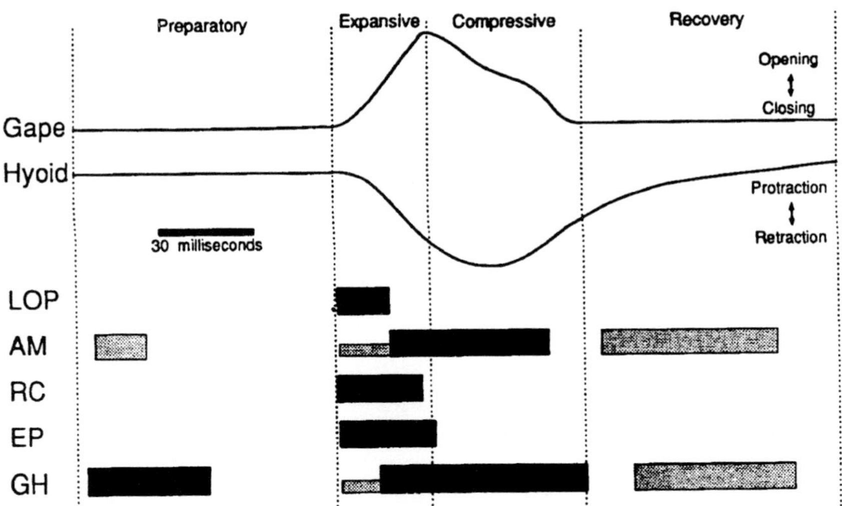

Figure 3. Schematic diagram of kinematic and motor patterns common to initial prey capture events in many ray-finned fishes. The names of phases associated with kinematic events are indicated at the top. Note that phase names differ in the fish and tetrapod literature. For example, in tetrapods the compressive phase is referred to as the closing (or fast closing) phase. The preparatory phase has only been observed in a few taxa to date. Black bars indicate times when muscles are consistently active whereas gray bars indicate activity that is only intermittently present. Modified from Lauder and Reilly (1994).

from jaw closure to the return of hyoid, suspensorial, and opercularelements to their initial positions) typically involves activity in the jaw, hyoid, and suspensorial adductor muscles. Finally, in some ray-finned fishes, a preparatory phase occurs prior to mouth opening in which the volume inside the mouth cavity is reduced by activity of jaw and hyoid adductors. This phase has primarily been observed in percomorph ray-finned fishes and has not been found in plesiomorphic taxa (Lauder, 1980a).

Intraoral Prey Transport

The process of moving prey from the jaws to the esophagus is referred to as prey transport. In many fishes, the process of transport involves two discrete components: hydraulic transport and pharyngeal

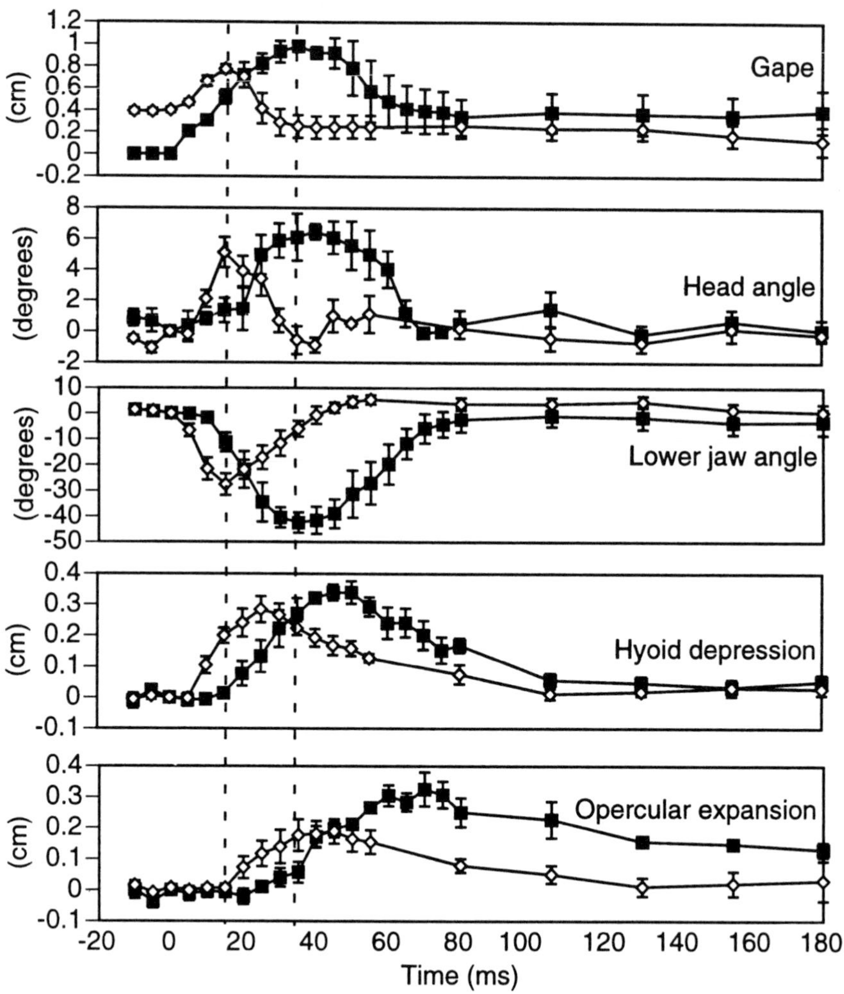

Figure 4. Comparison between prey capture and transport kinematics in *Lepomis macrochirus*. Prey transport is indicated by open symbols and initial capture events by solid symbols. The dashed lines indicate peak gape for transports and captures. After Gillis and Lauder (1995).

jaw transport. During pharyngeal jaw transport, fishes use active movements of the gill arches to grasp, manipulate, and move prey from the posterior region of the oral cavity directly to the esophagus (Liem, 1973; Liem and Greenwood, 1981). In order for pharyngeal jaw transport to occur, the prey must be located between the upper and lower pharyngeal jaws (Lauder, 1983b). To transport prey intothis location fish utilize suction (hydraulic transport) and create a current of water through the mouth that carries prey from the anterior jaws to the pharyngeal jaws posteriorly. Fishes, such as lungfishes and sharks, that do not have mobile tooth plates on the gill arches to manipulate prey use hydraulic transport exclusively to move prey to the esophagus (Bemis and Lauder, 1986).

Although the process of hydraulic prey transport is superficially similar to initial prey capture by suction feeding, recent results have shown that there can be substantial kinematic differences between prey capture and transport (Gillis and Lauder, 1995). The kinematics of hydraulic transport are illustrated in figure 4 and compared to the kinematics of prey capture. Prey caught between the jaws following the strike are moved posteriorly by a combination of jaw, hyoid, and opercular movements that are significantly more rapid than the motions used to capture prey initially. For example, the mean duration of prey capture in the *Lepomis macrochirus* studied by Gillis and Lauder (1995) was 65 msec, while hydraulic transport was accomplished in 36 msec. In addition, kinematic excursions during transport tend to be smaller than during prey capture.

The process of hydraulic prey transport is widespread among tetrapod out-group taxa and represents a general biomechanical strategy for manipulating prey in the aquatic medium. By creating patterns of water movement within the oral cavity, prey caught between the jaws may be moved into a position appropriate for swallowing. In this sense, hydraulic transport is the functional analog of the tetrapod tongue, and motor patterns associated with hydraulic manipulation in fishes may have played an important role in the evolution of tongue function in early tetrapods.

OUTGROUP PATTERNS: AMPHIBIANS

Amphibian taxa represent an important clade for understanding amniote feeding mechanisms. Within the Amphibia are species that exhibit aquatic feeding, terrestrial feeding, and (in some taxa) ontogenetic and/or ecological transitions between these two feeding modes. At first glance, it seems, one could hardly ask for a better out-group clade on which to conduct experimental analyses of feeding mechanisms. By studying aquatic feeding in amphibians, one can hold the environment constant and compare their feeding behaviors to those of fish out-group clades in order to examine which functional attributes in amphibians are likely to have been retained from ancestral patterns in fishes. Additionally, terrestrial feeding in amphibians can be compared to aquatic feeding in amphibians and fishes to better understand how the transition to land influenced feeding morphology and function. Furthermore, a longitudinal analysis of feeding across ontogenetic environmental transitions allows the effects of change in environment to be studied directly in the same individuals. Finally, by comparing terrestrial feeding in amphibians to that in amniotes one should be able to define amniote patterns that have been inherited directly from the terrestrial anamniotic feeding mechanism as well as those that appear to be novel for the clade.

Unfortunately, the promise of the Amphibia has yet to be completely fulfilled. Uncertainties in the phylogenetic relationships among and within the three extant clades (Canatella and Hillis, 1993; Larson and Dimmick, 1993; Trueb and Cloutier, 1991) make it difficult to determine which character states within extant clades are primitive for this group as a whole. This problem is complicated by the existence of numerous early amphibian fossil taxa that bear greatest resemblance in jaw morphology to only one of the three extant clades–salamanders (Carroll and Holmes, 1980). Also, in some amphibian clades such as caecilians, relatively few taxa have been studied functionally although recent results (O'Reilly, 1990; O'Reilly and Deban, 1991) will add considerably to current data. In addition, and despite considerable progress during the last five to six years in the comparative study of feeding in frogs (Anderson, 1993; Deban and Nishikawa, 1992; Gray and Nishikawa, 1995; Nishikawa *et al*., 1992;

Nishikawa and Canatella, 1991; Nishikawa and Roth, 1991; Trueb and Gans, 1983), salamanders (Beneski *et al.*, 1995; Elwood and Cundall, 1994; Findeis and Bemis, 1990; Larson *et al.*, 1996; Lauder and Reilly, 1990; Lauder and Schaffer, 1988; Maglia and Pyles, 1995; Miller and Larson, 1990; Reilly, 1995; Reilly and Lauder, 1988, 1989, 1990b, 1991a), and caecilians (Bemis *et al.*, 1993; Nussbaum, 1983; O'Reilly, 1990), we still lack data on many aspects of feeding behavior in this diverse taxonomic group. For the Amphibia as a whole, the process of prey transport has only been studied quantitatively in a few species, the metamorphosis of feeding function has received limited attention, and data on electromyographic patterns of muscle function are still very limited. Only in one species, for example, has the function of jaw musculature been studied across metamorphosis as well as during prey transport. Nonetheless, the diversity of taxa for which data are available is growing, and these data provide several important insights relevant to amniote feeding.

Aquatic Prey Capture

Based upon the phylogenetic distribution of suction feeding in non-tetrapod out-groups [such as Dipnoans (Bemis, 1987; Bemis and Lauder, 1986) and actinistians (Lauder, 1980b)] and on morphological correlates of suction feeding function in early tetrapod fossils (Carroll, 1988; Lauder and Reilly, 1994), it is likely that suction feeding is primitive for the class Amphibia. However, whereas each of the three extant clades of amphibians possess aquatic members, some utilize derived feeding mechanisms distinct from their suction feeding ancestors.

Aquatic and semiaquatic adult anurans are known to use their forelimbs to capture and help manipulate prey under water (O'Reilly and Deban, 1991). In addition, aquatic adult caecilians are not known to generate suction during feeding (O'Reilly, personal communication). Instead, like terrestrial caecilians examined to date, they utilize jaw prehension to capture prey. Interestingly, however, *Typhlonectes natans* (an aquatic South American caecilian) does possess certain kinematic features common to suction feeders (e.g., expansion of the buccal cavity during prey capture). Therefore, perhaps aspects of the ancestral suction feeding pattern have been retained in some adult aquatic caecilians whose derived morphologies preclude the production of adequate negative pressures to generate useful suction (O'Reilly, 1990; O'Reilly and Deban, 1991).

Although suction feeding is not retained in all aquatic amphibians, it is present within all three of the extant amphibian clades, being widespread among aquatic salamanders (larvae and adults), present in some aquatic anurans, and common to many aquatic larval caecilians (O'Reilly, 1990; personal communication). In addition, some tadpole species are known to use suction feeding (Wassersug and Hoff, 1982).

The most important general distinction to make concerning the diversity of suction feeding in amphibians is that taxa within this clade generally possess one of two fundamentally different feeding mechanisms: unidirectional systems in which water flows from anterior to posterior through the mouth cavity (as in fishes), and bidirectional systems in which water drawn into the mouth by suction during the initial phases of the strike must exit anteriorly as the mouth closes (Lauder and Shaffer, 1986). Among salamanders, this distinction is relevant to species that feed in the water as both larvae (with a unidirectional feeding system) and as adults (bidirectionally), and also to comparative analyses of aquatic adults, which possess only limited gill openings posteriorly. These taxa (e.g., *Cryptobranchus*) possess functionally bidirectional feeding mechanisms, and display features of the jaw movement during the strike that are different from taxa possessing unidirectional mechanisms either as larvae or as adults (Cundall *et al.* 1987; Elwood and Cundall, 1994; Reilly and Lauder, 1992).

Analyses of unidirectional suction feeding in salamanders have revealed many similarities with the suction feeding mechanisms of fishes (Lauder, 1985a). During the expansive phase (or fast opening phase in tetrapod terminology) cranial elevation and lower jaw depression both contribute to the gape, hyoid depression is a major effector of intraoral pressure reduction, and there is a distinct recovery phase that is similar to that of fishes. In addition, the fundamental sequence of peak excursions shown in figures 3 and 4 is retained during aquatic prey capture in salamanders, as is the onset of hyoid depression during the Expansive Phase. Hydraulic transport is used to manipulate prey within the oral cavity (Elwood and Cundall, 1994;

Gillis and Lauder, 1994), and electromyographic patterns of homologous muscles show general similarities to those of fishes (Lauder and Reilly, 1990; Lauder and Shaffer, 1985; Reilly, 1995; Shaffer and Lauder, 1985, 1988; Wainwright *et al.*, 1989). The morphological differences between salamanders and fishes (such as limited lateral suspensorial mobility and the lack of ossified opercular elements in salamanders) do not obviate the many kinematic similarities in the feeding mechanism.

The fundamental patterns described previously for aquatic prey capture in fishes thus are retained in many salamanders that feed in the water. These traits cannot then be considered unique to fishes and when similar traits are discovered in amniotes they cannot be regarded as amniote specializations.

Aquatic Prey Transport

Aquatic intraoral prey transport has been examined quantitatively in only one amphibian taxon to date–larvae of *Ambystoma tigrinum* (Gillis and Lauder, 1994). Suction-based transport in this larval salamander showed remarkable kinematic similarity to the suction-based transport utilized by bluegill sunfish. Kinematic traits shared by transport behaviors across taxa include similar timings of maximal gape, cranial elevation, and gape cycle duration; these behaviors cluster together in a multivariate analysis based on seven kinematic variables (Gillis and Lauder, 1995). Furthermore, suction-based transport behaviors in both sunfish and tiger salamander larvae, while similar to one another, exhibit consistent differences relative to the suction-based capture behaviors in both of these taxa. We suggest that the similarities between aquatic prey transport behaviors in sunfish and larval *A. tigrinum* reflect the retention of a suction-based transport behavior from a common ancestor, and we contend that the divergence between aquatic capture and transport behaviors may constitute a plesiomorphic feature of vertebrate feeding systems.

Terrestrial Prey Capture

The transition to land during vertebrate evolution required many substantial changes in the morphological and physiological components of organismal design. As amphibians represent the most primitive vertebrate class to have succeeded in making such a transition (all three extant clades have terrestrial representatives), they are an excellent group within which to examine terrestrial feeding. By comparing terrestrial feeding mechanisms in amphibians to those in aquatic amphibians and fishes, one can better appreciate the kinds of changes that evolved to facilitate feeding on land.

Due to the lower density and viscosity of air relative to water and prey, movement of the aerial medium itself is not a useful vehicle for bringing prey toward the jaws. Instead, the organism itself (or part of it) must move toward and capture the prey. Hence, terrestrial prey capture in many salamanders and frogs is generally associated with projection of the tongue out of the mouth toward the prey (Findeis and Bemis, 1990; Gans and Gorniak, 1982a,b; Nishikawa and Canatella, 1991; Nishikawa and Roth, 1991; Reilly and Lauder, 1989). Accordingly, concomitant with a transition to land in amphibians (be it developmental, ecological, or evolutionary) many structural components of the feeding mechanism are altered (Duellman and Trueb, 1988; Lauder and Reilly, 1990; Wassersug and Hoff, 1982). Many of these alterations facilitate tongue projection, such as osteological and myological modifications to the skull and associated muscles, the formation of a tongue and its intrinsic musculature, and the remodeling of gill arch elements to support the tongue.

Lingual-based terrestrial feeding in many amphibians thus contrasts sharply with the suction mechanism used during aquatic feeding by actively controlling and utilizing specialized musculature and skeletal designs during the protraction and retraction of a projectile tongue. In addition, in salamanders (Larsen *et al.*, 1996; Miller and Larsen, 1990) and frogs (Gray and Nishikawa, 1995; Nishikawa and Canatella, 1991; O'Reilly and Nishikawa, 1995) that lunge during prey capture (in addition to protracting their tongue), specializations in locomotor function may also be involved in prey capture. Even in terrestrial feeding systems in which prey are approached closely and the jaws are used to catch prey directly (thus obviating the need for tongue projection), as in terrestrial caecilians (Bemis *et al.*, 1983; Nussbaum, 1983; O'Reilly, 1990), specializations such as those of the jaw adduction mechanism can be present.

Interestingly, comparisons of salamander and frog feeding mechanisms suggest that these two clades possess fundamentally different systems of neural control of jaw musculature (Nishikawa *et al.*, 1992; Roth *et al.*, 1990). For example, in salamanders, very little muscle activity is present prior to mouth opening, whereas in *Bufo*, muscles such as the geniohyoideus and intermandibularis may be active for 100 to 200 msec prior to the start of the fast opening phase (Gans and Gorniak, 1982a). Frogs possess extensive sensory feedback mechanisms to modulate movements of the jaws during feeding (Anderson and Nishikawa, 1993; Nishikawa *et al.*, 1992), whereas salamanders appear to lack such mechanisms for altering the strike while it is in progress.

Analyses of frog feeding kinematics have shown that considerable diversity exists among taxa in the kinematic patterns used during prey capture, the underlying musculoskeletal mechanisms involved in prey acquisition, and the neural substrates of prey capture (Gans and Gorniak, 1982a; Gray and Nishikawa, 1995; Nishikawa *et al.*, 1992; Nishikawa and Canatella, 1991; Nishikawa and Roth, 1991; Nishikawa and Gans, 1992; Ritter and Nishikawa, 1995). However, only recently have terrestrial prey capture kinematics in salamanders been shown also to exhibit considerable diversity (Findeis and Bemis, 1990; Larsen and Beneski, 1988; Larson *et al.*, 1989; Lombard and Wake, 1977; Maglia and Pyles, 1995; Miller and Larsen, 1990).

Despite the variation in feeding kinematics, and differences in the neural control of jaw and tongue movements among frogs and salamanders, at least one generalization can be made regarding the kinematics of terrestrial prey capture in these taxa: jaw and tongue movements appear to be coordinated during the gape cycle. As a result, the gape profile of terrestrially feeding salamanders and frogs typically follows one of two general patterns, both of which are distinct from the bell-shaped profile seen during aquatic suction-based feeding (Fig. 3). In the first pattern, which we term a three-phase pattern (after Beneski *et al.*, 1995) the gape cycle consists of three distinct parts: first, there is a period of relatively rapid mouth opening (during which the tongue is raised from the floor of the mouth and begins to be protracted), second, there is a period of a relatively stable or slowly increasing gape (during which the tongue is protracted fully and begins to be retracted), and third, there is a period of rapid mouth closing (during which the tongue is brought back into the mouth while the jaws close on the prey item). This three phase gape cycle can be seen in figure 5 (panels A, B, and C) and is common to terrestrially feeding ambystomatid salamanders as well as many frogs studied to date (Beneski *et al.*, 1995; Deban and Nishikawa, 1992; Nishikawa and Canatella, 1991; Reilly and Lauder, 1989). In the four phase gape cycle pattern (Beneski *et al.*, 1995) (Fig. 5D-F), a second period of rapid mouth opening is inserted between the period of relatively stable or slowly increasing gape and the period of rapid mouth closing seen in the three-phase pattern. This second period of further gape opening occurs during tongue retraction and appears to accommodate the prey item being returned to the mouth. This four-phase gape cycle is exhibited by many terrestrial salamander clades examined to date (Findeis and Bemis, 1990; Larsen *et al.*, 1989; Miller and Larsen, 1990), including members of the most primitive terrestrial family, the Hynobiidae (Larsen *et al.*, 1996). Hence, the three-phase gape cycle during prey capture seen in ambystomatids is quite possibly a derived kinematic pattern for terrestrial salamanders, whereas the four-phase cycle is likely to be the primitive condition (Beneski *et al.*, 1995). Additionally, forward body lunges appear to be common to many terrestrially feeding frogs and salamanders (although not to ambystomatid salamanders), and this behavior is probably also primitive for terrestrial frogs and salamanders (e.g. Gray and Nishikawa, 1995).

To summarize briefly, projection of the tongue is nearly ubiquitous during prey capture in terrestrial frogs and salamanders. Although variability exists in the feeding mechanisms within and

Figure 5. Representative gape profiles from amphibian taxa showing the distinction between three-phase profiles (A-C) and four-phase profiles (D-F). Tongue protrusion is shown in relation to gape in panels C and F. A, *Hyla cinerea*, modified from Deban and Nishikawa (1992); B, *Spea multiplicata*, modified from O'Reilly and Nishikawa (1995); C, *Ambystoma tigrinum*, modified from Reilly and Lauder (1989); D, *Hynobius kimurae*, modified from Larsen *et al.* (1996); E, *Taricha torosa*, modified from Findeis and Bemis (1990), F, *Desmognathus fuscus*, modified from Larsen and Beneski (1988). Vertical dashed lines delimit the phases indicated in panels A and D. The scale bars indicate a time of 50 msec. ➔

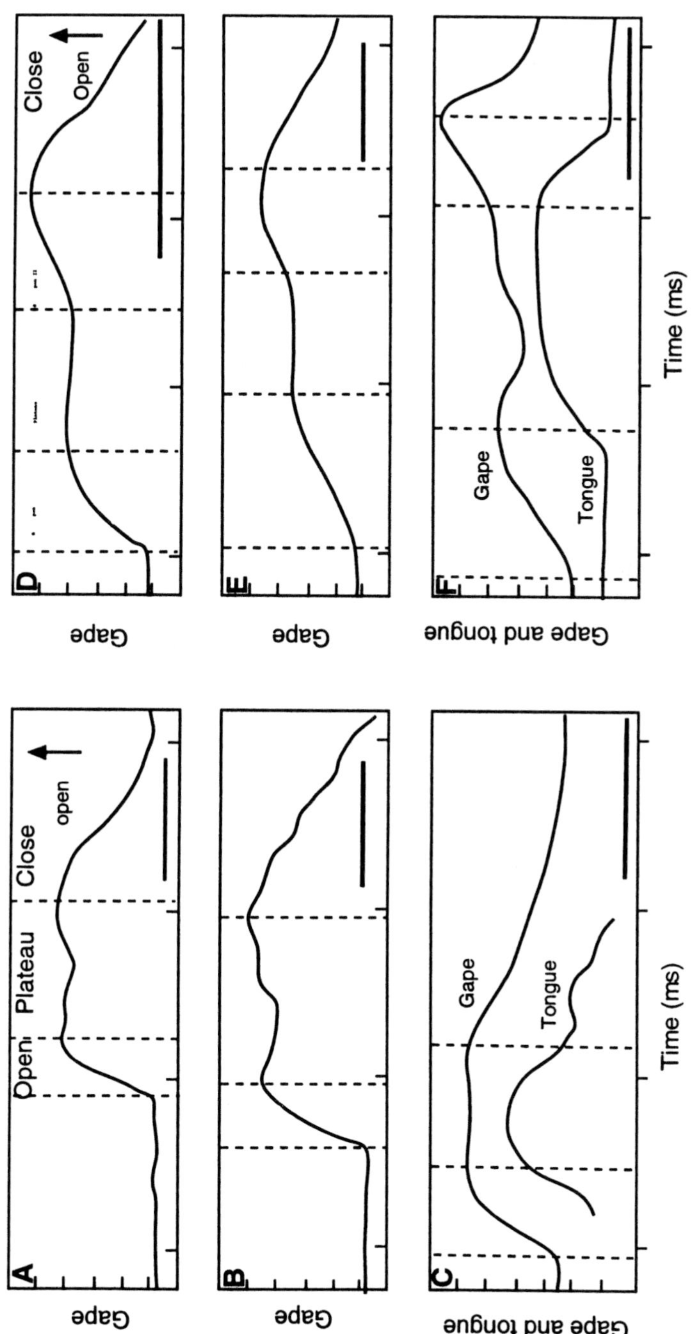
A
B
C
D
E
F
Gape
Gape
Gape and tongue
Gape
Gape
Gape and tongue
Open
Plateau
Close
open
Close
Open
Gape
Tongue
Gape
Tongue
Time (ms)
Time (ms)

among these taxa, lingual-based terrestrial prey capture shares several general patterns among amphibians, that distinguish it from the aquatic feeding behaviors present in aquatic ancestors. First, projection of the tongue during mouth opening involves protraction of the hyoid apparatus during this portion of the gape cycle, whereas during suction feeding the hyoid begins to be retracted during mouth opening in order to enlarge and reduce the pressure within the buccal cavity (Fig. 3). Second, a period of relatively stable or slowly increasing gape is present during tongue projection (the lack of lower jaw movement during this period presumably provides a more stable platform from which to project the tongue), whereas during suction-based aquatic feeding the gape profile is more bell shaped, lacking a relatively stable (plateau) phase (Fig. 3). Third, although movement of the hyoid apparatus has not generally been explicitly measured during terrestrial feeding in amphibians, in cases in which it has been quantified in terrestrial salamanders (Reilly and Lauder 1989; Findeis and Bemis, 1990), retraction of the hyoid continues through the mouth closing portion of the gape cycle (even after the tongue has been retracted fully back into the mouth), as well as after the mouth is closed. This contrasts with suction feeding wherein the hyoid is being protracted during the recovery phase of the gape cycle (Fig. 3).

Terrestrial Intraoral Prey Transport

Amphibians do not process their food intraorally in any significant manner beyond crushing or biting prey between the jaws following capture (Bemis *et al.*, 1983; DeVree and Gans, 1994; Elwood and Cundall, 1994; Erdman and Cundall, 1984; Schwenk and Wake, 1983). Prey held in the mouth may be pressed against vomerine teeth, but extensive processing or reduction of food into smaller pieces does not occur. A major function of the tongue following capture of prey is to transport food from the jaws to the esophagus, and use of the tongue in this manner has been documented in caecilians (Bemis *et al.*, 1983) and salamanders (Reilly and Lauder, 1990a). As is the case with initial prey capture in a terrestrial environment, prey that are much denser than the surrounding fluid require a non-medium dependent transport mechanism. As amphibians use their tongue to transport prey and are the most plesiomorphic clade of tetrapods, they represent an important clade for understanding the

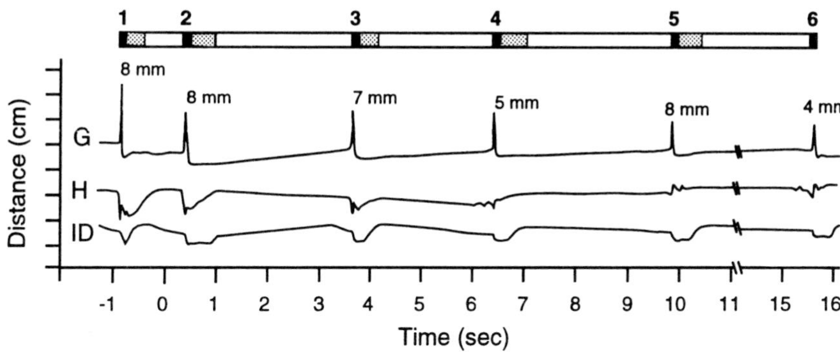

Figure 6. Gape and hyoid kinematic profiles for terrestrial prey transport in *Ambystoma tigrinum*. Six transport cycles following capture of an earthworm are shown, and at each cycle the prey is moved posteriorly toward the esophagus. Open bars at the top indicate the duration of the preparatory phase, black bars the transport gape cycles, and shaded bars the recovery phase. Distances shown above each gape cycle indicate the amount of prey transported posteriorly. Abbreviations: G, gape distance; H, hyoid depression; ID, intermandibular distance, which reflects movement of the tongue anteriorly between the mandibular rami. Modified from Reilly and Lauder (1990a).

use of tongue-based mechanisms for intraoral prey movement in amniotes.

Unfortunately, quantitative analyses of terrestrial prey transport have only been performed for one amphibian, *Ambystoma tigrinum*. An entire sequence of prey transport following capture of an earthworm is shown for *A. tigrinum* in figure 6. About 1 cm of a 5 cm long earthworm was captured between the jaws at the strike, and the remaining 4 cm was transported into the oral cavity and esophagus in a series of six transport cycles. The gape cycle of lingual-based preytransport closely resembles that of hydraulic transport in fishes and larval salamanders. Each cycle involves rapid mouth opening in a fast opening phase, followed by a closing phase, and between 4 and 8 mm of the prey is transported posteriorly with each cycle. Unlike the gape cycle of the initial strike, the gape profile of a transport is bell shaped (Fig. 7) with no plateau maintaining a near constant gape opening. The hyoid moves rapidly in a posteroventral direction during the fast opening phase, and this movement draws prey attached to the tongue posteriorly. During the recovery phase, the tongue is protracted anterodorsally sliding under the prey prior to the next transport cycle.

One key feature of the prey transport cycle is the extended preparatory phase which may last several seconds and has been divided into two parts (Reilly and Lauder, 1990a). During the first part (P1) the gape slowly increases by about 1 mm and the prey is compressed against the roof of the mouth by elevation and protraction of the hyoid. The second part of the preparatory phase (P2) is shorter and during this time just prior to the fast opening phase the gape is held constant. During prey transport in *A. tigrinum* the head and body do not move horizontally, and there is thus no inertial component of body movement relative to the prey: Posterior prey movement is entirely a consequence of posterior tongue movement.

Electromyographic analysis of jaw muscle function during prey transport (Fig. 7) has shown that muscle activity patterns used for prey transport differ significantly from those used to capture prey initially (Reilly and Lauder, 1991b). During prey transport, durations of muscle bursts tend to be shorter, intrinsic tongue muscles such as the genioglossus show very little activity, and the adductor mandibulae internus muscle reaches peak activity much earlier than during initial prey capture. Most surprisingly, the subarcualis rectus one muscle, the major tongue protractor, is strongly active in a single burst despite the lack of observed tongue projection during transport (Fig. 7).

Based on the general similarities of gape and hyoid kinematic profiles in terrestrial transport by *Ambystoma tigrinum* and hydraulic transport by fishes, Lauder and Reilly (1994; also see Reilly and Lauder, 1990a) hypothesized that these two behaviors are distinct from the process of terrestrial prey capture and that the kinematic and motor patterns used during terrestrial transport are derived from the plesiomorphic pattern used for hydraulic transport. Gillis and Lauder (1994) tested this hypothesis explicitly by comparing statistically the kinematic patterns of four behaviors in *A. tigrinum*: aquatic capture

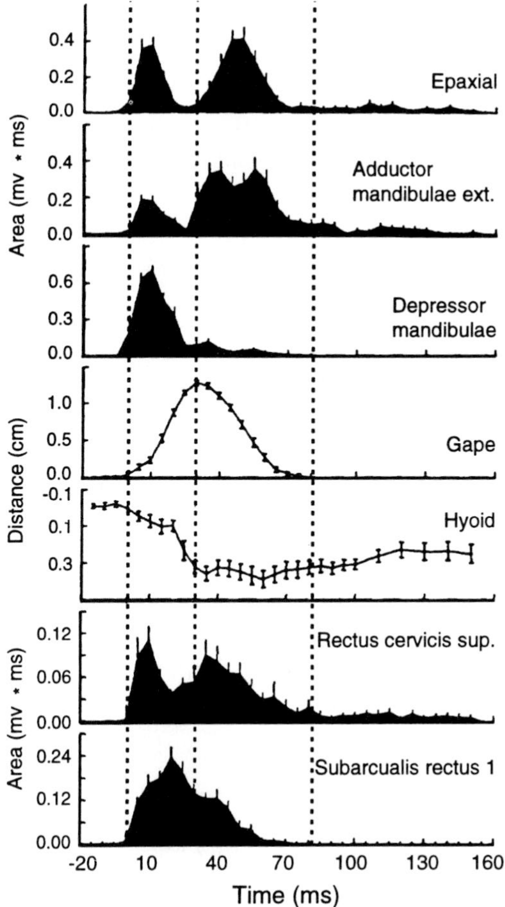

Figure 7. Kinematic and muscle activity profiles during prey transport in *Ambystoma tigrinum*. Solid curves represent the mean rectified integrated area at each time interval for each muscle. Short vertical lines indicate standard errors for all curves which are the mean of 17 transport events. All data were recorded simultaneously. Dashed vertical lines indicate the onset of mouth opening, peak gape, and mouth closing from left to right. Modified from Reilly and Lauder (1991b).

and transport and terrestrial capture and transport. Transport behaviors are indeed similar to one another and distinct from capture behaviors in that they occur significantly more rapidly and involve reduced excursions relative to prey capture. This suggests that many aspects of the lingual-based transport behavior seen in amphibians may have been inherited directly from the suction-based aquatic transport behaviors of aquatic ancestors. Further studies of prey transport in other amphibians are needed before such an hypothesis can be further supported.

SUMMARY OF OUTGROUP DATA

Based on these experimental results from extant out-group clades, ten key features of the tetrapod feeding mechanism can be hypothesized as having been present in the terrestrial anamniotic ancestors of amniotes. (1) Prior to the onset of the mouth opening, gape was held constant with the jaws closed or nearly so prior to initial prey capture, or held at a constant low value before mouth opening begins during the prey transport cycle. (2) Mouth opening occurred as a result of both cranial elevation (the product of epaxial muscle activity) and lower jaw depression (caused by activity in the rectus cervicis and depressor mandibulae muscles). (3) The presence of a fleshy tongue permitted lingual-based prey capture and transport. (4) During prey capture, protraction of the hyoid apparatus was used to project the tongue, and occurred as the mouth was opening, or during a period of relatively stable gape. (5) Retraction of the hyoid apparatus during prey capture returned the tongue to the mouth and continued even after the mouth was closed. (6) The prey capture gape cycle was characterized by four phases–a period of mouth opening, a period of stable or slowly increasing gape, a second period of further mouth opening, and a period of rapid mouth closing. (7) A forward lunge occurred concomitant with the prey capture gape cycle. (8) Terrestrial prey transports exhibited a bell-shaped gape profile and were distinct kinematically and electromyographically from the initial capture of prey. (9) Transport of prey within the mouth occurred by posteroventral movements of the hyoid apparatus during mouth opening (in a manner similar to aquatic prey transport). (10) During transport, the hyoid apparatus was protracted (moved anterodorsally) during a Recovery Phase, after the mouth was closed.

Several of these plesiomorphic patterns are modified significantly within amniotes, and yet without an understanding of the historical origin of amniote functional traits we would be unable to identify sequences of historical transformation in feeding function or to identify homologous functional attributes of feeding systems among tetrapods.

PRIMITIVE AMNIOTE FEEDING MECHANISMS

Although amniote taxa display considerable diversity in their feeding mechanisms, a number of authors have abstracted from the large base of comparative data several general features of amniote jaw function that are believed to be primitive for the clade as a whole (Bels *et al.*, 1994a; Bramble and Wake, 1985; Delheusy and Bels, 1992; Hiiemae and Crompton, 1985; Reilly and Lauder, 1990a; Schwenk and Throckmorton, 1989). Many of these characteristics of amniote feeding are well illustrated by the feeding systems in lizards that primitively use lingual prehension to capture prey (Herrel *et al.*, 1995; Schwenk and Throckmorton, 1989; Smith, 1984, 1988). Here we focus on lizard prey capture and manipulation as exemplifying many traits that may be representative of primitive amniote feeding mechanisms.

The transition to amniote feeding involves several significant morphological and functional changes from feeding systems described previously in amphibians and fishes. The gill arch elements that feature so prominently in fish and salamander feeding have been modified and greatly reduced. The tongue and supporting skeletal elements have become elaborated (Delheusy *et al.*, 1994; Smith, 1986), and the skull has a smaller number of independently mobile elements [the presence of various types of cranial kinesis notwithstanding (Frazetta, 1962; Smith and Hylander, 1985)]. Many amniote taxa utilize extensive intraoral food processing prior to swallowing prey, and several distinct behaviors associated with such prey manipulation have been described: for example, inertial feeding, "cleaning of teeth," chewing or reduction of prey, and pharyngeal packing. Here our focus will be on describing the function of the feeding mechanism during prey capture and transport, with the overall aim of assessing functional traits that might be primitive for the Amniota.

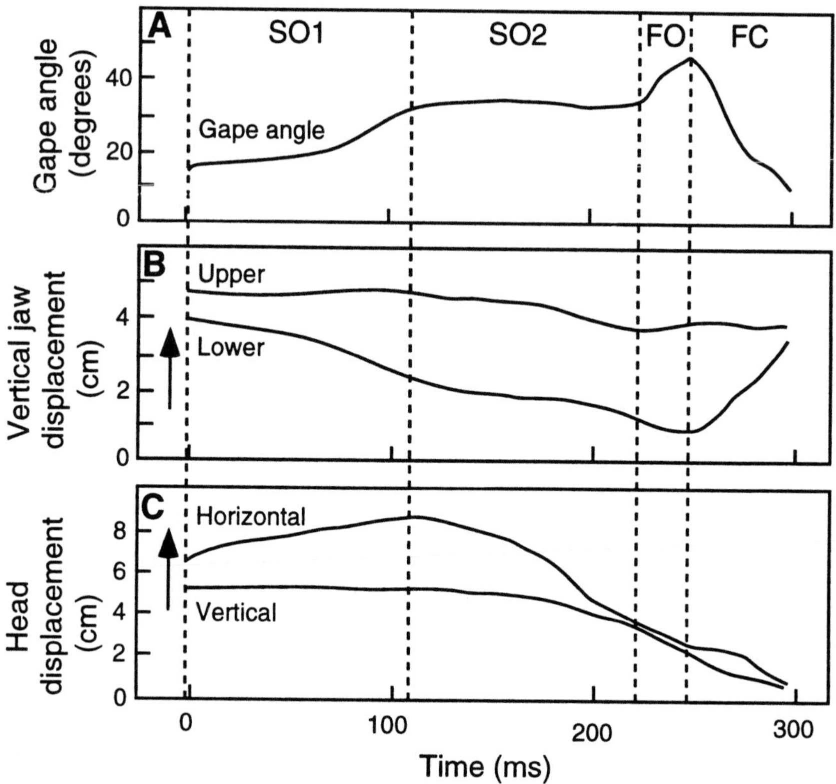

Figure 8. Gape distance (A) and vertical (B) and horizontal (C) jaw and head displacement plotted for prey capture in *Oplurus cuvieri.* Note the presence of a slow opening phase in the gape profile (SO1) and the relative lack of upper jaw movement. Arrows in B and C indicate dorsal and posterior movement respectively. Modified from Delheusy and Bels (1992).

Terrestrial Prey Capture

The gape profile of initial prey capture is illustrated for the iguanid lizard *Oplurus cuvieri* in figure 8 (Delheusy and Bels, 1992). The gape profile is divided into four distinct phases: an initial phase of slow opening (SO1), a plateau of relatively constant gape (SO2), fast opening (FO), and a fast closing phase (FC). This is somewhat similar to the four-phase cycle seen in many terrestrial salamanders, although the initial phase of opening in these out-group taxa is relatively rapid, whereas in lizards it can be quite slow. The skull moves posteriorly slightly during SO1 and reaches maximal posterior movement during the transition to SO2 (Figure 8) before moving anteriorly substantially during SO2, FO, and FC. Vertical motion of the skull is minimal during the SO1, but ventral movement begins during SO2 and continues through both the FO and FC phases. Gape is the result of movement of both the upper and lower jaws, however, the lower jaw tends to account for most of the gape changes observed during the gape cycle. For example, the gape increase during SO1 and FO is due mostly to lower jaw movement and much of the change in gape during Fast Close is due to elevation of the lower jaw while the upper jaw remains relatively immobile [Figure 2; (Delheusy and Bels, 1992)].

Anterior movement of the tongue begins during the SO1 phase but accelerates rapidly during SO2 as the tongue is extended toward the prey. Prey contact occurs at the end of SO2 (Delheusy and Bels, 1992), and during the FO phase the tongue and hyoid move posteroventrally to bring prey into the mouth. The kinematic pattern at this time is quite similar to that seen in terrestrial salamanders with a four-phase feeding cycle (Fig. 5).

Although the slow opening phases have been observed in prey capture events from taxa in all three major lineages of iguanid lizards (Herrel *et al.*, 1995; Schwenk and Throckmorton, 1989; Wainwright *et al.*, 1991), not all species show both these phases, and the occurrence of the SO phases may also depend on prey type (Bels and Grosse, 1990; Gorniak *et al.*, 1982; Urbani and Bels, 1995). *Anolis equestris*, for example, shows only a single SO phase with no distinct plateau (SO2) phase during prey acquisition (Bels and Delheusy, 1992). In addition, analysis of aquatic prey capture kinematics in amniotes such as the snapping turtle (*Chelydra serpentina*) show that there is no slow opening phase of the gape cycle (Lauder and Pendergast, 1992): Kinematic profiles during mouth opening resemble those of aquatic prey capture in aquatic anamniote ancestors.

Electromyographic studies of jaw muscle function during initial prey capture have shown that during the slow opening phases there is considerable activity in the hyoid and intrinsic tongue muscles (Herrel *et al.*, 1995) that continues into the fast opening phase. The onset of activity in the depressor mandibulae is coincident with the start of Fast Opening while the sternohyoideus muscle is active during fast opening and again during the closing phase [which consists of both fast close and slow close phases in *Agama*; (Herrel *et al.*, 1995]. The adductor and pterygoid muscles are strongly active during the fast opening and closing phases.

As noted by Herrel *et al.* (1995), jaw muscle activity patterns in *Agama* are very similar to those described previously for chameleons (Wainwright and Bennett, 1992), whereas there are significant differences with jaw motor patterns reported for *Sphenodon* (Gorniak *et al.*, 1982). However, given the paucity of comparative data on motor patterns during initial prey capture in lizards it is difficult to generalize on the causal bases for these differences.

Terrestrial Prey Transport

Following prey capture, lizards display a number of intraoral manipulatory behaviors including chewing, repositioning of prey, pharyngeal packing, cleaning, and transport (e.g. Deheusy and Bels, 1992; Kraklau, 1991; Smith, 1984, 1988). In order to make comparisons with nonamniote taxa it is important to base comparative analyses and evolutionary hypotheses on appropriate behavioral comparisons; in some cases it is difficult to discern from published figures which behavior is being analyzed, and without detailed kinematic studies or x-ray cinematography it is often difficult to know exactly how prey are being moved within the oral cavity. In an effort to describe comparable data for amniotes and anamniotes we will focus on terrestrial prey transport behavior, recognizing that jaw function in amniotes is associated with a wide diversity of postcapture behaviors.

Transport of prey from the anterior region of the oral cavity into the posterior portion for swallowing has been described for several taxa of lizards (e.g. Deheusy and Bels, 1992; Schwenk and Throckmorton, 1989; Smith, 1984; So *et al.*, 1992). Delheusy and Bels (1992) illustrate successive transport cycles in *Oplurus* and show that many of the phases of the gape cycle present during initial capture are also present during transport. The main consistent exception is the SO2 phase, which is absent. Similar results were obtained by So *et al.* (1992) (Fig. 9). Examination of transport gape profiles suggests that even the recognition of an SO phase in any form may be problematic

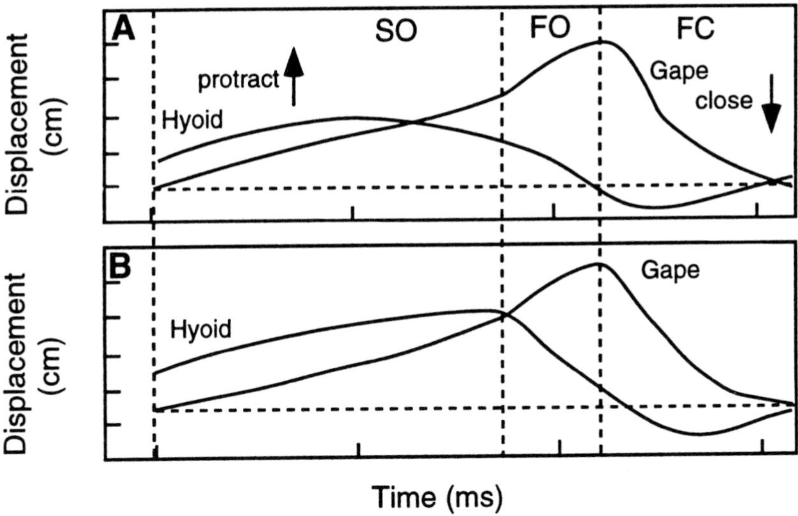

Figure 9. Schematic diagram of gape and hyoid kinematic profiles during prey transport in (A) *Chamaeleo jacksonii,* and (B) *Ctenosaura similis.* Note that the hyoid is protracted during the SO phase and retracted during FO. Modified from So *et al.* (1992); panel B after Smith (1984).

in many gape cycles, as there is only a slight change in slope of gape distance versus time (Fig. 9). Given that any increase in gape is likely to begin slowly, accelerate to a maximum rate of change, and finally decrease toward a maximum excursion, there must mathematically be an inflection point in the gape curve that could be identified as the end of slow opening and the start of fast opening. The existence of this inflection point need not be a reflection of any active neurological control or biomechanical feature of the feeding mechanism. The presence and extent of the SO phase is clearly highly variable both among transport cycles, among manipulatory behaviors, and among taxa.

Transport of prey following capture involves repeated cycles of hyoid protraction and retraction that move prey toward the esophagus. During the slow opening phase of the gape cycle the hyoid is protracted (Fig. 9). Retraction begins either just prior to or during fast opening and continues through the closing phase. This general pattern of gape and hyoid movements is superficially similar to that seen during the four-phase prey capture cycle seen in most terrestrial salamanders but contrasts substantially with that seen during prey transport in the only terrestrial salamander in which transport has been examined, *Ambystoma tigrinum*. In *A. tigrinum*, recall that the hyoid is retracted over the first portion of the gape cycle (during mouth opening) and protracted during a recovery phase after the gape cycle is finished (after mouth closing). Without further examination of terrestrial prey transport in other anamniotes, it will be hard to determine to what extent transport behavior in amniotes has diverged relative to that in terrestrial anamniotic ancestors.

Delheusy and Bels (1992) have conducted quantitative statistical analyses of transport behavior and compared kinematic transport patterns to jaw movements during chewing, initial capture, and cleaning. An analysis of variance showed chewing and transport cycles to differ significantly in duration and time to maximal lower jaw depression, and So *et al.* (1992) also found numerous significant differences between transport and chewing cycles in chameleons. A principal component analysis of these behaviors in *Oplurus* shows that cleaning behavior is the most distinctive and that there is considerable overlap between initial capture, transport, and reduction behaviors. These behaviors, although statistically distinct, nonetheless share a number of common kinematic patterns.

PLESIOMORPHIC AMNIOTE FUNCTIONAL TRAITS

Comparison of functional patterns in squamates to those described previously for amphibians and fishes suggests that several of the traits observed in amniotes are novel features of the feeding mechanism that are likely to have been present at the base of the amniote radiation. The diversity of intraoral processing behaviors in which the jaws, tongue, and hyoid are all involved (e.g., chewing, repositioning of prey, and cleaning) is a novel feature of the amniote feeding mechanism. In addition, the presence of a slow opening phase in which gape distance increases relatively slowly at the start of the gape cycle is an amniote trait. Associated with slow opening are seemingly unique patterns of hyoid muscle activity that result in hyoid and tongue protraction during slow opening. However, without further electromyographic studies of terrestrial prey capture in salamanders that utilize a four-phase cycle of feeding, it will be hard to determine whether these patterns of muscle of activity are really unique to amniotes or are simply not present in the ambystomatid salamanders (which have a three-phase cycle of prey capture) that have been studied to date.

It is tempting to view experimental data from amniotes as supporting the idea that a SO phase is required for the tongue-based feeding systems that characterize so many squamates. However, ambystomatid salamanders and many frogs possess well-developed tongue-based feeding systems that lack a SO phase.

SYNTHESIS: CONCLUSIONS AND UNRESOLVED ISSUES

Bramble and Wake (1985) presented a general model of kinematic and electromyographic patterns for tetrapod feeding mechanisms. This model has been of important heuristic value because it has provided a hypothesis against which empirical data from extant tetrapods can be tested. At present, Bramble and Wake's model has received some support (see, for example, Schwenk and Throckmorton, 1989). However, this model has also come under review where experimental results do not match predictions. Based on the results from analyses of individual taxa, various authors have examined specific predictions of this model (Deheusy and Bels, 1992; Reilly and Lauder, 1990a, 1991b; So *et al.*, 1992).

Althugh there is no doubt that, in amniotes, several general characteristics of the Bramble and Wake model do describe features of jaw function common to many amniote clades, the experimental results summarized above for amniotes and anamniote tetrapods also point out a number of complications that render their description of a "generalized tetrapod" functional pattern problematic.

First, the pattern of jaw movement and muscle function during prey transport in ambystomatid salamanders (the only anamniote taxon for which quantitative data are available on prey transport) is quite different than expected under the general tetrapod model. For example, there is no slow opening phase, hyoid protraction thus does not immediately precede fast opening, motion of the head and neck does not occur in the predicted manner (there is often very little horizontal skull movement), and there is no electrical activity in the depressor mandibulae and hyoid muscles just prior to the fast opening phase. In fact, many features of ambystomatid transport systems instead appear to be primitive traits inherited from aquatic ancestors. However, recall that ambystomatids show relatively distinct kinematic patterns during prey capture relative to most other salamanders, and it is indeed possible that this is true of their transport behavior as well. Further examination of transport behaviors in other urodeles is required before the accuracy of Bramble and Wake's model can be assessed relative to such primitive tetrapods.

Second, results from a variety of amniote taxa also suggest that kinematic and electromyographic data do not tightly fit predicted patterns. For example, in many transport and manipulation cycles there is no clear SO phase, and electromyographic patterns in the depressor mandibulae, sternohyoideus, adductor mandibulae, and pterygoideus show unpredicted patterns (Herrel *et al.*, 1995). In addition, comparing prey capture in iguanians to the proposed model has been done by several investigators (e.g. Kraklau, 1991; Schwenk and Throckmorton, 1989), and Delheusy and Bels (1992: p. 184) summarize their results by noting that "Our data do not support the model of Bramble and Wake or their speculation about the relationship between SO II duration and the size of the prey."

Given the limited experimental data available in 1985, it is perhaps not surprising that more recent results have called many of our previous concepts of amniote jaw function into question. However, even these additional data are insufficient to do more than suggest the outlines of a new view of amniote feeding function. Given the diversity of both anamniote and amniote taxa, the number of taxa for which we have both kinematic and electromyographic data is surprisingly few. We probably do not have a complete set of kinematic and electromyographic data for more than ten taxa of anamniote tetrapods and squamates. Furthermore, such data are rarely available for the full range of behavioral diversity exhibited by the feeding mechanism. In order to evaluate biomechanical models of jaw function and to produce evolutionary hypotheses of functional transformation, a much larger data set is needed. Such experimental data will permit a more quantitative assessment of the diversity of feeding system function in primitive amniotes and provide a better understanding of how plesiomorphic functional traits combined with novel features to form the basal amniote feeding mechanism.

ACKNOWLEDGMENTS

This work was supported by National Science Foundation Grants IBN 91-19502 and 95-07181 to George Lauder. We thank Steve Reilly for stimulating discussions on prey transport kinematics, Don Buth for his careful reading of the manuscript, and Jim O'Reilly for information on caecilian feeding.

LITERATURE CITED

Anderson, C. W. 1993. The modulation of feeding behavior in response to prey type in the frog *Rana pipiens*. *Journal of Experimental Biology,* 179:112.

Anderson, C. W., and K. C. Nishikawa. 1993. A prey-type dependent hypoglossal feedback system in the frog *Rana pipiens*. *Brain, Behavior, and Evolution* 42:89-96.

Bels, V. L., M. Chardon, and K. V. Kardong. 1994a. Biomechanics of the hyolingual system in Squamata. Pages 197-240 in: *Advances in Comparative and Environmental Physiology: Biomechanics of Feeding in Vertebrates* (V. L. Bels, M. Chardon, and P. Vandewalle, eds.). Berlin: Springer-Verlag.

Bels, V. L., Chardon, M., and P. Vandewalle. (Eds.). 1994b. *Advances in Comparative and Environmental Physiology 18. Biomechanics of Feeding in Vertebrates*. Berlin: Springer Verlag.

Bels, V. L. and V. Delheusy. 1992. Kinematics of prey capture in iguanid lizards: comparison between *Anolis equestris* (Anolinae) and *Oplurus cuverieri* (Oplurinae). *Belgium Journal of Zoology,* 122:223-234.

Bels, V. L., and V. Goosse. 1990. Comparative kinematic analysis of prey capture in *Anolis carolinensis* (Iguania) and *Lacerta viridis* (Scleroglossa). *Journal of Experimental Zoology,* 255:120-124.

Bemis, W. E. 1987. Feeding systems of living Dipnoi: anatomy and function. *Journal of Morphology, Supplement,* 1:249-275.

Bemis, W. E., and G. V. Lauder. 1986. Morphology and function of the feeding apparatus of the lungfish *Lepidosiren paradoxa* (Dipnoi). *Journal of Morphology,* 187:81-108.

Bemis, W. E., K. Schwenk, and M. H. Wake. 1983. Morphology and function of the feeding apparatus in *Dermophis mexicanus* (Amphibia: Gymnophiona). *Zoological Journal of the Linnean Society of London,* 77: 75-96.

Beneski, J. T., J. H. Larsen, and B. T. Miller. 1995. Variation in the feeding kinematics of mole salamanders (Ambystomatidae: *Ambystoma*). *Canadian Journal of Zoology,* 73353-366.

Bramble, D. M., and D. B. Wake. 1985. The feeding mechanisms of lower tetrapods. Pages 230-261in: *Functional Vertebrate Morphology* (M. Hildebrand, D. M. Bramble, K. F. Liem, and D. B. Wake, eds.). Cambridge, Massachusetts: Harvard University Press.

Cannatella, D. C., and D. M. Hillis. 1993. Amphibian relationships: phylogenetic analysis of morphology and molecules. *Herpetological Monographs,* 7:1-7.

Carroll, R. L. 1988. *Vertebrate Paleontology and Evolution.* San Francisco: W. H. Freeman and Company.

Carroll, R. L., and R. Holmes. 1980. The skull and jaw musculature as guides to the ancestry of salamanders. *Zoological Journal of the Linnean Society of London,* 68:1-40.

Cundall, D., J. Lorenz-Elwood, and J. D. Groves. 1987. Asymmetric suction feeding in primitive salamanders. *Experientia,* 43:1229-1231.

Deban, S. M., and K. C. Nishikawa. 1992. The kinematics of prey capture and the mechanism of tongue protraction in the green tree frog *Hyla cinerea. Journal of Experimental Biology*, 170:235-256.

Delheusy, V., and V. Bels. 1992. Kinematics of feeding behavior in *Oplurus cuvieri* (Reptilia: Iguanidae). *Journal of Experimental Biology,* 170:155-186.

Delheusy, V., G. Toubeau, and V. Bels. 1994. Tongue structure and function in *Oplurus cuvieri* (Reptilia: Iguanidae). *Journal of Morphology,* 238:263-276.

DeVree, F., and Gans, C. (1994). Feeding in tetrapods. Pages 93-118 in: *Advances in Comparative and Environmental Physiology: Biomechanics of Feeding in Vertebrates* (V. L. Bels, M. Chardon, and P. Vandewalle, eds.). Berlin: Springer-Verlag.

Duellman, W. E. and L. Trueb. 1986. *Biology of Amphibians.* New York: McGraw Hill.

Elwood, J. R., and D. Cundall. 1994. Morphology and behavior of the feeding apparatus in *Cryptobranchus alleganiensis* (Amphibia: Caudata). *Journal of Morphology,* 220:47-70.

Erdman, S., and D. Cundall. 1984. The feeding apparatus of the salamander *Amphiuma tridactylum*: morphology and behavior. *Journal of Morphology,* 181:175-204.

Findeis, E. K. and W. E. Bemis. 1990. Functional morphology of tongue projection in *Taricha torosa* (Urodela: Salamandridae). *Zoological Journal of the Linnean Society of London,* 99:129-157.

Frazzetta, T. H. (1962). A functional consideration of cranial kinesis in lizards. *Journal of Morphology,* 111:287-319.

Frazzetta, T. H. (1994). Feeding mechanisms in sharks and other elasmobranchs. Pages 31-57 in: *Advances in Comparative and Environmental Physiology: Biomechanics of Feeding in Vertebrates* (V. L. Bels, M. Chardon, and P. Vandewalle, eds.). Berlin: Springer-Verlag.

Gans, C., and G. C. Gorniak, G. C. 1982a. Functional morphology of lingual protrusion in marine toads (*Bufo marinus*). *American Journal of Anatomy,* 163:195-222.

Gans, C., and G. C. Gorniak. 1982b. How does the toad flip its tongue? Test of two hypotheses. *Science,* 216:1335-1337.

Gillis, G., B. and G. V. Lauder. 1994. Aquatic prey transport and the comparative kinematics of *Ambystoma tigrinum* feeding behaviors. *Journal of Experimental Biology,* 187:159-179.

Gillis, G. B., and G. V. Lauder. 1995. Kinematics of feeding in bluegill sunfish: is there a general distinction between aquatic capture and transport behaviors? *Journal of Experimental Biology,* 198:709-720.

Gorniak, G. C., H. I. Rosenberg, and C. Gans. 1982. Mastication in the tuatara, *Sphenodon punctatus* (Reptilia: Rhynchocephalia): structure and activity of the motor system. *Journal of Morphology,* 171:321-353.

Gray, L., and K. Nishikawa. 1995. Feeding kinematics of phyllomedusine tree frogs. *Journal of Experimental Biology,* 198:457-463.

Grobecker, D. B., and T. W. Pietsch. 1979. High-speed cinematographic evidence for ultrafast feeding in antennariid anglerfishes. *Science,* 205:1161-1162.

Hanken, J., and B. K. Hall. 1993. *The Vertebrate Skull.* Chicago: The University of Chicago Press.

Herrel, A., Cleuren, J. and F. De Vree. 1995. Prey capture in the lizard *Agama stellio. Journal of Morphology*, 224:313-329.

Hiiemae, K., and A. W. Crompton. 1985. Mastication, food transport, and swallowing. Pages 262-290 in: *Functional Vertebrate Morphology* (M. Hildebrand, D. M. Bramble, K. F. Liem, and D. B. Wake, eds.). Cambridge, Massachusetts: Harvard University Press.

Kraklau, D. 1991. Kinematics of prey capture and chewing in the lizard *Agama agama* (Squamata: Agamidae). *Journal of Morphology,* 210:195-212.

Larsen, J. H. and J. T. Beneski. 1988. Quantitative analysis of feeding kinematics in dusky salamanders (*Desmognathus*). *Canadian Journal of Zoology,* 66:1309-1317.

Larsen, J. H., J. T. Beneski, and B. T. Miller. 1996. Structure and function of the hyolingual system in *Hynobius* and its bearing on the evolution of prey capture in terrestrial salamanders. *Journal of Morphology,* 227:235-248.

Larsen, J. H., J. T. Beneski, and D. B. Wake. 1989. Hyolingual feeding systems of the Plethodontidae: comparative kinematics of prey capture by salamanders with free and attached tongues. *Journal of Experimental Zoology,* 252:25-33.

Larson, A., and W. W. Dimmick. 1993. Phylogenetic relationships of the salamander families: an analysis of congruence among morphological and molecular characters. *Herpetological Monographs*, 7:77-93.

Lauder, G. V. 1980a. Evolution of the feeding mechanism in primitive actinopterygian fishes: a functional anatomical analysis of *Polypterus, Lepisosteus,* and *Amia. Journal of Morphology,* 163:283-317.

Lauder, G. V. 1980b. The role of the hyoid apparatus in the feeding mechanism of the living coelacanth, *Latimeria chalumnae. Copeia,* 1980:1-9.

Lauder, G. V. 1980c. The suction feeding mechanism in sunfishes (*Lepomis*): an experimental analysis. *Journal of Experimental Biology*, 8:49-72.

Lauder, G. V. 1983a. Food capture. Pages 280-311 in: *Fish Biomechanics* (P. W. Webb and D. Weihs, eds.). New York: Praeger Publishing Company.

Lauder, G. V. 1983b. Functional design and evolution of the pharyngeal jaw apparatus in euteleostean fishes. *Zoological Journal of the Linnean Society,* 77:1-38.

Lauder, G. V. 1983c. Prey capture hydrodynamics in fishes: experimental tests of two models. *Journal of Experimental Biology*, 104:1-13.

Lauder, G. V. 1985a. Aquatic feeding in lower vertebrates. Pages 210-229 in: *Functional Vertebrate Morphology* (M. Hildebrand, D. M. Bramble, K. F. Liem, and D. B. Wake, eds.). Cambridge, Massachusetts: Harvard University Press.

Lauder, G. V. (1985b). Functional morphology of the feeding mechanism in lower vertebrates. Pages 179-188 in: *Functional Morphology in Vertebrates* (H.-R. Duncker and G. Fleischer, eds.). New York: Gustav Fischer Verlag.

Lauder, G. V., and T. Prendergast. 1992. Kinematics of aquatic prey capture in the snapping turtle *Chelydra serpentina. Journal of Experimental Biology,* 164:55-78.

Lauder, G. V., and S. M. Reilly. 1990. Metamorphosis of the feeding mechanism in tiger salamanders (*Ambystoma tigrinum*): the ontogeny of cranial muscle mass. *Journal of Zoology*, 222:59-74.

Lauder, G. V., and S. M. Reilly. 1994. Amphibian feeding behavior: comparative biomechanics and evolution. Pages 163-195 in: *Advances in Comparative and Environmental Physiology: Biomechanics of Feeding in Vertebrates* (V. L. Bels, M. Chardon, and P. Vandewalle, eds.). Berlin: Springer-Verlag.

Lauder, G. V., and H. B. Shaffer. 1985. Functional morphology of the feeding mechanism in aquatic ambystomatid salamanders. *Journal of Morphology,* 185:297-326.

Lauder, G. V., and H. B. Shaffer. 1986. Functional design of the feeding mechanism in lower vertebrates: Unidirectional and bidirectional flow systems in the tiger salamander. *Zoological Journal of the Linnean Society of London*, 88:277-290.

Lauder, G. V., and H. B. Shaffer. 1988. Ontogeny of functional design in tiger salamanders (*Ambystoma tigrinum*): Are motor patterns conserved during major morphological transformations? *Journal of Morphology,* 197:249-268.

Lauder, G. V., and H. B. Shaffer. 1993. Design of feeding systems in aquatic vertebrates: Major patterns and their evolutionary interpretations. Pages

113-149 in: *The Skull, Volume 3: Functional and Evolutionary Mechanisms* (J. Hanken and B. K. Hall, eds.). Chicago: University of Chicago Press.

Liem, K. F. 1970. Comparative functional anatomy of the Nandidae (Pisces: Teleostei). *Fieldiana Zoology,* 56:1-166.

Liem, K. F. 1973. Evolutionary strategies and morphological innovations: cichlid pharyngeal jaws. *Systematic Zoology,* 22:425-441.

Liem, K. F. 1984. Functional versatility, speciation and niche overlap: are fishes different? Pages 269-305 in: *Trophic Interactions Within Aquatic Ecosystems* (D. G. Myers and J. R. Strickler, eds.). AAAS Selected Symposium No. 85, Washington, D.C.

Liem, K. F., and P. H. Greenwood. 1981. A functional approach to the phylogeny of pharyngognath teleosts. *American Zoologist,* 21:83-101.

Lombard, R. E., and D. B. Wake. 1977. Tongue evolution in the lungless salamanders, family Plethodontidae, II. Function and evolutionary diversity. *Journal of Morphology,* 153:39-80.

Maglia, A. M., and R. A. Pyles. 1995. Modulation of prey-capture behavior in *Plethodon cinereus* (Green) (Amphibia: Caudata). *Journal of Experimental Zoology,* 272:167-183.

Miller, B. T., and J. H. Larsen. 1990. Comparative kinematics of terrestrial prey capture in salamanders and newts (Amphibia: Urodela: Salamandridae). *Journal of Experimental Zoology,* 256:135-153.

Moss, S. A. 1977. Feeding mechanisms in sharks. *American Zoologist,* 17:355-364.

Motta, P. J., R. E. Hueter, and T. C. Tricas. 1991. An electromyographic analysis of the biting mechanism of the lemon shark, *Negaprion brevirostris*: Functional and evolutionary implications. *Journal of Morphology,* 210:55-69.

Nishikawa, K., C. W. Anderson, S. M. Deban, and J. O'Reilly, J. 1992. The evolution of neural circuits controlling feeding behavior in frogs. *Brain, Behavior, and Evolution,* 40:125-140.

Nishikawa, K., and D. C. Cannatella. 1991. Kinematics of prey capture in the tailed frog, *Ascaphus truei* (Anura: Ascaphidae). *Zoological Journal of the Linnean Society of London,* 103:289-307.

Nishikawa, K., and G. Roth. 1991. The mechanism of tongue protraction during prey capture in the frog, *Discoglossus pictus*. *Journal of Experimental Biology,* 159:217-234.

Nishikawa, K. C., and C. Gans. 1992. The role of hypoglossal sensory feedback during feeding in the marine toad, *Bufo marinus*. *Journal of Experimental Zoology,* 264:245-252.

Norton, S. F., and E. L. Brainerd. 1993. Convergence in the feeding mechanics of ecomorphologically similar species in the Centrarchidae and Cichlidae. *Journal of Experimental Biology,* 176:11-29.

Novacek, M. 1993. Patterns of diversity in the mammalian skull. Pages 438-545 in: *The Skull. Volume 2. Patterns of Structural and Systematic Diversity* (J. H. Hanken and B. K. Hall, eds.). Chicago: The University of Chicago Press.

Nussbaum, R. A. 1983. The evolution of a unique dual jaw-closing mechanism in caecilians (Amphibia: Gymnophiona) and its bearing on caecilian ancestry. *Journal of Zoology, London,* 199:545-554.

Nyberg, D. W. 1971. Prey capture in the largemouth bass. *American Midland Naturalist,* 86:128-144.

O'Reilly, J. C. 1990. Aquatic and terrestrial feeding in caecilians (Amphibia: Gymnophiona): a possible example of phylogenetic constraint. *American Zoologist,* 30:140A.

O'Reilly, J. C., and S. M. Deban. 1991. The evolution of aquatic prey capture in amphibians: phylogenetic constraints and exaptations. *American Zoologist,* 31:17A.

O'Reilly, S. R., and K. C. Nishikawa. 1995. Mechanism of tongue protraction during prey capture in the spadefoot toad *Spea multiplicata* (Anura: Pelobatidae). *Journal of Experimental Zoology,* 273:282-296.

Reilly, S. M. 1995. The ontogeny of aquatic feeding behavior in *Salamandra salamandra*: stereotypy and isometry in feeding kinematics. *Journal of Experimental Biology,* 198:701-708.

Reilly, S. M., and G. V. Lauder. 1988. Ontogeny of aquatic feeding performance in the eastern newt *Notophthalmus viridescens* (Salamandridae). *Copeia,* 1988:87-91.

Reilly, S. M., and G. V. Lauder. 1989. Kinetics of tongue projection in *Ambystoma tigrinum*: quantitative kinematics, muscle function, and evolutionary hypotheses. *Journal of Morphology,* 199:223-243.

Reilly, S. M., and G. V. Lauder. 1990a. The evolution of tetrapod prey transport behavior: kinematic homologies in feeding function. *Evolution,* 44:1542-1557.

Reilly, S. M., and G. V. Lauder. 1990b. The strike of the tiger salamander: quantitative electromyography and muscle function during prey capture. *Journal of Comparative Physiology*, 167:827-839.

Reilly, S. M., and G. V. Lauder. 1991a. Experimental morphology of the feeding mechanism in salamanders. *Journal of Morphology*, 210:33-44.

Reilly, S. M., and G. V. Lauder. 1991b. Prey transport in the tiger salamander: quantitative electromyography and muscle function in tetrapods. *Journal of Experimental Zoology*, 260:1-17.

Reilly, S. M., and G. V. Lauder. 1992. Morphology, behavior, and evolution: comparative kinematics of aquatic feeding in salamanders. *Brain, Behavior, and Evolution,* 40:182-196.

Ritter, D., and and K. Nishikawa. 1995. The kinematics and mechanism of prey capture in the African pig-nosed frog (*Hemisus marmoratum*): description of a radically divergent anuran tongue. *Journal of Experimental Biology,* 198:2025-2040.

Roth, G., K. Nishikawa, D. B. Wake, U. Dicke, and T. Matsushima. 1990. Mechanics and neuromorphology of feeding in amphibians. *Netherlands Journal of Zoology,* 40:115-135.

Russell, A. P., and J. J. Thomason. 1993. Mechanical analysis of the mammalian head skeleton. Pages 345-383 in: *The Skull. Volume 2. Patterns of Structural and Systematic Diversity* (J. H. Hanken and B. K. Hall, eds.). Chicago: University of Chicago Press.

Schwenk, K., and G. S. Throckmorton. 1989. Functional and evolutionary morphology of lingual feeding in squamate reptiles: phylogenetics and kinematics. *Journal of Zoology, London,* 219:153-175.

Schwenk, K., and D. B. Wake. 1993. Prey processing in *Leurognathus marmoratus* and the evolution of form and function in desmognathine salamanders (Plethodontidae). *Biological Journal of the the Linnean Society,* 49:141-162.

Shaffer, H. B., and G. V. Lauder. 1985. Aquatic prey capture in ambystomatid salamanders: patterns of variation in muscle activity. *Journal of Morphology,* 183:273-326.

Shaffer, H. B., and G. V. Lauder. 1988. The ontogeny of functional design: metamorphosis of feeding behavior in the tiger salamander (*Ambystoma tigrinum*). *Journal of Zoology, London,* 216:437-454.

Smith, K. K. 1984. The use of the tongue and hyoid apparatus during feeding in lizards (*Ctenosaura similis* and *Tupinambis nigropunctatus*). *Journal of Zoology, London,* 202:115-143.

Smith, K. K. 1986. Morphology and function of the tongue and hyoid apparatus in *Varanus* (Varanidae, Lacertilia). *Journal of Morphology,* 187:261-287.

Smith, K. K. 1988) Form and function of the tongue in agamid lizards with comments on its phylogenetic significance. *Journal of Morphology,* 196:157-171.

Smith, K. K. 1993. The form of the feeding apparatus in terrestrial vertebrates: studies of adaptation and constraint. Pages 150-196 in: *The Skull. Volume 3. Functional and Evolutionary Mechanisms* (J. H. Hanken and B. K. Hall, eds.). Chicago: The University of Chicago Press.

Smith, K. K., and W. L. Hylander. 1985. Strain gauge measurement of mesokinetic movement in the lizard *Varanus exanthematicus. Journal of Experimental Biology,* 114:53-70.

So, K.-K. J., P. C. Wainwright, and A. F. Bennett. 1992. Kinematics of prey processing in *Chamaeleo jacksonii*: conservation of function with morphological specialization. *Journal of Zoology, London,* 226:47-64.

Trueb, L., and R. Cloutier. 1991. A phylogenetic investigation of the inter- and intrarelationships of the Lissamphibia (Amphibia: Temnospondyli). Pages 223-313 in: *Origins of the Higher Groups of Tetrapods: Controversy and Consensus* (H.-P. Schultze and L. Trueb, eds.). Ithaca: Comstock Publishing Associates.

Trueb, L., and C. Gans. 1983. Feeding specializations of the Mexican burrowing toad, *Rhinophrynus dorsalis* (Anura: Rhinophrynidae). *Journal of Zoology, London,* 199:189-208.

Urbani, J.-M., and V. L. Bels. 1995. Feeding behavior in two scleroglossan lizards: *Lacerta viridis* (Lacertidae) and *Zonosaurus laticaudatus* (Cordylidae). *Journal of Zoology, London,* 236:265-290.

Wainwright, P., C. P. Sanford, S. M. Reilly, and G. V. Lauder. 1989. Evolution of motor patterns: aquatic feeding in salamanders and ray-finned fishes. *Brain, Behavior, and Evolution,* 34:329-341.

Wainwright, P. C. and A. F. Bennett. 1992. The mechanism of tongue projection in chameleons. I. Electromyographic tests of functional hypotheses. *Journal of Experimental Biology,* 168:1-21.

Wainwright, P. C., D. M. Kraklau, and A. F. Bennett. 1991. Kinematics of tongue projection in *Chamaeleo oustaleti*. *Journal of Experimental Biology,* 159:109-133.

Wassersug, R. J., and K. Hoff. 1982. Developmental changes in the orientation of the anuran jaw suspension. Pages 223-246 in: *Evolutionary Biology* (M. K. Hecht, B. Wallace, and G. T. Prance, eds.). New York: Plenum Press.

Weijs, W. A. 1994. Evolutionary approach of masticatory motor patterns in mammals. *Advances in Comparative and Environmental Physiology,* 18:281-320.

Westneat, M. W., and P. C. Wainwright. 1989. Feeding mechanism of *Epibulus insidiator* (Labridae: Teleostei): evolution of a novel functional system. *Journal of Morphology*, 202:129-150.

CHAPTER 7

AMNIOTE ORIGINS AND THE DISCOVERY OF HERBIVORY

Nicholas Hotton III

Everett C. Olson

Richard Beerbower

INTRODUCTION

The adoption of herbivory was a critical aspect of amniote differentiation and diversification not only for the herbivores but for predatory amniotes; it has also played critical role in the evolution of terrestrial plants and of other terrestrial consumers from bacteria and fungi to insects. The early history of tetrapod herbivory has long been a subject of inquiry, but knowledge is limited by the scarcity and equivocality of the fossil record and by dependence on indirect evidence. The Late Devonian-Early Carboniferous record is particularly scanty, and even the later Paleozoic chronicle is biased and far from complete. Further, we have no direct evidence about diet from either tetrapod gut contents or coprolites during this interval. Any inferences based on phylogenetic affinities with modern tetrapods require extreme extrapolations because Permo-Carboniferous tetrapods lack close modern relatives.

We propose a model for interpretation of diet from skeletal characteristics based primarily on biomechanical arguments but tested against recent studies of extant tetrapod herbivores; we also propose a general model for analysis of the critical abiotic and biotic factors affecting the evolution of tetrapod diets. To be sufficient, such models should cover all the critical processes, factors, and products involved in herbivory and should be "universal" in that they derive from the fundamental rules of mechanics, chemistry, and biological organization. Although most knowledge of these factors and processes derives from extant mammals (Chivers *et al.*, 1994; Van Soest, 1994), the same set of variables appear to determine the patterns of herbivory in other extant tetrapods (including birds, turtles, and lizards). Finally, in analysis and interpretation of our results we have mapped inferred diets and circumstances onto current interpretations of early tetrapod phylogeny and have applied comparative methods wherever appropriate.

Contrary to currently widespread views (e.g., DiMichele *et al.* 1992, p. 230), the consumption of plant (tracheophyte) tissues probably originated in the earliest Carboniferous (355-340 mybp), and was associated with synapomorphies established in the initial divergence and differentiation of the batrachosaur clade (sensu Laurin and Reisz, this volume), which includes the amniotes as a derived subclade. These transformations yielded animals that could augment primary diets of terrestrial invertebrates with exaptive consumption of plant material of of low-fiber content (readily digestible, nutrient-rich and cellulose-poor) . By the Mid-Carboniferous (320-310 mybp) a modest variety of amniotes and some representatives of other batrachosaur subclades probably had primary diets of low-fiber plant tissues–evidently derived exaptively and adaptively from omnivorous ancestors. Finally, before the end of the Late Carboniferous (~290 mybp), obligate high-fiber herbivores, dependent on microbial endosymbiotic fermentation of cellulose-rich, poor quality, plant tissues, had evolved independently in several amniote and other cotylosaurian subclades (*sensu* Laurin and Reisz, this volume)– seemingly exaptively and adaptively from low-fiber herbivores.

During this interval, herbivory moved up a "fiber curve," from consumption of relatively nutritious but rare low-fiber foodstuffs to one of less nutritious but much more abundant high-fiber material.

The evolution of herbivory in amniotes is also associated with transformations in circumstances, particularly by increases in: (1) the availability of small invertebrates, particularly insects, as prey; (2) the availability of plant items as nutritious as invertebrates; and (3) the availability and potential nutritional value of high-fiber plant tissues. These changes reflect the evolution of new kinds of plants and animals and the shifts in the characteristics and distribution of plant associations driven by climatic trends. The correlations between transformations in circumstances and evolution of herbivory not only generate a chronicle for the origins of amniote herbivory, but also provide a sufficient and persuasive explanation for the early evolution of amniote herbivory: from insectivory through omnivory and low-fiber herbivory to high-fiber herbivory with changes in both phylogenetic and ecologic opportunities and constraints. The independent evolution among several amniote clades of functionally comparable aptitudes for omnivory and herbivory in comparable circumstances supports such a causal interpretation as do the similarities and differences in dietary evolution and associated circumstances among the various out-groups. Finally, comparison of the fossil record with patterns of dietary and morphologic variation within extant analogs implies a gradualistic pattern at the species or even subspecific level.

DIAGNOSING TETRAPOD HERBIVORY

Requirements of Herbivory

Because extrinsic circumstances provide the context for interpretation of intrinsic functional capabilities, we begin with a consideration of the environmental factors most significant in herbivory. These obviously focus on food sources (Table 1, columns 1 and 2), but they also encompass other resources as well as "disturbance", that is, the sources of disability and death. In general, low-fiber components, such as buds, shoots, young leaves, and seeds are easily digested and nourishing. Most, however, are small and relatively rare, widely dispersed in space and time, and typically are protected by thin, tough or hard coverings. In contrast, mature leaves and stems are relatively large, very abundant, and uniformly and widely distributed. However, they are much less digestible, have relatively low nutritional value, and tend to be tough. Systematic variation of physical and chemical characteristics and of abundance and distribution within and among plant associations determines in large part the kinds of aptitudes necessary for utilization of potential food items and thus the principal modes of tetrapod herbivory (Hladik and Chivers, 1994; Langer and Chivers, 1994; Lucas, 1994; Chivers, *et al.*, 1994).

Other critical resources include water, heat sources and sinks, and refuges from disturbance; they are significant to herbivores not only because their availability affects acquisition of food and exposure to disturbance but also because they determine the characteristics of plant associations. The principal sources of disturbance subsume predation, desiccation, and thermal variation. They affect herbivores as they influence the places and times available for foraging and collection of food and the latter as well as the costs of regulation and replacement. For example, the abundance and diversity of reptilian omnivores and the proportion of plant tissue in their diets vary with circumstances (Brown and Pérez-Mellado, 1994). They are restricted primarily to tropical and subtropical climates presumably because high ambient temperatures are correlated with the high levels of physiological performance in food processing as well as with the availability of food and the minimization of thermal extremes. Omnivorous and herbivorous lizards are primarily terrestrial–even those that forage primarily in upper littoral sites–and are most abundant and diverse where predation by mammals is relatively low, such as, on small islands and in deserts. Further, the proportion of plant tissues consumed tends to vary inversely with the availability of suitable animal prey. Among extant tetrapods, relatively low metabolic and growth rates are associated with ecological success primarily where required costs of regulation and replacement are low relative to availability of resources (Zimmerman and Tracy, 1989).

Table 1. Patterns in Tetrapod Omnivory and Herbivory. ➔

1. TYPE OF DIET	2. CHARACTERISTICS OF FOOD	3. REQUIRED CAPABILITIES	4. SATISFYING REQUIREMENTS	5. DIAGNOSTIC SKELETON FEATURES	6. REFERENCES ON EXTANT ANALOGS
A. DUROPHAGOUS OMNIVORY HEAVILY SCLERITIZED INVERTEBRATES AND PLANT MATERIALS (SEEDS, ETC.)	Readily digestible, high value Small size; quite hard but brittle; resist collection; strongly resist fracturing . Mobility low or nil but rare and scattered	Digestion of starches as well as fats, proteins, chitin . Hold-and-crush processing Seize-and-squeeze collection Search over wide area.	Amylase plus lipase, proteinase, chitinase Maximum force near occlusion; fracture propagation critical. Modest mobility.	Infer from extant clades. Narrow, S-P jaws; near-isodont (possibly short fangs); teeth bluntly ogival, gouged/ pitted; Limbs longish; body relative short; skull small.	Dalrymple, 1979; Riepple and Labhardt, 1979; Greene, 1982; Auffenberg, 1988, and Estes and Williams, 1984.
B. NON-DUROPHAGOUS OMNIVORY LIGHTLY SCLERITIZED INVERTEBRATES and PLANT MATERIALS	Readily digestible, high value Small size; moderately hard, brittle; avoid and/or resist collection; some resistence to fracturing Mobility considerable to nil; moderately rare and scattered	Digestion of starches as well as fats, proteins, chitin . Hold/crush processing. Seize/squeeze collection. Search over wide area.	Amylase plus lipase, proteinase, chitinase Maximum force near occlusion; fracturing critical. Considerable mobility.	Infer from extent clades. Narrow, S-P jaws; near-isodont dentition (possibly short fangs); teeth ogival, unworn, pitted or faceted. Limbs long; body short; skull small.	Hotton, 1955; Greene, 1982; Burquez, et al., 1986; Rocha, 1989; Castilla, et al., 1991; Nunez, et al., 1992; Van Sluys, 1993; Schall and Resssell, 1991; Paulissen and Walker, 1994; Valido and Nogales, 1994.
C. LOW-FIBER HERBIVORY LIGHTLY SCLERITIZED PLANT MATERIALS (SEEDS, ETC.)	Readily digestible, high value Small size, moderately hard, brittle; resist shearing forces during collection and fracturing during mechanical processing Mobility nil but moderately rare and scattered	Digestion of starches as well as fats, proteins Hold/crush processing. Seize/squeeze collection Search over wide area	Amylase plus lipase, proteinase Maximum force near occlusion; fracturing critical. Considerable mobility	Infer from extent clades. Narrow, S-P jaws; near-isodont dentition (rarely fangs); teeth ogival to blades; unworn or faceted, Moderately long legs; short body; skull small.	Hotton, 1955; Schall and Resssell, 1991; Dearing and Schall, 1993;
D. HIGH-FIBER HERBIVORY: LIGHTLY SCLERITIZED BUT TOUGH, FIBROUS PLANT MATERIALS (MATURE LEAVES AND SHOOTS	High in indigestible cellulose, low in digestible energy and protein and enclosed by refractory layers of cellulose or cutin; include digestive inhibitors and toxins. Typically small to moderately large; soft but tough; resist shearing and fracturing forces during collection and processing. Mobility nil but patchy distribution in spite of relatively great abundance.	Digestion of starches plus fats, proteins. Fermentation of cellulose and deactivation of inhibitors and toxins Hold/crush processing. Seize/squeeze collection Search over considerable area	Amylase plus lipase, proteinase. Endosymbiotic cellulysis and breakdown of inhibitors and toxins Maximum force near occlusion; fracturing critical. At least modest	Infer from extant clades Typically large with expanded torso as mark of endosymbiosis. S-P jaws; typically anisodont with chisel-form cropping mortar and pestle crushing teeth; smooth or micro-striated wear facets. Moderately long legs; body short; skull small.	Hotton, 1955; Pough, 1973; Warren and Lee, 1973; Nagy and Shoemaker, 1975; Greer, 1976; Skoczylas, 1978; Rand, 1978; Bjorndal, 1979; Auffenberg, 1982; Case, 1982; Guard, 1980; Greene, 1982; Iverson, 1982; McBee and McBee, 1982; Nagy, 1982; Van Devender, 1982; Wiewandt, 1982; Christian, et al., 1984; Estes and Williams, 1984; Troyer, 1984a, 1984b; Smits, 1985; Mautz and Nagy, 1987; Zimmerman and Tracy, 1989; Bjorndal, et al., 1990; Rand, et al., 1990; Foley, et al., 1992; Bjorndal and Bolton, 1992, 1993.
E. REFERENCES CHARACTERISTICS OF FOOD, REQUIRED CAPABILITIES, SATISFYING REQUIREMENTS, DIAGNOSTIC SKELETAL	Janzen, 1973; Bennett and Gorman, 1979; Brown and Perez-Mellado, 1994; Lucas, 1994; Vincent, 1990, 1991; Bjorndahl et al., 1990; Moir, 1994; Demment and Van Soest, 1983, 1985; Hladik and Chivers, 1994; Langer and Chivers, 1994; Chivers et al., 1994; Van Soest, 1994; Chivers, Langer, et al. 1994.	**Tooth Wear:** Walker, et al., 1978; Lucas and Corlett, 1991. **Tooth Form and Mechanics:** Lucas, 1994; Lucas and Corlett, 1991; Lucas and Luke, 1984; Vincent, 1990, 1991; Chivers and Langer, 1994; Bjorndahl, et al., 1990; Chivers, Langer, et al. 1994. **Gut Volume and Structure:** Parra, 1978; Demment and Van Soest, 1983, 1985; Alexander, 1991; Penry And Jumars, 1987; Chivers And Langer, 1994; Van Soest, 1994; Chivers, Langer, et al. 1994. Chivers, Langer, et al. 1994. **Production of Enzymes**: Stevens, 1988; Van Soest, 1994; Chivers, Langer, et al. 1994. **Endosymbiotic Microbiota of Gut:** Parra, 1978; Bauchop, 1977; Savage, 1977; Clarke, 1977; Prins, 1977; McBee, 1977; McBee and McBee, 1983; Van Soest, 1994; Chivers, Langer, et al. 1994. **Bite Mechanics:** Sokol, 1967; Olson, 1961b; Riepple and Labhardt, 1979; Chivers, Langer, et al. 1994.. **Size**: Pough, 1973, and Wilson and Lee, 1974; Van Soest, 1994; Chivers, Langer, et al. 1994.			

Establishing Aptitudes and Behavior

The presence of aptitudes for utilization of plants as nutritional sources suggests which fossil tetrapods could have been facultative herbivores and which, obligate herbivores of low- or high-fiber persuasion. Measurement of these aptitudes relies on biomechanical analyses of skeletal form and on analogies with modern animals–particularly those most nearly comparable in terms of size and physiological requirements and, ideally, from a variety of clades, both sister and out-group (Reif, 1982). Biomechanical analyses derive from consideration of the mechanical and chemical factors and processes operating in extant animals independent of specific ecological as well as genealogic affinities (Niklas, 1992; Reif, 1982; Signor, 1982).

The essential capabilities for tetrapod herbivory include not only those involved in chemical and mechanical processing of plant tissues but also those related to collection and to search and discovery (Table 1). Discovery is a function in large part of mobility (and agility) as they delimit the search for food (foraging) and thus the likelihood of encountering food items for a particular expenditure of time and energy. Mobility is reflected in body dimensions, particularly in overall size and limb lengths and proportions that suggest at one extreme relatively wide foraging and at the other localized foraging. Collection also depends on mobility and agility in the capture of a particular item with a range from "sit-and-wait" to pursuit predation as well as on the mechanical characteristics of the skull, jaws and teeth as they determine bite sizes, rates, and forces for collection and mechanical processing. Relatively wide, lightly built skulls and jaws with kinetic-inertial jaw mechanics that produce maximal bite forces at full gape (Olson, 1961b) facilitate grab-and-gulp feeding. The kinetic-inertial jaw mechanism involves very rapid jaw closure and food ingestion but supplies little force for collection and mechanical processing. Conversely, relatively narrow, strongly built skulls and jaws with static-pressure mechanics suggest seize-and-squeeze feeding in which the maximal bite forces are exerted near occlusion (Olson, 1961b) to provide a powerful bite for collection and mechanical processing. Grab-and-gulp feeding facilitates the capture and rapid ingestion of relatively small, nonresisting items; seize-and-squeeze, the capture of relatively large resistant ones and the detachment of pieces for mechanical processing and ingestion.

Biomechanically, acutely conical teeth serve well for piercing and holding relative soft material; chisel-edged ones, for shearing and fracturing hard and/or tough tissues, particularly in mortar and pestle combinations. Predictably, acutely conical teeth would best serve grab-and-gulp feeding patterns; the inclusion of relatively large fangs would contribute to capture of relatively large, slippery, and/or strong prey. In contrast, near-isodont dentitions of chisels or blades with a modest admixture of other kinds of teeth would best serve seize-and-squeeze feeders; the optimum form for the latter would vary with diet ranging from caniniform in specialized carnivores to incisiform and mortar and pestle-like in specialized herbivores. Finally, differences in the kinds of food collected and modes of collection and mechanical processing should leave their marks in tooth wear: the fracturing of very hard items chips, pits, and scars tooth surfaces; the shearing of tough, fibrous ones between adjacent teeth grinds facets cross-cutting enamel and dentine, surface texture of the facets reflecting both the characteristics of the food and the patterns of jaw movement.

The principal factors in chemical processing of plant materials are (1) the mechanical preparation for such processing, (2) the enzymes that assist digestion, (3) the endosymbiotic fermentation of plant cellulose, (4) the temperature, and (5) the duration of the process. Skull, jaw and tooth dimensions and jaw mechanics register these capacities; static-pressure bites and wear-faceted mortar-and-pestle dentitions point toward extensive fracturing and shearing and thus a primary diet of cellulose-rich plant tissues.

The enzymes critical for effective and efficient herbivory include amylases necessary for digestion of starch and other soluble plant carbohydrates; the presence (or absence) of such enzymes in fossil taxa has to be inferred from their distribution among extent sister clades and outgroups. Similarly, effective and efficient utilization of cellulose-rich (high-fiber) plant tissues depends on endosymbiotic cellulysis in practically all extant animals, vertebrate and invertebrate. As with enzymatic production, the distribution of this process among extent sister clades and out-groups provides a partial measure of its likelihood in extinct taxa. Body proportions provide a more direct measure because they reflect original gut volume–as cellulysis is very slow process it is typically associated with a relatively voluminous gut and thus relatively large body sizes and disportionally bulky torsos. (Specialized processing systems in some tetrapods, for example rodents, minimize those requirements but do not eliminate them.) Additionally, body mass affects cellulysis as far as it provides a disproportionaly voluminous gut that allows retention of the digesta for long periods and as it favors a relatively high, constant temperature (inertial endothermy) that increases the rate of microbial fermentation.

Modesto (1992) has argued that amniotic reproductive modes are necessary for transmission of cellulytic endosymbionts to neonates and therefore for the evolution of high-fiber herbivory. (Lee and Spencer (this volume) employ the converse of this argument in referring the Diadectomorpha to the Amniota because of the evidence for high-fiber herbivory with endosymbiotic fermentation.) Although we agree that this could be contributing factor, endosymbiotic cellulysis is known in marine fish, and some endosymbionts–though apparently not cellulytic ones–occur in frogs (Stevens and Hume, 1995). Moreover, there seems no biomechanical necessity of such a limit as many lissamphibian larvae and juveniles feed at sites where adults defecate (Van Soest, 1994), and the neonates and juveniles of high-fiber herbivores among lizards can survive and grow on an exclusive diet of insects and low-fiber plant tissues (Pough, 1973).

To the extent that the morphologies of early amniotes (and their inferred physiologies) parallel those of extant lizards (Carroll, 1991) and turtles, these groups can provide particularly useful analogs for analysis of early tetrapod herbivory. In addition, examples of herbivory among modern amphibian larvae and actinopterygian fish as well as extant mamals and birds permit out-group comparisons for analysis of early amniote herbivory (Stevens and Hume, 1995; Duellman and Trueb, 1986; Van Soest, 1994). A survey of current information on feeding patterns in lizards and turtles suggests four end members in the consumption of plant tissues: (1) durophagous omnivory, (2) non-durophagous omnivory, (3) low-fiber herbivory and (4) high-fiber herbivory (Table 1, Column 1). In general, the skeletal morphology of the taxa in these categories is consistent with that predicted from biomechanical analysis as shown in the table and with the dietary patterns observed in other extant vertebrates (both sister amniote and lissamphibian and piscine out-groups (e.g., Stevens and Hume, 1995). The diagnostic characteristics cited in Table 1 provide a reasonable guide for analysis of diet in fossil tetrapods given their consistency with biomechanical requirements and observation of extent tetrapods, but such diagnoses are of likelihood rather than certainty.

Among modern amniotes in each category, the level of specialization for herbivory appears correlated with minimum levels of aptitude necessary to satisfy nutritional requirements given resources and disturbance (Greene, 1982), and thus varies considerably with circumstances. Levels of specialization are low where the animals in question have relatively low physiological requirements or consume a low proportion of plant tissues. They also tend to be low where nutrients are readily accessible from plant tissues and where exposure to disturbance is relatively low. In particular, the aptitudes for omnivory and low-fiber herbivory correspond so closely to those for predation that capabilities for one are sufficient for the other given otherwise favorable circumstances. As discussed earlier, reptilian omnivores vary in abundance and diversity with circumstances, as does the proportion of plant tissue in their diets (Brown and Pérez-Mellado, 1994). In consequence, extant tetrapods exercising the first option are difficult to distinguish morphologically from pure molluscivores, and those exploiting the second, from pure insectivores. Thus, although early amniotes are commonly interpreted as insectivores because of their similarities to small, extant, ostensibly insectivorous lizards, some lizards indistinguishable from confirmed insectivores in dentition and osteology are nevertheless partly or fully herbivorous (Greene, 1982).

In contrast, all reptilian high-fiber herbivores manifest comparable, distinctive characteristics associated with utilization of tough, cellulose-rich leaves and shoots, but some difficulties appear in evaluation of tooth form and wear. The diagnosis of herbivory in lizards has been based traditionally on complex patterns of dental crown sculpture, and Pough (1973) and Wilson and Lee (1974) have argued that herbivory among reptiles must be limited to relatively large animals. However, recent observations demonstrate that some extant reptiles attain a considerable degree of herbivory without having evolved either complex dental crowns or large size (e.g., Greene, 1982). Although they all have closely spaced isodont teeth with laterally compressed blades, none has capacities for microcomminution of food comparable to those provided by the closely occluding, mortar-and-pestle teeth of mammals. In addition, none of their teeth show any significant wear despite consumption of abrasive material. But, given their relatively low nutrition requirements, lizards do not require the high rates of chemical processing possible with micro-comminution; they simply retain the ingested fiber for a longer period to maximize yield. In contrast, the much greater nutritional requirements typical of mammals and birds presumably demand micro-communition to ensure high rates of fermentation. Wear on lizard teeth is presumably minimal because (1) their total food intake is relatively low relative to the rate of tooth replacement and (2) their teeth need not contact each other in occlusion. Thus, the absence of mammal-like molar teeth and wear patterns does not preclude high-fiber herbivory, but, from the biomechanical point of view, their presence in ancient tetrapds was almost certainly associated with a high-fiber diet–particularly where linked with the capacities for endosymbiotic cellulysis associated with relatively bulky, barrel-shaped torsos and large adult sizes.

DIETS INFERRED

Phylogenetic Infrastructure

Exploration of evolution of omnivory and herbivory among early amniotes requires consideration of phylogenetic affinities and the variation in nutritional requirements within the clade and among the several tetrapod out-groups (Figure 1). Recent analyses agree that the amniote clade encompasses two subclades, Synapsida and Reptilia (e.g., Gauthier, 1994; Laurin and Reisz, 1995; Lee and Spencer, this volume). The synapsids comprise seven "pelycosaurian-grade" families: the Eothyrididae, Caseidae, Varanopsidae, Ophiacodontidae,

---

Figure 1. Phylogeny and dietary styles; phylogeny based on Laurin and Reisz (1995), Laurin and Reisz (this volume), Carrroll (1995), and Hopson (1994). ➔

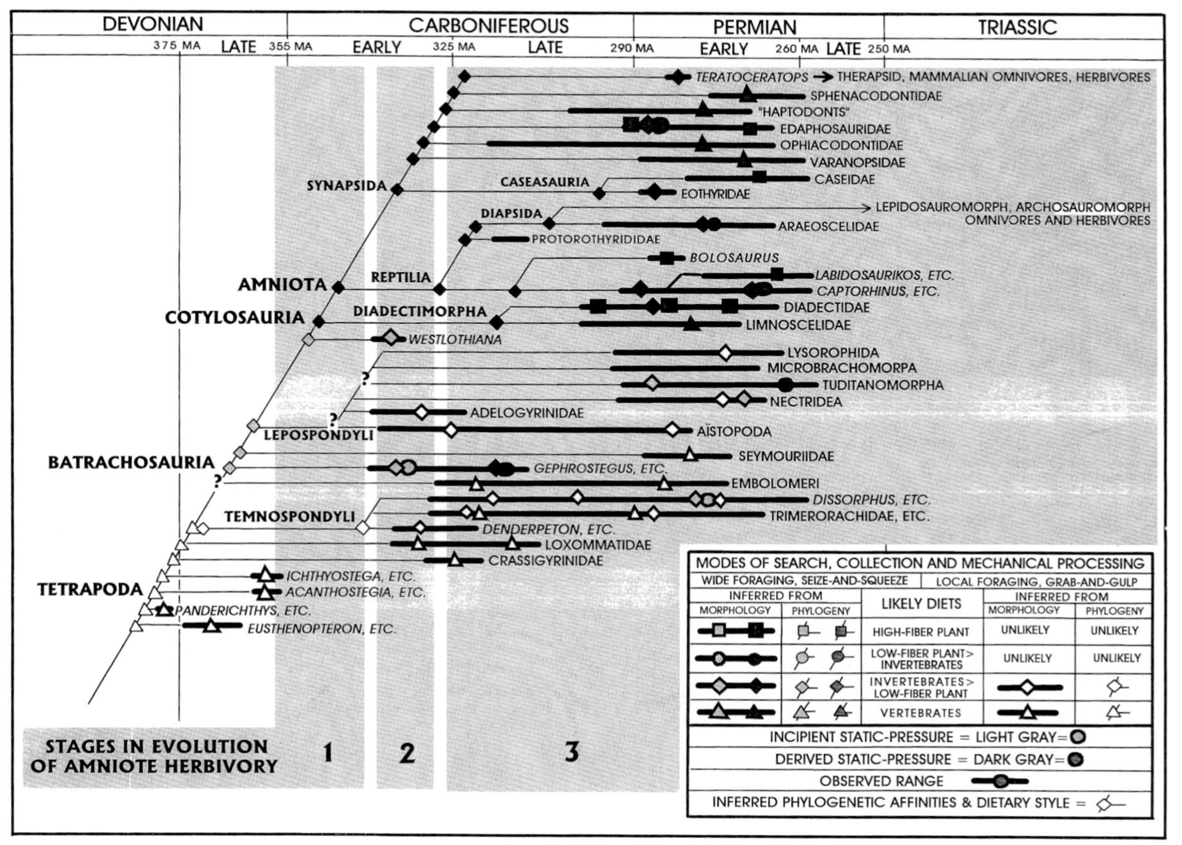

DEVONIAN
CARBONIFEROUS
PERMIAN
TRIASSIC
375 MA
LATE
355 MA
EARLY
325 MA
LATE
290 MA
EARLY
260 MA
LATE
250 MA
TERATOCERATOPS → THERAPSID, MAMMALIAN OMNIVORES, HERBIVORES
SPHENACODONTIDAE
"HAPTODONTS"
EDAPHOSAURIDAE
OPHIACODONTIDAE
VARANOPSIDAE
CASEIDAE
SYNAPSIDA
CASEASAURIA
EOTHYRIDAE
DIAPSIDA
→ LEPIDOSAUROMORPH, ARCHOSAUROMORPH OMNIVORES AND HERBIVORES
ARAEOSCELIDAE
PROTOROTHYRIDIDAE
BOLOSAURUS
LABIDOSAURIKOS, ETC.
AMNIOTA
REPTILIA
CAPTORHINUS, ETC.
DIADECTIMORPHA
DIADECTIDAE
COTYLOSAURIA
LIMNOSCELIDAE
WESTLOTHIANA
LYSOROPHIDA
MICROBRACHOMORPHA
TUDITANOMORPHA
NECTRIDEA
ADELOGYRINIDAE
LEPOSPONDYLI
AÏSTOPODA
BATRACHOSAURIA
SEYMOURIIDAE
GEPHROSTEGUS, ETC.
EMBOLOMERI
DISSORPHUS, ETC.
TEMNOSPONDYLI
TRIMERORACHIDAE, ETC.
DENDERPETON, ETC.
LOXOMMATIDAE
CRASSIGYRINIDAE
ICHTHYOSTEGA, ETC.
TETRAPODA
ACANTHOSTEGIA, ETC.
PANDERICHTHYS, ETC.
EUSTHENOPTERON, ETC.
STAGES IN EVOLUTION OF AMNIOTE HERBIVORY
1
2
3
MODES OF SEARCH, COLLECTION AND MECHANICAL PROCESSING
WIDE FORAGING, SEIZE-AND-SQUEEZE
LOCAL FORAGING, GRAB-AND-GULP
INFERRED FROM
MORPHOLOGY
PHYLOGENY
LIKELY DIETS
HIGH-FIBER PLANT
LOW-FIBER PLANT> INVERTEBRATES
INVERTEBRATES> LOW-FIBER PLANT
VERTEBRATES
UNLIKELY
INCIPIENT STATIC-PRESSURE = LIGHT GRAY=
DERIVED STATIC-PRESSURE = DARK GRAY=
OBSERVED RANGE
INFERRED PHYLOGENETIC AFFINITIES & DIETARY STYLE =

Edaphosauridae, "Haptodontidae" (a polyphyletic and paraphyletic cluster of taxa), and Sphenacodontidae (Hopson, 1994). The early reptiles encompass two subclades: the Captorhinidae and probably the Bolosauridae in one and the Protorothyrididae and Araeoscelidae in the other (Gauthier, 1994, Laurin and Reisz, 1995). Most authors view the diadectomorphs (including the Limnoscelidae and Diadectidae) as the immediate sister group of amniotes, forming with them the Cotylosauria (Gauthier, 1994; Carroll, 1995; Laurin and Reisz, this volume) although some authors (e.g., Berman *et al.*, 1992; Lee and Spencer, this volume) argue for their placement within the Amniota. One step further out, is an Early Carboniferous clade represented by *Westlothiana* (Carroll, 1995).

Beyond this point the relationships are less clear. Cladistic analyses by Carroll (1995) and Laurin and Reisz (this volume) place a monphyletic lepospondylian clade (subsuming the Aistopoda, Adelogyrinidae, Nectridea, Tuditanomorpha, Microbrachomorpha, Lysorophia, and Lissamphibia) as the cotylosaur sister group. Others have argued, however, for the polyphyletic attachment of these groups to the temnospondyles as outlined below. The seymouriamorphs share sufficient synapomorphies with cotylosaurs to justify their co-assignment (possibly along with lepospondyls) to a common clade, the Batrachosauria, encompassing this whole portion of the tetrapod evolutionary tree (Gauthier, 1994; Carroll, 1995; Laurin and Reisz, this volume). Finally, the various animals traditionally designated as "anthracosaurs" appear to have their closest affinities with the batrachosaurs possibly as a monophyletic sister clade (as suggested by both Gauthier and Carroll) or in a polyphyletic arrangement with one of the two well-know "anthracosaurian" groups, the gephyrostegids, included in the Batrachosauria leaving the embolomeres, as the proximal out-group (as argued by Laurin and Reisz and illustrated in Fig. 1).

The immediate sister clade of the batrachosaurs appears to be the Temnospondyli (Ahlberg and Milner, 1994; Carroll, 1995; Laurin and Reisz, this volume), a node bracketing a considerable number and variety of subclades ranging from the relatively plesiomorphic (e.g., the Dendrepetontidae) through the modestly derived (e.g. the Eryopidae), to the highly derived Dissorophidae (Milner, 1993). Some authors have also argued for the attachment of some of the "lepospondyl" groups to the temnospondyls: the Lissamphibia to the dissorophids (Milner, 1993) and the microsaurs and nectridians as successive out-groups on the temnospondyl branch (Ahlberg and Milner, 1994). Finally, the base of the tetrapod clade encompasses a heterogeneous, largely unresolved cluster of disparate, fish-like "proto-tetrapods" (the best known, *Ichthyostega* and *Acanthostega*) with panderichtyian fish as the closest out-group (Ahlberg and Milner, 1994; Carroll, 1995; Laurin and Reisz, this volume).

Analysis of dietary patterns in amniotes and other tetrapods also requires reconstruction of fundamental, shared nutritional requirements relevant to adoption of omnivory and herbivory. So far as requirements are concerned, all early tetrapods appear to have shared with living lizards and turtles a physiology that minimized dietary needs. Their levels of metabolism were low, growth rates low and episodic, and capabilities for thermoregulation limited, as indicated by lamellar-zonal patterns of primary bone growth (Peabody, 1961; Enlow, 1969; Ricqlés, 1975, 1976, 1978; Chinsamy, 1993; Chinsamy and Dodson, 1995). In addition, their sprawling limb posture and intervertebral articulations indicate laterally sinuous terrestrial locomotion, which would have restricted lung ventilation and thus metabolic performance (Carrier, 1987). Shortness of the limbs indicates low enough running speeds that even a sustained maximum could have been supported by relatively low rates of aerobic metabolism. Finally, their relatively small brains would have required correspondingly low levels of oxygen transport and would have functioned adequately with low levels of blood circulation and aerobic metabolism. Thus, they might suffice their caloric and other nutritional demands from consumption of plant tissues with relatively limited aptitudes for herbivory–as exemplified in extant tetrapods with relatively low metabolic rates, particularly where required costs of regulation and replacement are low relative to availability of resources (Zimmerman and Tracy, 1989).

Early tetrapods seem to have derived from predatory ancestors (Carroll, 1995) and thus certainly carried the ontogenetic patterns necessary for development of piercing teeth and for synthesis of the proteinases and lipases utilized in digestion of animal tissues. Such teeth could also be utilized, albeit rather inefficiently, in collection and mechanical processing of plant tissues, and such enzymes, for digestion of the proteins and fats in plant tissues. However, any tetrapod that was going to prosper primarily on plant food must have been capable of producing enzymes for digestion of carbohydrates. As at least some members of all extant tetrapod clades produce carbohydrases (Stevens and Hume, 1995), their common Late Devonian or Early Carboniferous ancestor had at least the evolutionary potential for their synthesis. Conversely, because only amniotes among extant tetrapods are capable of endosymbiotic cellulysis, that ability cannot be taken as a universal tetrapod capacity.

CANDIDATES FOR OMNIVORY

Amniotes

Among the early amniotes, sundry captorhinids, protorothyridids, araeoscelids, and primitive ophiacodonts and the edaphosaurians are likeliest candidates for omnivory, consuming nutrient-rich plant materials along with invertebrates (Fig. 1). Some of their shared characteristics, such as relatively short coupled bodies, robust limbs, and big feet, indicate a primitive emphasis on wide foraging in terrestrial habitats. Their jaw mechanics embody a derived form of the static-pressure bite marked by the addition of the pterygoid flanges and associated musculature. This arrangement added a strong kinetic-inertial "snap" that facilitated capture of small, agile terrestrial prey such as insects and small tetrapods to the powerful "squeeze" from the temporal adductors that facilitated the seizure, fracturing, and shearing of tough items such as indurated arthropod skeletons, seeds and other low-fiber plant materials). Their teeth are conical, ogival, or slightly cuspate, near-isodont (although four to six marginal teeth in the front half of the tooth row may be slightly taller than the rest), and commonly lack evidence of wear. In their jaw articulation, the glenoid is not appreciably longer than the quadrate condyle and fits the condyle fairly closely, imparting a measure of stability to the joint and thus constraining lateral jaw slippage during collection and mechanical processing of tough items. Jaw motion was primarily orthal, although a measure of fore-and-aft motion is possible. These aptitudes are comparable to those in extant insectivorous and omnivorous lizards (Table 1).

Feeding patterns apparently varied considerably among the members of this group. The relatively short legs of the protorothyridids, captorhinids, and synapsids suggest limited running speeds and a combination of slow-stalking and lurk-and-lunge tactics; the longer limbs and gracile bodies of araeoscelids show that they could have foraged more widely and operated as pursuit predators. The jaw adductor chambers of protorothyridids and araeoscelids are of modest size implying relatively moderate bite forces at occlusion. This limitation coupled with the presence of short fangs and a considerable battery of sharp albeit small palatal teeth suggest a rather generalized diet focused on some mix of small, lightly scleritized and relatively brittle invertebrates, seeds, buds, and shoots–perhaps with an emphasis on insects rather than plants. This interpretation is supported by the absence of scarification or wear facets on the cheek teeth. The captorhinids and synapsids diverge from this pattern in the greater size of the chambers housing the medial and posterior adductor jaw muscles and in presence of chisel-like teeth bearing wear marks. The former indicates a corresponding enlargement of the adductor muscles indicating that the static-pressure bite was very powerful; the latter, utilization of harder or more abrasive items.

Variation in body size and dental charcteristics among captorhinids apparently maps variation in diets within and among species comparable to that observed in some clades of modern lizards. The primitive genera (*Romeria*, *Protocaptorhinus*, *Labidosaurus*, and *Rhiodenticulatus*) are relatively small and their dentitions have a single row of maxillary and dentary teeth (Heaton, 1979; Gaffney and McKenna, 1979). Typically, the cheek teeth are smaller than the front teeth, all much of a similar size, and ogival with simple cones in *Romeria* and *Protocaptorhinus* and pinched crowns in *Labidosaurus*. Tooth-to-tooth wear on crowns of the simple, conical teeth of *Romeria* generated a series of narrow, longitudinal blades, effectively chisels, all along the tooth row (Heaton, 1979), and the pinched form of the cones in *Labidosaurus* provides a comparable row of chisel-like cheek teeth accentuated by wear facets. Such dental patterns are biomechanically apt for fracturing and shredding of relatively tough insect exoskeletons and plant tissues and are characteristic of smaller extant iguanid lizards, a group that includes a number of omnivores and facultative low-fiber herbivores.

Other early but more specialized members of the captorhinid clade couple somewhat larger size with further enlargement of the jaw adductors and modifications in dentition. The premaxillary teeth of *Labidosaurus* and *Captorhinus* are relatively large and narrowly spatulate, and the premaxilla is recurved so that the upper teeth are recumbent, behind and above which the anterior dentary teeth, also large, are procumbent. The front teeth, upper and lower, are worn where they occlude with one another; the wear facets bevel enamel and dentine smoothly and microstriated parallel to the long axis of the teeth (Figure 2B.1 in contrast in wear pattern with carnivorous sphenacodont in Figure 2A). The front of the first and largest premaxillary tooth also bears a facet, its striations paralleling the long axis of the tooth, typically worn through the enamel and into the dentine. This facet is peculiar in the apparent lack of occlusion with other teeth that might produce such wear and presumably resulted from contact with abrasive materials during the collection of food, for example, by probing into litter and soil, galleried wood, or plant strobili and cones.

The cheek teeth in both genera have laterally pinched crowns and thus approaching occlusion present opposing parallel rows of chisels at occlusion. *Labidosaurus* has one such row on each side of the upper and lower jaws and *Captorhinus* has one to three (Gaffney and McKenna, 1979). Alhough wear of any kind on cheek teeth of *Captorhinus* is most uncommon (observed in only a few of the 30 or so dentaries and maxillae of *Captorhinus* examined from the Dolese Brothers quarry at Fort Sill, Oklahoma), the pattern of wear, where observed, is more varied than in less specialized forms. Typically, it consists mostly of pitting that blunts rather than sharpens the ogival crowns (Figure 2B.2) and presumably reflects the contact of one tooth with another and with very hard items as highly indurated invertebrates or seeds were crushed. In one example however (Figure 2B.3), the cheek teeth bear striated facets of the kind produced by attrition as one tooth slides over another during fore-and-aft motion of jaw while processing fibrous, grit-covered plant materials. These trends in form and diet are of special interest as they anticipate the aptitudes discussed below for high-fiber herbivory exhibited by still more specialized captorhinids from the uppermost Lower Permian and Upper Permian assemblages.

The synapsids in this group of potential omnivores (or low-fiber herbivores) are also of special interest because of their genealogical connection with likely high-fiber herbivores. The Early Permian eothyrids, the most primitive known synapsids (and the sister clade of the caseids), are small animals with dentitions dominated by a relatively uniform series of relatively large conical teeth, of two to four pairs of anteriad, marginal fangs (Langston, 1965). In addition, the relatively primitive, mid-Carboniferous ophiacodont, *Archaeothyris*, presents a slightly modified version of this pattern (Reisz, 1972, 1975, 1986). Alhough Reisz argues these features suggest an emphasis on predation, they could well have consumed at least a modest portion of plant materials perhaps anticipating of the adoption of high-fiber herbivory in their sister clades as discussed below. Among primitive edaphosaurids, the Late Carboniferous (Stephanian) *Ianthosaurus* has a near-isodont dentition of slightly compressed acutely conical, slightly recurved teeth; the mid-Early Permian (Sakmarian) *Glaucosaurus*, has a fully isodont dentition of ogival, laterally compressed, sharp-edged teeth. The characteristics of the first are compatible with insectivory as Reisz and Berman (1986) suggest but are equally consonant with omnivory; those of the second suggest a diet of hard-bodied arthropods to Modesto (1994) but appear to us more compatible with omnivory or low-fiber herbivory as a precedent to the high-fiber herbivory inferred below for the more derived *Edaphosaurus*.

The dietary modes of other early ("pelycosaurian-grade") synapsids are also relevant to analysis and interpretation of early tetrapod herbivory. They are clearly variations on a single theme–a synapsid version of the wide foraging, seize-and-squeeze pattern–but fall into two distinct groups clearly related with differences in diet. The variations in one set focus primarily on aspects of functional morphology related to discovery, collection, and mechanical and chemical processing of tough, fibrous, cellulose-rich plant tissues; they were almost certainly primarily high-fiber herbivores utilizing somewhat different kinds of such tissues in somewhat different ways. Those in the other set are just as clearly related to discovery, collection, and processing of animals though they range from a relatively primitive emphasis on insectivory to highly derived ones for carnivory (for examples see interpretations in Romer and Price, 1940; Reisz, 1986).

Out-group Dietary Patterns

Potential aptitudes for omnivory and even high-fiber herbivory also appear in diadectomorphs on the other branch of the cotylosaur clade. They share with the amniotes relatively small heads and short-coupled bodies mounted on robust limbs demonstrating capacities for wide foraging on land. They also share features indicating static-pressure jaw mechanics bite including a transverse process on the pterygoid although the latter has a longitudinal rather than transverse orientation. Thus, they were well fitted for wide overland foraging and for seize-and-squeeze collection and mechanical processing. The early Permian diadectomorph *Tseajaia* (Moss,1972), is comparable in size and dental characteristics to the eothyrid synapsids described previously and like them is best interpreted primarily as a predator on invertebrates and very small vertebrates but with significant capacities for collection and processing of low-fiber plant tissues. These capacities for the use of plant tissues anticipate those of the more

Figure 2. Tooth wear in sphenacododontid pelycosaurs and captorhinid reptiles. (A) sphenacodont, cf. *Dimetrodon*, apparent wear facet on cheek tooth; note the fine striations oriented more or less vertically on enamel surface at lower right, and lack of striations on exposed dentine at upper left. Also note ragged edge of enamel. (B.1) *Captorhinus aguti*, wear facet on anterior aspect of premaxillary tooth; worn surface did not come into contact with other teeth. (B.2) *Captorhinus aguti*, ogival cheek teeth with crowns irregularly pitted and gouged as though by pounding of scleritized material. (B.3) *Captorhinus aguti*, ogival cheek teeth with transverse wear facets comparable to those of extant herbivorous mammals; microstriations indicate propalinal motion, and wear facets have sharp edges. (C.1) *Labidosaurikos*, ogival cheek tooth with pitted wear facet on crown and flank of tooth. (C.2) *Labidosaurikos*, ogival cheek tooth with short microstriae, preferred orientation parallel to long axis of jaw. Note enamel rim faired into dentary surface. (D.1) *Rothianiscus*, cheek tooth of worn flat, original shape not determined, pitted but smoothed after pitting. (D.2) Same as D.1. but higher magnification, heavily pitted but pits subdued, surface very smooth and polished. ➔

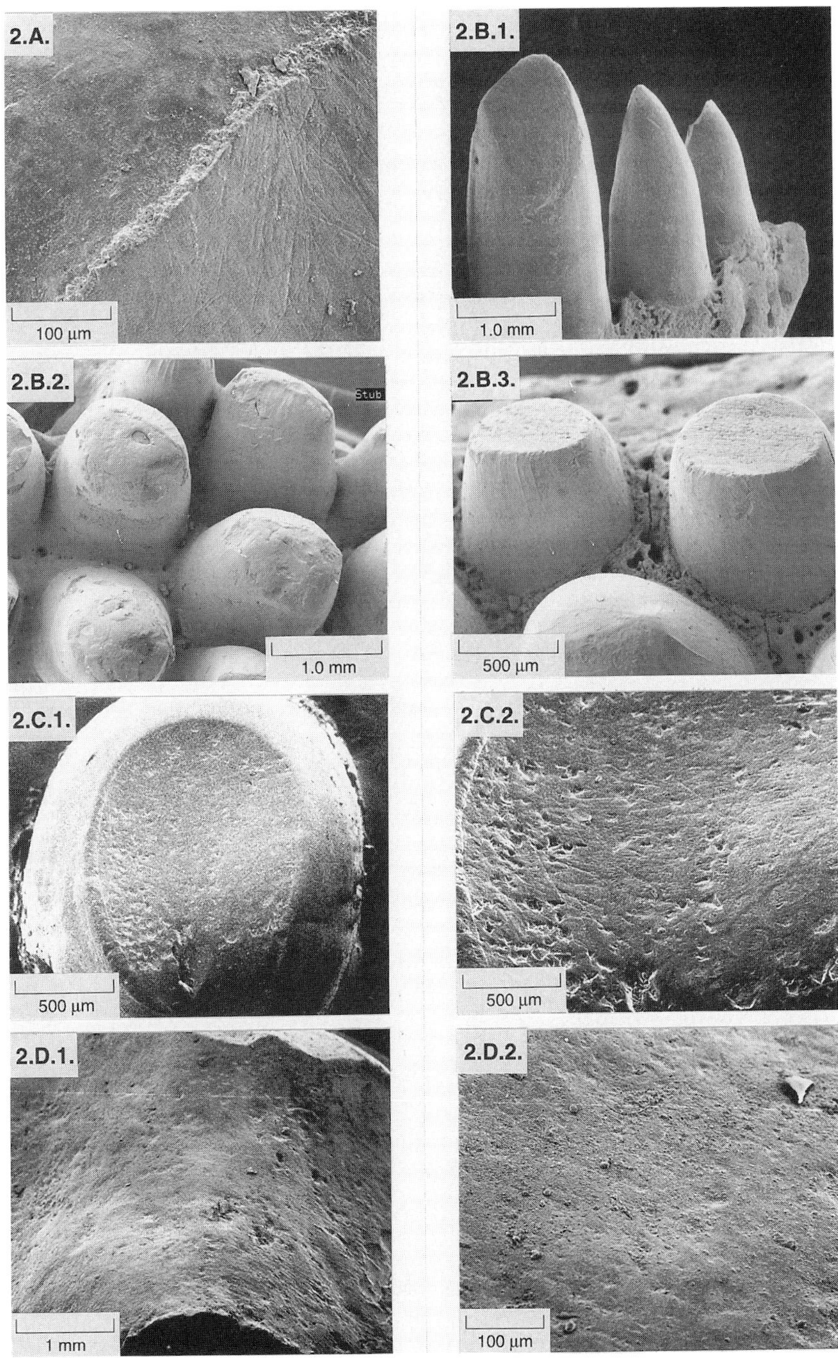
2.A.
100 μm
2.B.1.
1.0 mm
2.B.2.
Stub
1.0 mm
2.B.3.
500 μm
2.C.1.
500 μm
2.C.2.
500 μm
2.D.1.
1 mm
2.D.2.
100 μm

derived diadectomorphs, the diadectids *Desmatodon* and *Diadectes*. They were almost certainly high-fiber herbivores.

The recently discovered Early Carboniferous tetrapod, *Westlothiana* (Smithson *et al.*, 1994) shares sufficient derived characteristics with cotylosaurs to suggest a proximal shared ancestry. Carroll (1996) suggests that their small, elongate bodies with short limbs represent secondary specializations associated primarily with life as foragers on small food items in dense terrestrial vegetation where snake-like, laterally-sinuous locomotion would have facilitated extensive foraging. Nothing in Carroll's arguments preclude utilization of densely vegetated upper littoral habitats, in the transition between shallow, near shore aquatic (the "upper littoral zone") and near-shore terrestrial (the "lower supra-littoral"). The structure of their jaws and an isodont battery of small, sharply conical teeth connote incipient static-pressure, "hold-and-squeeze" collection and mechanical processing, and imply that they may have been primarily nondurophagous predators or even omnivores. In the absence of a pterygoid flange, they might not have had enough of a "snap" in jaw closure to capture small, quickly moving prey, but their "power bite" would have permitted them to capture and crush medium-sized, armored or muscular but relatively sluggish terrestrial invertebrates such as millipedes and, in combination with their likely agility in water, small aquatic tetrapods (Carroll, 1994).

The lepospondyls, the next out-group on the batrachosaur branch (following Carroll, 1995b; Laurin and Reisz, this volume), comprise a heterogeneous set of clades with a variety of feeding patterns. Only the keraterpetontids and tuditanomorphs have features associated with a significant potential for omnivory and low-fiber herbivory. The keraterpetontids though they had incipient static-pressure jaw mechanisms (A.R. Milner, 1980) possessed only limited capacities for overland foraging, but the tuditanomorphs parallel cotylosaurs in skeletal and dental features suggesting a substantial potential for omnivory and low-fiber herbivory. The tuditanomorphs, which first appear in mid-Carboniferous assemblages (e.g., Schultze and Bolt, 1996; Carroll, *et al.*, 1991), were animals of small to moderate size with relatively short bodies and stout, moderately long legs. These features, together with the absence of lateral line canals on adult skulls, suggest primarily terrestrial habits as foragers, with potential for pursuit of prey as well as for lurk-and-lunge collection. Tuditanomorph skulls and jaws are proportionately large and robust, and their form indicates a powerful bite at occlusion, one approaching the static-pressure pattern (Thomson and Bossy, 1970). Dentitions are isodont or near isodont; their teeth are ogival (bluntly conical) and in some specimens show considerable wear (Carroll and Gaskill, 1978). These features parallel those of many extant lizards as well as some early amniotes; they suggest considerable aptitude for collection and mechanical processing of hard or tough items, potentially including seeds and other nutrient-rich plant tissues. They differ sufficiently in details from the cotylosaurian pattern, however, to demonstrate an independent origin somewhere within the lepidospondyl clade from an ancestor that had a rather different feeding pattern than that inferred for the ancestral cotylosaurs. Unfortunately the primitive state of the lepospondyls is obscure. If, as Carroll suggests, these small elongate forms represent relatively highly derived states, they provide little direct information about locomotor and feeding patterns in their common ancestor though by interpolation from their placement within the batrachosaurs that ancestor was presumbly intermediate in character between the primitive anthracosaurids (discussed below) and *Weslothiana*.

In the remaining clades of batrachosaurs (seymourids and gephrostegids along with their primitive relatives) the structure of skull and jaws indicates incipient static-pressure bite mechanics departing from the more derived cotylosaur condition only in the absence of a transverse process on the pterygoid. The seymourids seem to have been rather specialized medium sized, aquatic predators; their size plus the presence of relatively large, acutely conical cheek teeth, a battery of even larger caniniform teeth, and a set of sharply conical, albeit short palatal fangs suggests that as adults they were more thoroughly terrestrial and likely carnivores, preying on other vertebrates, than insectivores or omnivores. The characteristics of the gephrostegids and their immediate relatives suggest rather different life styles than that of *Seymouria*. The best known member of the clade is the relatively late (Westphalian) and relatively derived *Gephrostegus* (Carroll, 1970). Its close-coupled body with limbs of moderate length and a smallish head indicates its capacities for extensive foraging in terrestrial habitats and for lurk-and-lunge collection. The association of a static-pressure bite (albeit primitive) with an isodont dentition of smallish, slightly recurved or ogival cones, the absence of caniniform teeth, and the shortness of the palatal fangs suggests a diet of small, weakly indurated invertebrates and potentially also of small low-fiber plant materials. *Eldeceeon* (Smithson, 1994) and *Sylvanerpeton* (Clack, 1994) from Lower Carboniferous strata (Visean) probably represent the primitive pattern for gephrostegids. *Eldeceeon* was a small, gracile, terrestrial animal and thus potentially a wide-ranging forager. The single skeleton known of *Sylvanerpeton* is only partially ossified; that condition suggests a possible aquatic habit but might equally reflect a juvenile rather than an adult condition. Incomplete preservation of their skulls, jaws, and dentitions obscures jaw and dental mechanics, but their relatively deep and narrow skulls and batteries small ogival or sharply conical teeth are features commonly associated with a static-pressure bite. Thus, though the evidence is not conclusive, both may have had an incipient seize-and-squeeze feeding style for a diet of small invertebrates and possibly even some low-fiber plant tissues.

The earliest and most primitive members of the embolomere clade, the proximal outgroup to batrachosaurs, have characteristics (e.g., Holmes, 1980, 1984) that suggest descent from ancestors somewhat like gephrostegids. They are small to medium-sized animals with short-coupled bodies, stout legs, and relatively small, narrow heads. Furthermore, although the structure of their skulls and jaws hints at kinetic-inertial jaw mechanics, the mechanical plan also appears compatible with a seize-and-squeeze bite albeit a relatively weak one (Clack, 1987). Such a bite, in addition to an isodont marginal dentition of moderately large, recurved conical teeth and large palatal fangs, suggests a diet primarily of large but relatively brittle invertebrates and/or relatively small vertebrates; omnivory seems unlikely but not impossible. Overall, such a primitive pattern would seem akin functionally to that inferred for the the most primitive batrachosaurs.

In contrast to the prevalence of wide-foraging/seize-and-squeeze feeding patterns among batrachosaurs, most temnospondyls appear to have had rather inconsiderable foraging capacities and to have been limited to grab-and-gulp collection and minimal mechanical processing. A version of this pattern appears even in relatively primitive forms from lower Carboniferous strata (*Dendrepeton* and *Balanerpeton* per Milner and Sequeira, 1994). Given these characteristics, the larger animals probably fed principally on small and medium-sized vertebrates; the smaller ones, on small vertebrates and unarmored invertebrates. The aptitudes of such animals for even facultative omnivory were probably very low, and their consumption of plant materials must have been minimal–although we cannot not rule out incidental utilization of small, unscleritized, readily digestible items (e.g., megaspores and seeds), encountered in the course of foraging as observed in extant grab-and-gulp feeders, including fish as well as amphibians.

Given this emphasis in dietary patterns, the exceptional temnospondyls with features implying omnivory and low-fiber herbivory provide valuable comparative data for analysis and interpretation of such dietary patterns in amniotes and other batrachosaurs. Such exceptions appear in some relatively late and highly derived members of the dissorophid subclade, for example *Cacops*; these animals appear to be the most highly terrestrialized temnospondyls as they have long legs and small skulls as well as short-coupled bodies. The structure of their skulls and jaws suggests an incipient static-pressure bite, one functionally comparable to that of stem batrachosaurs (e.g., *Seymouria* per Olson, 1961b), and they also resemble primitive batrachosaurs in the relatively small size of their palatal fangs. Overall, their features suggest capture of relatively small prey; their long limbs could have contributed to discovery and capture of particularly agile invertebrates and small tetrapods as well as escape from large carnivorous tetrapods. Thus, although their jaw and dental mechanics suggest only modest capabilities for collection and processing of plant material, their aptitudes for omnivory and low-fiber herbivory would have been considerably greater than those of other temnospondyls, even the most primitive, and would have approximated those in primitive batrachosaurs and even some amniotes.

None of the remaining clades of tetrapods appear to have possessed any consequential aptitudes for herbivory, but are significant in that they might reveal something of the primitive tetrapod feeding style from which that of the batrachosaurs (and temnospondyls) evolved. The loxommatids and crassigyrinids share numerous features related to feeding style with the Devonian "proto-tetrapods" grab-and-gulp feeding on large prey.

HIGH-FIBER HERBIVORY

High-fiber herbivory has probably been entirely limited to the cotylosaur clade and is widely distributed among the various subclades, carrying to the extreme the adoption of omnivory and low-fiber herbivory outlined earlier. A diadectid, *Desmatodon*, provides the oldest evidence for primary high-fiber diets with its presence in mid-Upper Carboniferous assemblages (probably Stephanian B) from Pennsylvania and southwestern North America (Berman and Sumida, 1995); another, more derived representative of this clade, *Diadectes*, first appears in the uppermost Carboniferous assemblages from those areas (Hook, 1989). The earliest known amniotes with comparable aptitudes are from the Stephanian C (uppermost Carboniferous) or Autunian A (lowermost Permian) terrestrial assemblages in the southwestern United States and include a synapsid, *Edaphosaurus*, and a reptile, *Bolosaurus*, with likely affinities to the captorhinids. The remaining, likely high-fiber herbivores (i.e., the latest, largest, most derived captorhinids and the synapsid caseids) make much later entries, appearing first (and very rarely) in upper Lower Permian assemblages (upper Artiniskian and lower Kungurian) from Texas and Oklahoma but becoming the most diverse and abundant tetrapods in the overlying Mid-Permian assemblages (uppermost Kungurian and/or lowermost Kazanian). Beyond the Early Permian the phylogenetic roster expands to include a considerable variety of diapsids as well as many therapsid grade synapsids.

Common Themes

As adults, all of these animals (except *Bolosaurus*) were of considerable to very large size (20 to 600 kg), and their relatively short-tailed, short-coupled, barrel-shaped, small-headed bodies were mounted on robust albeit short legs. These features indicate a primarily terrestrial habitus although any of them could probably also have foraged in near-terrestrial aquatic habitats. Their conspicuously bulky torsos betoken disproportionally large guts for their size, and large size in itself makes cellulysis of high-fiber plant tissue more feasible. Large size also confers other benefits that can outweigh the large capital investment in time, energy, and material implied by such bulk: It contributes to the effectiveness and efficiency of endomicrobial cellulysis, increases retention of metabolic heat and thus body temperatures and, all else equal, magnifies capabilities for regulation of disturbance (e.g., increasing tolerance of low temperatures and starvation as well as resistance to predation) (Peters, 1983, Calder, 1984).

Although the relatively small size of their heads along with their short muzzles suggests anomalously low rates of food collection relative to body mass and gut volume, such rates would be compatible with relatively low metabolic and growth levels and passage of food through the gut, slow enough to allow extensive fermentation of cellulose by the microflora of the gut. On the other hand, their relatively small skulls and jaws would have allowed selective collection of small, immature shoots and leaves, relatively low-fiber, high-nutrient items (Jarman and Sinclair, 1979; Janis and Ehrhardt, 1988). Although dental patterns differ among these animals, in all cases tooth form and arrangement indicate an emphasis on fracturing, tearing, and compacting relatively tough, fibrous materials such as plant leaves and stems. The wear patterns resemble those in extant herbivores: they are smoothly faceted by tooth-to-tooth wear and the facets are microstriated, presumably by abrasive grains of silt and dust present on low-growing plants (Walker *et al.*, 1978). Moreover, though none have multiridged cheek teeth, the linear patterns of striation suggest patterns of jaw motion that would have maximized mammal-like comminution and grinding of plant tissues and cells. Conversely, none of these animals possess the caniniform and sectorial teeth that would have facilitated capture, dismemberment, and ingestion of large prey such as other tetrapods.

Although the limb and body proportions of these animals limited foraging speeds and agility, given their overall size they still could have foraged over considerable areas and reached food items a half a meter or more above the substrate. Conversely, their limited mobility and agility would have limited food collection to plant materials, to slow-moving invertebrates such as snails, and to animal carcasses. A cost-effectiveness analysis suggests that expenditures in time and energy for discovery and collection of relatively rare, small, scattered food items, namely insects, seeds, and snails, would have been high relative to their needs (see Pough, 1973, on diets of extant lizards). Thus their primary food sources had to be large or abundant items such as large tetrapods or high-fiber plant tissues, though it does not rule out secondary consumption of small or rare items as sources of protein or other nutrients. Overall, their absurd proportions virtually require them to have been walking fermentation vats, as such bulky bodies would have significantly hindered their other activities. They would, however, have permitted efficient albeit slow processing of considerable volume of high-fiber plant foliage along with endosymbiotic recycling of nitrogen and synthesis of other nutrients as in extant mammalian megaherbivores (Van Soest, 1994).

Variations on the Theme

Beyond these common elements, differences among these animals in dental pattern and in body size and proportions suggest differences in collection and mechanical processing of food and probably also in diet. Bolosaurs are at the lower end of the size range with likely adult body masses of less than 1 kg and might seem too small even as adults to be high-fiber specialists with endomicrobial fermentation. They were larger, however, than the extant herbivorous iguanid lizard *Dipsosaurus* that does have a gut flora capable of dealing with cellulose (Mautz and Nagy, 1987). Therefore we conclude that *Bolosaurus* could have hosted such endosymbionts. Though small they have a gracile build and relatively long, well-ossified legs that indicate considerable capacities for long-distance foraging as well as for escape from tetrapod predators. In addition, given their small size and gracile bodies, they might have scrambled up into low-growing, bushy trees for food and refuge.

Bolosaurs couple expanded, complex cheek teeth with relatively simple but procumbent premaxillary and anterior dentary teeth. Their maxillary cheek teeth are bulbous on the labial side, each supporting a stout conical cusp that curves toward the lingual side of the tooth. Lingual to the base of the cusp the crown is rimmed by a robust cingulum, forming a shallow basin between the cingulum and the base of the cusp. Dentary cheek teeth are mirror images, bulbous and cusped on the lingual side and bearing a cingulum on the labial side. Watson (1954) concluded that they might have been insectivores; however, he lacked detailed information on the tooth wear pattern: The maxillary cheek teeth are heavily worn on the lingual side and dentary cheek teeth on the labial side (Fig. 3C). In both cases, although the enamel coating is thick, the cingulum is obliterated and dentine is exposed on the cusp, with the enamel surface surviving only in the bottom of the basin. Coarse striations, with a preferred orientation parallel to the long axis of the jaw, mark the exposed dentine surfaces and continue onto the enamel edges surrounding the dentine. This pattern suggests a fore-and-aft jaw motion consonant with the greater length of the inferior component of the jaw articulation and implies tooth-to-tooth comminution of tough plant tissues interlaced with silt. Despite of their small size, bolosaurs apparently possessed the essential capabilities for discovery, collection, and mechanical processing of fibrous plant tissues–although like modern small, high-fiber herbivores they probably selected the lowest fiber materials readily available and consumed small animals when encountered.

The list of possible high-fiber herbivores also includes a number of mid-sized captorhinids, *Labidosaurikos*, *Captorhinikos*, *Kahneria*, and *Rothianiscus*, from Mid-Permian (upper Kungurian or lower Kazanian) assemblages in Texas and Oklahoma (Olson, 1955, 1971; Dodick and Modesto, 1995); Dodick and Modesto suggest that they represent one or more distinct clades, possibly derived from a common ancestor with *Captorhinus*. Adults are considerably larger than *Bolosaurus* as well as earlier captorhinids with skull lengths from 12.5 to 45 cm as compared to 5.7 to 15 cm for the earlier forms, but because their skulls are relatively large in proportion to body size (Heaton and Reisz, 1986) they are still considerably smaller than most Permo-Carboniferous, high-fiber herbivores. Body proportions, as far as known, are comparable to those of the smaller, primitive taxa: short-coupled torsos of rather gracile build and moderately long legs that conjointly imply sufficient mobility and agility for extensive foraging including ascension of large shrubs and bushy trees.

As with *Bolosaurus*, their modest sizes and relatively slender torsos argue against a specialized diet of high-fiber tissues and for some sort of low-fiber diet or omnivory. On the other hand, the characteristics of their skulls, jaws, and teeth connote substantial aptitudes for collection and mechanical processing of fibrous leaves and tough stems as well as more nutritious items. They have relatively narrow muzzles but a greatly expanded temporal region with large adductor chambers, an association that combines a potential for selective feeding on small items with a very powerful static-pressure bite for collection and mechanical processing. Thus, they could have probed into masses of foliage for the more nutritious young leaves and shoots but also exerted strong forces on hard or tough items; their dental characteristics are certainly consistent with such varied feeding. This potential is complimented in the multiple, isodont rows of marginal cheek teeth: The number of rows varies, in proportion to size, from four to seven, and the form of the teeth and the patterns of wear, where known, are similar. The individual cheek teeth are blunt, laterally pinched ogives and are typically wear-faceted so that each row forms a row of small chisels parallel to the jaw margins. The wear facets are marked by both pits and striations in at least two taxa, *Labidosaurikos* (Fig. 2C.1 and 2C.2) and *Rothianiscus* (Fig. 2D.1 and 2D.2), but are marked by both pits and striations (although subsequent polishing tends to obscure the pits, especially in *Rothianiscus*). The pitting suggests they crushed strongly indurated but small items; the striations, which parallel the long axis of the jaw, confirm the

---

Figure 3. Tooth wear in *Diadectes*, *Bolosaurus* and *Edaphosaurus* (A.1) *Diadectes*, posterior aspect of incisiform tooth, striations on wear facet parallel with long axis of tooth. (A.2) Same as A.1 further enlarged. (B.1) *Diadectes*, crown of "molariform" cheek tooth, striations on wear facet predominantly parallel to long axis of jaw. (B.2) same as B.1, further enlarged. (C.1) *Bolosaurus*, unworn crown of "molariform" cheek tooth. (C.2) *Bolosaurus*, worn crown of "molariform" cheek tooth, striations on wear facet parallel to long axis of jaw. (D.1) *Edaphosaurus*, ogival buccal teeth. (D.2) Same as D.1, enlarged views of tooth. Note development of wear facets on flanks as well as crown. Microstriations parallel to long axis of jaw and enamel rim faired into dentine. ➔

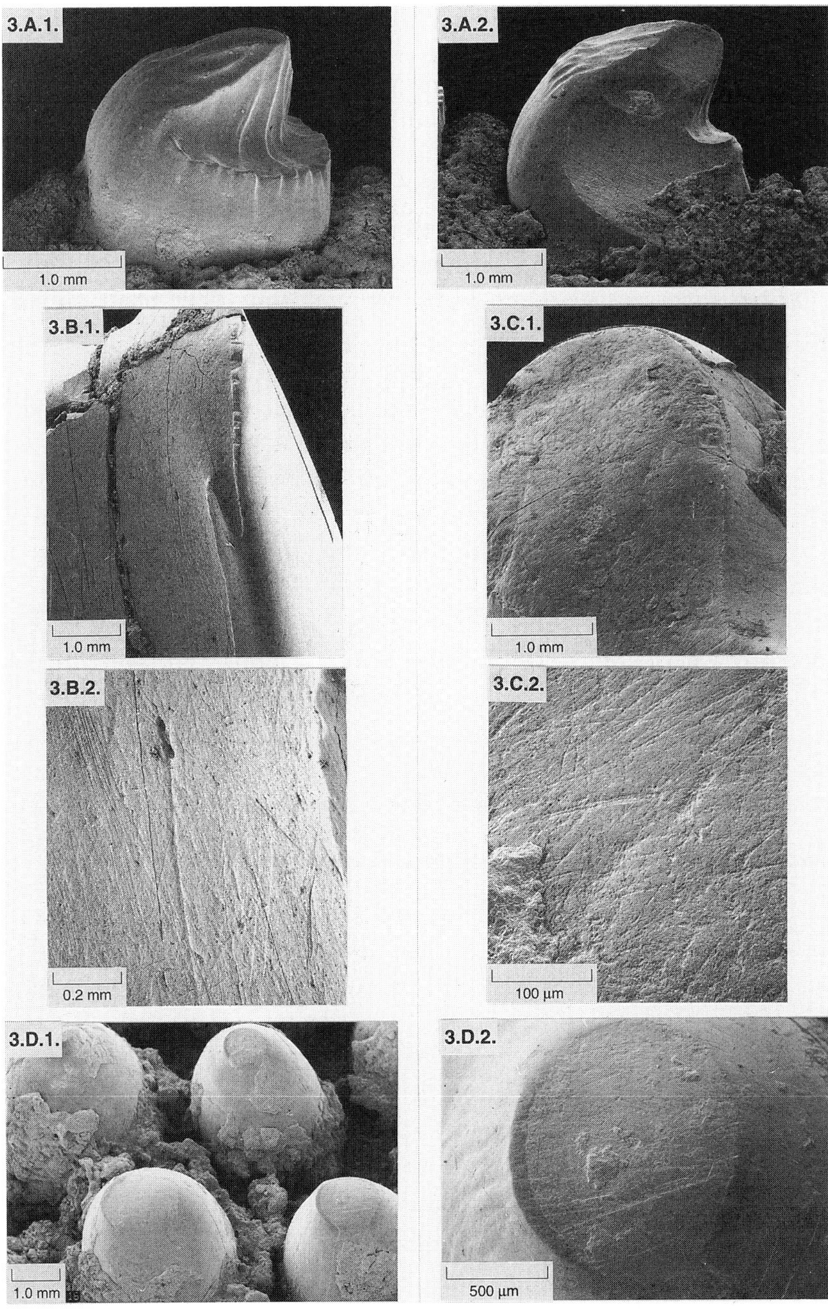
3.A.1.
1.0 mm
3.A.2.
1.0 mm
3.B.1.
1.0 mm
3.C.1.
1.0 mm
3.B.2.
0.2 mm
3.C.2.
100 µm
3.D.1.
1.0 mm
3.D.2.
500 µm

argument advanced by Doddick and Modesto (1995) for propalinal shearing and crushing during collection and processing of gritty items. Finally, the variation in tooth form and wear suggests a varied diet. Clearly, these animals could have collected high-fiber plant tissue and processed them effectively even though they lack the advantages of large size. In addition their bulk was great enough to demand a diet of large and abundant food items (Pough, 1973) and provide sufficient gut volume for meaningful fermentation. Thus, they probably took high-fiber plant tissues along with indurated, low-fiber plant organs such as megasporangia, strobili, and seeds–and possibly armored invertebrates.

In contrast to *Bolosaurus* and captorhinids, adult diadectids were large animals with bulky, rotund bodies, and with major differences in jaw and tooth morphology and likely mechanics. They have relatively short but deep faces, deep, massive mandibles, and a distinctive backward deflection of the pterygoid flange. They possess procumbent, bluntly spatulate incisiform teeth and large, blunt, anteroposteriorly compressed cheek teeth, the latter typically much eroded by wear (Romer, 1956; Olson, 1947, 1966b). The wear facets on the incisiform teeth display vertical microstriations that evince orthal jaw movements (Figure 3A.1 and 3A.2)–presumably for grasping compact food items and for severing tough stems and leathery leaves. In contrast, the striations on the cheek teeth (Figures 3B.1 and 3B.2) parallel the long axis of the jaw, implying backward (propalinal) motion of the lower jaw at occlusion, movement consistent with the structure of the jaw articulation and the pterygoid flange. This doubtless fractured and tore the cell walls in tough, resistant items such as stems, stiff fronds, and leaves although they probably did not contribute much to comminution of such tissues. Finally, the fore-and-aft ridges on skull above the cheek teeth connote muscular cheeks that could have held food between the cheek teeth permitting extended mastication. Diadectids could have collected and processed a wide variety of plant tissue, but the relative massiveness of diadectid teeth suggests a special facility for the collection and processing of large, hard seeds and even relatively large, heavily armored arthropods and thin-shelled snails. All these would have augmented the proteins and other nutrients in short supply in a high-fiber diet.

Edaphosaurs are also of considerable size, and their teeth and jaws, although morphologically distinct from those of diadectids, were functionally comparable. Their anterior marginal teeth are isodont and rather small and presumably they contributed to grasping and detaching small food items. Their cheek teeth are isodont, slightly swollen with laterally compressed, finely serrated tips with the plane of each blade oblique to the longitudinal axis of the tooth row as detailed by Modesto (1995) and in a pattern comparable to that in the modern herbivorous lizard *Iguana*. In addition, successive teeth are increasingly laterally-directed back along the maxillary row, and opposing dentary teeth, increasingly medially directed. Tooth replacement appears to have been relatively rapid, but some teeth show a marked wear facet on their lingual surfaces, a pattern presumably incurred in tooth-to-tooth severing of tough, fibrous tissues.

Further mechanical breakdown of the plant material probably involved interaction between the posterior cheek teeth and between those on the palate and the buccal margin of the jaw. The latter consist of multiple rows of small, closely spaced ogives that could have pierced and held leaves and stems as they were ripped from the plant; the sharp-edged wear facets on these teeth must have contributed to fracturing and shearing plant cuticle and fiber. Microstriations on these facets parallel the long axis of the jaw (Figures 3D.1 and 3D.2) and demonstrate backward movement of the lower jaw at occlusion as a means of macerating and compacting plant tissues held between the tips of upper and lower marginal teeth. This interpretation of jaw function is further supported by the form of the jaw articulation (facilitating retraction of the lower jaw at occlusion) and by inclination of the posterior cheek teeth (minimizing interference of the marginal teeth with such movement at occlusion while still holding food within the mouth). The overall consequences as described in detail by Modesto (1995: pp. 232-233) would have extensive fracturing, tearing, and compaction if not comminution and breakage of cell walls. This is similar to that in diadectids though accomplished in a different fashion. Edaphosaurs, like diadectids, could have consumed a wide variety of plant tissue. However, the relative delicacy of edaphosaur teeth suggests utilization of somewhat more frangible items; their jaw and dental mechanics also appear sufficient for processing less indurated seeds and of megaspores, sporangia, and strobili, adding protein and readily digested carbohydrates to a high-cellulose diet.

Evidently, caseid pelycosaurs differed markedly from both diadectids and edaphosaurs in feeding patterns because they have quite different jaw and tooth characteristics, and different body proportions (Romer and Price, 1940; Stovall *et al.*, 1966; Olson, 1968, 1976). They have large conical "incisors" and spatulate cheek teeth, and very small (albeit numerous) palatal teeth. The incisiform teeth presumably served to grasp and hold potential food items; the cheek teeth have compressed, tuberculate edges that presumably held tough, fibrous materials in place while they were severed, broken, and torn during occlusion The lack of wear facets on these teeth testifies to a lack of tooth-to-tooth occlusion, a deficiency that would have limited collection and processing of hard items or even woody stems. In addition, their lack of any crushing teeth would have very much circumscribed their capacities for fracturing plant cuticle and cell walls. On the other hand, their large, robust hyoid structures indicate the presence of a large, powerful tongue that could have compressed and fractured food into a compact bolus against the small teeth that covered the palatal surface. These differences from diadectids and edaphosaurs in faculties for mechanical processing suggest differences in primary food sources–a conjecture supported by their association with a different set of plants. Caseids may have offset these apparent deficiencies in mechanical processsing by greater capacities for chemical processing and for collection of more palatable food. Their torsos are grotesquely enlarged implying very high gut volume relative to body size compensating for inadequacies in mechanical processing. Further, their disproportionally large and powerful forelimbs, particularly the distal elements, suggest capabilities for digging up rhizomes and roots and for uprooting shrubs. Further, the structure of the posterior trunk and the sacral vertebrae and ribs also suggests that they might have been able to raise their forequarters to reach up into the larger shrubs and tear down higher branches. The latter capabilities would have extended of their vertical feeding range into the upper shrub strata, providing greater access to higher quality food such as immature, low-fiber shoots and foliage.

Overall, the characteristics of these early, high-fiber herbivores suggest life styles consistent with their aptitudes for herbivory. Although low metabolic and growth levels would have their reduced capabilities for acquisition of resources, regulation of disturbance and replacement of losses, so too would they lessen those necessary for discovery, collection, and processing of food. For example, their low metabolic levels would have permitted even large animals to obtain sufficient food and water in relatively small, food-rich areas over considerable periods of time and, conversely, to tolerate starvation for relatively long intervals. The large size and compact shape of the adults–as it increased and stabilized body temperatures–also accelerated digestion and fermentation of refractory, cellulose-dominated food. Food supply could have been increased and expenditures for foraging areas reduced insofar as large size permitted knocking over, breaking down, and uprooting shrubby plants and digging up rhizomes and roots. Such feeding techniques might also have provided access to higher quality food items, ones with lower fiber and higher protein content. Their small heads and jaws would have permitted selective collection of relatively small but high-quality items–indeed, the crushing teeth observed in *Diadectes* and *Desmatodon* and the palatal tooth plates of *Edaphosaurus* suggest that they may have fed on indurated, low-fiber items such as lycopsid and sphenopsid megaporangia and strobili as well as large pteridospem seeds. Such a diet would include a larger supply of protein and allow higher rates of chemical processing and thus a larger income of nutrients. Finally, their modest food requirements would have allowed nearly continuous concealment in patches of dense vegetation, minimizing exposure to predation as well as thermal stress and desiccation.

Replacement would, of course, have been hindered not only by the low rates of growth and maturation inherent with low metabolic levels but also by the high rates of juvenile mortality associated with ovipary and production of small neonates, all of which in the absence of parental care would have led to high rates of mortality for embryos, hatchlings, and juveniles. On the other hand, large adult size must have maximized adult survival and the period of reproductive activity. Finally, large body masses along with low metabolic levels would have permitted allocation of a large portion of nutritional income to reproduction even when food was relatively scarce. Such a capability for repeated production of large numbers of small eggs could offset embryo, neonate and early juvenile mortality, but if predation were intense and/or if physical circumstances were particularly unfavorable, population densities would have been low. (We would note in this regard that the extant, giant, herbivorous tortoises are quite successful on food-poor desert islands–in the absence of mammalian predators–but disappear rapidly when such animals are introduced.)

ANALYSIS AND DISCUSSION

Analysis of the distribution of dietary aptitudes among amniotes and other early tetrapod clades in relation to abiotic and biotic circumstances (Table 2 and Fig. 1) provides the primary basis for an explanation of the evolution of those aptitudes. Obviously, the incidence and modes of herbivory varied among the several tetrapod clades: most of the likely omnivores are batrachosaurians and high-fiber herbivory is limited to cotylosaurs. Conversely, omnivory was clearly extremely rare among non-batrachosaurs, and none of the latter show any of the features bespeaking high-fiber herbivory. Further, the patterns of predation differ on the two sides of the genealogic divide: most batrachosaurians apparently functioned as wide-foraging, seize-and-squeeze predators and most non-batrachosaurs as lurk-and-lunge, grab-and-gulp ones. What of the common ancestor of batrachosaurs and temnospondyls? The characteristics of remaining outgroups argue for a shared, primitive grab-and-gulp mechanics–and primitive lurk-and-lunge foraging–in the ancestry of both batrachosaurs and temnospondyls. So far as known, both the pretetrapods (e.g., *Eusthenopteron* and *Panderichthys*), the primitive tetrapods (e.g., *Acanthostega*, *Ichthostega*, loxommatids and crassigyrinds) have very primitive versions of kinetic-inertial jaw mechanics and primitive body proportions implying that they were ambush, grab-and-gulp predators. Evidently, these characteristics represent the primitive condition for the entire tetrapod clade and their immediate piscine antecedents and thus both the batrachosaurs and temnospondyls had such a source, albeit modestly modified.

Even the principal anomalies tend to support these generalizations for they are associated with and probably consequent on adoption of anomalous life styles and living places. If the lepospondyls are batrachosaurs, they appear to have specialized for predation on small animals in densely vegetated, upper littoral and/or lower supra-littoral habitats (Carroll, 1996). If the diverse lepospondyl subclades are close sister clades of temnospondyls then that particular aberration disappears. In either case tuditanomorphs reversed the general lepospondyl trend and developed aptitudes for wide-foraging, seize-and-squeeze feeding analogous to those of primitive cotylosaurs. Similarily, those dissorphophids that have analogous capacities for such feeding appear to have been the only members of the temnospondyl clade to have both lived and fed primarily in upper supralittoral, mesic forests.

The establishment of high-fiber herbivory in amniotes and diadectids (constituting the cotylosaurian clade) and its limitation to that clade are, in turn, clearly associated with the circumstances that promoted omnivory and low-fiber herbivory. The probable connections are clearest in the edaphosaurids and captorhinids: from *Ianthasaurus* through *Glaucosaurus* to *Edaphosaurus*; the first is a likely insectivore-omnivore and the last a high-fiber herbivore (Modesto, 1994, 1995). A comparable spectrum of aptitudes appears in the phylogenetic succession of captorhinids from the most plesiomorphic members of the group such as *Romeria* to the most derived such as *Rothianiscus* (Gaffney and McKenna, 1979) and demonstrates an analogous shift from predominately insectivorous to predominately herbivorous. A comparable pattern appears among edaphosaurians, and diadectid and caseasaurian phylogenies also suggest similar (although less clear-cut) sequences. These phylogenetic patterns demonstrate that high-fiber herbivory evolved independently in several different clades including independent adoption of endosymbiotic cellulytic fermentation. This view is consistent with the later, iterative evolution of fermentation-based herbivory among various subclades of therapsid and mammalian-grade synapsids as well as among parareptiles and diapsids.

The localization of high-fiber herbivory within the cotylosaurian clade, its iteration within that clade, and the phylogenetic evidence for transitional stages in its evolution suggest exaptation. Conversely, the various, shared analogous specializations for high-fiber herbivory demonstrate adaptation whether driven entirely by selection or produced by a incremental series of exaptative and adaptive steps. The pattern of dietary variation among species of the whiptail lizard *Cnemidophorus* (Family Teiidae) provides an intriguing model for the adoption of herbivory. Although representatives of this clade are traditionally presented as an archetypal example of an active predator particularly apt for insectivory (Pianka, 1970), members of some species consume significant amounts of plant material (Schall and Ressel, 1991), and on the Caribbean island of Bonaire, the diet of *C. murinus* varies locally from low- to high-fiber herbivory. Low-fiber feeders augment their diet with minor components of high-fiber plant tissues and insects whereas high-fiber consumers incorporate significant admixtures of flowers, fruit, and insects (Dearing and Schall, 1993). Adults are relatively small and lack overt morphological specializations for collection or mechanical processing of plant tissues. Their guts, however, are longer than those of primarily insectivorous species of *Cnemidophorus*, and some observations even suggest endosymbiotic fermentation (Dearing, 1993). Thus, the *Cnemidophorus* clade in itself encompasses the entire spectrum of physiological, behavioral, ecological, and evolutionary possibilities from insectivory to high-fiber herbivory. It is important to note, however, that in these animals the observed dietary differences are not reflected in skeletal proportions nor in dental characteristics and so would not be evident even in complete, well-preserved fossils.

The incidence and modes of early tetrapod herbivory varied considerably among different habitats and ecosystems and with their transformations (Table 2). Although some portion of this variation represents sampling biases, a clear pattern remains even after these biases are taken into account. Thus the association of likely omnivores and low-fiber herbivores with supralittoral habitats is almost certainly of biological significance: Their considerable aptitudes for terrestrial locomotion suggest primarily supralittoral habits as does their near absence from primarily aquatic assemblages despite the higher probabilities of preservation and discovery in aquatic sites. Likewise, the increase in the relative abundance and diversity of likely omnivores and herbivores from Early Carboniferous to Early Permian almost certainly reflects biological reality rather than sampling biases because it is observed in all comparable assemblages, even littoral ones, regardless of age. These correlations assuredly reflect differences in resources and disturbance between various littoral and supralittoral plant associations and trophic systems and in the evolutionary potential of the various clades of tetrapods utilizing these sites.

The scarcity of primary omnivores and herbivores among the aquatic tetrapods presumptively reflects not only the availability of suitable food but also the requirements of aquatic predation. As for food supply, no fully aquatic tracheophytes are known from Permo-Carboniferous, continental fossil assemblages (Collinson and Scott, 1987). On the other hand, the mechanics of food collection in water seem unlikely to favor evolution of static-pressure jaw mechanics. For example, extant herbivorous fish that utilize mechanically resistant plant materials such as seeds have added auxiliary, static-pressure "pharyngeal jaws" to supplement their primitive kinetic-inertial, mandibular bite. Extant crocodilians have largely lost their ancestral static-pressure bite and reverted to a kinetic-intertial pattern (Olson, 1961b). Any general deficiency in the capacities of aquatic tetrapods for chemical processing of plant tissues seems extremely unlikely as a potential for digestion of the primary constituents of low-fiber plant tissues appears to be a fundamental character of all vertebrates. [Indeed, the aquatic larvae of most living frogs and toads feed largely or exclusively on starch-rich algae, and some ray-fin fishes only distantly related to tetrapods consume some variety of low-fiber tissues (Sibbing, 1991).]

Conversely, potential seize-and-squeeze omnivores and herbivores are relatively abundant and diverse where the assemblages sample primarily supralittoral ecosystems (Table 2). These positive correlations probably reflect not only the mechanical feasibilty of static-pressure jaw mechanics in subaerial feeding but also the greater availability of small, tough, and hard arthropods and low-fiber plant materials in those systems. That abundance, combined with the scattered distribution of these items, would encourage adoption of wide-foraging, seize-and-squeeze tactics and ultimately an abundance of omnivores and low-fiber herbivores. In addition, the characteristics of Permo-Carboniferous plant associations and likely analogies with modern associations in comparable circumstances suggest that the most favorable locations for omnivory and herbivory would have been in dense patches of bushy or shrubby pteridosperms within warm, subhygric to submesic forests. Such thickets would have produced not only a relatively large amount of readily accessible, high-quality plant tissue but also a large biomass of insects supported by that resource. In addition, they would have provided easy access to water, to heat sources and sinks, and to refuges from disturbance. Conversely, the least favorable sites would presumably have been (1) in lycopsid-dominated swamp forests where primary production was evidently quite low and refuges from predators rare and (2) in highly seasonal, relatively open, subxeric, coniferous woodland and shrub where again primary production was probably quite low and access to water and refuges quite limited, except during brief rainy periods. The very large size of Mid-Permian caseids may reflect their need for toleration of starvation and desiccation and for resistance to predation in what appear to be subxeric plant associations.

The distribution of likely omnivores and herbivores also appears correlated with long-term changes in the characteristics of supralittoral ecosystems. First, the greater abundance and phylogenetic diversity of likely tetrapod insectivores, omnivores, and low-fiber herbivores in Mid-Carboniferous (Westphalian) fossil assemblages than in earlier ones (Visean through Westphalian) coincides with an apparent increase in the supply of insects and large seeds. It was also associated with an increase in the abundance and

Table 2. Chronicle of Tetrapod Omnivory and Herbivory. References: Paleobiogeography: Carroll, 1994. Paleogeography and Paleoclimatology: Van der Zwan,, *et al.*, 1985; Witzke, 1990; Ziegler, 1990; Calder, 1994; DiMichele, *et al.*, 1985; Frakes, *et al.*, 1992; Cecil, 1990; Dulong, *et al.*, 1994; Kvale, *et al.*, 1994; Wright, 1990; Miller and West, 1993; Tandon and Gibling, 1994; plants: Ziegler *et al.*, 1981; Allen and Dinely, 1988; Retallack and Germain-Heins,1994; DiMichele, Hook *et al.*, 1992; DiMichele and Aronson, 1992. ➔

PERIOD	SERIES	STAGE	CLIMATE	HABITATS AND SIGNIFICANT ECOSYSTEM FEATURES	OMNIVORY AND HERBIVORY
PERMIAN	UPPER	KAZANIAN 260 MA	SEASONAL WARM\COOL AND WET/DRY	➢ DOMINATELY UPPER LITTORAL TO UPPER SUPRALITTORAL. ➢ ?SUB-MESIC WOODS AND SHRUBLAND. ➢ PTERIDOSPERMS AND CONIFERS DOMINATE ➢ INSECTS ABUNDANT; TOP CARNIVORES RELATIVELY RARE.	➢ INSECTIVORES, OMNIVORES, LOW-FIBER HERBIVORES VERY ABUNDANT, DIVERSE, DISPARATE ➢ HIGH-FIBER HERBIVORES VERY ABUNDANT, DIVERSE, AND DISPARATE
	LOWER	KUNGURIAN 270 MA	WARM, MODERATELY WET/DRY BECOMING DRIER TOWARD TOP	➢ PRIMARILY UPPER LITTORAL AND LOWER SUPRALITTORAL WITH MESIC TO SUBXERIC HINTERLANDS AT TOP. ➢ HYGRO- TO MESOPHYLOUS FORESTS EXCEPT SUB-MESOPHYLOUS NEAR TOP. ➢ PTERIDOSPERMS DOMINATE EXCEPT IN WETTEST SITES. ➢ INSECTS ABUNDANT; TOP CARNIVORES ABUNDANT EXCEPT AT TOP	➢ HIGH-FIBER HERBIVORES PRESENT BUT RELATIVELY RARE EXCEPT VERY ABUNDANT LOCALLY NEAR TOP ➢ OMNIVORES AND HERBIVORES ABUNDANT, DIVERSE
		ARTINSKIAN 275 MA SAKMARIAN 280 MA ASSELIAN 290 MA	WARM, WET TO MODERATELY WET/DRY	➢ VARIETY OF UPPER LITTORAL AND SUPRALITTORAL SITES WITH MESIC HINTERLANDS. ➢ HYGRO- TO MESOPHYLOUS FORESTS ➢ PTERIDOSPERMS DOMINATE EXCEPT IN WETTEST SITES ➢ INSECTS ABUNDANT; TOP CARNIVORES RELATIVELY ABUNDANT	➢ INSECTIVORES, OMNIVORES, LOW-FIBER HERBIVORES ABUNDANT, DIVERSE ➢ HIGH-FIBER HERBIVORES PRESENT BUT RELATIVELY RARE ➢ LATEST LIKELY ORIGIN OF CASEID AND CAPTORHIND HIGH-FIBER HERBIVORES
UPPER CARBONIFEROUS	STEPHANIAN 300 MA		RELATIVELY WARM, MODERATELY WET	➢ PREDOMINATELY UPPER LITTORAL AND LOWER SUPRALITTORAL BUT SOME UPPER SUPRALITTORAL SITES. ➢ HYGRO- TO MESOPHYLOUS FORESTS ➢ PTERIDOSPERMS DOMINATE EXCEPT IN WETTEST SITES ➢ INSECTS ABUNDANT; TOP CARNIVORES RELATIVELY ABUNDANT	➢ FIRST RECORDS OF HIGH-FIBER HERBIVORY: DIADECTIDS, POSSIBLY EDAPHOSAURS ➢ LATEST LIKELY ORIGIN OF DIADECTIDS, EDAPHOSAURS NEAR BEGINNING
	WESTPHALIAN 310 MA		RELATIVELY COOL, WET	➢ UPPER LITTORAL AND LOWER SUPRALITTORAL BUT ONLY A VERY FEW UPPER SUPRALITTORAL SITES. ➢ HYGRO- TO SUB-HYGROPHYLOUS FORESTS ➢ FURTHER INCREASE IN ABUNDANCE, DIVERSITY OF PTERIDOSPERMS; LARGE SEEDS COMMON ➢ INSECTS ABUNDANT; TOP CARNIVORES RELATIVELY ABUNDANT	➢ INSECTIVORES, OMNIVORES/ LOW-FIBER HERBIVORES MODERATELY DIVERSE AND ABUNDANT ➢ EARLIEST LIKELY ORIGIN OF DIADECTIDS, EDAPHOSAURS
	NAMURIAN		RELATIVELY WARM, WET	➢ POORLY SAMPLED; UPPER LITTORAL AND LOWER SUPRALITTORAL SITES VERY RARE. ➢ HYGRO- TO SUB-HYGROPHYLOUS FORESTS ➢ PTERIDOSPERMS MORE ABUNDANT AND DIVERSE; SOME LARGER SEEDS ➢ APPEARANCE OF INSECTS; TOP CARNIVORES RELATIVELY ABUNDANT	➢ INSECTIVORES, OMNIVORES, LOW-FIBER HERBIVORES VERY RARE ➢ EARLIEST LIKELY ORIGIN OF REPTILE, SYNAPSID, DISSOROPHID, TUDITANOMORPH CLADES
LOWER CARBONIFEROUS	VISEAN 340 MA		RELATIVELY WARM, MODERATELY DRY	➢ POORLY SAMPLED, UPPER LITTORAL TO SUPRALITTORAL. ➢ HYGRO- TO SUB-MESOPHYLOUS FORESTS ➢ PTERIDOSPERM DISTRIBUTION, ABUNDANCE, DIVERSITY LIMITED; SEEDS VERY SMALL ➢ FEW TERRESTRIAL INVERTEBRATES; TOP CARNIVORES ABUNDANT	➢ INSECTIVORES VERY RARE; POSSIBLY NO OMNIVORES OR LOW-FIBER HERBIVORES ➢ LATEST LIKELY ORIGIN OF COTYLOSAUR, LEPOSPONDYL GEPHYROSTEGID CLADES
	TOURNAISIAN 355 MA		RELATIVELY WARM AND MODERATELY DRY	➢ NO SIGNIFICANT SAMPLES OF TETRAPODS YET DESCIBED SAMPLED ➢ HYGRO- TO SUB-MESOPHYLOUS FORESTS ➢ PTERIDOSPERM DISTRIBUTION, ABUNDANCE, DIVERSITY LIMITED; SEEDS VERY SMALL ➢ FEW TERRESTRIAL INVERTEBRATES	➢ NO SIGNIFICANT TETRAPOD RECORD YET DESCRIBED ➢ LATEST LIKELY ORIGIN OF BATRACHOSAUR, TEMNOSPONDYL CLADES

References: Paleobiogeography: Carroll, 1994. Paleogeography and Paleoclimatology: Van der Zwan *et al.*, 1985; Witzke, 1990; Ziegler, 1990; Calder, 1994; DiMichele *et al.*, 1985; Frakes *et al.*, 1992; Cecil, 1990; Dulong *et al.*, 1994; Kvale *et al.*, 1994; Wright, 1990; Miller and West, 1993; Tandon and Gibling, 1994. Plants: Ziegler *et al.*, 1981; Allen and Dinely, 1988; Retallack and Germain-Heins, 1994; DiMichele, Hook *et al.*, 1992; DiMichele and Aronson, 1992.

diversity of shrubby pteridosperms and thus, presumably, the production of low-fiber tissues. Second, the genesis, initial appearance, and diversification of high-fiber herbivores coincides with the further diversification of shrubby seed-ferns through the later Carboniferous and Early Permian (Stephanian through Artinskian) in both lower and upper supralittoral assemblages. Third, the increase in relative abundance and diversity of high-fiber herbivores during the latest Early Permian and the Late Permian (Kungurian and Kazanian) is associated with a further shift from mesic forest to submesic or even subxeric woodland and shrub and accompanies a decrease in the relative abundance of large terrestrial carnivores. These correlations between phylogenetic and ecological patterns argue for the major roles of food supply and predation in the early evolution of tetrapod omnivory and herbivory. This conclusion is intuitively and deductively satisfying. It is strengthened inductively by coincident, analogous transformations in several different clades in comparable circumstances and similar events in similar circumstances at different times (not only among amniotes but also in variously distant out-groups).

INTERPRETIVE SUMMARY AND CONCLUSIONS

Analysis of the historical sequence suggests that the evolution of herbivory among early amniotes proceeded in three more or less distinct steps dependent on particular configurations of circumstance, exaptation, and adaptation–stages and patterns paralleled in other primitive tetrapod clades (Fig. 1, Table 2). The initial transformations came in the differentiation of the batrachosaur ancestry from the primitive tetrapod stock during the latest Devonian and earliest Carboniferous (360-340 mybp); it encompassed basic innovations in functional morphology and, presumably, in physiology and behavior for a primitive version of a wide-foraging, seize-and-squeeze pattern of feeding in terrestrial habitats. The second set of transformations was accomplished during the Mid-Carboniferous (340-320 mybp) with the genesis of the cotylosaur clade. This event established an advanced version of the wide-foraging, seize-and-squeeze feeding style, one that increased considerably aptitudes for insectivory, omnivory, and low-fiber herbivory. The third generated in parallel the specialized functional aptitudes necessary for effective and efficient high-fiber herbivory among a large number of cotylosaur subclades; it began with diadectids, captorhinds, caseids, and edaphosaurids clades in the Late Carboniferous (320 mybp onward) and has continued to the present.

The first stage provided a modest potential for discovery, collection, and mechanical processing of small, scattered, tough food items in supralittoral habitats. Thus it also provided at least a minimal potential for insectivory and omnivory as documented by the appearance of primitive batrachosaurs (including the near-cotylosaur *Westlothiana*) in the Mid-Early Carboniferous (Visean). The paucity of tetrapod remains from older strata constrains reconstruction of specific environmental factors and possible exaptive and adaptive events, but the evolution of these aptitudes was certainly correlated with invasion of supralittoral habitats. It appears exaptive so far as it is related to primitive tetrapod capacities for a diet of small prey, and adaptive so far as it transformed them for more effective and efficient utilization of new kinds of prey in new circumstances. (This interpretation is supported by the chronicle of analogous transformations among dissorophid temnospondyls and kerapetontid and tuditanomorph lepospondyls during the latter part of the Carboniferous.)

The second step subsumed further enhancement of the primitive batrachosaurian aptitudes for insectivory, omnivory, and low-fiber herbivory–some of which could even have contributed to use of high-fiber plant tissues. The paucity of upper littoral and lower supralittoral tetrapod assemblages from the lowest Mid-Carboniferous (Namurian and lower Westphalian) strata restricts direct information on this stage, but it is documented by shared, derived features that appear first in early Late Carboniferous amniotes. However, because homologous traits also occur in diadectomorphs, they were evidently present in the common ancestor of the various cotylosaurian subclades in the mid-Early Carboniferous (late Visean or early Namurian). Moreover, analogous aptitudes evolved independently in other batrachosaurian clades (the gephrostegids and the more derived tuditanomorphs) at about the same time in similar circumstances. These coincidences in time and circumstance suggest that all were triggered by changes in circumstances; their restriction to groups that already had some aptitudes for wide-foraging, seize-and-squeeze feeding implies that their response to that change depended on exaptation as well as adaptation. Seemingly then, changing circumstances, exaptive possibilities and adaptive responses conspired to initiate and direct this phase in the evolution of amniote herbivory.

The third step, the establishment of the essential functional capacities for high-fiber herbivory, occurred in parallel among diadectids, reptilian captorhinids and bolosaurs, and synapsid edaphosaurids and caseids beginning in Late Carboniferous. The early history of these transformations is apparently obscured by the bias in upper Mid-Carboniferous (upper Westphalian) assemblages toward wetland habitats and hydrophilous plant association, but in all these cases, diadectid to synapsid, their later history is associated with a shift toward drier climates and mesic to submesic forests. Furthermore, though clearly evolved independently in each clade, these capabilities all derive from the shared cotylosaurian aptitudes for wide-foraging, seize-and-squeeze omnivory and low-fiber herbivory variously modified for specialized high-fiber herbivory. Thus, the third phase in the evolution of amniote herbivory also represents an amalgam of circumstances, exaptation, and adaptation.

A two-phase scenario is sugggested for the initiation and early evolution of high-fiber herbivory among diadectids and amniotes; the first primarily exaptation, and the second adaptation. Initially, the very considerable aptitudes of primitive cotylosaurs for omnivory suggest a potential for use of low-fiber items as a supplement to animal prey but as the primary food source given shortages of animal prey. So far as low-fiber materials were relatively abundant, as they probably were in open, mesic, pteridosperm forests, behavioral shifts toward low-fiber herbivory would have been relatively likely, followed by morphological and physiological adaptation. In turn, specialization for consumption of low-fiber materials might favor the ingestion of high-fiber items, perhaps incidentally at first but subsequently as supplements to and replacements for low-fiber items. Such transitions to a mixed diet would seem most likely if there was a relative abundance of high-quality, high-fiber material as in the lower strata of open mesic and sub-mesic pteridospermous forest and woodland. There the rates of vegetative production were probably relatively high, providing a considerable biomass of immature, relatively nutritious shoots and leaves. Thus, the dietary aptitudes for omnivory appear exaptive for adoption of low-fiber herbivory and adaptation for low-fiber herbivory exaptive for mixed diet or even limited kinds of high-fiber diets. Such patterns appear in extant tetrapods, perhaps most strikingly in mammals and birds, but also among lizards such as the variation in the diet of *Cnemidophorus* from entirely insectivorous to predominately herbivorous described earlier.

The second phase in evolution of high-fiber herbivory presumably involved adaptation in the establishment of two functionally related innovations. One of these, cellulytic endosymbiosis (as marked by a disproportionally voluminous torso) must have required coevolution of tetrapod and microbial morphology and physiology. The other, marked by establishment of special aptitudes for collection and mechanical processing of cellulose-rich materials, probably represents an evolutionary response to the demands and possibilities of endosymbiotic cellulysis. The independent origin of high-fiber herbivores in several clades of cotylosaurs argues for independent acquisition of cellulytic endosymbionts in each clade and for relative ease of such events. On the other hand, symbiont populations are very similar in composition across the spectrum of extant, phylogenetically independent, high-fiber herbivores–symbiont populations tend to vary more with food intake than with host phylogeny (Clarke, 1977; Stevens and Hume, 1995). In addition, though these populations include a considerable variety of taxa with very different phylogenetic sources, nearly all have their closest affinities with various groups of anaerobic decomposers. These patterns argue for independent genesis of endosymbiosis in each clade and thus again for relative ease of such coevolutionary events.

Therefore, cellulytic endosymbiosis could very likely have begun as a sort of commensalism as appears to have happened in the evolution of extant, fungus-farming ants (Chapela, *et al.*, 1994). Animals foraging in forest litter would have picked up a wide variety of microbes, including decomposers. Those microorganisms that survived residence in the tetrapod gut were predisposed for endo-commensalism so far as the herbivores ingested plant material useful to their gut flora and as endomicrobial fermentation produced volatile fatty acids, proteins, sugars, and vitamins that could be digested and assimilated by the host. Given a supply of relatively high quality though high-fiber plant tissues, natural selection would likely tend to increase effectiveness and efficiency on both sides, transforming incidental commensalism to obligatory symbiosis. For tetrapods, such enhancements would have reasonably involved selection for morphological and physiological transformations that would maximize the rate and efficiency of fermentation relative to the herbivores's nutritional requirements. The possibilities include, obviously enough, those features that characterize extant high-fiber herbivores: (1) more thorough mechanical processing of food, (2) increases in body size and in gut length relative to body size, (3) compartmentalization of the gut, and (4) changes in gut chemistry. Further, such changes in the characteristics of herbivore guts would enhance invasion by new kinds of potential endosymbionts, drawn from a variety of clades, fungal and protist as well as bacterial. In turn these features would have maximized the reproductive success of the endosymbionts–so long as they did not reduce that of the host. For endosymbionts it would have include such changes in morphology, physiology, and biochemistry as would maximize survival and productivity within the gut and successful infection of other guts, to say nothing of the reproductive success of the host.

In addition, such "infectious strains" once established in one tetrapod group would be likely to spread to other tetrapod clades wherever and whenever behavior, aptitudes, and circumstances were favorable--that is, where browsing was easy. This argument is supported by the chronicle: The first appearance of a high-fiber herbivore, the diadectid, *Desmatodon*, in Late Carboniferous (Early Stephanian) terrestrial assemblages, was followed shortly by establishment of similar diets and functional capacities, not only in closely related forms such as *Diadectes*, but also in more distantly related animals, such as edaphosaurs, bolosaurs, caseids, and captorhinids, all apparently frequenting relatively browse-rich plant associations. The characteristics of the extant *Cnemidophorus murinus* (Dearing, 1993) suggest an early stage in the process of infection with the endosymbionts assuming a role in host physiology and ecology–and thus in its potential evolution.

Finally, the activities of early high-fiber herbivores may have contributed to the expansion of that feeding guild as they "re-engineered" plant associations through intensification of disturbance by mechanical damage as well as consumption of plants and accelerated nutrient recycling. Such transformations could have had a considerable impact on competition among plants as well as on their survival and growth and would have favored rapidly growing herbs and/or shrubs and thus an expansion in the number and size of food-rich, refuge-rich patches for the herbivores. One would also expect high-fiber herbivory to be a regular component of terrestrial trophic systems thereafter because of the continued potential for infection of new groups of omnivores. The abundance and diversity of high-fiber herbivores would become a function primarily of the availability of critical resources and intensity of disturbance relative to their capabilities for acquisition, regulation, and replacement rather than initiation of endosymbiosis from scratch.

ACKNOWLEDGMENTS

In this project, the senior authors, N. Hotton and E. C. Olson, had primary responsibility for the review of the fossil record, and the junior author, R. Beerbower, for an analysis of the physiological ecology of early tetrapod herbivory. The latter also bears immediate responsibility for the information on the history of terrestrial arthropods, plants, plant associations, and environments. Messrs. Hotton and Beerbower bear full responsibility for the text since E. C. Olson's death in November 1993 precluded his participation in its completion. In 1987 we benefited greatly from discussions of early tetrapod herbivory with Bill DeMichele, Bob Hook, Jürgen Boy, Bob Gastaldo, Tom Phillips, Steve Scheckler, Bill Shear and Hans-Dieter Sues, the members of the "Paleozoic Working Group" at the Smithsonian conference on the Evolution of Terrestrial Ecosystems. Gary Harmon brought us a new perspective on the origins of endosymbiotic cellulysis, and Ernest Lundelius provided a fresh insight into the mechanics and patterns of tooth wear. We are especially indebted to those who provided access to material in their custody and permitted additional preparation: Richard Cifelli, Oklahoma Museum of Natural History; John R. Bolt, Field Museum of Natural History; and Philip D. Gingrich, Museum of Paleontology, University of Michigan. Finally, as we presented our preliminary results at various professional meetings, we gathered useful responses from colleagues in paleobotany as well as vertebrate paleontology.

LITERATURE CITED

Ahlberg, P. E., and A. C. Milner. 1994. The origin and early diversification of tetrapods. *Nature*, 368:507-514.

Alexander, R. McN. 1991. Optimization of gut structure and diet for higher vertebrate herbivores. *Philosophical Transactions of the Royal Society of London* B, 333:249-255.

Allen, K. C., and D. L. Dineley. 1988. Mid-Devonian to mid-Permian floral and faunal regions and provinces. Pages 531-548 in: *The Caledonian-Appalachian Orogen* (A.L. Harris and D.J. Fettes, eds.). Geological Society of London. Special Paper No. 38.

Auffenberg, W. 1988. *Gray's Monitor Lizard.* Gainsville: University of Florida Press.

Bauchop, T. 1977. Foregut fermentation. Pages 223-251 in: *Microbial Ecology of the Gut* (R. T. J. Clarke and T. Bauchop, eds.). New York: Academic Press.

Barrington, E. J. W. 1977. The alimentary canal and digestion. Pages 109-161 in:, *The Physiology of Fishes* (M. E. Brown, ed.). New York: Academic Press.

Bennett, A. F., and G. C. Gorman. 1979. Population density and energetics of lizards on a tropical island. *Oecologia*, 42:339-358.

Berman, D. S, and S. S. Sumida. 1995. New cranial material of the rare diadectid *Desmatodon hesperis* (Diadectimorpha) from the Late Pennsylvania of central Colorado. *Annals of the Carneigie Museum*, 64:315-336.

Berman, D. S, S. S. Sumida, and R. E. Lombard. 1992. Reinterpretation of the temporal and occipital regions in *Diadectes* and the relationships of the diadectomorphs. *Journal of Paleontology,* 66:481-499.

Bjorndal, K. A. 1979. Cellulose digestion and volatile fatty acid production in the green turtle, *Chelonia mydas. Comparative Biochemistry and Physiology*, 63A:127-129.

Bjorndal, K. A., and A. B. Bolton. 1992. Body size and digestive efficiency in a herbivorous freshwater turtle: advantages of small bite size. *Physiological Zoology*, 65:1028-1039.

Bjorndal, K. A., and A. B. Bolton. 1993. Digestive efficiencies in herbivorous and omnivorous freshwater turtles on plant diets: Do herbivores have a nutritional advantage? *Physiological Zoology*, 66:384-395.

Bjorndal, K. A., A. B. Bolton, and J. E. Moore. 1990. Digestive fermentation in herbivores: effect of food particle size. *Physiological Zoology*, 63:710-721.

Brooks, D. R., and D. A. McLennan. 1991. *Phylogeny, Ecology, and Behavior.* Chicago: The University of Chicago Press.

Brown, R. P., and V. Pérez-Mellado. 1994. Ecological energetics and food acquisition in dense Menorcan islet populations of the lizard Podarcis lilfordi. *Functional Ecology*, 8:427-434.

Burquez, A., O. Flores-Villela and A. Hernandez. 1986. Herbivory in a small iguanid lizard, *Sceloporus torquatus torquatus*. *Journal of Herpetology*, 20: 262-264.

Calder, J. H. 1994. Duration and periodicity of controls on coal formation; A Carboniferous perspective. Geological Society of America, Abstracts, Annual Meeting, 1994, p. A-95,

Calder, W. A., III. 1984. *Size, Function, and Life History.* Cambridge, Massachusetts: Harvard University Press.

Calow, P. and C. R. Townsend. 1981. Energetics, ecology and evolution. Pages 3-19 in: *Physiological Ecology: An Evolutionary Approach to Resource Use* (C.R. Townsend and P. Calow, eds.). Sunderland, Massachusetts: Sinauer Associates .

Carrier, D. R., 1987. The evolution of locomotor stamina in tetrapods: circumventing a mechanical constraint. *Paleobiology*, 13:326-341.

Carroll, R. L. 1964. Early evolution of the dissorophid amphibians. *Bulletin of the Museum of Comparative Zoology*, 131:161-250.

Carroll, R. L. 1969. A Middle Pennsylvanian captorhinomorph and the interrelationships of primitive reptiles. *Journal of Paleontology*, 43:151-170.

Carroll, R. L. 1970. The ancestry of reptiles. *Philosophical Transactions, Royal Society of London*, B 257:267-308.

Carroll, R. L. 1991. The origin of reptiles. Pages 331-353 in: *Origins of the Higher Groups of Tetrapods* (H.-P. Schultze and L. Trueb, eds.). Ithaca: Cornell University Press.

Carroll, R. L. 1992. The primary radiation of terrestrial vertebrates. *Annual Review of Earth Science 1992*, 20:45-84.

Carroll, R. L. 1994. Evaluation of geologic age and environmental factors in changing aspects of the terrestrial vertebrate fauna during the Carboniferous. *Transactions of the Royal Society of Edinburgh, Earth Sciences,* 84:427-431.

Carroll, R. L. 1995. Problems of the phylogenetic analysis of Paleozoic choanates. *Bulletin du Museum National d'histoire Naturelle, Section C: Sciences de la terre*, 17:389-445.

Carroll, R. L. 1996. Elongate early tetrapods: Why "lizzie" had little limbs. *Special Papers in Palaeontology*, 52:139-148.

Carroll, R. L., and J. Chorn. 1995. Vertebral development in the oldest microsaur and the problem of "lepospondyl" relationships. *Journal of Vertebrate Paleontology*, 15:37-56.

Carroll, R. L., P. Bybee, and W. D. Tidwell. 1991. The oldest microsaur (Amphibia). *Journal of Paleontology*, 65:314-322.

Carroll, R. L., and Gaskill, P. 1978. The order Microsauria. *American Philosophical Society Memoir,* 126:1-211.

Case, T. J. 1982. Ecology and evolution of the giantic chuckwallas, *Sauromalis hispidus* and *Sauromalus varius*. Pages 184-212 in: *Iguanas of the World* (G.M. Burghardt and A.S. Rand, eds.). Park Ridge, New Jersey: Noyes Publications.

Castilla, A. M., D. Bauwens, and G. A. Llorente. 1991. Diet composition of the lizard *Lacerta lepida* in central Spain. *Journal of Herpetology,* 25:30-36.

Cecil, C. B. 1990. Paleoclimate controls on stratigraphic repetition of chemical and siliciclastic rocks. *Geology*, 18:533-536.

Chapela, I. H., S. A. Rehner, T. R. Schultz, and U.G. Mueller. 1994. Evolutionary history of the symbiosis between fungus-growing ants and their fungi. *Science*, 266:1691-1696.

Chinsamy, A. 1993. Image analysis and the physiological implications of the vascularization of femora in archosaurs. *Modern Geology*, 19:101-108.

Chinsamy, A., and P. Dodson. 1995. Inside a dinosaur bone. *American Scientist*, 83: 174-180.

Chivers, D. J., and P. Langer. 1994. Gut form and function: variations and terminology. Pages 3-8 in: *The Digestive System in Mammals: Food, Form and Function* (D.J. Chivers and P. Langer, eds.). Cambridge, UK: Cambridge University Press.

Chivers, D. J., and P. Langer. 1994. Food, form, and function: interrelationships and future needs. Pages 411-430 in: *The Digestive System in Mammals: Food, Form and Function* (D. J. Chivers and P. Langer, eds.). Cambridge, UK: Cambridge University Press.

Christian, K. A., C. R. Tracy, and W. P. Porter. 1984. Diet, digestion, and food preferences of Galapagos land iguanas. *Herpetologica*, 40:205-212.

Clack, J. A. 1987. *Pholiderpeton scutigerium* Huxley, an amphibian from the Yorkshire Coal Measures. *Philosophical Transactions of the Royal Society of London* , B 318:1-103.

Clack, J. A. 1994. *Silvanerpeton miripedes*, a new anthracosauroid from the Visean of East Kirkton, West Lothian, Scotland. *Transactions of the Royal Society of Edinburgh*, 84:369-376.

Clarke, R. T. J. 1977. The gut and its micro-organisms. Pages 223-251 in: *Microbial Ecology of the Gut* (R. T. J. Clarke and T. Bauchop, eds.). New York: Academic Press.

Collinson, M. E., and A. C. Scott. 1987. Implications of vegetational changes through the geologic record on models for coal-forming swamp environments. Pages 67-85 in: *Coal and Coal-Bearing Strata: Recent*

Advances (A.C. Scott, ed.). *Geological Society of London, Special Publication*, No. 32.

Dalrymple, G. H. 1979. On the jaw mechanism of the snail-crushing lizards, *Dracaena daudin* 1802 (Reptilia, Lacertilia, Teiidae). *Journal of Herpetology*, 13:303-311.

Dearing, M. D. 1993. An alimentary specialization for herbivory in the tropical whiptail lizard *Cnemidophorus murinus*. *Journal of Herpetology*, 27:111-114.

Dearing, M. D., and J. J. Schall. 1993. Testing models of optimum diet assembly by the generalist herbivorous lizard *Cnemidophorus murinus*. *Ecology*, 73: 845-858.

DeMar, R. 1968. The Permian labyrinthodont amphibian *Dissorophus muticinctus*, and adaptations and phylogeny of the family Dissorophidae. *Journal of Paleontology*, 42:1210-1242.

Demment, M. W., and P. J. Van Soest. 1983. *Body size, Digestive Capacity, and Feeding Strategies of Herbivores.* Morrilton, Arkansas: Winrock International Livestock Research and Training Center.

Demment, M. W., and P. J. Van Soest. 1985. A nutritional explanation for body size patterns of ruminant and non-ruminant herbivores. *American Naturalist*, 125:641-672.

DiMichele, W. A., and R. B. Aronson. 1992. The Pennsylvanian-Permian vegetational transition: a terrestrial analog to the onshore-offshore hypothesis. *Evolution*, 46:807-824.

DiMichele, W. A., R. W. Hook, R. Beerbower, J. A. Boy, R. A. Gastaldo, N. Hotton III, T. L. Phillips, S. E. Scheckler, W. A. Shear, and H.-D. Sues. 1992. Paleozoic terrestrial ecosystems. Pages 205-325 in: *Terrestrial Ecosystems through Time* (A. K. Behrensmeyer, J. D. Damuth, W. A. DiMichele, R. Potts, H.-D. Sues, and S. L. Wing, eds.). Chicago: The University of Chicago Press.

DiMichele, W. A., T. L. Phillips, and R. A. Peppers. 1985. The influence of climate and depositional environment on the distribution and evolution of Pennsylvanian coal-swamp plants. Pages 223-256 in: *Geological Factors and the Evolution of Plants* (B. H. Tiffney, ed.). New Haven: Yale University Press.

Doddick, J. T., and S. P. Modesto, 1995. The cranial anatomy of the captorhinid reptile *Labidosaurikos meachami* from the Lower Permian of Oklahoma. *Palaeontology*, 38: 687-711.

Duellman, W. E. and L. Trueb. 1986. *Biology of Amphibians*. New York: McGraw-Hill.

Dulong, F. T., C. B. Cecil, and B. R. Wardlaw. 1994. Allogenesis of a Middle Pennsylvanian sedimentary cycle. *Geological Society Of America, Annual Meeting Abstracts*, 1994, p. A-241

Enlow, D. H. 1969. The bone of reptiles. Pages 45-80 in: *The Biology of the Reptilia, 1* (C. Gans and A.d'A. Bellairs, eds.). New York: Academic Press.

Estes, R., and E. W. Williams. 1984. Ontogenetic variation in the molariform teeth of lizards. *Journal of Vertebrate Paleontology*, 4:96-107.

Foley, W. J., A. Bouskila, A. Shkolnik, and I. Choshniak. 1992. Microbial digestion in the herbivorous lizard *Uromastyx aegyptius* (Agamidae). *Journal of the Zoological Society of London*, 226:387-398.

Frakes, L. A., J. E. Francis, and J. I. Syktus. 1992. *Climate Modes of the Phanerozoic.* Cambridge, UK: Cambridge University Press.

Gaffney, E. S., and M. C. McKenna. 1979. A Late Permian captorhinid from Rhodesia. *Novitates*, no. 2688, 15 pages.

Gauthier, J. A. 1994. The diversification of the amniotes. Pages 129-159 in: *Major Features of Vertebrate Evolution* (R. S. Spencer, ed.), *Short Courses in Paleontology*, Number 7.

Gould, S. J., and R. C. Lewontin. 1979. The spandrels of San Marco and the Panglossian paradigm: a critique of the adaptationist programme. *Philosophical Transactions, Royal Society of London*, B205:581-598.

Gould, S. J. and E. S. Vrba. 1982. Exaptation -- a missing term in the science of form. *Paleobiology*, 8:4-15

Greene, H. W. 1982. Dietary and phenotypic diversity in lizards: why are some organisms specialized? Pages 107-128 in: *Environmental Adaptation and Evolution* (D. Mossakowski and G. Roth, eds.). Stuttgart: Gustav Fischer.

Greene, H. W. 1994. *Homology: The Hierarchic Basis of Comparative Biology.* New York: Academic Press.

Greer, A. E. 1976. On the evolution of the giant Cape Verde scincid lizard *Macroscincus coctei. Journal of Natural History*, 10:691-712.

Guard, C. L. 1980. The reptilian digestive system: general characteristics. Pages 43-52 in: *Comparative Physiology: Primitive Mammals* (K. Schmidt-Nielsen, ed.). Cambridge, UK: Cambridge University Press.

Harvey, P. H., and M. D. Pagel. 1991. *The Comparative Method in Evolutionary Biology.* Oxford: Oxford University Press.

Heaton, M. J. 1979. Cranial anatomy of primitive captorhinid reptiles from the Late Pennsylvania and Early Permian, Oklahoma and Texas. *Oklahoma Geological Survey Bulletin*, 127:1-84.

Heaton, M. J., and R. R. Reisz. 1986. Phylogenetic relationships of captorhinomorph reptiles. *Canadian Journal of Earth Sciences*, 23:402-418.

Hladik, C. M., and D. J. Chivers. 1994. Food and the digestive system. Pages 65-73 in: *The Digestive System in Mammals: Food, Form and Function* (D. J. Chivers and P. Langer, eds). Cambridge, UK: Cambridge University Press.

Hook, R. W. 1989. Stratigraphic distribution of tetrapods in the Bowie and Wichita Groups, Permo-Carboniferous of North-Central Texas. Pages 47-53 in: *Permo-Carboniferous Vertebrate Paleontology, Lithostratigraphy, and Depositional Environments of North-Central Texas* (R. W. Hook, ed.). Field Trip Guidebook No. 2, 49th Annual Meeting., Society of Vertebrate Paleontology.

Hopson, J. A. 1994. Synapsid evolution and the radiation of non-eutherian mammals. Pages 190-219 in: *Major Features of Vertebrate Evolution* (R. S. Spencer, ed.), *Short Courses in Paleontology*, Number 7.

Hotton, N., III. 1955. A survey of adaptive relationships of dentition to diet in the North American Iguanidae. *American Midland Naturalist*, 53:88-114.

Iverson, J. B. 1982. Adaptations to herbivory in iguanine lizards. Pages 49-59 in: *Iguanas of the World* (G. M. Burghardt and A. S. Rand, eds.). Park Ridge, New Jersey: Noyes Publications.

Janis, C. M., and D. Ehrhardt. 1988. Correlation of relative muzzle width and relative incisor width with dietary preferences in ungulates. *Zoological Journal of the Linnean Society*, 92:267-284.

Janzen, D. H. 1973. Sweep samples of tropical foliage insects: effects of seasons, vegetation types, elevation, time of day, and insularity. *Ecology*, 54:687-708.

Jarman, P. J., and A. R. E. Sinclair. 1979. Feeding strategy and the patterning of resource partition in ungulates. Pages 130-163 in: *Serengeti. the Dynamics of an Ecosystem* (A. R. E. Sinclair and M. Norton-Griffiths,eds.). Chicago: The University of Chicago Press.

Kvale, E. P., G. S. Fraser, A. W. Archer, A. Zawistoski, N. Kemp, and P. McGough. 1994. Evidence of seasonal precipitation in Pennyslvanian sediments of the Illinois basin. *Geology*, 22:331-334.

Labandeira, C. C., and B. S. Beale. 1990. Arthropod terrestriality. Pages 214-256 in: *Arthropod Paleobiology* (S. J. Culver, ed.), *Short Courses in Paleontology*, Paleontological Society.

Langer, P., and D. J. Chivers. 1994. Classification of foods for comparative analysis of the gastro-intestinal tract. Pages 74-86 in: *The Digestive System in Mammals: Food, Form and Function* (D. J. Chivers and P. Langer, eds.). Cambridge, UK: Cambridge University Press.

Langston, W. Jr. 1965. *Oedaleops campi* (Reptilia, Pelycosauria), new genus and species from the Lower Permian of New Mexico, and the family Eothyrididae. *Bulletin of the University of Texas Museum*, Number 9.

Laurin, M., and R. R. Reisz. 1995. A reevaluation of early amniote phylogeny. *Zoological Journal of the Linnean Society,* 113:165-223.

Levins, R., and R. Lewontin. 1985. *The Dialectic Biologist.* Cambridge, Massachusetts, Harvard University Press.

Lucas, P. W. 1994. Categorization of food items relevant to oral processing. Pages 197-218 in: *The Digestive System in Mammals: Food, Form and Function* (D. J. Chivers and P. Langer, eds.). Cambridge, UK: Cambridge University Press.

Lucas, P. W., and R. T. Corlett. 1991. Quantitative aspects of the relationship between dentitions and diets. Pages 93-121 in: *Feeding and the Texture of Foods* (J. F. V. Vincent and P. J. Lilliford, eds.). Cambridge, UK: Cambridge University Press.

Lucas, P. W., and D. A. Luke. 1984. Chewing it over: Basic principles of food breakdown. Pages 283-299 in: *Food Acquisition and Processing in*

Primates (D. J. Chivers, B. A. Wood, and A. Bilsborough, eds.). London: Plenum Press.

Mautz, W. J., and K. A. Nagy. 1987. Ontogenetic changes in diet, field metabolic rate, and water flux in the herbivorous lizard *Dipsosaurus dorsalis*. *Physiological Zoology*, 60:640-658.

McBee, R.H. 1977. Fermentation in the hindgut. Pages 185-222 in: *Microbial Ecology of the Gut* (R. T. J. Clarke and T. Bauchop, eds.). New York: Academic Press.

McBee, R. H., and V. H. McBee. 1982. The hindgut fermentation in the Green Iguana, *Iguana iguana*. Pages 77-83 in: *Iguanas of the World.* (G. M. Burghardt and A. S. Rand, eds.). Park Ridge, New Jersey: Noyes Publications.

McLellan, B. N., and F. W. Hovey. 1995. The diet of grizzly bears in the Flathead River drainage of southeastern British Columbia. *Canadian Journal of Zoology*, 73:704-712.

Miller, K. B., and R. R. West. 1993. Reevaluation of Wolfcampian cyclothems in northeastern Kansas: Significance of subaerial exposure and flooding surfaces. *Kansas Geology Survey, Bulletin*, 235:1-26.

Milner, A. R. 1980. The tetrapod assemblage from Nyrany, Czechoslovakia. Pages 439-496 in: *The Terrestrial Environment and the Origin of Land Vertebrates* (A.L. Panchen, ed.). Systematics Association, Special Volume 15. New York: Academic Press.

Milner, A. R. 1987. The Westphalian tetrapod fauna: some aspects of its geography and ecology. *Journal of the Geological Society, London*, 144: 495-506.

Milner, A. R. 1993. The Paleozoic relatives of the lissamphibians. *Herpetological Monographs*, 6:8-27.

Milner, A. R. 1994. Biogeography of Paleozoic tetrapods. Pages 324-353 in: *Paleozoic Vertebrate Biostratigraphy and Biogeography* (J.A. Long, ed.). Baltimore: Johns Hopkins University Press .

Milner, A. R. and S. E. K. Sequeira. 1994. The temnospondyl amphibians from the Visean of East Kirton, West Lothian, Scotland. *Transactions of the Royal Society Of Edinburgh*, 84:331-361.

Milner, A. C. 1980. A review of the Nectridea (Amphibia). Pages 377-405 in: *The Terrestrial Environment and the Origin of Land Vertebrates* (A.L. Panchen, ed.). Systematics Association, Special Volume 15. New York: Academic Press.

Modesto, S. P. 1992. Did herbivory foster early amniote diversification? *Journal of Vertebrate Paleontology*, 12:44A.

Modesto, S. P. 1994. The Lower Permian synapsid *Glaucosaurus* from Texas. *Palaeontology*, 37:51-60.

Modesto, S. P. 1995. The skull of the herbivorous synapsid *Edaphosaurus boanerges* from the Lower Permian of Texas. *Palaeontology*, 38:213-239.

Modesto, S. P., and R. R. Reisz. 1990a. A new skeleton of *Ianthasaurus hardestii*, a primitive edaphosaur (Synapsida, Pelycosauria) from the Upper Pensylvanian of Kansas. *Canadian Journal Of Earth Sciences*, 27:834-844.

Modesto, S. P., and R. R. Reisz. 1992. Restudy of the Permo-Carboniferous synapsid *Edaphasaurus novomexicanus* Williston and Case, the oldest known herbivorous amniote. *Canadian Journal of Earth Sciences*, 29:2653-2662.

Moir, R. J. 1994. The "carnivorous" herbivores. Pages 87-102 in: *The Digestive System in Mammals: Food, Form and Function* (D. J. Chivers and P. Langer, eds.). Cambridge, UK: Cambridge University Press.

Moss, J. L. 1972. The morphology and phylogenetic relationships of the Lower Permian tetrapod *Tseajaia campi* Vaughn (Amphibia: Seymouriamorpha). *University of California, Publications in Geological Science*, 98:1-72.

Nagy, K. A., and V. H. Shoemaker. 1975. Energy and nitrogen budgets of the free-living desert lizard *Sauromalus obesus*. *Physiological Zoology*, 48:252-262.

Niklas, K. J. 1992. *Plant Biomechanics*. Chicago: The University of Chicago Press.

Nunez, H., J. Supan, H. Torres, J. H. Carrothers, and F. M. Jaksic. 1992. Autecologic observation on the endemic central Chilean Lizard *Pristidactylus volcanenis*. *Journal of Herpetology*, 26:228-230.

Olson, E. C. 1955. Parallelism in the evolution of the Permain vertebrate faunas of the old and new world. *Fieldiana: Geology*, 37:385-401.

Olson, E. C. 1947. The family Diadectidae and its bearing on the classification of reptiles. *Fieldiana: Geology*, 11:1-53.

Olson, E. C. 1961a. Food chains and the origin of mammals. International Colloquium on the Evolution of Lower and Unspecialized Mammals. *Koninllijke Vlaammse Academie Voor Wetenshappen, Letteren en Schne Kunsten von Belgie*, pt. 1:97-116.

Olson, E. C. 1961b. Jaw mechanisms: rhipidistians, amphibians, reptiles. *American Zoologist*, 1:205-215.

Olson, E. C. 1966a. Community evolution and the origin of mammals. *Ecology*, 47: 291-308.

Olson, E. C. 1966b. The relationships of *Diadectes*. *Fieldiana: Geology*, 14:199-227.

Olson, E. C. 1968. The family Caseidae. *Fieldiana: Geology*, 17:225-349.

Olson, E. C. 1971. *Vertebrate Paleozoology*. New York: Wiley-Interscience.

Olson, E. C. 1975. Permo-Carboniferous paleoecology and morphotypic series. *American Zoologist*, 15:371-389.

Olson, E. C. 1976. The exploitation of the land by early tetrapods. Pages 1-30 in: *Morphology and Biology of Reptiles* (A. d'A. Bellairs and C.B. Cox, eds.). Linnean Society, Symposium Series, no.3.

Panchen, A. L. 1972. The skull and skeleton of *Eogyrinus attheyi*. *Philosophical Transactions, of the Royal Society of London* , B 263:279-326.

Panchen, A. L., and T. R. Smithson, 1988. The relationships of the earliest tetrapods. Pages 1-31 in: *The Phylogeny and Classification of the Tetrapods, Volume 1: Amphibians, Reptiles, Birds* (M. J. Benton, ed.). Systematics Association, Special Volume No. 35A. Oxford: Clarendon Press.

Parra, R. 1978. Comparison of foregut and hindgut fermentation in herbivores. Pages 205-230 in: *The Ecology of Arboreal Folivores* (G. G. Montgomery, ed.). Washington, D.C.: Smithsonian Institution Press.

Paulissen, M. A., and J. M. Walker, 1994. Diet of the insular whiptail lizard *Cnemidophorus nigricolor* (Teiidae) from Grand Rocques Island, Venezuela. *Journal Of Herpetology*, 28:524-526.

Peabody, F. E. 1961. Annual growth zones in living and fossil vertebrates. *Journal of Morphology*, 108:11-62.

Penry, D. L., and P. A. Jumars. 1987. Modeling animal guts as chemical reactors. *American Naturalist*, 129:69-96.

Peters, R. H. 1983. *The Ecological Implications of Body Size*. Cambridge, UK: Cambridge University Press.

Pianka, E. R. 1970. Comparative autecology of the lizard *Cnemidophorus tigris* in different parts of its geographic range. *Ecology*, 51:703-720.

Pianka, E. . 1981. Resource acquisition and allocation among animals. Pages 300-314 in: *Physiological Ecology: An Evolutionary Approach to Resource Use* (C.R. Townsend and P. Calow, eds.). Sunderland, Massachusetts: Sinauer Associates .

Pough, F. H. 1973. Lizard energetics and diet. *Ecology*, 54:837-844.

Prins, R. A. 1977. Biochemical activities of gut micro-organisms. Pages 73-183 in: *Microbial Ecology of the Gut* (R. T. J. Clarke and T. Bauchop, eds.). New York: Academic Press.

Rand, A. S. 1978. Reptilian arboreal herbivores. Pages 115-122 in: *The Ecology of Arboreal Folivores* (G. G. Montgomery, ed.). Washington, D.C.: Smithsonian Institution Press.

Rand, A. S., B. A. Dugan, H. Monteza, and D. Vianda. 1990. The diet of a generalized folivore: *Iguana iguana* in Panama. *Journal of Herpetology*, 24:211-214.

Reif, W.-E. 1982. Functional morphology on the procrustean bed of the neutralism-selectionism debate–notes on the constructional morphlogy approach. Pages 46-59 in: *Studies in Paleoecology* (A. Seilacher, W.-E. Reif,and F. Westphal, eds.), *Neus Jahrbuch für Geologie und Paleontogie*, Abhandlungen 164 (Heft 1/2).

Reisz, R. R. 1972. Pelycosaurian reptiles from the Middle Pennsylvanian of North America. *Bulletin of the Museum of Comparative Zoology*, 144:27-62.

Reisz, R. R. 1975. Pennsylvanian pelycosaurs from Linton, Ohio, and Nyrany, Czechoslovakia. *Journal of Paleontology*, 49:522-527.

Reisz, R. R. 1986. Pelycosauria. Pages 1-102 in: *Handbuch der Palaoherpetology*, Volume 17 (P. Wellnhofer, ed.). Stuttgart: Fischer.

Reisz, R. R., and D. S Berman. 1986. *Ianthasaurus hardestii* n.sp., a primitive edaphosaur (Reptilia, Pelycosauria) from the Upper Pennsylvanian Rock Lake Shale near Garnett, Kansas. *Canadian Journal of Earth Sciences*, 23: 77-91.

Retallack, G. J., and J. Germain-Heins. 1994. Evidence from paleosols for the geological antiquity of rain forest. *Science*, 265:499-502.

Rieppel, O., and L. Labhardt. 1979. Mandibular mechanics in *Varanus niloticus* (Reptilia: Lacertilia). *Herpetologica*, 35:158-163.

Ricqlés, A. J. de. 1975. Recherches paléohistologiques sur les os longes des tétrapodes. VII. Sur la classification, la signication functionelle et l'histoire des tissus osseux des tétrapodes. Première partie: structures. *Annales de Paléontologie (Vertebres)*, 61:49-129.

Ricqlés, A. J. de. 1976. Recherches paléohistologiques sur les os longes des tétrapodes. VII. Sur la classification, la signication functionelle et l'histoire des tissus osseux des tétrapodes. Deuxième partie: Fonctions. *Annales de Paléontologie (Vertébrés)*, 62:71-126.

Ricqlés, A.J. de. 1978. Recherches paléohistologiques sur les os longes des tetrapodes. VII. Sur la classification, la signication functionelle et l'histoire des tissus osseux des tétrapodes. Troisième partie: evolution. *Annales de Paléontologie (Vertébrés)*, 64:85-111.

Rocha, C. F. D. 1989. Diet of a tropical lizard (*Liolaemus lutzae*) of southeastern Brazil. *Journal of Herpetology*, 23:292-294.

Romer, A. S. 1946. The primitive reptile *Limnoscelis* restudied. *American Journal of Science*, 244:149-188.

Romer, A. S. 1956. *Osteology of the Reptiles*. Chicago: University of Chicago Press.

Romer, A. S., and L. I.. Price. 1940. Review of the Pelycosauria. *Geological Society of America, Special Paper*, No. 28, 538 pages

Savage, D. C. 1977. Interactions between the host and its microbes. Pages 277-310 in: *Microbial Ecology of the Gut* (R. T. J. Clarke and T. Bauchop, eds.). New York: Academic Press.

Shear, W. 1991. The early development of terrestrial ecosystems. *Nature*, 351:283-289.

Schall, J. J., and S. Ressel. 1991. Toxic plant compounds and the diet of the predominately herbivorous lizard *Cnemidophorus arubensis*. *Copeia*, 1991: 111-119.

Scheckler, S. E. 1986. Floras of the Devonian-Mississippian transition. Pages 81-96 in: *Land Plants: Notes for a Short Course* (T. W. Broadhead, ed.). *University of Tennessee, Department of Geological Science, Studies in Geology*, 15.

Schultze, H.-P. 1991. A comparison of controversial hypotheses on the origin of tetrapods. Pages 29-67 in: *Origins of the Higher Groups of Tetrapods, Contoroversy and Consensus* (H.-P. Schultze and L. Trueb, eds.). Ithaca: Cornell University Press .

Schultze, H.-P., and J. R. Bolt. 1996. The lungfish *Tranodis* and the tetrapod fauna from the Upper Mississippiann of North America. Pages 31-54 in: *Studies on Carboniferous and Permian Vertebrates* (A. R. Milner, ed.). *Special Papers in Palaeontology*, 52.

Signor, P. W. 1982. A critical re-evaluation of the paradigm method of functional inference. Pages 59-63 in: *Studies in Paleoecology* (A. Seilacher, W.-E.

Reif,and F. Westphal, eds.), *Neus Jahrbuch für Geologie und Paleontogie*, Abhandlungen 164 (Heft 1/2).

Sibbing, F. A. 1991. Food capture and oral processing. Pages 377-412 in: *Cyprinid Fishes: Systematics, Biology and Exploitation* (I. J. Winfield and J. S. Nelson, eds.). London: Chapman and Hall.

Skoczylas, R. 1978. Physiology of the digestive tract. Pages 589-717 in: *Biology of the Reptilia, Volume 8, Physiology B* (C. Gans, ed.). New York: Academic Press.

Smithson, T. R. 1994. *Eldeceeon rolfei*, a new reptilomorph from the Visean of East Kirton, West Lothian, Scotland. *Transactions of the Royal Society of Edinburgh, Earth Sciences*, 84:377-382.

Smithson, T. R., R. L. Carroll, A. L. Panchen, and S. M. Andrews. 1994. *Westlothiana lizziae* from the Visean of East Kirkton, West Lothian, Scotland, and the amniote stem. *Transactions of the Royal Society of Edinburgh, Earth Sciences*, 84:383-412.

Smits, A. W. 1985. Behavioral and dietary responses to aridity in the chuckwalla, *Sauromalus hispidus. Journal Of Herpetology*, 19:441-449.

Sokol, O. M. 1967. Herbivory in lizards. *Evolution*, 21:192-194.

Sparks, R. E. 1995. Need for ecosystem management of large rivers and their floodplains. *Biosciences*, 45:168-182.

Stevens, C. E., and I. D. Hume. 1996. *Comparative Physiology of the Vertebrate Digestive System,* 2nd Edition. Cambridge, UK: Cambridge University Press.

Stovall, J. W., L. I. Price, and A.S. Romer. 1966. The postcranial skeleton of the giant Permian pelycosaur *Cotylorhynchus romeri. Bulletin of the Museum of Comparative Zoology*, 135:1-30.

Tandon, S. K., and M. R. Gibling. 1994. Calcrete and coal in late Carboniferous cyclothems of Nova Scotia, Canada: climate and sea-level changes linked. *Geology*, 22:755-758.

Thomson, K. S. 1993. The origin of tetrapods. *American Journal of Science*, 293-A: 33-62.

Thomson, K. S. 1994. The origin of tetrapods. Pages 85-107 in: *Major Features of Vertebrate Evolution* (R. S. Spencer, ed.). *Short Courses in Paleontology*, Number 7.

Thomson, K. S., and K. H. Bossy. 1970. Adaptive trends and relationships in early Amphibia. *Forma et Functio*, 3:7-31.

Throckmorton, G. 1973. Digestive efficiency in the herbivorous lizard *Ctenosauria pectinata. Copeia*, 1973:431-435.

Toft, C. A. 1981. Feeding ecology of Panamanian litter anurans: patterns in diet and foraging mode. *Journal of Herpetology*, 15:139-144.

Troyer, K. 1984a. Structure and function of digestive tract of a herbivorous lizard, *Iguana iguana. Physiological Zoology*, 57:1-8

Troyer, K. 1984b. Diet selection and digestion in *Iguana iguana*: the importance of age and nutrient requirements. *Oecologia*, 61:201-207.

Van der Zwan, C. J., M. C. Bouldter, and R. N. L. B. Hubbard. 1985. Climatic change during the lower Carboniferous in Euramerica, based on multivariate statistical analysis of palynological data. *Palaeogeography, Palaeo-climatology, Palaeoecology*, 52:1-20.

Van Devender, R. W. 1982. Growth and ecology of spiny-tailed and green iguanas in Costa Rica, with comments on the evolution of herbivory and large body size. Pages 162-183 in: *Iguanas of the World* (G.M. Burghardt and A.S. Rand,eds.). Park Ridge, New Jersey: Noyes Publications.

Van Sluys, M. 1993. Food habits of the lizard *Tropidurus itambere* (Tropiduridae) in southeastern Brazil. *Journal Of Herpetology*, 27:347-351.

Van Soest, P. J. 1994. *Nutritional Ecology of the Ruminant*, 2nd Edition. Ithaca: Cornell University Press.

Vincent, J. F. V. 1990. Fracture properties of plants. *Advances in Botanical Research*, 17:235-287.

Vincent, J. F. V. 1991. Texture of plants and fruits. Pages 19-34 in: *Feeding and the Texture of Foods* (J. F. V. Vincent and P. J. Lilliford, eds.). Cambridge, UK: Cambridge University Press.

Vitt, L. J., and J. P. Caldwell. 1994. Resource utilization and guild structure of small vertebrates in the Amazon forest leaf litter. *Journal of Zoological Society of London*, 234:463-476.

Waldschmidt, S. R., S. M. Jones, and W. P. Porter. 1988. Reptilia. Pages 553-619 in: *Animal Energetics, Volume 2, Bivalvia through Tetrapoda* (T. J. Pandian and F. J. Vernberg, eds.). New York: Academic Press.

Walker, A., H. N. Hoeck, and L. Perez. 1978. Microwear of mammalian teeth as an indicator of diet. *Science*, 201:908-910.

Watson, D. M. S. 1954. On *Bolosaurus* and the origin and classification of reptiles. *Bulletin of the Museum of Comparative Zoology*, 111:297-449.

Wiewandt, T. A. 1982. Evolution of nesting patterns in iguanine lizards. Pages 119-141 in: *Iguanas of the World* (G. M. Burghardt and A. S. Rand, eds.). Park Ridge, New Jersey: Noyes Publications.

Wilson, K. J., and A. K. Lee. 1974. Energy expenditure of a large herbivorous lizard. *Copeia*, 1974:338-348.

Witzke, B. J. 1990. Palaeoclimatic constraints for Paleozoic latitudes of Laurentia and Euramerica. Pages 57-74 in: *Palaeozoic Palaeogeography and Biogeography* (W. S. McKerrow and C. R. Scotese, eds.). *Geological Society of London*, Memoir No. 12

Wright, V. P. 1990. Equatorial aridity and climatic oscillation during the early Carboniferous, southern Britain. *Journal of the Geological Society of London*, 147:359-363.

Ziegler, A. M. 1990. Phytogeographic patterns and continental configurations during the Permian Period. Pages 363-382 *Palaeozoic Palaeogeography and Biogeography* (W. S. McKerrow and C. R. Scotese, eds.). *Geological Society of London*, Memoir No. 12

Ziegler, A. M., R. K. Bambach, J. T. Parrish, S. F. Barrett, E. H. Gierlowski, W. C. Parker, A. Raymond, and J. J. Sepkosi, Jr. 1981. Paleozoic biogeography and climatology. Pages 231-266 in: *Paleobotany, Paleoecology and Evolution, Volume 2* (K. J. Niklas, ed.). New York: Praeger.

Zimmerman, L. C., and C. R. Tracy. 1989. Interactions between the environment and ectothermy and herbivory in reptiles. *Physiological Zoology*, 62:374-409.

CHAPTER 8

EVOLUTION OF THE AMNIOTE EGG

Mary J. Packard

Roger S. Seymour

INTRODUCTION

A major difference between modern amphibians and the remainder of the tetrapods is the nature of the egg produced by the former: Amphibians lay anamniotic-eggs, whereas all other tetrapods produce amniotic eggs. The distinction in egg types actually rests on the presence of a suite of extraembryonic membranes, the amnion (which gives the "amniote" egg its name), chorion, allantois, and yolk sac. However, evolution of the amniotic egg involved a number of important changes that may have-been unrelated to the appearance of extraembryonic membranes. These included, but were not necessarily limited to, the replacement and/or modification of egg envelopes, the elimination of barriers to diffusion of gases, an increase in egg size, changes in-patterns of development, and alterations in how nutrients are supplied during embryogenesis.

Our purpose in this chapter is to present a plausible, but necessarily speculative, scenario for the evolution of the amniote egg from an anamniotic predecessor. In so doing, we will not ascribe a particular egg type to a particular fossil group or lineage. Fossil forms that are assumed to be amniotes are identified partly on the basis of similarity to extant amniotes, that is, tetrapods known to produce an amniote egg. However, morphological similarities among tetrapods do not provide insight into the egg produced by those organisms. Fossil forms that may clearly be "anamniotes" might have laid an egg that would be characterized as amniotic just as forms that would be classified as amniotes might have produced eggs that lacked the extraembryonic structures characteristic of eggs of extant amniotes. Eggs produced by the predecessor to the amniotes and extraembryonic structures within such eggs are unlikely to be fossilized; therefore, insight into the transition in egg form and function must come primarily from a consideration of eggs and embryos of contemporary anamniotes and amniotes.

EVOLUTION OF THE AMNIOTE EGG

One view for the evolution of amniotes from anamniotes (Carroll, 1969, 1970, 1982, 1988) concludes that the transition was made by species of relatively small adult size that presumably laid eggs similar in many respects to the direct developing eggs of a variety of extant amphibians. The oocyte produced by the predecessor to amniotes, like the oocyte produced by contemporary amphibians

---

Figure 1. Highly schematic representation of evolution of the early amniote egg from an anamniotic predecessor. (A) The anamniotic predecessor presumably was similar to eggs of contemporary amphibians and consisted of a relatively large mass of yolk surrounded by the vitelline membrane and the jelly layers. Collectively, the latter two components comprise the egg capsule. (B) The perivitelline space, the space in which the embryo develops, was formed by the movement of water from the environment across the capsule. This process thinned the capsule and reduced, but did not eliminate, the barrier to gas exchange. (C) The elaboration of a fibrous shell membrane external to the jelly layers marked the onset of the transition from the anamniotic to the amniotic state. This additional egg envelope provided structural support that the capsule alone could not provide. (D-F). The movement of water from the jelly into the yolk compartment of the preamniote/early amniote egg eliminated the jelly as a barrier to diffusion and helped to dry the spaces between the fibers of the shell membrane. Both processes facilitated gas exchange and contributed to the increase in egg size that characterizes the anamniote amniote transition. ➔

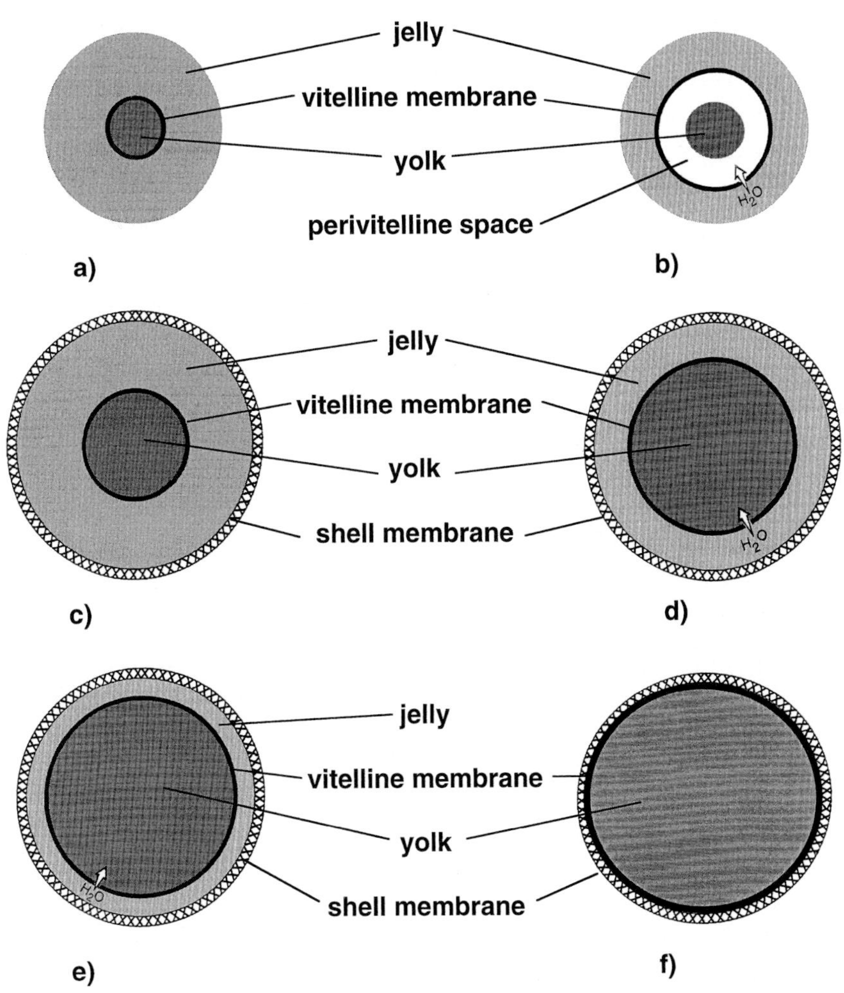
jelly
vitelline membrane
yolk
perivitelline space
H_2O
a)
b)
jelly
vitelline membrane
yolk
shell membrane
H_2O
c)
d)
jelly
vitelline membrane
yolk
shell membrane
H_2O
e)
f)

(Dumont and Brummett, 1985), presumably lacked an eggshell but was delimited by at least two egg envelopes: the vitelline membrane and one or more jelly layers. These egg envelopes were elaborated external to the plasma membrane that surrounded the oocyte (Dumont and Brummett, 1985) and formed a combined structure called the egg capsule (Fig. 1A). A consideration of eggs of contemporary amphibians indicates that the egg capsule probably provided some support and protection to the egg but was not an effective barrier to water loss (Taigen *et al.*, 1984). The ancestral egg probably was oviposited in terrestrial environments and thus may not have been in contact with free standing water (Carroll, 1969). However, the extreme permeability of the egg capsule to water vapor required that eggs be deposited in very wet settings or that one of the parents provide the eggs with water (Taigen *et al.*, 1984). After a relatively protracted incubation-compared to that of amphibian eggs that give rise to larvae (Bradford, 1990), small, fully formed, precocial young emerged.

The predecessor of the amniote egg, like the direct developing eggs of some contemporary amphibians, was large relative to eggs laid by species with a larval periods (but still was small compared to eggs of the vast majority of amniotes (Carey *et al.*, 1980; Carroll, 1970; Seymour, 1979). We view the increase in egg size that occurred as part of the transition from the anamniotic to the amniotic state as critical to the evolutionary success of the amniote egg. However, increases in egg size required that conflicts and constraining factors be overcome and also have important implications with regard to metabolic intensity, the exchange of respiratory gases, and the mobilization and transport of nutrients. Recent investigations of these conflicts, constraints, and implications now permit us to examine the quantitative and qualitative differences between eggs and embryos of living anamniotes and amniotes and draw inferences concerning the evolution of the amniote egg (Seymour and Bradford, 1995).

Egg Size and Embryonic Respiration

Evolution of chorioallantoic gas exchange (Szarski, 1968) clearly is an important factor that contributed to evolution of the vast range of egg sizes characterizing contemporary amniotes. However, we view functional replacement of the gelatinous egg capsule by a fibrous shell membrane, rather than evolution of extraembryonic membranes, to be among the key innovations leading to evolution of the amniote egg. An appreciation of the importance of this innovation comes from an understanding that the capsule of eggs of extant amphibians severely limits gas exchange between the egg and its environment and thus limits embryonic respiration and size of embryos. This barrier to gas transfer had to be eliminated before size of eggs could increase substantively.

The embryo of contemporary amphibians develops within the perivitelline space, a fluid filled compartment located just beneath the capsule (Fig. 1B). Early in development the osmotic uptake of water causes the capsule to expand and lift away from the embryo (Salthe, 1965). These changes permit adequate space to accommodate embryonic growth and also increase the surface area and decrease the thickness of the capsule, both of which facilitate gas exchange (Seymour and Bradford, 1987, 1995). These changes notwithstanding, however, the capsule remains a critical barrier to the exchange of gases by embryos of extant amphibians, and diffusion of oxygen through the jelly layers is about 75% that for diffusion through pure water and less than 1/300,000 that of diffusion in air (Seymour, 1994).

Several lines of evidence indicate that the barrier to diffusive movement of oxygen through the jelly constrains the metabolism of embryos of contemporary amphibians late in development, particularly in large eggs or in eggs naturally incubated at higher temperatures (Seymour, 1994; Seymour and Bradford, 1995). For example, oxygen uptake often increases greatly when the capsule is breached at hatching, and lowering the external oxygen level stimulates hatching. In addition, the allometric relationship between hatching stage oxygen consumption and ovum mass for amphibian embryos has a low scaling exponent (0.52 at 30° C) compared to that of (reptilian (0.82) and avian (0.74) embryos, an observation indicating that oxygen diffusion through the capsule becomes progressively more limiting in larger embryos. In contrast, the allometric relationship between incubation time and ovum mass for amphibian embryos has a high scaling exponent (0.44) in comparison to reptiles (0.12) and birds (0.22). Long incubation time and slow rate of development are known to be caused by oxygen limitation (Bradford and Seymour, 1988). Lastly, oxygen limitation can be demonstrated in eggs incubated at higher temperatures because metabolic oxygen demand increases with temperature, and diffusion cannot keep pace with the requirements of the embryo.

The preceding observations indicate that modern amphibians are straining the capacity of the capsule to supply sufficient oxygen, especially in the large eggs that characterize direct developing species. In addition, long development times resulting from oxygen limitation increase the energetic cost of maintenance and the dangers of dehydration or attack by parasites or predators. From the standpoints of gas exchange and rate of development, therefore, the capsule of the preamniote egg, like that of eggs of extant amphibians, was a potential liability. Evolution of larger eggs that would accommodate larger, more metabolically active embryos could not occur unless this liability were circumvented or eliminated.

An additional consideration with regard to increasing egg size is the requirement for physical support of the egg. It is a general principle that the size of supporting tissues increases disproportionately with the size of the objects they support. For example, the mass of supporting structures for both invertebrates and vertebrates as well as the mass of mollusc shells and avian eggshells increases with body mass raised to approximately the 1.1 power (Anderson *et al.*, 1979). Consequently, the supporting tissues of large animals represent a greater proportion of total mass than in small animals. In the case of fluids held within a spherical shell, such as the amphibian egg capsule, a Laplace principle requires that walls become thicker or stronger if the internal volume increases. These observations indicate that larger eggs ought to have thicker supporting walls, but this requirement conflicts with the need for adequate diffusion of respiratory gases as egg and embryo size increases. The requirement for additional support with increases in egg size is even more critical for eggs deposited in terrestrial environments because eggs laid in such settings lack the support provided by an aquatic medium. Under such circumstances, gravity and surface tension would tend to deform the egg and the enclosed embryo unless the strength of the capsule were able to resist these forces.

Seymour and Bradford (1995) evaluate the conflicting requirements of physical support and gas transfer as a function of egg size and conclude that oxygen would become limiting in eggs with capsules about 12 mm diameter and 0.5 mm thickness. Such a large, thin capsule would be extremely delicate and would enclose a space of only 850 μl. No living amphibian is known to produce a capsule as large and as thin, and the reason may be that such eggs would be structurally unsound.

We suggest that a fibrous shell membrane, similar to those characterizing eggs of contemporary oviparous amniotes (Board, 1982; Packard and DeMarco, 1991; Packard and Hirsch,1986; Packard *et al.*, 1982), was the evolutionary innovation that enabled eggs to overcome the constraints imposed by the egg capsule (Fig. 1C). Deposited over the surface of the egg in the oviduct, the tough, proteinaceous shell membrane would provide structural support that the vitelline membrane and jelly layers alone could not provide. However, acquisition of this egg envelope would not have been effective without two additional changes in the ancestral egg: The jelly layers between the embryo and the shell membrane had to be eliminated, and the spaces between fibers of the shell membrane had to be free of water. These requirements could have been met by movement of water from the jelly layers into the yolk compartment (Figs. 1D-F). This process would cause the yolk to swell and thus would bring the embryo into close proximity to the shell membrane and the site for gas exchange and also would facilitate drying of the shell membrane.

Transfer of water from albumen to yolk, swelling of the yolk compartment, and drying of the shell membrane occur early in development in eggs of oviparous reptiles (Packard and Packard, 1988). The hydrated shell membrane of freshly laid eggs gives the eggshell a translucent appearance. However, as water moves from the albumen into the yolk and the shell membrane dries, the eggshell becomes increasingly white and opaque, and the permeability of the egg to gases increases (Deeming and Thompson, 1991). These movements of water occur only in fertile eggs and result from a shift in the balance between the osmotic tension of the egg contents and the capillary tension developed in the water occupying the spaces among fibers of the shell membrane (Seymour and Piiper, 1988).

Although both egg envelopes are formed in the oviduct (Dumont and Brummett, 1985), it is not clear that the jelly of the amphibian egg capsule and the albumen of amniote eggs are homologous. Nonetheless, they are analogous in location and structure, and the possibility exists that modification of the jelly layers gave rise to the albumen of amniote eggs. Moreover, movements of water similar to those known to occur in eggs of modern oviparous reptiles may have evolved during the anamniote–amniote transition. Indeed, osmotic removal of water from the albumen and shell membrane of reptilian eggs may have antecedents in the osmotic uptake of environmental water into the perivitelline space that occurs in fertile eggs of contemporary amphibians (Salthe, 1965).

Released from the constraints imposed by the capsule, the preamniote egg could have increased to a size considerably larger than eggs produced by all extant amphibians. Even at this stage, the additional respiratory capacity provided by the chorioallantoic epithelium may not have been essential for the increase in egg size. Cutaneous gas exchange alone satisfies the requirements of adult plethodontid salamanders weighing more than 10 g (Gatten *et al.*, 1992); therefore the skin and gills may well have been sufficient to supply very large embryos with oxygen, once the constraint of the jelly had been overcome. Elaboration of accessory respiratory structures such as bell gills or highly vascularized tails, which characterize embryos of a variety of modern direct developing amphibians (Elinson, 1987), might also have provided additional respiratory capacity. However, these structures do not eliminate the constraint to gas exchange by the jelly layers among eggs of extant amphibians and could not relieve the constraint to egg size inherent in the anamniotic state.

The preceding observations do not provide insight into the timing of the evolution of chorioallantoic gas exchange. This epithelium could have evolved prior to, during, or subsequent to the appearance of the shell membrane and elimination of the jelly layers. However, if the epithelium arose prior to the changes outlined previously, we assume that its respiratory function, like that of the

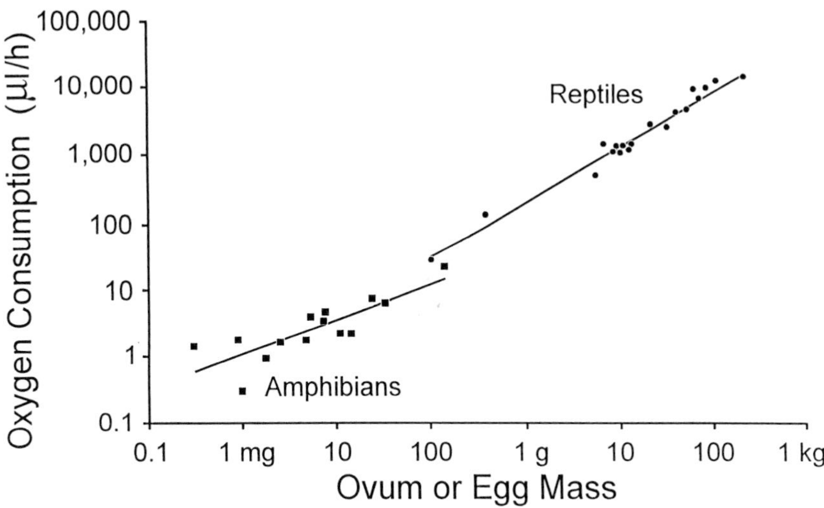

Figure 2. Relationship between the rate of oxygen consumption in hatching stage embryos of amphibians (Seymour and Bradford, 1995) and reptiles (Vleck and Hoyt, 1991). The equation for amphibians is $VO_2 = 1.06\ M^{0.52}$ and that for reptiles is $VO_2 = 0.68\ M^{0.82}$ in which rate of oxygen consumption (VO_2) has units of μl/hr and egg mass (M) has units of mg. Some of the data for amphibian eggs were corrected to 30°C assuming a Q_{10} of 1.5. The data for reptilian eggs were collected at temperatures between 29 and 31°C. The data points for the smallest reptilian eggs, those of *Menetia greyii*, came from unpublished work of M. B. Thompson and was not used to derive the regression.

accessory respiratory structures in eggs of modern amphibians, would have been constrained by the presence of the jelly layers. Thus, chorioallantoic gas exchange could not support the full range of sizes that characterize eggs of contemporary amniotes in the absence of other changes.

The respiratory consequences of the transition from jelly capsules to shell membranes and evolution of chorioallantoic gas exchange are obvious in an allometric comparison of hatching stage oxygen consumption rates in embryonic amphibians (Fig. 2). The regression for reptiles has a significantly higher slope than for amphibians, illustrating that the larger eggs of reptiles are not limited by diffusion through a capsule. Even where the range of masses overlaps, reptilian eggs have higher metabolic rates. For example, the largest amphibian eggs for which we have data are those of the salamander, *Ensatina eschscholtzii*, with a 137 mg ovum and a hatching stage rate ofoxygen consumption of 11.7 μl/hr at 14° C, or 22.4 μl/hr when corrected to 30° (Bradford, 1984), the highest point for amphibians shown on figure 2. It is important to emphasize, however, that this species is unlikely to survive incubation at 30°C for reasons including the capsule's limitation to oxygen diffusion. A 137 mg reptilian egg would be expected to consume oxygen at 38.2 μl/hr, and higher still if oxygen consumption could be calculated from ovum (yolk) mass. The smallest reptilian eggs for which data on oxygen consumption data are available are produced by the skink, *Menetia greyii*, with a 104 mg fresh egg, 93 mg of yolk, and a maximum rate of oxygen consumption of 27 μl/hr at 29°C (M. B. Thompson, unpublished data). At 30°C, an amphibian embryo from a 93 mg ovum would be expected to consume oxygen at a rate of 11μl/hr. It is clear that when the capsule of anamniote eggs was replaced by shell membranes, the rate of aerobic metabolism and the potential for embryonic growth were released from the constraints of oxygen diffusion.

In summary, our view of evolution of the amniote egg to this point includes replacement of the egg capsule with a fibrous shell membrane and elimination of the diffusive barrier presented by the jelly layers. These modifications enabled major increases in egg size and could have occurred prior to evolution of extraembryonic membranes. However, the additional respiratory capacity afforded by the chorioallantois was critical to further increases in egg size.

Provision of Nutrients and Calcium to Embryos

The transition from the anamniotic to the amniotic condition required a number of developmental changes in addition to those outlined previously. For example, the evolution of a large egg that gives rise to a relatively large precocial young necessitates a substantial increase in the nutrient reserves of the egg, (i.e., an increase in the mass of yolk). Increases in the size of the yolk introduced a suite of problems with regard to acquision of yolk reserves during embryogenesis.

Eggs of contemporary amphibians undergo holoblastic cleavage so that the entire mass of yolk is partitioned among individual cells (Elinson, 1987). Consequently, each daughter cell generally receives a supply of yolk platelets during division, and this complement of yolk is sufficient to meet the energy demands of cells in eggs giving rise to larvae (Karasaki, 1963; Kielbowna, 1975; Selman and Pawsey, 1965). Incubation is relatively short and yields young that essentially complete development during the larval stage (Bradford, 1990; Duellman and Trueb, 1986).

Evolution of the amniote egg necessitated changes in this basic mechanism for nutrient provision. As egg size increased, the amount of yolk relative to the amount of cytoplasm increased, and cleavage of the large body of yolk became increasingly difficult (Elinson, 1989). Indeed, among contemporary amphibians with direct development, holoblastic cleavage may form two populations of cells: Those at the embryonic area are yolk poor, whereas those cleaved from the bulk of the yolk-laden oocyte are relatively yolk rich (del Pino and Escobar, 1981; Elinson and del Pino, 1985). These observations indicate that provision of nutrients to the developing embryo in the preamniote egg could not have been met simply by provision of all cells with a complement of yolk platelets during cleavage. This challenge was met via two important changes in developmental pattern during evolution of the amniote egg: Abandonment of holoblastic cleavage in favor of meroblastic cleavage and development of a highly vascularized yolk sac (Elinson, 1987, 1989).

Meroblastic cleavage divides cells in the embryonic area but leaves the yolk uncleaved. Early in development, when cells of the embryonic area are in intimate contact with the mass of yolk, nutrients probably are obtained simply by engulfing yolk. However, growth of the embryo leads to increasing isolation of cells from the nutrient reserve of the yolk, and nutrients are supplied by the cellular epithelium of the yolk sac membrane and the developing vascular network. Some extant amphibians exhibit an "amniote like" pattern of cleavage, but none exhibits true meroblastic cleavage (Elinson, 1987). A vascularized yolk sac forms in direct developing species, but this structure is not homologous to the yolk sac of amniote embryos (Elinson, 1987, 1989; Elinson and del Pino, 1985). Nonetheless, the problem of nutrient provision as egg size increased was met in a similar way among anamniotes and amniotes.

The modifications outlined above–elaboration of a fibrous shell membrane, elimination of the barrier to gas exchange provided by the jelly layer, increases in nutrient reserves, the abandonment of holoblastic cleavage, and the formation of the full complement of extraembryonic membranes–complete the transition from the anamniotic to the amniotic state. However, the primitive amniote egg probably lacked calcareous material on the outer surface of the shell membrane. In that regard, the primitive amniote egg may have resembled the eggs of modern monotremes (Hughes, 1984). In contrast, shells of eggs of most contemporary, oviparous amniotes contain crystalline material ranging from isolated deposits or a thin crust in eggs of some lizards and snakes to the thick, compact, highly organized calcareous layer of archosaurian eggs (Board, 1982; Ferguson, 1982; Packard and DeMarco, 1991). Thus, few extant amniotes lay an egg characterized only by the basic features exhibited by eggs of early amniotes.

Increases in egg size during evolution of the amniote egg probably necessitated proportionately greater increases in some nutrients than others. For example, one critical alteration in developmental pattern accompanying the evolution of direct development in amphibians is the change in the timing of skeletal ossification (Elinson, 1990). A clear dichotomy in onset of ossification characterizes species with a larval stage and those with direct development. Larvae may exhibit extensive ossification prior to completion of metamorphosis or may concentrate ossification in the post metamorphic period, depending on the species, but ossification does not occur during embryogenesis (Atkinson, 1981; Dodd and Dodd, 1976; Duellman and Trueb, 1986; Hanken and Hall, 1984, 1988; Wiens, 1989). In contrast, ossification is initiated during embryogenesis in species that undergo direct development (Elinson, 1990; Hanken *et al.*, 1992; Lynn, 1942). This change in the timing of ossification presumably requires that direct developing eggs contain considerably more calcium than eggs that will give rise to a cartilaginous larva. A change in the timing of ossification presumably would have necessitated an increase in calcium content of eggs of direct developing amphibians even if size of eggs had not increased appreciably.

Sources of calcium for development and patterns of mobilization and deposition of calcium among embryos of contemporary amphibians have been studied in only one species of direct developing anuran, *Eleutherodactylus coqui* (Packard *et al.*, 1996). A consideration of these patterns in this species and among contemporary amniotes and of the general differences in developmental pattern among anamniotes and amniotes allows the formulation of a plausible sequence of events to characterize the transition from the anamniotic to the amniotic state with respect to mobilization and deposition of calcium.

In the absence of other sources of calcium (e.g., a calcareous eggshell), all the calcium required to support development must be invested in the oocyte if an egg is to give rise to precocial young. Thus, the direct developing egg of ancestral amphibians and the derivative amniote egg must have contained considerably more calcium than characterized eggs that gave rise to larvae. In addition, the pattern of mobilization and deposition of calcium by direct-developing embryos of ancestral amphibians and embryos of early amniotes probably was similar (Fig. 3). Calcium content of yolk probably did not change appreciably during the differentiation phase of development but presumably declined rapidly once embryos entered the growth phase (Fig. 3). This transfer of calcium caused calcium content of yolk to decline and that of the carcass of the embryo to increase. However, the total quantity of calcium within the egg did not change during incubation (Fig. 3). Thus, freshly laid eggs and eggs near the end of incubation contained similar quantities of calcium.

The calcium required for ossification of larvae of extant amphibians is obtained from the diet and via transport across the gills and skin and may be sequestered in endolymphatic deposits prior to its mobilization to support ossification (Burggren and Just, 1992;

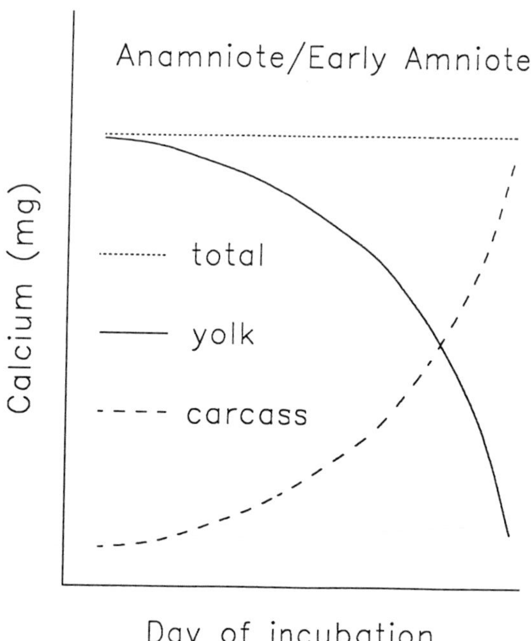

Figure 3. Presumed pattern of mobilization and deposition of calcium by embryos of anamniotes and early amniotes, prior to appearance of a calcareous layer on the outer surface of the shell membrane. Embryos rely exclusively on the yolk as a source of calcium. Calcium is transferred from yolk to carcass, but the total quantity of calcium inside the egg does not change during incubation.

Guardabassi, 1960; Pilkington and Simkiss, 1966). A somewhat similar process seemingly occurs in some extant amphibians with direct development in that distinct endolymphatic deposits form during embryogenesis well before the onset of ossification (Townsend and Stewart, 1985). The sacs enlarge and then regress coincident with the beginning of ossification (Townsend and Stewart, 1985). These observations indicate that calcium is transferred from yolk to endolymphatic deposits and then to the developing skeleton. Some yolk calcium may be used for direct support of ossification, but this possibility cannot be evaluated.

The ancestral preamniote probably exploited a similar strategy with regard to yolk calcium mobilization: Yolk calcium may have been deposited in the endolymphatic sacs and then mobilized from those deposits to support ossification. An analogous pattern may also have characterized early amniotes. However, at some point during evolutionary modification of the amniote egg, the pattern of utilization of yolk calcium shifted. As a result, more and more of the calcium mobilized from the yolk was used in direct support of ossification. This shift may have come about as a result of the increase in size of eggs and embryos. The quantity of calcium that can be sequestered in endolymphatic deposits may be too small to make deposition of yolk calcium therein and its subsequent mobilization a viable strategy for embryos larger than some critical size. In this regard, the pattern of mobilization and deposition of calcium characterizing anamniote-preamniote eggs may have constrained embryo size just as the diffusive properties of the egg capsule did.

The basic pattern of calcium mobilization and deposition of the early amniote egg at this point probably consisted of a transfer of calcium from the yolk compartment to the carcass of the developing embryo, and the cellular epithelium of the yolk sac controlled transport of calcium from the yolk to the vascular supply. The chorioallantoic epithelium was not involved in calcium mobilization or transport and thus functioned largely as a respiratory organ.

A critical modification to the basic plan of the early amniote egg involved deposition of calcium salts on the enveloping shell membrane. Unfortunately, uncalcified or poorly calcified eggshells are unlikely to be preserved in the fossil record (Hirsch and Packard, 1987), and the point at which this modification arose cannot be determined. However, the shells of ancestral eggs that first exhibited this modification may have resembled the shells of eggs of a variety of extant lizards and snakes (Packard *et al.*, 1982; Packard and DeMarco, 1991; Packard and Hirsch, 1986). These eggshells are characterized by a very thick, fibrous shell membrane with widely spaced, isolated calcareous deposits or a thin crust of crystalline material on the outer surface (Packard and Hirsch, 1986).

The appearance of calcareous material on the shell membrane made available a new source of calcium for embryos. Exploiting this

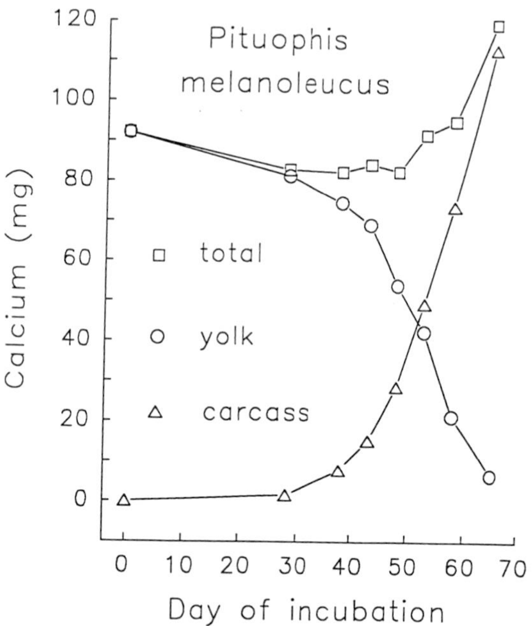

Figure 4. Pattern of mobilization and deposition of calcium by embryos of the bullsnake *Pituophis melanoleucus*. Embryos rely primarily on the yolk as a source of calcium and obtain only a small proportion of their calcium from the eggshell. Calcium mobilized from the shell is deposited in the carcass of the embryo, and yolk is depleted of essentially its entire complement of calcium. Data are from M. J. Packard and Packard (1988). Eggs were incubated at 26°C. The last data point for each line is plotted on the average day of hatching.

resource presumably required parallel modifications to the cellular epithelium of the chorioallantoic membrane that heretofore had functioned primarily in a respiratory capacity. Calcium sequestered in the eggshell cannot be used by a developing embryo unless the calcium can be solubilized and then transported into the vasculature (Packard and Clark, 1996). An epithelium that mediates gas exchange is unlikely to be able to mediate these additional processes without modifications to its physiological capabilities.

The capacity to use calcium sequestered in the eggshell resulted in a change in the pattern of mobilization and deposition of calcium during embryogenesis. Amniote embryos still relied primarily on the yolk as a source of calcium but began to obtain a small fraction of their calcium requirement from the eggshell. The influx of calcium from the eggshell eventually resulted in a net increase in the quantity of calcium contained within eggs (i.e., within the body of the embryo and the yolk sac) so that contents of near term eggs came to contain more calcium than freshly laid eggs. This pattern of mobilization and deposition of calcium probably was similar to that characterizing eggs of contemporary snakes (Fig. 4).

Yolk of eggs of extant snakes contains most of the calcium available to support development, and the eggshell contains relatively small amounts of calcium (Packard, 1994). The quantity of calcium in yolk does not change appreciably during the differentiation phase of development but declines precipitously during the growth phase as embryos draw on the calcium reserve of the yolk to support skeletogenesis (Fig. 4). During much of development, calcium is simply transferred from the yolk to the carcass of the growing embryo, and the total quantity of calcium within the egg remains constant in a pattern similar to that characterizing amniote eggs prior to the appearance of calcareous material in the eggshell. However, mobilization of calcium from the eggshell late in development causes a net increase in the quantity of calcium within the egg (Fig. 4). As a result, the total quantity of calcium within the egg (i.e., the amount in the yolk and the body of the embryo) increases between oviposition and hatching (Fig. 4). Calcium mobilized from both the yolk and the eggshell is deposited in the carcass of the embryo, and yolk calcium is severely depleted (Fig. 4). Consequently, residual yolk contains too little calcium to support skeletogenesis of neonates (Packard, 1994).

The pattern of mobilization and deposition of calcium described above raises the possibility that early amniotes could have exhibited substantial plasticity in the degree to which they relied on the eggshell as a source of calcium. The large quantity of calcium in the yolk and the small percentage of calcium obtained from the shell indicates that embryos probably could have relied exclusively on the yolk for their calcium even after the evolution of a crystalline component in the eggshell. However, embryos that did not or could not, for whatever reason, exploit the calcium available in the eggshell may have emerged from incubation smaller than those that did exploit the shell as a source of mineral.

Morphology of eggshells and patterns of mobilization and deposition of calcium among extant oviparous, amniotic vertebrates (Board, 1982; Packard, 1994; Packard and DeMarco, 1991; Packard and Hirsch, 1986; Packard *et al.*, 1982) have diverged considerably from the presumed ancestral condition. The trend exemplified by contemporary amniotes is toward more heavily calcified eggshells with less reliance on the yolk as a source of calcium and greater reliance on the shell (Packard, 1994). Squamate reptiles have eggshells and patterns of mobilization and deposition of calcium more similar to the presumed ancestral condition than is the case for turtles, crocodilians, or birds (Packard, 1994). Embryonic snakes place more reliance on the yolk as a source of calcium than do embryonic lizards, but the yolk is the primary source of calcium in both groups. The general pattern of mobilization and deposition of calcium for embryonic turtles differs from that of embryonic squamates primarily in the degree of reliance on the shell as a source of calcium. Turtle embryos obtain a much larger proportion of their calcium from the eggshell (Packard, 1994). Nonetheless, squamate and turtle embryos deplete the yolk of most of its calcium so that residual yolk withdrawn into the body cavity just prior to hatching contains too little calcium to support skeletal growth in neonates (Packard, 1994).

The greatest divergence from the presumed ancestral pattern of mobilization and deposition of calcium occurs among crocodilians and birds (Packard, 1994), members of the archosaurian lineage (Carroll, 1988). Embryos of precocial species in this lineage obtain the bulk of their calcium from the eggshell, and most of the calcium mobilized from the shell is deposited in the carcass of the developing embryo as is the case for embryonic squamates and turtles. However, some of the calcium mobilized from the shell is deposited in the yolk. As a result, calcium content of residual yolk is similar to or greater than that of yolk of fresh eggs (Fig. 5), and residual yolk has the potential to support skeletogenesis in neonates.

The pattern of mobilization and deposition of calcium exhibited by embryos of altricial birds diverges in varying degrees from that described for archosaurians in general (Packard, 1994).

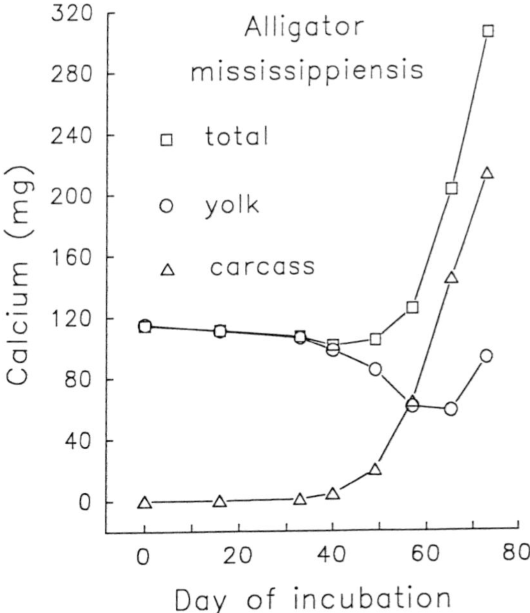

Figure 5. Pattern of mobilization and deposition of calcium by embryos of the American alligator, *Alligator mississippiensis*. Embryos rely primarily on the shell as a source of calcium. Most of the calcium mobilized from the eggshell is deposited in the carcass of the embryo. Yolk is depleted of calcium early in incubation, but some of the calcium removed from the eggshell is deposited in the yolk late in development. Note that the total quantity of calcium inside eggs is considerably higher at the end of incubation than at oviposition. Data are from M. J. Packard and Packard (1989). Eggs were incubated at 30°C. The last data point for each line is plotted on the average day of hatching.

Patterns of mobilization and deposition of calcium in pigeons are similar to those of embryos of domestic fowl and crocodilians (Hart *et al.*, 1992). However, embryonic yellowheaded blackbirds obtain more than 90% of the calcium required for development from the eggshell and deposit most of that calcium in the yolk rather than the carcass (Packard and Packard, 1991). Thus, the carcass of the hatchling contains only a small quantity of calcium, whereas residual yolk contains considerably more calcium than the carcass and much more than was present in yolk of fresh eggs (Packard, 1994). Unfortunately, too little information is available from altricial birds to identify evolutionary patterns and trends, and we cannot ascertain whether the patterns exhibited by yellowheaded blackbirds are common among species with altricial development or more limited in distribution.

An important aspect of the anamniote-amniote transition that has yet to be addressed concerns the physiology and morphology of the epithelia involved in respiration, nutrient transfer, and calcium mobilization and transport. For example, the partitioning of yolk into individual cells in direct developing eggs of extant amphibians means that nutrients contained in the yolk are not in direct contact with the epithelium of the yolk sac. This arrangement is likely to affect the mechanism of nutrient transfer from the yolk to the vascular supply of the yolk sac. Some transfer of yolk during embryogenesis in domestic fowl occurs as a result of endocytosis of yolk by cells of the yolk sac membrane (Juurlink and Gibson, 1973; Lambson, 1970; Mobbs and McMillan, 1979; Romanoff, 1960). However, a similar mechanism of yolk uptake seems unlikely when the cells of the yolk sac are not in direct contact with their substrate.

The chorioallantoic membrane of chicken eggs exhibits morphological and physiological characteristics that seemingly are related to the role of this epithelium in mediating mobilization of calcium from the eggshell and transport of calcium into the vasculature (Akins and Tuan, 1993a,b; Anderson *et al.*, 1981; Gay *et al.*, 1981; Rieder *et al.*, 1980). For example, the influx of calcium into the cells of the chorioallantoic membrane occurs at a much higher rate than that which characterizes other calcium transporting cells (Akins and Tuan, 1993b). Thus, the characteristics of this membrane may vary depending on the degree to which the developing embryo relies on calcium from the eggshell to support skeletogenesis. Unfortunately, morphological and physiological studies of calcium mobilization and transport have concentrated on eggs of domestic fowl. As a result, these and other questions related to the anamniote-amniote transition and evolution of the amniotic egg cannot be addressed at this time.

We consider that evolution of a fibrous shell membrane and elimination of the jelly layers as a barrier to diffusion were key steps in the evolution of the amniote egg from an anamniotic precursor.

These innovations opened the door to critical increases in egg size and metabolic intensity of embryos that could not be supported by the anamniotic condition. The suite of extraembryonic membranes characteristic of amniote eggs could have evolved at any time during the transition from the anamniotic to amniotic state. However, chorioallantoic gas exchange could not support the wide range of sizes characterizing eggs of extant amniotes until the diffusive barrier of the jelly layers had been eliminated. Other key changes include increases in nutrient reserves, evolution of holoblastic cleavage, and deposition of calcium salts on the outer surface of the fibrous shell membrane. The latter change has culminated in wide variation in structure of shells of eggs of oviparous amniotes and increasing reliance on the shell as a source of calcium. Mobilization of calcium from the eggshell presumably necessitated modifications to the physiological capabilities of the chorioallantoic epithelium. Unfortunately, comparative studies of form and function of this epithelium are not available.

SUMMARY

The amniote egg presumably evolved from an anamniote egg that gave rise via direct development to fully formed, precocial young. The precursor egg, like eggs of extant amphibians, was enclosed by a capsule composed of the vitelline membrane and jelly layers. Studies of eggs of contemporary amphibians indicate that the capsule probably greatly restricted diffusive gas exchange and thus limited the range of egg sizes that could be realized. The key innovations that permitted the evolution of larger-reptilian eggs were the envelopment of the egg by a fibrous shell membrane external tothe jelly layers and transfer of water from the jelly layers into the yolk compartment. The shell membrane provided physical support that the capsule alone could not supply and the transfer of water from the jelly to the yolk functionally eliminated the jelly as a barrier to gas exchange and also facilitated drying of the spaces between the fibers of the shell-membrane, both of which permitted rapid diffusion of gases between the egg and its environment. Only after these changes had taken place could the chorioallantois become an effective gas exchange organ.

The anamniotic predecessor of the amniote egg presumably lacked a calcareous layer, and embryos relied on the yolk for all the calcium required to support skeletogenesis. The evolution of the shell membrane provided a site for deposition of calcium salts and thus a new source of calcium for developing young. Eggs of oviparous lepidosaurians, the closest contemporary parallel to the early amniote egg, have a calcium rich yolk and calcium poor shell, and embryos rely on the yolk for most of their calcium. Eggs of other contemporary oviparous amniotes have diverged substantially from the basic plan assumed to characterize early amniote eggs.

Eggshells are calcium rich, yolk is calcium poor, and embryos rely primarily on the shell for calcium. Considerable variability in the pattern of deposition of shell calcium also has emerged. Shell calcium is deposited in the carcass of embryonic lepidosaurians and chelonians, in both the carcass and yolk of archosaurs with precocial development, and primarily in the yolk of some archosaurs with altricial development.

ACKNOWLEDGMENTS

We thank K. L. M. Martin and S. Sumida for inviting us to contribute to this volume and G. F. Birchard and G. C. Packard for constructive criticisms of several drafts of this chapter. Our research was supported in part by grants IBN87 18191 and IBN94 07136 (MJP) from the U. S. National Science Foundation and the Australian Research Council.

LITERATURE CITED

Akins, R. E., and R. S. Tuan. 1993a. Transepithelial calcium transport in the chick chorioallantoic membrane. I. Isolation and characterization of chorionic ectoderm cells. *Journal of Cell Science*, 105:369-379.

Akins, R. E., and R. S.Tuan. 1993b. Transepithelial calcium transport in the chick chorioallantoic membrane. II. Compartmentalization of calcium during uptake. *Journal of Cell Science*, 105:381-388.

Anderson, J. F., H. Rahn, and H. D. Prange. 1979. Scaling of supporting tissue mass. Quarterly Review of Biology, 54:139-148.

Anderson, R. E., C. V.Gay, and H. Schraer. 1981. Ultrastructural localization of carbonic anhydrase in the chorioallantoic membrane by immunocytochemistry. Journal of Histochemistry and Cytochemistry, 29:1121-1127.

Atkinson, B. G. 1981. Biological basis of tissue regression and synthesis. Pages 397-444 in: *Metamorphosis: A Problem in Developmental Biology*, (L. I. Gilbert and E. Frieden, eds.). New York: Plenum Press.

Board, R. G. 1982. Properties of avian egg shells and their adaptive value. *Biological Review of the Cambridge Philosophical Society*, 57:1-28.

Bradford, D. F. 1984. Physiological features of embryonic development in terrestrially breeding plethodontid salamanders. Pages 87-98 in: *Respiration and Metabolism of Embryonic Vertebrates* (R. S. Seymour, ed.). Junk: Dordrecht.

Bradford, D. F. 1990. Incubation time and rate of embryonic development in amphibians: the influence of ovum size, temperature, and reproductive mode. *Physiological Zoology*, 63:1157-1180.

Bradford, D. F., and R.S. Seymour. 1988. Influence of environmental PO_2 on embryonic oxygen consumption, rate of development, and hatching in the frog *Pseudophryne bibroni*. *Physiological Zoology*, 61:475 482.

Burggren, W. W., and J. J. Just. 1992. Developmental changes in physiological systems. Pages 467-530 in: *Environmental Physiology of the Amphibians* (M. E. Feder and W. W. Burggren, eds.). Chicago: The University of Chicago Press.

Carey, C. , H. Rahn, and P. Parisi. 1980. Calories, water, lipid and yolk in avian eggs. *Condor*, 82:335-343.

Carroll, R. L. 1969. Problems of the origin of reptiles. *Biological Reviews*, 44:393-432.

Carroll, R. L. 1970. Quantitative aspects of the amphibian-reptilian transition. *Forma et Functio*, 3:165-178.

Carroll, R .L. 1982. Early evolution of reptiles. *Annual Review of Ecology and Systematics*, 13:87-109.

Carroll, R. L.1988. *Vertebrate Paleontology and Evolution*. New York: W. H. Freeman and Company.

Deeming, D. C., and M. B. Thompson. 1991. Gas exchange across reptilian eggshells. Pages 277-284 in: *Egg Incubation: Its Effects on Embryonic Development in Birds and Reptiles* (D. C. Deeming and M. J. W. Ferguson, eds.). Cambridge, UK: Cambridge University Press.

del Pino, E. M., and B. Escobar. 1981. Embryonic stages of *Gastrotheca riobambae* (Fowler) during maternal incubation and comparison of development with that of other egg brooding hylid frogs. *Journal of Morphology*, 167:277-295.

Dodd, M. H. I., and J. M. Dodd. 1976. The biology of metamorphosis. Pages 467-599 in: *Physiology of the Amphibia, Volume III* (B. Lofts, ed.). New York: Academic Press.

Duellman, W. E., and L. Trueb. 1986. *Biology of Amphibians*. New York: McGraw Hill .

Dumont, J. N., and A. R. Brummett. 1985. Egg envelopes in vertebrates. Pages 235-288 in: *Developmental Biology, Volume 1* (L. Browder, ed.). New York: Plenum Publishing Corporation.

Elinson, R. P.1987. Change in developmental patterns: embryos of amphibians with large eggs. Pages 1-21 in: *Development as an Evolutionary Process* (R. A. Raff and E. C. Raff, eds.). New York: Alan R. Liss.

Elinson, R. P.1989. Egg evolution. Pages 251-262 in: *Complex Organismal Functions:Integration and Evolution in Vertebrates* (D. B. Wake, and G. Roth, eds.). Chinchester: John Wiley Sons.

Elinson, R. P. 1990. Direct development in frogs: wiping the recapitulationist slate clean. *Seminars in Developmental Biology*, 1:263-270.

Elinson, R. P., and E. M. del Pino. 1985. Cleavage and gastrulation in the egg-brooding, marsupial frog, *Gastrotheca riobambae. Journal of Embryology and Experimental Morphology*, 90:223-232.

Ferguson, M. W. J. 1982. The structure and composition of the eggshell and embryonic membranes of *Alligator mississippiensis. Transactions of the Zoological Society of London*, 36:99-152.

Gatten, R .E. Jr., K. Miller, K. and R. J. Full. 1992. Energetics at rest and during locomotion. Pages 314-377 in: *Environmental Physiology of the Amphibians* (M. E. Feder and W. W. Burggren, eds.). Chicago: The University of Chicago Press.

Gay, C. V., H. Schraer,D. J. Sharkey,and E. Rieder. 1981. Carbonic anhydrase in developing avian heart, blood and chorioallantoic membrane. *Comparative Biochemistry and Physiology*, 70A:173-177.

Guardabassi, A. 1960. The utilization of the calcareous deposits of the endolymphatic sacs of *Bufo bufo bufo* in the mineralization of the skeleton: Investigations by means of Ca45. *Z. Zellforsch. Mikrosk. Anat.* 51:278-282.

Hanken, J., and B. K. Hall. 1984. Variation and timing of the cranial ossification sequence of the oriental fire bellied toad, *Bombina orientalis* (Amphibia, Discoglossidae). *Journal of Morphology*, 182:245-255.

Hanken, J., and B. K. Hall. 1988. Skull development during metamorphosis: I. early development of the first three bones to form–the exoccipital, the parasphenoid, and the frontoparietal. *Journal of Morphology*, 195:247-256.

Hanken, J., M. W. Klymkowsky, C. H. Summers, D. W. Seufert,and N. Ingebrigtsen, N. 1992. Cranial ontogeny in the direct developing frog, *Eleutherodactylus coqui* (Anura: Leptodactylidae), analyzed using whole mount immunohistochemistry. *Journal of Morphology*, 211:95-118.

Hart, L. E., V. Ravindran, and A. Young. 1992. Accumulation of calcium and phosphorus in pigeon (*Columba livia*) embryos. Journal of Comparative Physiology B, 162:535-538.

Hirsch, K. F., and M. J. Packard. 1987. Review of fossil eggs and their shell structure. *Scanning Microscopy*, 1:383-400.

Hughes, R. L. 1984. Structural adaptations of the eggs and fetal membranes of monotremes and marsupials for respiration and metabolic exchange. Pages

389-421 in: *Respiration and Metabolism of Embryonic Vertebrates* (R. S. Seymour, ed.). Junk: Dordrecht.

Juurlink, B. H. J., and M. A. Gibson. 1973. Histogenesis of the yolk sac in the chick. *Canadian Journal of Zoology*, 51:509-519.

Karasaki, S. 1963. Studies on amphibian yolk 5. Electron microscopic observations on the utilization of yolk platelets during embryogenesis. *Journal of Ultrastructure Research*, 9:225-247.

Kielbowna, L. 1975. Utilization of yolk platelets and lipid bodies during the myogenesis of *Xenopus laevis* (Daudin). *Cell Tissue Research*, 159:279-286.

Lambson, R. O. 1970. An electron microscope study of the entodermal cells of the yolk sac of the chick during incubation and after hatching. *American Journal of Anatomy*, 129:1-20.

Lynn, W. G. 1942. The embryology of *Eleutherodactylus nubicola*, an anuran which has no tadpole stage. *Contributions in Embryology, Carnegie Institute of Washington Publication 541*, 30:27-62.

Mobbs, I. G., and D. B. McMillan. 1979. Structure of the endodermal epithelium of the chick yolk sac during early stages of development. *American Journal of Anatomy*, 155:287-309.

Packard, G. C., and M. J. Packard. 1988. The physiological ecology of reptilian eggs and embryos. Pages 523-605 in: *Biology of the Reptilia*, (C. Gans, and R. B. Huey, eds.).New York: Alan R. Liss.

Packard, M. J. 1994. Patterns of mobilization and deposition of calcium in embryos of oviparous, amniotic vertebrates. *Israel Journal of Zoology*, 40:481 492.

Packard, M. J., and N. B. Clark. 1996. Aspects of calcium regulation in embryonic lepidosaurians and chelonians and a review of calcium regulation in embryonic archosaurians. *Physiological Zoology*, in press.

Packard, M. J., and V. G. DeMarco. 1991. Eggshell structure and formation in eggs of oviparous reptiles. Pages 53-69 in: *Egg Incubation: Its Effects on Embryonic Development in Birds and Reptiles* (D. C. Deeming and M. W. J. Ferguson, eds.). Cambridge, UK: Cambridge University Press.

Packard, M. J., and K. F. Hirsch. 1986. Scanning electron microscopy of eggshells of contemporary reptiles. *Scanning Electron Microscopy*, 1986/4:1581-1590.

Packard, M. J., J. H. Jennings, and J. Hanken. 1996. Growth and uptake of mineral by embryos of the direct developing frog *Eleutherodactylus coqui*. *Comparative Biochemistry and Physiology*, in press.

Packard, M. J., and G. C. Packard. 1988. Sources of calcium and phosphorus during embryogenesis in bullsnakes (*Pituophis melanoleucus*). *Journal of Experimental Zoology*, 246:132-138.

Packard, M. J., and G. C. Packard. 1989. Mobilization of calcium, phosphorus, and magnesium by embryonic alligators (*Alligator mississippiensis*). *American Journal of Physiology*, 257:R1541 R1547.

Packard, M. J., and G. C. Packard. 1991. Patterns of mobilization of calcium, magnesium, and phosphorus by embryonic yellow headed blackbirds (*Xanthocephalus xanthocephalus*). *Journal of Comparative Physiology* B, 160:649-654

Packard, M. J., G. C. Packard, and T. J. Boardman. 1982. Structure of eggshells and water relations of reptilian eggs. *Herpetologica*, 38:136-155.

Pilkington, J. B., and K. Simkiss, K. 1966. The mobilization of the calcium carbonate deposits in the endolymphatic sacs of metamorphosing frogs. *Journal of Experimental Biology*, 45:329-341.

Rieder, E., C. V. Gay, and H. Schraer. 1980. Autoradiographic localization of carbonic anhydrase in the developing chorioallantoic membrane. *Anatomy and Embryology*, 159:17-31.

Romanoff, A. L.1960. *The Avian Embryo*. New York: The Macmillan Company.

Salthe, S. N. 1965. Increase in volume of the perivitelline chamber during development of *Rana pipiens* Schreber. *Physiological Zoology*, 38:80 98.

Selman, G. G., and /g. J. Pawsey. 1965. The utilization of yolk platelets by tissues of *Xenopus* embryos studied by a safranin staining method. *Journal of Embryology and Experimental Morphology*, 14:191-212.

Seymour, R. S. 1979. Dinosaur eggs: Gas conductance through the shell, water loss during incubation and clutch size. *Paleobiology*, 5:1 11.

Seymour, R.S. 1994. Oxygen diffusion through the jelly capsules of amphibian eggs. *Israel Journal of Zoology*, 40:493-506.

Seymour, R. S., and D. F. Bradford. 1987. Gas exchange through the jelly capsule of the terrestrial eggs of the frog, *Pseudophryne bibroni*. *Journal of Comparative Physiology*, 157B:477-481.

Seymour, R .S., and D. F. Bradford. 1995. Respiration of amphibian eggs. Physiological Zoology, 68:1-25.

Seymour, R. S., J and Piiper. 1988. Aeration of the shell membranes of avian eggs. *Respiration Physiology*, 71:101-116.

Szarski, H. 1968. The origin of vertebrate foetal membranes. *Evolution*, 22:211-213.

Taigen, T. L., F. H. Pough, and M. M. Stewart. 1984. Water balance of terrestrial anuran (*Eleutherodactylus coqui*) eggs: importance of partental parental care. *Ecology*, 65:248-255.

Townsend, D. S., and M. M. Stewart. 1985. Direct development in *Eleutherodactylus coqui* (Anura: Leptodactylidae): A staging table. *Copeia* 1985:423-436.

Vleck, C. M., and D. F. Hoyt. 1991. Metabolism and energetics of reptilian and avian embryos. Pages 285-306 in: Egg Incubation: Its Effects on Embryonic Development in Birds and Reptiles (D. C. Deeming and M .W. J. Ferguson, eds.). Cambridge, UK: Cambridge University Press.

Wiens, J. J. 1989. Ontogeny of the skeleton of *Spea bombifrons* (Anura: Pelobatidae). *Journal of Morphology*, 202:29-51.

CHAPTER 9

MORPHOLOGY AND EVOLUTION OF THE EGG OF OVIPAROUS AMNIOTES

James R. Stewart

"The Whole Egg is the Embryo" - Agassiz, 1857

INTRODUCTION

The evolutionary significance of the amniote egg has long been recognized. Indeed, Romer (1967) elevated this evolutionary event to primacy among vertebrate innovations. Conceptually, the amniote egg is a "terrestrial" egg; an achievement of embryonic emancipation from the aquatic realm, replete with specializations for an independent existence (Needham, 1931; Romer, 1957). These specializations include an eggshell, oviductal proteins (albumen), yolk (enclosed in a yolk sac), amnion, chorion, and allantois. The yolk sac, amnion, chorion, and allantois endow the eggs of amniote vertebrates with an elaborate interface between the developing embryo and its immediate environment. These structures compartmentalize the egg, by enclosing specific egg and embryonic elements, and regulate exchange between the egg and the external environment and among compartments within the egg (Simkiss, 1980). The conventional term for these tissues, "extraembryonic membranes," refers to their development as extensions of the body of the embryo and to their fleeting role in the life of the organism (Patton, 1951). Neither of these distinctions is entirely accurate and the terminology tends to obscure the independent function, development and history of these four structures. Nonetheless, the "extraembryonic membranes" are derived characteristics that confer unique structural and functional attributes to the eggs of amniotes.

The development of the extraembryonic membranes has been described in relatively few species of amniotes and functional characteristics are even less well known. Uneven representation among taxa has limited understanding of developmental variation and constrained analysis of the evolution of the egg of amniotes. Mammalian species, especially Eutheria, are the most widely known (Mossman, 1937, 1987; Luckett, 1977, 1980). The pattern of development and structure of the extraembryonic membranes is known for only a few turtles, crocodilians, and birds. Descriptions are more common for lizards and snakes; most are of viviparous species. However, because of the apparent conservative nature of the development of the extraembryonic membranes, the general pattern of variation among amniote taxa may be evident from the available data. However, there is considerable variation among amniotes in the interaction between the egg and the environment and likely variation in the functional role specific structures play in the development of embryos. It remains to be determined whether the few "models" for amniote taxa are actually representative.

This chapter is not intended to be an exhaustive review of amniote extraembryonic membranes and have attempted to synthesize material primarily on Reptilia (*sensu* Gauthier et al., 1988b), with consideration of Monotremata as a basal mammalian taxon (Novacek *et al.*, 1988). Detailed reviews of mammalian extraembryonic membranes can be found in Luckett (1977), Mossman (1987), and Tyndale-Biscoe and Renfree (1987). This review follows the excellent treatment of the topic by Luckett (1977). The systematics and taxonomy recognized in this paper are those of Gauthier *et al.* (1988a, 1988b). Thus, to analyze the pattern of egg morphology among extant Amniota, I consider Amniota to include the sister taxa, Mammalia and Reptilia. Reptilia includes the terminal taxa, Chelonia, Lepidosauria, and Archosauria. The following discussion considers the information available to address three questions: (1) What are the characteristics of the amniote egg? (2) What is the pattern of variation among extant amniotes? and (3) what can we surmise about the origin of the amniote egg?

MORPHOLOGY OF THE EXTRAEMBRYONIC MEMBRANES

Synopsis and Terminology

The development of the yolk sac, chorion, amnion, and allantois is remarkably uniform for all extant Reptilia (Luckett, 1977; Mossman, 1987) and suggests a high degree of evolutionary constraint. Reptilian eggs undergo meroblastic (partial) cleavage. The yolk, which is generally substantial, does not divide. Cleavage divisions produce a flattened aggregation of cells, the blastodisc, that rests on the surface of the yolk mass (Romanoff, 1960; Ewert, 1985). The perimeter of the blastodisc rests directly on the yolk but the central region is separated from the yolk by a space, the subgerminal cavity. That portion of the blastodisc overlying the subgerminal cavity, the area pellucida, contributes to the embryo. The perimeter of the blastodisc, termed the area opaca, does not form a blastocoel and lies directly on the yolk. The area opaca essentially is the first extraembryonic structure to form. Cells at the periphery of the area opaca migrate circumferentially over the surface of the yolk as the first yolk sac membrane. This structure, which contains elements of two germ layers, ectoderm and endoderm, is termed the bilaminar omphalopleure (Hill, 1897). Mesoderm initially appears between the ectoderm and endoderm within the area pellucida and inner margin of the area opaca and gradually extends outward between the two layers of the bilaminar omphalopleure. Blood islands within this extraembryonic mesoderm are the first sites of hematopoeisis and a network of blood vessels develops in the yolk sac. These vessels grow both centrally and peripherally and form the primordia of the vitelline circulation. The concentric growth of this vascular region, the area vasculosa, produces a second stage in the formation of the yolk sac, a vascularized trilaminar omphalopleure–the choriovitelline membrane

(Mossman, 1937). The choriovitelline membrane gradually extends around the yolk mass to form a vascularized interface between the egg and the external environment. This structure is transitory in all Reptilia. The mesodermal layer of the choriovitelline membrane subsequently splits to form a cavity, the extraembryonic coelom. The vascularized mesoderm remains with the yolk sac, forming the yolk sac splanchnopleure, while the nonvascularized mesoderm remains associated with the outer ectodermal layer to contribute to the somatopleure, or chorion. The extraembryonic coelom initially appears adjacent to the embryo and gradually extends around the yolk mass.

Reptilian amniogenesis occurs by the formation of folds or ridges of the blastoderm that grow over the embryo gradually enclosing it in a fluid-filled cavity, the amniotic vesicle (Fisk and Tribe, 1949). The fold in the head region initially consists of ectoderm and endoderm and has the form of a horseshoe that opens toward the tail of the embryo. The definitive amnion (somatopleure) and chorion (somatopleure) are formed when mesoderm becomes associated with the folds of ectoderm and an extraembryonic coelom separates the amnion from the chorion. The region of closure of the amniochorion in the trunk region of the embryo forms a band of tissue, the seroamniotic connection, anchoring the inner ectodermal layer to the outer ectodermal layer. The seroamniotic connection is invaded by mesoderm, which separates to form an extraembryonic coelom, in most (Fisk and Tribe, 1949) but not all (Mitsukuri, 1891) Reptilia. The tail region of the embryo is enclosed by an amniochorion consisting of ectoderm and mesoderm (Fisk and Tribe, 1949). In some species (Mitsukuri, 1891; Dendy, 1899; Romanoff, 1960), a posterior amniotic tube extends beyond the body of the embryo to open into the space under the vitelline membrane. The posterior amniotic tube is transitory.

The allantois, which is splanchnopleuric (endoderm and mesoderm), forms as an extension of the endodermal layer of the hindgut of the embryo. The allantoic vesicle gradually expands and eventually the outer allantoic membrane contacts the chorion. Fusion of the mesodermal layers of the chorion and allantois produces the chorioallantoic membrane. As long as the choriovitelline membrane is intact, the allantois is restricted to the embryonic hemisphere of the egg. The allantoic vesicle of all Reptilia eventually expands to fill the extraembryonic coelom that develops as a cavity formed by the separation of mesodermal layers of the choriovitelline membrane.

CHARACTERISTICS OF SPECIFIC TAXA

Monotremata

Griffiths (1978) provides a detailed summary of monotreme reproduction and development. Luckett (1977) and Hughes (1993) are excellent descriptions of the development of the extraembryonic membranes of echidna, *Tachyglossus aculeatus*, and platypus, *Ornithorhynchus anatinus*, that incorporate original observations into classical literature accounts. There is great similarity in the development of the extraembryonic membranes of these two species. An intrauterine phase of development is followed by oviposition and incubation of the egg within a pouch. The length of incubation for *Tachyglossus* is 10-10.5 days (Griffiths *et al.*, 1969; Griffiths, 1978). A thin (17-30 μm) mucoid coat and eggshell are deposited around the egg by the female reproductive tract (Hughes et al., 1975; Griffiths, 1978; Hughes, 1984). The eggshell of *Ornithorhynchus* is composed of three distinct zones (Hughes and Carrick, 1978). The composition of the eggshell is largely protein (Hughes, 1977). During intrauterine development the volume of the egg contents increases dramatically through absorption of uterine secretions (Flynn and Hill, 1939; Hughes, 1984). Recently oviposited embryos contain approximately 19 somites. A choriovitelline membrane occupies the immediate periphery of the embryo at oviposition and the yolk sac is enclosed by a bilaminar omphalopleure. Neither a chorion nor an allantois is present, but early stages of amniogenesis are evident.

In *Ornithorhynchus* embryos of 35 or 36 somites the allantois is a small outgrowth of the hindgut (Luckett, 1977). The choriovitelline membrane extends beyond the equator of the yolk sac. The abembryonic pole of the yolk sac is a bilaminar omphalopleure. The histology of the two yolk sac membranes differs. The outer epithelium of the choriovitelline membrane is squamous, and that of the bilaminar omphalopleure is columnar. The endodermal layers are

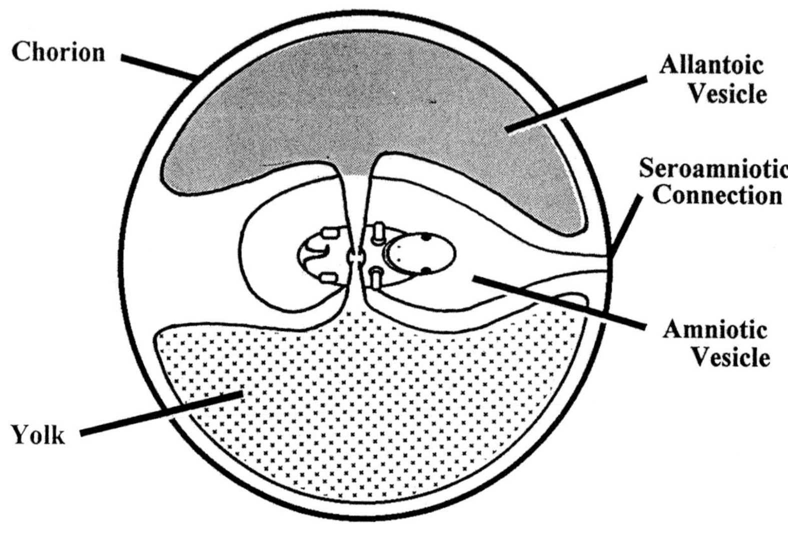

Figure 1. Extraembryonic membranes of Monotremata, based on *Ornithoryhnchus anatinus* (Semon, 1894; Luckett, 1977).

columnar in the choriovitelline membrane and squamous in the bilaminar omphalopleure.

The following description of a platypus embryo (Fig. 1), estimated to be 3.5-4.5 days prior to hatching, from Hughes (1993) is similar to details provided by Luckett (1977) for late incubation stage from Semon (1894). The allantoic vesicle fills the embryonic hemisphere of the egg and is fused to the chorion to form a chorioallantoic membrane. The allantoic contribution to the chorioallantoic membrane is highly vascularized and the chorion consists of two layers, an outer cuboidal epithelium overlies a layer of squamous cells. The choriovitelline membrane fills the abembryonic hemisphere of the egg. This structure is well vascularized and overlain by a thin, squamous epithelium. The amnion fits tightly over the embryo. The chorion occupies a zone between the chorioallantoic membrane and the trilaminar omphalopleure and there is a well-defined seroamniotic connection dorsal to the fetus in this region. A small exocoelom surrounds the tail of the embryo in late incubation (Luckett, 1977).

Chelonia

The stage of embryos of most turtle species at oviposition is late gastrula (Ewert, 1985). The embryo rests on the yolk mass and is surrounded by the vitelline membrane, albumen, shell membrane, and the mineral layer of the eggshell (Ewert, 1985). The mass of albumen is roughly equal to that of the yolk (Ewert, 1979). Very early in development in *Mauremys japonica* (as *Clemmys japonica*), the albumen overlying the blastodisc disperses or is absorbed and the blastodisc becomes adherent to the shell membrane (Mitsukuri, 1891). Amniogenesis in *M. japonica* and *Trionyx sinensis* (as *Trionyx japonicus*) (Mitsukuri, 1891) and *Chelydra serpentina* and *Chrysemys picta* (as *Chrysemys marginata*) (Fisk and Tribe, 1949) proceeds as folds of ectoderm begin to extend over the embryo. The lateral portions of the folds are invaded by mesoderm. The sheet of mesoderm is disrupted and the amnion, chorion, and intervening extraembryonic coelom are formed. The ectodermal folds that converge over the trunk region of the embryo are not invaded by mesoderm and a seroamniotic connection develops. The seroamniotic connection anchors the embryo and persists throughout incubation. An additional event associated with amniogenesis in these four species is the development of a posterior amniotic tube. The lateral folds extend posterior to the embryo and form a tube that connects the amniotic vesicle to the cavity between the yolk sac and the vitelline membrane.

The choriovitelline membrane is clearly evident in embryos of *C. serpentina* with 8 somites (Yntema, 1968). A rudimentary vitelline circulation is functional at 19 somites. Agassiz (1857) provides

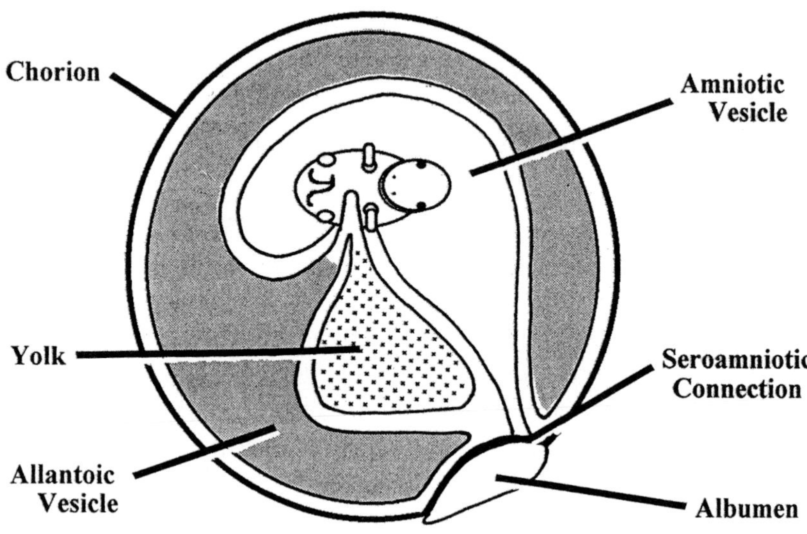

Figure 2. Extraembryonic membranes of Chelonia, based on *Trionyx sinensis* (Mitsukuri, 1891).

detailed drawings of later development of the yolk sac membranes in *C. serpentina*. The choriovitelline membrane gradually extends over the yolk but is likewise gradually supplanted as the expanding allantois follows. This pattern continues as both membranes grow toward the abembryonic pole. The choriovitelline membrane vascularizes the abembryonic pole of the yolk before being replaced by the allantois. In terminal stages of development, the allantois encloses the entire egg contents.

The allantois of *M. japanica* and *T. sinensis* (Mitsukuri, 1891) also gradually expands to surround the yolk sac. Late in development (Fig. 2), the remaining albumen of the egg is concentrated at the base of the yolk sac and the lobes of the allantois converge from both sides and extend between the yolk sac and the mass of albumen. The ectodermal layer of the chorioallantois contains columnar cells where there is contact with the albumen. The allantois thus entirely encloses the egg contents with the exception of the seroamniotic connection dorsal to the embryo.

Lepidosauria

Sphenodontida.–Descriptions of the development of the extraembryonic membranes of tuatara, *Sphenodon punctatus*, relate mostly to amniogenesis (Dendy, 1899; Fisk and Tribe, 1949; Moffat, 1985). The amniotic cavity is formed by the growth of folds that extend dorsally over the embryo. The head region of the embryo is enclosed by a fold consisting of ectoderm and endoderm; lateral folds over the trunk region consist solely of ectoderm (Dendy, 1899; Fisk and Tribe, 1949). The extension of ectoderm over the trunk region forms a seroamniotic connection (Fisk and Tribe, 1949). Dendy (1899) reported that the seroamniotic connection persisted in late-stage embryos, however, Fisk and Tribe (1949) could find no trace of this structure in two earlier embryos and concluded that the chorion and amnion achieve complete separation in *S. punctatus*. Subsequent to the formation of the amniotic cavity, a posterior amniotic tube develops in the tail region of the embryo (Dendy, 1899). The roof of the posterior amniotic tube is an extension of the seroamniotic connection (Fisk and Tribe, 1949). The posterior amniotic tube opens on the surface of the blastoderm and eventually achieves a length greater than the body of the embryo before disappearing.

The allantois is initially visible as a small outgrowth in the tail region at embryonic stage L (Dendy, 1899), which corresponds to stage 28 of Dufaure and Hubert (1961) for *Lacerta vivipara* (Moffat, 1985). The allantoic vesicle enlarges to surround the embryo and most of the yolk sac, such that in late stages of incubation, the chorioallantoic membrane encloses most of the egg (Dendy, 1899). The development of the yolk sac has not been described. Dendy (1899) does report that the yolk sac splanchnopleure is richly vascularized and that the "serous envelope" has not separated from the yolk sac at the abembryonic pole of late stage embryos.

Squamata.–The prevalence of viviparity among lizards and snakes has stimulated interest in placental structure among Squamata. As a result, the development of the extraembryonic membranes is known for a variety of species, most of which are viviparous. Major features of the pattern of development do not differ for the oviparous species that have been studied.

The development of the yolk sac consists of the three phases characteristic of other Reptilia. The primary yolk sac is a bilaminar omphalopleure. A trilaminar omphalopleure is formed as a layer of mesoderm proliferates between the ectodermal and endodermal layers of the bilaminar omphalopleure. The nonvascular trilaminar omphalopleure is vascularized shortly thereafter to form the choriovitelline membrane. Transformation of the yolk sac is progressive as modifications associated with each phase spread outward over the yolk mass. The concentric growth of the choriovitelline membrane over the surface of the yolk is delimited by a vascular ring, the sinus terminalis.

In contrast to all other Reptilia, the course of mesodermal growth in advance of the choriovitelline membrane deviates from the bilaminar omphalopleure and extends into the yolk (Stewart, 1993). This intravitelline mesoderm most commonly forms a sheet that is several cell layers thick and which continues to grow to ultimately form a continuous membrane within the yolk across the entire abembryonic pole. These cells undergo a morphogenetic process similar to that of the mesoderm of the choriovitelline membrane and an extraembryonic coelom, the yolk cleft, is formed. The yolk cleft is lined by splanchnopleure. The outer compartment of yolk that is separated from the yolk mass by this process is termed the isolated yolk mass (Fig. 3). The isolated yolk mass is lined internally by splanchnopleure and externally by the bilaminar omphalopleure. The pattern of growth of the extraembryonic mesoderm has an important effect on the early distribution of blood vessels at the abembryonic pole of the egg. Blood vessels that subsequently form in the abembryonic region of the yolk sac do so in association with intravitelline cells, whereas the bilaminar omphalopleure remains

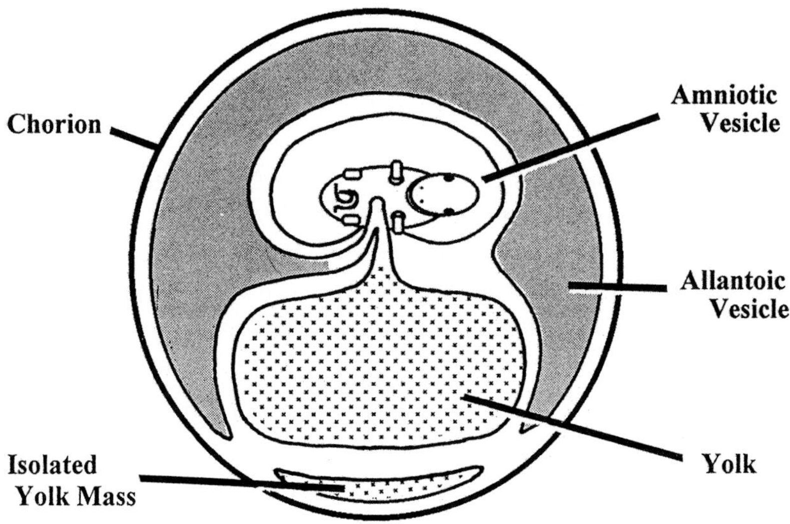

Figure 3. Extraembryonic membranes of Squamata, based on *Elgaria multicarinata* (Stewart, 1985).

intact and nonvascular at the abembryonic pole. The formation of an isolated yolk mass has been documented in all species of Squamata that have been studied (Stewart, 1993). The subsequent development of this structure following its formation varies among species, but this structural variation has not been linked to specific functional characteristics (Stewart, 1993).

Development of the region of the yolk sac that transforms into a choriovitelline membrane is similar to other Reptilia. The mesodermal layer separates and an extraembryonic coelom lined externally by somatopleure and internally by splanchnopleure is formed. The vitelline vessels remain associated with the yolk sac splanchnopleure following disruption of the choriovitelline membrane.

The few studies of amniogenesis among squamates are reviewed by Fisk and Tribe (1949). The most detailed work is on several species of *Chamaeleo* that are unusual compared to other reptiles in that amniogenesis proceeds concentrically and early in development. The composition and development of the amniochorion for the few species of snakes and lizards that have been described is similar to that of *Sphenodon* (Fisk and Tribe, 1949).

The allantois of squamates expands to fill the extraembryonic coelom and its outer membrane fuses with the chorion (Fig. 3). Growth of the allantois toward the abembryonic pole of the egg is dependent on the fate of the isolated yolk mass (Stewart, 1993). In some species, the allantois expands as the isolated yolk mass regresses and eventually surrounds the entire yolk mass. In other species, the allantois extends into the yolk cleft.

Archosauria

Crocodylia.–Embryos of *Crocodylus johnsoni, C. porosus* and *Alligator mississippiensis*, have 9-20 pairs of somites (mean = 12) at oviposition (Ferguson, 1985). The embryo lies free on top of the yolk. A layer of albumen, roughly equal in volume to the yolk, lies between the yolk and the shell membrane. Shortly after oviposition, in embryos with 21-25 pairs of somites, the albumen overlying the embryo dissipates and the blastoderm attaches to the shell membrane (Ferguson, 1982). This connection persists throughout incubation (Ferguson, 1985). The albumen layer gradually loses water to the yolk sac and, as the volume decreases, becomes restricted to the ends of the egg.

Details of amniogenesis are unknown. Fisk and Tribe (1949) describe a single specimen, referred to as crocodile material. The amnion and chorion are separated by an extraembryonic coelom over the embryo, but a short seroamniotic connection is present posteriorly. The extraembryonic coelom extends between the amnion and chorion dorsal to the embryo in 13 to 14-day embryos (Ferguson, 1985). There is no evidence of a posterior amniotic tube (Fisk and Tribe, 1949).

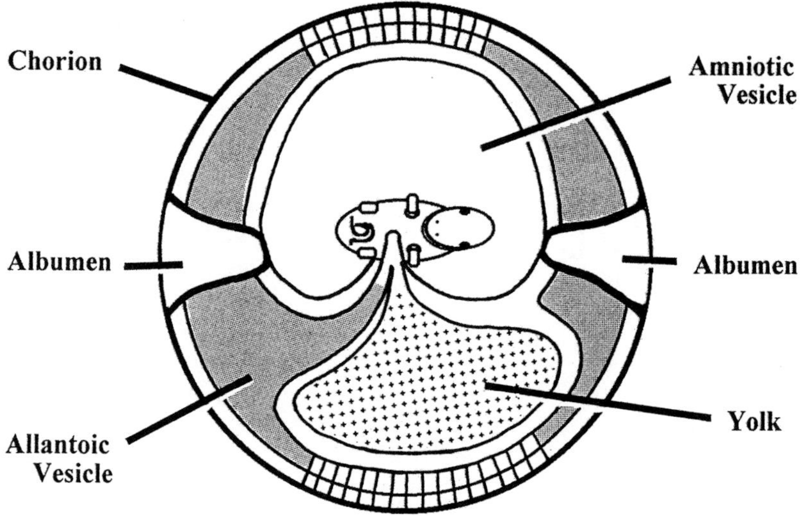

Figure 4. Extraembryonic membranes of Crocodylia, based on *Alligator mississippiensis* (Ferguson, 1982, 1985).

The allantoic bud is distinct in embryos with 36-40 pairs of somites and the allantois is fused to the chorion in 9-day-old embryos (Ferguson, 1985). The allantoic vesicle fills the extraembryonic coelom to form a circumferential band about the central axis of the egg in 15-day-old embryos.

Ferguson (1982, 1985) provided a detailed description and diagram of the distribution of fetal membranes for embryos of *A. mississippiensis* aged 40 days postoviposition (Fig. 4). Albumen is concentrated at either end of the equatorial axis of the egg. The

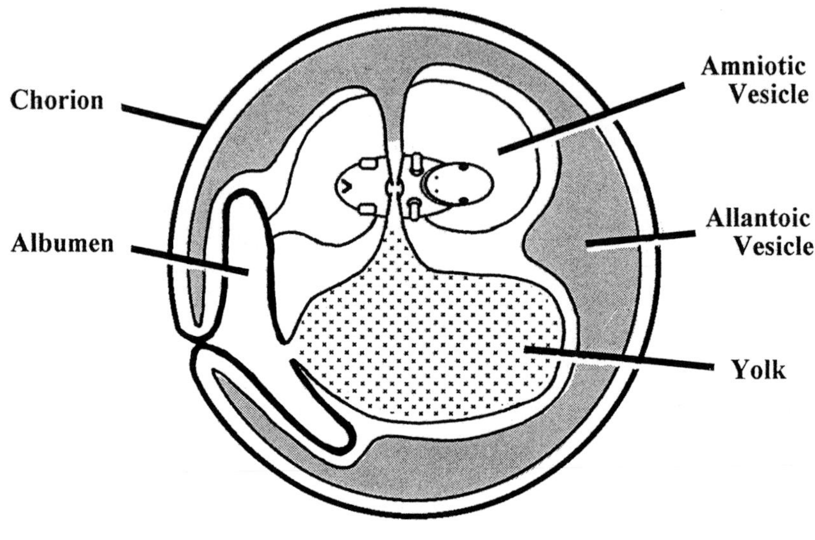

Figure 5. Extraembryonic membranes of Aves, based on *Gallus gallus* (Hamilton, 1952).

expanded allantoic vesicle surrounds the embryo, which is attached to the yolk sac via a yolk stalk, and fills the central region of the egg between the two masses of albumen. The chorioallantoic membrane thus forms a band that contacts the shell membrane about the latitudinal axis of the egg. An elliptical-shaped "fusion zone" is present at the two poles of the egg (embryonic and abembryonic). These are regions of fusion of the eggshell membrane, chorion, outer and inner limbs of the allantois, and either amnion or yolk sac which anchor the embryo and yolk sac within the eggshell. The yolk sac is withdrawn into the body of the embryo prior to hatching.

Aves.–The chicken, *Gallus gallus*, has been one of the "model" amniote species for experimental embryology and development of the extraembryonic membranes is known in detail. There are several excellent syntheses of the embryology of this species, including Hamilton (1952) and Romanoff (1960). The following account is drawn largely from these two sources (Fig. 5). The recently ovulated egg contains a preprimitive streak blastodisc resting on a large yolk mass. The blastoderm and yolk are surrounded by a vitelline membrane. A large mass of albumen surrounds the vitelline membrane and is in turn surrounded by the shell membrane and eggshell. Development of the yolk sac is sequential, consisting of three phases, and proceeds rapidly such that by the third day of incubation all three phases are in progress. The bilaminar omphalopleure (area vitellina) is visible extending over the dorsal surface of the yolk mass during the first day of incubation. The choriovitelline membrane (area vasculosa) is clearly present during the second day of incubation. The choriovitelline membrane extends around the yolk as vascularized mesoderm invades the bilaminar omphalopleure. As the bilaminar omphalopleure is transformed into the choriovitelline membrane, the trailing edge of the choriovitelline membrane is progressively disrupted by separation of the layer of mesoderm to form an extraembryonic coelom enclosed by the outer somatopleure (chorion) and the inner splanchnopleure of the yolk sac. The splanchnopleuric yolk sac gradually replaces the choriovitelline membrane.

The amnion encloses the embryo by 72 hr incubation (39 somites) except for a small amniotic navel. At this stage the seroamniotic connection is present over the trunk region of the embryo. The amniotic navel of *G. gallus* does not extend beyond the tail region of the embryo and closes at Days 3 or 4 of incubation. The posterior region of the amnion varies among Aves, and in some species a posterior amniotic tube develops (Romanoff, 1960). The ectoderm of the seroamniotic connection is gradually invaded by a layer of mesoderm. This process begins at Day 4 and is complete on Days 9 or 10. The resulting structure, composed of ectoderm, mesoderm, and ectoderm, develops perforations shortly after formation. As a result of this patency the amniotic cavity is confluent with the egg compartment containing the albumen.

The allantois is the last of the extraembryonic membranes to appear. During the fourth day the allantois grows out on the right side of the embryo and contacts the chorion shortly thereafter. By Day 6 the allantois is about the same size as the embryo and reaches the equator of the yolk sac at Days 7 or 8. The allantoic vesicle completely surrounds the egg contents by Days 11 or 12. The chorioallantois extends beyond the yolk sac and surrounds the mass of albumen at the abembryonic pole. The resulting structure is the albumen sac, which has a flask shape. The growth of the allantois encloses the connection between the amniotic cavity and the albumen sac to form an albumen filled tubular channel, the albumen duct. Albumen enters the amniotic cavity via the perforations in the seroamniotic connection.

EVOLUTIONARY PATTERNS OF EGG MORPHOLOGY

The eggs of Reptilia and Monotremata undergo meroblastic (partial) cleavage and are surrounded by tertiary shell membranes that are proteinaceous secretions of the oviduct (Hughes, 1977). The large yolk mass is enclosed by a choriovitelline membrane as a stage in yolk sac formation, amniogenesis proceeds by a process of folding, and a large allantoic vesicle develops. These characteristics are likely ancestral for amniotes and represent a combination of primitive and derived traits. The tertiary shell membranes (Gray, 1928; Packard and Packard, 1980) and choriovitelline membrane (Mossman, 1987; Elinson, 1989) have possible antecedents among anamniotes. For example, the yolk mass of some amphibians is enclosed and vascularized (Elinson, 1989). The addition of glycoproteins to the oviductal egg is common among vertebrates (Hughes, 1977), and the eggs of Amphibia are enclosed by proteinaceous layers, the egg capsules, secreted by the oviduct (Salthe, 1963). An amniochorion and a large allantoic vesicle are derived amniote characteristics that have no clear antecedent among modern Amphibia.

Meroblastic cleavage has evolved five times among modern vertebrates and is correlated with large yolk quantity in all lineages except teleosts (Collazo *et al.*, 1994). Although some Amphibia ovulate moderately large eggs (Elinson, 1987; Del Pino, 1989), all have holoblastic cleavage, i.e., the egg is completely divided into blastomeres. Meroblastic cleavage, in which the yolk does not divide, is derived among amniotes and likely was associated with production of a large yolked egg. Meroblastic cleavage is an exaptation for increased egg size because of constraints imposed on morphogenesis by holoblastic cleavage (Elinson, 1989). The eggs of Reptilia and Monotremata cleave meroblastically, whereas those of Marsupialia (Tyndale-Biscoe and Renfree, 1987) and Eutheria (Balinsky, 1975) undergo a similar cleavage pattern although little or no yolk is present.

Like all organisms, eggs are defined by distinct functional characteristics that couple requirements for growth and metabolism to sources of these materials. One aspect of variation among the eggs of amniotes is related to differences in the source of materials, or conceptually, variation in the degree of independent existence. Differences in the pattern of exchange between the egg and the environment are a well-studied aspect of comparative egg physiology among amniotes (Packard and Packard, 1988). Expressed as a dichotomy, eggs either exhibit considerable flux of water, gases, and perhaps other molecules during development or they exchange principally gases. The latter pattern Needham (1931) termed "cleidoic". For example, the eggs of birds undergo minimal development prior to oviposition, are enclosed in a thick calcareous eggshell, and require only respiratory exchange and suitable temperature for successful incubation, whereas the eggs of monotremes are retained *in utero*, are enclosed in a thin shell membrane, and absorb uterine secretions all prior to oviposition.

Provisions to the egg are supplied by vitellogenesis prior to ovulation or by the maternal gonoduct following ovulation (Blackburn *et al.*, 1985). If eggs are to be oviposited, oviductal secretions influence the subsequent pattern of exchange between the egg and the incubation environment. These secretions include albumen, proteins of the eggshell membranes, and inorganic minerals of the eggshell (Packard and DeMarco, 1991; Palmer and Guillette, 1991). Eggshells are composed of an inner, proteinaceous shell membrane and an outer inorganic layer (Packard and DeMarco, 1991). Both layers are variable among species of amniotes. Eggshells of oviparous species can be classified as either flexible or rigid based on the relative differences in thickness of the two layers and the degree of compactness of mineral units in the outer layer (Packard and Demarco, 1991).

Water movement into and out of the egg is correlated with the degree of mineralization of the eggshell (Ewert, 1985; Packard and Packard, 1988; Ackerman, 1991; Packard, 1991). The monotreme eggshell is not mineralized (Hughes, 1977) and undergoes exchange, at least during intrauterine development (Flynn and Hill, 1939; Hughes, 1984). Eggshell structure varies considerably among turtles (Ewert, 1985; Packard and Packard, 1988), but based on physical characteristics, eggshells have been defined as either pliable, hard-expansible, or brittle (Ewert, 1985). Using the dichotomy, rigid vs. pliable, Iverson and Ewert (1991) compared variation in eggshell characteristics to a cladogram for turtles. The resulting pattern revealed that eggshell rigidity is not phylogenetically constrained because few clades exhibit only a single shell type. Further, the most likely explanation for the pattern of eggshell distribution is that pliable shelled eggs are primitive and that rigid shelled eggs have evolved independently at least five times among turtles (Iverson and Ewert, 1991).

The eggshells of squamates also vary, but all are flexible except for members of two subfamilies of gekkonid lizards (Packard and Packard, 1988). Unlike other squamates, in which the degree of mineralization is slight, these species have rigid eggshells that are heavily mineralized (Packard and Hirsch, 1989). The tuatara eggshell is flexible but differs structurally from most squamates (Packard et al., 1982). All crocodilians and birds produce heavily mineralized, rigid eggshells (Packard and DeMarco, 1991).

The distribution of flexible and rigid eggshell types among modern amniotes is most easily explained if the primitive amniote eggshell was flexible (Figure 6). This condition is present in monotremes and primitive lepidosaurs and likely is primitive for turtles (Iverson and Ewert, 1991). A mineralized eggshell membrane is a synapomorphy for Reptilia (Packard, 1994). A heavily mineralized, rigid eggshell evolved independently among archosaurs, some gekkonid lizards and at least five times among turtles.

In addition to contributing a substantial layer of minerals to the egg, the oviduct of archosaurs secretes a large mass of albumen which surrounds the yolk (Romanoff and Romanoff, 1949; Ferguson, 1982, 1985). The hydrophilic albumen traps and binds water and thus serves as a water reservoir during passage of the egg through the oviduct. This water is transferred to the yolk following oviposition (Romanoff and Romanoff, 1949; Hamilton, 1952; Ferguson, 1985). In contrast, the eggs of monotremes (Hughes, 1977; Hughes and Carrick, 1978) and lepidosaurs (Packard *et al.*, 1977; Packard and Packard, 1988) contain little albumen, but see Tracy and Snell (1985) for an alternative interpretation for lepidosaurs. Recently, the first detailed study of protein during egg development in a lepidosaur, *Anolis pulchellus*, demonstrated that the oviduct does not contribute an albumen layer to the egg and water contributed by the oviduct is stored in yolk (Cordero-Lopez and Morales, 1995). This pattern may be typical for lepidosaurs with flexible eggshells. The calcareous eggs of geckos have not been studied. The amount of albumen supplied to eggs of archosaurs, lepidosaurs, and monotremes, is apparently correlated with the water flux characteristic of the eggs; eggs with rigid eggshells that do not take up water from the substratum contain a large mass of albumen. One possible hypothesis is that provision of a large amount of albumen evolved in heavily mineralized, rigid eggshells for water storage.

Turtle eggs provide a test for this hypothesis because all are provisoned with a substantial amount of albumen, yet they vary widely in water flux physiology and eggshell composition (Ewert, 1979; Packard and Packard, 1988). Ewert (1979) provides data for 15 species of turtles for which eggshell type and yolk and albumen quantities are known. For these species, albumen constitutes approximately one-half ($X = 0.53 \pm 0.06$ SD) of the total mass of yolk plus albumen. Species with brittle shells ($N = 7$) do not differ significantly from species with parchment shells ($N = 8$) ($F = 0.68$, df 1,14, $P = 0.42$). Among these turtles, quantity of albumen is not correlated with eggshell type. If pliable eggshells are primitive for

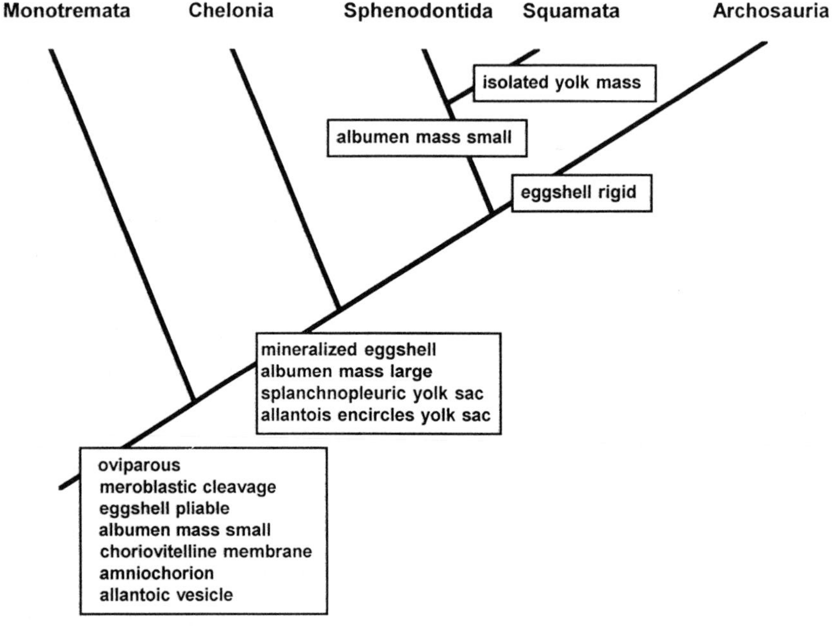

Figure 6. Characteristics of the eggs of extant amniote taxa. Characters at the base of the cladogram are presumed ancestral for Amniota. Possible transformations are indicated for individual clades. The cladogram is based on Gauthier *et al.* (1988b) and Novacek *et al.* (1988).

turtles (Iverson and Ewert, 1991), the evolution of large albumen stores must have preceeded the evolution of rigid eggshells.

The phylogenetic distribution of quantity of albumen provision to eggs is most simply explained by the evolution of a large albumen provision within pliable shelled eggs of the earliest Reptilia (Fig. 6). Reduction in albumen provision is derived for Lepidosauria and the primitive condition is retained among Chelonia and Archosauria. An alternative hypothesis, which results in an evolutionary sequence only slightly more complex, is that a small albumen provision is primitive for Reptilia. This hypothesis requires that Lepidosauria retained the primitive condition (Packard and Packard, 1980) and Chelonia and Archosauria independently evolved large albumen stores.

Development of the yolk sac of Reptilia proceeds in three stages, each producing a distinct structure (Romanoff, 1960; Ewert, 1985; Ferguson, 1982, 1985; Stewart, 1993). The primary yolk sac is an extension of the area opaca that grows over the surface of the yolk to form a bilaminar omphalopleure (ectoderm and endoderm) (Hill,1897). The bilaminar omphalopleure is converted to a vascularized trilaminar omphalopleure, the choriovitelline membrane (Mossman, 1937), by the insertion of mesoderm between the layers of ectoderm and endoderm. The choriovitelline membrane is the secondary yolk sac. The choriovitelline membrane is transitory because it is disrupted by the formation of a cavity, the extraembryonic coelom, within the mesodermal layer (Agassiz, 1857; Romanoff, 1960; Stewart, 1985). Development of the extraembryonic coelom produces the terminal yolk sac, the yolk sac splanchnopleure (endoderm and mesoderm), and the outer somatopleure (mesoderm and ectoderm). This sequence of events occurs in all Reptilia. In addition to the three stages outlined previously, a unique yolk sac membrane develops among Squamata (Weekes, 1935; Yaron, 1985; Stewart and Blackburn, 1988; Stewart, 1993). This tissue is formed from mesoderm that proliferates into the yolk mass (intravitelline mesoderm) and cavitates to form the yolk cleft. The inner margin of the yolk cleft is lined by splanchnopleure that is continuous with the yolk sac splanchnopleure originally associated with the choriovitelline membrane. The outer boundary of the yolk cleft, also splanchnopleure, forms the inner margin of the isolated yolk mass. The bilaminar omphalopleure lies at the perimeter of the isolated yolk mass. The isolated yolk mass is a synapomorphy for Squamata, or possibly Lepidosauria; the development of the yolk sac has not been studied in *Sphenodon*. The function of this structure is unknown, but as a result of its formation the outer boundary of the yolk sac at the abembryonic pole of squamates is unlike that of any other Reptilia. Blood vessels associated with the yolk sac are separated from the shell membrane by a nonvascular bilaminar omphalopleure, a narrow mass of yolk, and an extraembryonic coelom.

In the terminal stage of incubation, the yolk sac of monotremes consists of two structurally distinct regions (Hughes, 1993). The dorsal surface of the yolk sac, which faces the amnion, is splanchnopleuric, whereas the choriovitelline membrane surrounds the remainder of the perimeter of the yolk and is in apposition to the shell membrane. This pattern is similar to a stage in development of the yolk sac of Reptilia. Development of the extraembryonic coelom adjacent to the embryo in Reptilia occurs during amniogenesis (Fisk and Tribe, 1949; Romanoff, 1960) and this cavity subsequently extends within the choriovitelline membrane about the perimeter of the yolk. Because there are no structural equivalents among Amphibia, the polarity of yolk sac structure is uncertain. Either monotremes are paedomorphic for this character or they represent the condition of the ancestral amniote egg. In marsupials, both a bilaminar omphalopleure and a choriovitelline membrane invest different regions of the yolk (Tyndale-Biscoe and Renfree, 1987). The terminal stage in yolk sac development among eutherians is basically one of three different conditions, a choriovitelline membrane, a yolk sac splanchnopleure, or a novel structure, an inverted yolk sac (Mossman, 1987).

The amniochorion is a synapomorphy for Amniota and the process of development by folding is similar in all Reptilia (Fisk and Tribe, 1949; Romanoff, 1960) and most Mammalia (Luckett, 1977). Details of amniogenesis are known for relatively few species of Reptilia, complicating any attempt to discern phylogenetic patterns. Two basic patterns of amniogenesis occur among amniotes–by the formation of folds and by cavitation. An analysis of the evolution of amnion formation by cavitation among Mammalia (Luckett, 1976, 1977) indicates that this developmental pattern is derived and the formation of amniotic folds, characteristic of Reptilia, is primitive. Amniogenesis by folding occurs as folds of the blastoderm in the head and trunk region of the embryo grow to extend over the embryo. The folds in the trunk region are composed entirely of ectoderm and constitute the seroamniotic connection (Fisk and Tribe, 1949). The seroamniotic connection is apparently a feature of amniogenesis in all amniotes in which amniotic folds develop. It is reported in monotremes (Hughes, 1984), turtles (Mitsukuri, 1891), lepidosaurs (Dendy, 1899; Hrabowski, 1926; Fisk and Tribe, 1949), and archosaurs (Fisk and Tribe, 1949; Romanoff, 1960). A permanent seroamniotic connection has been reported in monotremes (Hughes, 1984), turtles (Mitsukuri, 1891) and tuatara (Dendy, 1899). However, Fisk and Tribe (1949) examined specimens of tuatara analyzed by Dendy (1899) and found no evidence of a permanent seroamniotic connection. The seroamniotic connection is invaded by mesoderm in lizards and this new structure persists throughout embryonic development (Hrabowski, 1926). Among birds, the seroamniotic connection is invaded by mesoderm and then develops perforations that connect to the albumen sac (Romanoff, 1960).

An additional reported variation in amniogenesis is the pattern of closure of the amnion over the tail region of the embryo. In some species, the folds posterior to the seroamniotic connection of the trunk region form a circular opening to the amniotic cavity, the amniotic navel (Fisk and Tribe, 1949; Romanoff, 1960). In turtles (Mitsukuri, 1891), tuatara (Dendy, 1899; Fisk and Tribe, 1949), lizards (Fisk and Tribe, 1949), and birds (Romanoff, 1960) an extension of the seroamniotic connection forms a long tube posterior to the body of the embryo. This tube, the posterior amniotic tube, connects the amniotic cavity to the cavity under the vitelline membrane.

Amniogenesis by folding was likely a process characteristic of the ancestral amniote egg. In all extant amniotes, a dorsal sheet of ectoderm, the seroamniotic connection, develops during amniogenesis. The development of an extraembryonic coelom adjacent to the embryo occurs in all extant amniotes during later stages of amniogenesis. Based on developmental pattern, Fisk and Tribe (1949) suggested that the seroamniotic connection is the homolog of the ancestral amniochorion. The possibility that this structure is retained in monotremes (Hughes, 1984, 1993) and turtles (Mitsukuri, 1891) but lost in lepidosaurs (Hrabowski, 1926; Fisk and Tribe, 1949) and archosaurs (Romanoff, 1960) distinguishes Diapsida from Anapsida and Monotremata.

The allantois is a synapomorphy for Amniota and an expanded allantoic vesicle is characteristic of monotremes and Reptilia (Luckett, 1977; Mossman, 1987). The pattern of distribution of the allantois is related to the development of the yolk sac. In monotremes, the permanence of the choriovitelline membrane restricts the allantois to the embryonic pole (Hughes, 1993). Among turtles (Agassiz, 1857), crocodilians (Ferguson, 1985), and birds (Romanoff, 1960), yolk sac splanchnopleure develops and the allantois expands into the extraembryonic coelom to enclose the entire egg contents. Among squamates (Stewart, 1993), expansion of the allantois follows two patterns: The allantois extends about the periphery of the egg as the isolated yolk mass is absorbed, or it extends into the yolk cleft and does not contact the perimeter of the egg at the abembryonic pole. The greatest variation in the development and structure of the allantois is associated with viviparity (Luckett, 1977; Mossman, 1987; Blackburn, 1993).

In addition to secretions contributing to albumen and eggshell formation, the oviduct of amniotes may contribute to embryonic nutrition by providing other organic and inorganic molecules (Blackburn *et al.*, 1985). The degree of nutrient provision, or matrotrophy, would be expected to be related to the length of egg retention and thus most highly developed among viviparous species. The prevalence of matrotrophy among oviparous species has not been studied, but the absorption of uterine secretions by the eggs of monotremes (Flynn and Hill, 1939; Hughes, 1984) indicates that this pattern of embryonic nutrition is not restricted to viviparous species. Extended egg retention and viviparity result in a relationship between the egg and the developmental environment that may differ substantially from that of eggs oviposited at early embryonic stages. The pattern of water provision differs, respiratory constraints may differ, and the source of nutrition may differ, all of which may require embryonic modifications to access new sources or alter acquisition strategies. As a result, we might expect the greatest differences among amniote eggs and embryos to be correlated with reproductive mode. The development and structure of the extraembryonic membranes of viviparous amniotes (marsupials, eutherians, and squamates) are impressively diverse and the evolution of amniote placentation has been analyzed in numerous previous publications (Weekes, 1935; Mossman, 1937, 1987; Amoroso, 1952; Bauchot, 1965; Luckett, 1974, 1975, 1976, 1977, 1980; Yaron, 1985; Blackburn, 1985, 1993;

Stewart, 1993). A detailed review of placentation is beyond the scope of this work except to emphasize the general categories of diversity as they relate to the extraembryonic membranes. The primary placental exchange tissues of amniotes include allantoplacentation and several types of yolk sac placentation. In marsupials, squamates, and some eutherians, the choriovitelline membrane develops a functional interaction with the uterus, the choriovitelline placenta (Mossman, 1987; Tyndale-Biscoe and Renfree, 1987; Stewart, 1993). The choriovitelline placenta is permanent in marsupials and a few eutherians, but transitory in Squamata and most Eutheria. Many Eutheria (for example, rodents, lagomorphs, some Insectivora, and Chiroptera) have an inverted yolk sac placenta (Mossman, 1987). This structure results from the collapse of the yolk sac such that the opposing yolk sac membranes are placed in proximity. In some species, the splanchnopleuric yolk sac is apposed to a bilaminar omphalopleure; in others, the bilaminar omphalopleure is absent. In either condition, the yolk sac splanchnopleure adjacent to the embryo provides circulatory support for the embryonic side of the placenta. Two additional forms of yolk sac placentation occur among squamates and are associated with the isolated yolk mass–the omphaloplacenta and omphalallantoic placenta (Stewart, 1993). The allantois does not participate in placentation in most marsupials (Tyndale-Biscoe and Renfree, 1987), but does contribute to a diverse array of placental forms among squamates and eutherians (Mossman, 1937, 1987; Luckett, 1977; Blackburn, 1993).

The distribution of reproductive mode among modern amniotes indicates that oviparity is primitive and that viviparity has evolved once among Mammalia and numerous times among Lepidosauria. The clades Chelonia and Archosauria are oviparous and lecithotrophic. Matrotrophy evolved among Mammalia and independently among several lineages of Squamata.

ORIGIN OF THE AMNIOTE EGG

The absence of recognizable antecedents to most structural components of the egg of amniotes, particularly the extraembryonic membranes, has forced speculation on the origin of the amniote egg to focus on functional analogs among modern anamniotes and on the ecological and physiological advantages of specific structures to modern amniotes. The resulting scenarios have emphasized three questions: (1) Which came first, the amniote or the amniote egg (Romer, 1957; Carroll, 1970); (2) what was the order of appearance and functional characteristic of the extraembryonic membranes (Mitsukuri, 1891; Hubrecht, 1912; Szarski, 1968); and 3) what was the reproductive mode of the first amniotes (Hubrecht, 1912; Mossman, 1987).

In contrast to Amphibia, amniotes are terrestrial organisms and reproductive traits are among their specializations for terrestrial life. Romer (1957) considered that all stages of the life cycle likely did not evolve terrestrial specializations simultaneously; that based on the fossil record, the adults of the earliest amniotes were aquatic, and that predation pressure and risk of drying in unstable aquatic habitats would exert high risk on aquatic eggs. Given these three assumptions, Romer (1957, 1967) argued that the amniote egg evolved among aquatic organisms as a response to high egg mortality in aquatic environments. Tihen (1960) concurred on the sequence, but noted that the most likely source of egg mortality was predation, not drought. An alternative suggestion for the ecological conditions selecting for the ancestral amniote egg was proposed by Goin and Goin (1962). They argued that terrestrial breeding habits, specifically internal fertilization, large egg size, and oviposition in terrestrial sites are correlated with humid, montane habitats in modern Amphibia and, by analogy, similar ecological conditions could have selected for the amniote egg. This scenario followed the observation by Lutz (1948) that, based on conditions in Anura in which terrestrial habits are associated with large egg size, an increase in yolk reserves followed by gradual loss of the larval stage of development likely were critical steps in the evolution of the reptilian egg. Elinson (1989) noted that large size is a significant feature of the amniote egg, but because the pattern of cleavage limits egg size the evolution of meroblastic cleavage, followed by the evolution of large vascularized surfaces, must precede increases in egg size.

Carroll (1970) offered an imaginative argument to link reproductive characteristics of the early amniotes to body size and cranial morphology. Because size of terrestrial eggs is limited by respiratory constraints and adult body size of modern Caudata is correlated with egg size, the ancestors of amniotes must have had small body size (<80 mm snout-vent). Larger adult size could evolve only after the evolution of embryonic respiratory specializations released the size limitation on eggs. The evolution of small adult body size thus preceeded the evolution of the terrestrial amniote egg. Carroll's (1970) scenario was founded on an hypothesis that Captorhinomorpha is the sister taxon to all other amniotes and that the amniote egg evolved among primitive captorhinomorphs with small adult body size. Alternative phylogenetic hypotheses (see Lombard and Sumida, 1992) have placed Captorhinomorpha within Amniota. Further, according to recent phylogenetic analyses, the likely sister groups to amniotes are either Diadectomorpha (Gauthier *et al.*, 1988b; Laurin and Reisz, this volume) or Seymouriamorpha (Berman *et al.*, 1992; Lee and Spencer, this volume), both of which are relatively large bodied organisms. If the amniote egg evolved in organisms with small body size as Carroll (1970) suggested, this "terrestrial" egg must have preceeded the evolution of the Amniota (sensu Gauthier et al., 1988a, 1988b).

The amniochorion and allantois are synapomorphies for Amniota for which no clearly intermediate structures are known among Amphibia. Speculation on the evolutionary order of the appearance of these structures has been based on assessment of a possible preeminent functional significance to the ancestral amniote egg. Mitsukuri (1891) considered the frequent development of nonspecific folds in the spreading blastoderm, the universal occurrence of an amniotic headfold, the development of a seroamniotic connection, and the tendency of the head region of the embryo to sink into the large yolk mass to provide clues to both the formation and the function of the earliest amnion. In his view, the amnion arose as a consequence of the mechanics of growth of the blastoderm over the yolk mass and functioned to prevent the embryonic head region from sinking into the yolk. Primitively, the amnion consisted of ectoderm. The mesodermal contribution arose in association with a secondary event–the growth of a respiratory organ, the allantois. Fisk and Tribe (1949) also cite the great tendency for growth of extraembryonic ectoderm and support the view that the ancestral amnion was entirely ectodermal; the seroamniotic connection is the homolog of this structure. However, the evolutionary sequence proposed by Fisk and Tribe (1949) arises from a different functional concern. They argue that a shell membrane, and perhaps eggshell, preceeded the evolution of the amnion that arose to protect the developing embryo from adhering to the outer investment of the egg.

The evolution of the extraembryonic coelom and its relationship to the yolk sac, particularly the transition from a choriovitelline membrane to a splanchnopleuric yolk sac, was a primary event in the origin of the amniote egg (Mossman, 1987). Mossman (1987) also noted that the allantois, originally an embryonic urinary bladder, was the last of the extraembryonic membranes to appear. Szarski (1968) likewise placed primary emphasis on the evolution of the allantois as an excretory organ, but differed in the sequence of membrane appearance by suggesting that the evolution of an amnion and chorion were later events. In both scenarios, the respiratory function of the allantois was secondarily derived.

For Hubrecht (1912), the primitive amniote was viviparous and the key innovation was a chorion-like embryonic envelope. This structure anchored the eggs in the maternal oviduct and absorbed fluids from the uterine cavity. An embryonic vascular system such as the allantois developed later, and the amnion evolved still later as a protective water jacket. Mossman (1987) also believed that the amniote egg evolved as a specialization for viviparity but for different reasons. As in other scenarios (Lutz, 1948; Elinson, 1989), Mossman (1987) emphasized that large yolked eggs could not evolve in the absence of sufficient respiratory support. If a respiratory function for the allantois evolved prior to the advent of large eggs, then the products of these small eggs must have been immature, altricial young, thus requiring parental care. These reproductive characteristics would be most likely to occur in a viviparous species.

These scenarios offer much for consideration but, as reconstructions of historical events, most understandably lack the capacity to be constructed as testable hypotheses. This does not make them any less interesting nor thought provoking but does foster a sense of frustration that such a significant event in the evolution of vertebrates is so elusive. Based on modern species, the structure of the amniote egg differs markedly from that of anamniotes, but even similar characteristics such as oviductal secretions have not been carefully compared. Luckett (1977) predicted characteristics of the predecessor to the egg of amniotes based on the distribution of traits among modern species, and similarly, the pattern of character transformation among extant amniotes can be predicted as phylogenetic relationships become more clearly defined. Such a method was used by Iverson and Ewert (1991), for example. However, there is much that is not known about embryonic development and egg physiology for virtually every major amniote group. For example, details of the development of the nominal structure for the taxon, the amnion, are known for very few species. Reproductive and developmental characters can contribute importantly to phylogenetic analyses (Luckett, 1977, 1993). Unfortunately, the greatest impediment to this approach is the lack of detailed studies of egg development among amniotes.

DISCUSSION

Although a comprehensive analysis of the pattern of egg evolution among extant amniotes is compromised by limited data, a robust phylogenetic approach will be possible ultimately as information on egg morphology and physiology of modern species increases. However, answers to fundamental questions concerning the origin of the egg and the polarity of character transformation will be more elusive because of the absence of living representatives of a suitable outgroup. The closest outgroup to modern Amniota, Lissamphibia, is a phylogenetically distant taxon with specialized reproduction and development.

Recent phylogenetic analyses identify Synapsida as the sister group to all other Amniota, which are represented by Reptilia among extant species (Gauthier *et al.*, 1988b; Lombard and Sumida, 1992). Based on this phylogenetic hypothesis, the egg of extant oviparous Reptilia is distinct from that of extant oviparous Synapsida, represented by Monotremata (Fig. 6). The structural characteristics of the eggs of Monotremata, (1) meroblastic cleavage, (2) proteinaceous shell membrane, (3) choriovitelline membrane as the definitive yolk sac, (4) amniochorion formed by folding, and (5) expanded allantoic vesicle confined to the embryonic pole of the egg, are likely plesiomorphic for Amniota. Monotremes share oviparity, and oviductal secretions of glycoprotein and keratinous protein with modern Amphibia (Hughes, 1977). Meroblastic cleavage, a derived trait, is shared with Reptilia. Two of the characters shared between Reptilia and Monotremata, an amniochorion and allantoic vesicle, are synapomorphies for Amniota.

The distribution of characters among extant taxa (Fig. 6) indicates that the ancestral condition for Reptilia was an oviparous, lecithotrophic egg. The egg cleavage pattern was meroblastic. Oviductal contributions were a proteinaceous and mineralized shell membrane and a large mass of albumen. The eggshell was pliable. The sequence of yolk sac development included a bilaminar omphalopleure, a choriovitelline membrane and a splanchnopleure. The amniochorion was formed from folds and an expanded allantoic vesicle surrounded the egg contents in terminal stages of incubation. The reptilian egg contains four derived characters not present in monotremes, a mineralized eggshell membrane, a large mass of albumen, the presence of a yolk sac that is entirely splanchnopleuric and an allantoic vesicle that surrounds the egg contents.

SUMMARY

The megalecithal eggs of oviparous amniotes are incubated in terrestrial environments. This life history pattern is contingent on a unique combination of specializations that include the pattern of cleavage, oviductal secretions, and elaborations of the developing embryo. During passage through the oviduct the egg is enclosed by glycoprotein (albumen) and an eggshell. The eggshell has an organic (protein) matrix and a variable inorganic component. A mineralized eggshell is characteristic of Reptilia and heavily mineralized, rigid eggshells have evolved on multiple occasions. Extensions of the embryo, the extraembryonic membranes, regulate embryonic development. The development of the extraembryonic membranes, which include the yolk sac, amnion, chorion, and allantois, is surprisingly uniform among oviparous amniotes. The amniochorion develops by a process of folding and results in an extraembryonic coelom in the embryonic hemisphere of the egg. The allantois forms as an outgrowth of the hindgut of the embryo and quickly expands to fill the extraembryonic coelom that forms during amniogenesis. Subsequent development of the allantois is influenced by the pattern of yolk sac formation. All oviparous amniotes develop a choriovitelline membrane (vascular trilaminar omphalopleure) which encloses the large yolk mass. The extraembryonic coelom of Reptilia expands to surround the yolk mass by splitting of the choriovitelline membrane. This process extends the somatopleure (chorion) and produces the definitive splanchnopleuric yolk sac of Reptilia. The extraembryonic coelom accomodates expansion of the allantoic vesicle, which grows to encircle the yolk sac. The distribution of characters among modern taxa suggests that the primitive amniote egg had a vascularized trilaminar yolk sac, an amniochorion, and a large allantoic vesicle. The outer investments of this egg consisted of a thin layer of glycoprotein surrounded by a proteinaceous shell membrane. This combination of characters is found among modern Monotremata. The outer investments likely were derived from the egg investments of anamniotes. The amniochorion and allantoic vesicle are synapomorphies for Amniota, with no clear antecedents among modern anamniotes. The egg of Reptilia evolved a thick layer of albumen, a mineralized shell membrane, a splanchnopleuric yolk sac, and an allantoic vesicle that surrounds the yolk sac.

ACKNOWLEDGMENTS

I am grateful to S. S. Sumida and K. L. M. Martin for the invitation to contribute to this volume. Thanks to R. A. Pyles for help with the figures. I also thank R. L. Reeder, S. S. Sumida and M. B. Thompson for critical comments.

LITERATURE CITED

Ackerman, R. A. 1991. Physical factors affecting the water exchange of buried reptile eggs. Pages 193-211 in: *Egg Incubation: Its Effects on Embryonic Development in Birds and Reptiles* (D. C. Deeming and M. W. J. Ferguson, eds.). Cambridge, UK: Cambridge University Press.

Agassiz, L. 1857. Contributions to the natural history of the United States of America. Volume II, Part III. Embryology of the turtle, pages 451-643. Boston: Little, Brown and Company.

Amoroso, E. C. 1952. Placentation. Pages 127-311 in:, *Marshall's Physiology of Reproduction* (A. S. Parkes, ed.). London: Longman's, Green and Company, Ltd.

Bauchot, R. 1965. La placentation chez les reptiles. *Annee Biologique*, 4:547-575.

Balinsky, B. I. 1975. *An Introduction to Embryology*. Philadelphia: W. B. Saunders.

Berman, D. S, S. S. Sumida, and R. E. Lombard. 1992. Reinterpretation of the temporal and occipital regions in *Diadectes* and the relationships of diadectomorphs. *Journal of Paleontology*, 66:481-499.

Blackburn, D. G. 1985. The evolution of viviparity and matrotrophy in vertebrates with special reference fo reptiles. Unpublished Ph.D. Dissertation. Cornell University, Ithaca, New York.

Blackburn, D. G. 1993. Chorioallantoic placentation in squamate reptiles: structure, function, development, and evolution. *Journal of Experimental Zoololgy*, 266:414-430.

Blackburn, D. G., H. E. Evans, and L. J. Vitt. 1985. The evolution of fetal nutritional adaptations in viviparous vertebrates. Pages 437-439 in: *Functional Morphology in Vertebrates* (H.R. Duncker and G. Fleischer, eds.). Stuttgart: Gustav Fischer Verlag.

Carroll, R. L. 1970. Quantitative aspects of the amphibian-reptilian transition. *Forma et Functio*, 3:165-178.

Collazo, A., J. A. Bolker, and R. Keller. 1994. A phylogenetic perspective on teleost gastrulation. *American Naturalist*, 144:133-152.

Cordero-Lopez, N., and M. H. Morales. 1995. Lack of proteins of oviductal origin in the egg of a tropical anoline lizard. *Physiological Zoology*, 68:512-523.

Del Pino, E. M. 1989. Modifications of oogenesis and development in marsupial frogs. *Development*, 107:169-187.

Dendy, A. 1899. Outlines of the development of the tuatara, *Sphenodon (Hatteria) punctatus*. *Quarterly Journal of Microscopical Science*, 42:1-87.

Dufaure, J.P., and J. Hubert. 1961. Table de developpement du lezard vivipare: *Lacerta (Zootica) vivipara* Jacquin. *Archives d'Anatomie Microscopique et de Morphologie Experimentale*, 50:309-328.

Elinson, R. P. 1987. Change in developmental patterns: embryos of amphibians with large eggs. Pages 1-21 in: *Development as an Evolutionary Process* (R.A. Raff and E.C. Raff, eds). New York: Alan R. Liss, Incorporated.

Elinson, R. P. 1989. Egg evolution. Pages 251-262 in: *Complex Organismal Functions: Integration and Evolution in Vertebrates* (D. B. Wake and G. Roth, eds.). Chichester: John Wiley and Sons Ltd.

Ewert, M. A. 1979. The embryo and its egg: development and natural history. Pages 333-413 in: *Turtles: Perspectives and Research* (M. Harless and H. Morlock, eds.). New York: Wiley and Sons.

Ewert, M. A. 1985. Embryology of turtles. Pages 75-267 in: *Biology of the Reptilia. Vol. 14, Development A* (C. Gans, F. Billett, and P. F. A. Maderson, eds.). New York: John Wiley and Sons.

Ferguson, M. W. J. 1982. The structure and composition of the eggshell and embryonic membranes of *Alligator mississippiensis*. *Transactions of the Zoological Society of London*, 36:99-152.

Ferguson, M. W. J. 1985. The reproductive biology and embryology of the crocodilians. Pages 329-421 in: *Biology of the Reptilia. Vol. 14, Development A* (C. Gans, F. Billett, and P. F. A. Maderson, eds.). New York: John Wiley and Sons.

Fisk, A., and M. Tribe 1949. The development of the amnion and chorion of reptiles. *Proceedings of the Zoological Society of London*, 119:83-114.

Flynn, T. T., and J.P. Hill. 1939. The development of the Monotremata. Part IV. Growth of the ovarian ovum, maturation, fertilisation, and early cleavage. *Transactions of the Zoological Society of London*, 24:445-623.

Gauthier, J., A. G. Kluge, and T. Rowe. 1988a. Amniote phylogeny and the importance of fossils. *Cladistics*, 4:105-209.

Gauthier, J. A., A. G. Kluge, and T. Rowe. 1988b. The early evolution of the Amniota. Pages 103-155 in: *The Phylogeny and Classification of the Tetrapods, Vol. 1: Amphibians, Reptiles, Birds* (M.J. Benton, ed.). Oxford: Clarendon Press.

Goin, O. B., and C. J. Goin. 1962. Amphibian eggs and the montane environment. *Evolution*, 16:364-371.

Gray, J. 1928. The role of water in the evolution of the terrestrial vertebrates. *British Journal of Experimental Biology*, 6:26-31.

Griffiths, M. 1978. *The Biology of Monotremes.* New York: Academic Press.

Griffiths, M., D. L. McIntosh, and R. E. A. Coles. 1969. The mammary gland of the echidna, *Tachyglossus aculeatus* with observations on the incubation of the egg and on the newly-hatched young. *Journal of Zoology, London*, 158:371-386.

Hamilton, H. L. 1952. *Lillie's Development of the Chick.* New York: Henry Holt and Company.

Hill, J. P. 1897. Contributions to the embryology of the Marsupialia. I. The placentation of *Perameles*. *Quarterly Journal of Microscopical Science*, 40:385-446.

Hrabowski, H. 1926. Das dotterorgan der eidechsen. *Zeitschrift für Wissenschaftliche Zoologie*, 128:305-382.

Hubert, J. 1985. Embryology of the Squamata. Pages 1-34 in: *Biology of the Reptilia. Vol. 15, Development B* (C. Gans and F. Billett, eds.). New York: John Wiley and Sons.

Hubrecht, A. A. W. 1912. The foetal membranes of the vertebrates. *Proceedings of the Seventh International Zoological Congress*, pages 426-434.

Hughes, R. L. 1977. Egg membranes and ovarian function during pregnancy in monotremes and marsupials. Pages 281-291 in: *Reproduction and Evolution* (J. H. Calaby and C. H. Tyndale-Biscoe, eds.). Canberra: Australian Academy of Science.

Hughes, R. L. 1984. Structural adaptations of the eggs and the fetal membranes of monotremes and marsupials for respiration and metabolic exchange. Pages

389-421 in: *Respiration and Metabolism of Embryonic Vertebrates* (R. S. Seymour, ed.). Dordrecht: Dr W. Junk Publishers.

Hughes, R. L. 1993. Monotreme development with particular reference to the extraembryonic membranes. *Journal of Experimental Zoology*, 266:480-494.

Hughes, R. L., and F. N. Carrick. 1978. Reproduction in female monotremes. Pages 233-253 in: *Monotreme Biology* (M. L. Augee, ed.). Royal Zoolological Society of New South Wales.

Hughes, R. L., F. N. Carrick, and C. D. Shorey. 1975. Reproduction in the platypus, *Ornithorhynchus anatinus*, with particular reference to the evolution of viviparity. *Journal of Reproduction and Fertility*, 43:374-375.

Iverson, J. B., and M. A. Ewert. 1991. Physical characteristics of reptilian eggs and a comparison with avian eggs. Pages 87-100 in: *Egg Incubation: Its Effects on Embryonic Development in Birds and Reptiles* (D. C. Deeming and M. W. J. Ferguson, eds.). Cambridge, UK: Cambridge University Press.

Lombard, R. E., and S. S. Sumida. 1992. Recent progress in understanding early tetrapods. *American Zoologist*, 32:609-622.

Luckett, W. P. 1974. Comparative development and evolution of the placenta in Primates. Pages 142-234 in: *Reproductive Biology of the Primates* (W. P. Luckett, ed.). Basel: S. Karger.

Luckett, W.P. 1975. Ontogeny of the fetal membranes and placenta. Their bearing on primate phylogeny. Pages 157-182 in: *Phylogeny of the Primates* (W. P. Luckett and F. S. Szalay, eds.). New York: Plenum Press.

Luckett, W. P. 1976. Cladistic relationships among primate higher categories: evidence of the fetal membranes and placenta. *Folia Primatologica*, 25:245-276.

Luckett, W. P. 1977. Ontogeny of amniote membranes and their application to phylogeny. Pages 439-516 in: *Major Patterns in Vertebrate Evolution* (M. K. Hecht, P. C. Goody, and B. M. Hecht, eds.). New York: Plenum Press.

Luckett, W. P. 1980. Monophyletic of diphyletic origins of Anthropoidea and Hystricognathi. Evidence of the fetal membranes. Pages 347-368 in: *Evolutionary Biology of New World Monkeys and Continental Drift* (R. L. Ciochon and A.B. Chiarelli, eds.). New York: Plenum Press.

Luckett, W. P. 1993. Uses and limitations of mammalian fetal membranes and placenta for phylogenetic reconstruction. *Journal of Experimental Zoology*, 266:514-527.

Lutz, B. 1948. Ontogenetic evolution in frogs. *Evolution*, 2:29-39.

Mitsukuri, K. 1891. On the foetal membranes of Chelonia. *The Journal of the College of Science, Imperial University, Japan*, 4:1-53.

Moffat, L. A. 1985. Embryonic development and aspects of reproductive biology in the tuatara, *Sphenodon punctatus*. Pages 493-521 in: *Biology of the Reptilia. Vol. 14, Development A* (C. Gans, F. Billett, and P. F. A. Maderson, eds.). New York: John Wiley and Sons.

Mossman, H. W. 1937. Comparative morphogenesis of the fetal membranes and accessory uterine structures. *Contribibutions to Embryology Carnegie Institution*, 26:129-246.

Mossman, H. W. 1987. *Vertebrate Fetal Membranes*. New Brunswick, New Jersey: Rutgers University Press.

Needham, J. 1931. *Chemical Embryology*. Cambridge: Cambridge Univ. Press.

Novacek, M. J., A. R. Wyss, and M. C. McKenna. 1988. The major groups of eutherian mammals. Pages 31-71 in: *The Phylogeny and Classification of the Tetrapods, Vol. 2: Mammals* (M. J. Benton, ed.). Oxford: Clarendon Press.

Noble, R. C. 1991. Comparative composition and utilisation of yolk lipid by embryonic birds and reptiles. Pages 17-28 in: *Egg Incubation: Its Effects on Embryonic Development in Birds and Reptiles* (D. C. Deeming and M. W. J. Ferguson, eds.). Cambridge, UK: Cambridge University Press.

Packard, G. C. 1991. Physiological and ecological importance of water to embryos of oviparous reptiles. Pages 213-228 in: *Egg Incubation: Its Effects on Embryonic Development in Birds and Reptiles* (D. C. Deeming and M. W. J. Ferguson, eds.). Cambridge, UK: Cambridge University Press.

Packard, G. C., and M. J. Packard. 1980. Evolution of the cleidoic egg among reptilian antecedents of birds. *American Zoologist*, 20:351-362.

Packard, G. C., and M. J. Packard. 1988. The physiological ecology of reptilian eggs and embryos. Pages 523-605 in: *Biology of the Reptilia. Vol. 16, Ecology B* (C. Gans and R. B. Huey, eds.). New York: Alan R. Liss.

Packard, G. C., C. R. Tracy, and J. J. Roth. 1977. The physiological ecology of reptilian eggs and embryos, and the evolution of viviparity within the class Reptilia. *Biological Reviews*, 52:71-105.

Packard, M. J. 1994. Patterns of mobilization and deposition of calcium in embryos of oviparous, amniotic vertebrates. *Israel Journal of Zoology*, 40:481-492.

Packard, M. J., and K. F. Hirsch. 1989. Structure of shells from rigid-shelled eggs of the geckos *Gekko gecko* and *Phelsuma madagascarensis*. *Canadian Journal Zoology*, 67:746-758.

Packard, M. J., K. F. Hirsch, and V. B. Meyer-Rochow. 1982. Structure of the shell from eggs of the tuatara, *Sphenodon punctatus*. *Journal of Morphology*, 174:197-205.

Packard, M. J., and V.G. DeMarco 1991. Eggshell structure and formation in eggs of oviparous reptiles. Pages 53-69 in: *Egg Incubation: Its Effects on Embryonic Development in Birds and Reptiles* (D. C. Deeming and M. W. J. Ferguson, eds.). Cambridge, UK: Cambridge University Press.

Palmer, B. D., and L. J. Guillette, Jr. 1991. Oviductal proteins and their influence on embryonic development in birds and reptiles. Pages 29-46 in: *Egg Incubation: Its Effects on Embryonic Development in Birds and Reptiles* (D. C. Deeming and M. W. J. Ferguson, eds.). Cambridge, UK: Cambridge University Press.

Patton, B. M. 1951. *Early Embryology of the Chick, 4th Edition*. Philadelphia: The Blakiston Company.

Richards, M. P., and N. C. Steele 1987. Trace element metabolism in the developing avian embryo: A review. *Journal of Experimental Zoology, Supplement*, 1:39-51.

Romanoff, A. L. 1960. *The Avian Embryo*. New York: Macmillan and Company.

Romanoff, A. L., and A. J. Romanoff. 1949. *The Avian Egg*. New York: John Wiley and Sons.

Romer, A. S. 1957. Origin of the amniote egg. *Scientific Monthly*, 85:57-63.

Romer, A. S. 1967. Major steps in vertebrate evolution. *Science*, 158:1629-1638.

Salthe, S. N. 1963. The egg capsules in the Amphibia. *Journal of Morphology*, 113:161-171.

Semon, R. 1894. Die embryonalhullen der monotremen und marsupialier. *Denkschriften der Medicinisch-Naturwissenschaftlichen Gesellschaft*, 2:19-58.

Simkiss, K. 1980. Water and ionic fluxes inside the egg. *American Zoologist*, 20:385-393.

Stewart, J. R. 1985. Placentation in the lizard *Gerrhonotus coeruleus* with a comparison to the extraembryonic membranes of the oviparous *Gerhonotus multicarinatus* (Sauria, Anguidae). *Journal of Morphology*, 185:101-114.

Stewart, J. R. 1993. Yolk sac placentation in reptiles: structural innovation in a fundamental vertebrate fetal nutritional system. *Journal of Experimental Zoology*, 266:431-449.

Stewart, J. R. and D. G. Blackburn. 1988. Reptilian placentation: structural diversity and terminology. *Copeia*, 1988:838-851.

Szarski, H. 1968. The origin of vertebrate foetal membranes. *Evolution*, 22:211-214.

Tihen, J. A. 1960. Comments on the origin of the amniote egg. *Evolution*, 14:528-531.

Tracy, C. R., and H. L. Snell. 1985. Interrelations among water and energy relations of reptilian eggs, embryos, and hatchlings. *American Zoologist*, 25:999-1008.

Tyndale-Biscoe, H., and M. Renfree. 1987. *Reproductive Physiology of Marsupials*. Cambridge, UK: Cambridge University Press.

Weekes, H. C. 1935. A review of placentation among reptiles with particular regard to the function of the placenta. *Proceedings of the Zoological Society of London*, 1935:625-645.

Williams, J. 1967. Yolk utilization. Pages 341-382 in: *The Biochemistry of Animal Development* (R. Weber, ed.). New York: Academic Press.

Yaron, Z. 1985. Reptilian placentation and gestation: structure, function, and endocrine control. Pages 527-603 in: *Biology of the Reptilia. Vol. 15, Development B* (C. Gans and F. Billett, eds.). New York: John Wiley and Sons.

Yntema, C. L. 1968. A series of stages in the embryonic development of *Chelydra serpentina*. *Journal of Morphology*, 125:219-252.

CHAPTER 10

THE ROLE OF THE SKIN IN THE ORIGIN OF AMNIOTES: PERMEABILITY BARRIER, PROTECTIVE COVERING AND MECHANICAL SUPPORT

Larry M. Frolich

> He was a very small frog with wide, dull eyes. And just as I looked at him, he slowly crumpled and began to sag. The spirit vanished from his eyes as if snuffed. His skin emptied and drooped...I watched the taut, glistening skin on his shoulders ruck, and rumple, and fall. Soon, part of his skin, formless as a pricked balloon, lay in floating folds like bright scum on top of the water; it was a monstrous and terrifying thing.
>
> –Annie Dillard, 1974

INTRODUCTION

In her description of a dying frog, Annie Dillard, as acute nature observer, recognizes the preeminence of the skin in maintaining the structural integrity of the animal. Ironically, and perhaps unfortunately to the study of amniote origins, for most comparative vertebrate anatomists, the skin typically has been what is ripped off quickly to get to the structures of interest underneath. Physiological and functional study of the skin, meanwhile, has focused primarily on the local, and often microscopic, properties of isolated pieces. If, however, considered both micro- and macro-scopically as an integrated structural-functional system, this largest of organs is revealed to be a crucial element in virtually every organismal function important to understanding the origin of amniotes. Skin serves as a protective covering against trauma, disease, heat, cold and other environmental threats; as a permeability barrier that regulates fluid and gas exchange with the environment; and as a mechanical support for the muscles and other skeletal elements involved in breathing and locomotion.

The unfortunate aspect of the comparative anatomist's tradition of quickly ripping the skin off an interesting specimen is that the mode in which the skin has been least studied–functional analysis of macroscopic properties of the whole organ–potentially provide the most fruitful contributions to the understanding of amniote origins. When all the skin's diverse functions–permeability membrane, protective covering and mechanical support–are considered simultaneously in reference to the origin of amniotes, the following three key points emerge:

1. Regarding permeability function: Despite years of textbook indoctrination to the contrary, the osmoregulatory role of the skin is insignificant in the origin of amniotes. Terrestriality does not require a waterproof integument.

2. Regarding protective function: Bony scales are a heavy way to provide protection in the terrestrial environment. Thus, speedy terrestrial locomotion requires replacing bony dermal scales. Keratinized epidermal scales are a good substitute for providing protection, but skeletal support is lost.

3. Regarding mechanical support function: The bony scales of amniote ancestors also served as mechanical support and attachment for axial muscles. With the origin of amniotes, the dermal collagen evolved into a dense mat to serve these functions during locomotion. A densely woven mat of dermal fibers is the most consistent and

significant evolutionary skin innovation associated with the origin of amniotes.

Before examining the biomechanical and physiological data pertinent to the three main facets of skin function (protective covering, permeability membrane and mechanical support--these form the main sections for the heart of this review), some attempt must be made to wrangle out those vertebrate groups, both extant and fossil, for which a review of the available data on skin might prove most profitable. Elsewhere in this volume (Laurin and Reisz; Lee and Spencer), thorough coverage is given to the phylogenetic issues that confound the origin of amniotes. However, the following section teases out a few of these issues that are most important relative to the review of skin physiology that follows.

FOSSIL SKIN, LISSAMPHIBIAN SPECIALIZATION AND THE LACK OF AN EXTANT SURROGATE FOR EARLY AMNIOTES.

Fossil skin is extremely rare and rather limited in what it reveals about the function of the skin in the living organism. Virtually nothing can be discerned about permeability. If well preserved scales are present, they might reveal something about the skin's protective role, especially against trauma. In addition, although not often considered in paleontological analyses, the pattern revealed by a patch of scales discovered *in situ* can reveal something about the mechanical support provided by the overall skin organ. In general however, we must turn to living groups in order to understand the comparative physiology of the skin as it relates to amniote origins.

Selecting an appropriate out-group to amniotes presents a serious problem (Benton, 1991; Carroll, 1982), especially in reference to analyzing skin features. Lissamphibians (frogs, salamanders and caecilians) are the obvious natural out-group for discerning derived amniote features (see Fig. 3). However, for several reasons, lissamphibian skin exhibits a highly derived suite of traits relative to the early tetrapods from which amniotes evolved. Lissamphibians are inveterate skin breathers (Feder and Burggren, 1985) and their skin is highly specialized to facilitate gas exchange with the environment.

They have a bimodal life cycle with frequent aquatic/terrestrial transitions and the fluid permeability of their skin reflects this unusual (and quite likely derived relative to most early tetrapods) mode of life. Although the predominant environment of amniote ancestors is not an easy matter to discern (Carroll, 1991), it seems unlikely that they filled a niche anything at all like that of modern lissamphibians. Lissamphibians are small and generally fossorial or arboreal and their protective needs are quite different from the large, ground-dwelling early tetrapods which gave rise to amniotes. The diminutive size and lack of a rib cage in lissamphibians also results in mechanical support needs from the skin that are quite different from those of the bulky, stout-ribbed amniote sister groups.

If Lissamphibians are of questionable value as an out-group for amniotes, then to what modern vertebrate group can we turn? Unfortunately, looking beyond Lissamphibians for an out-group means looking beyond land vertebrates (Fig. 3). If Lissamphibians are "over-specialized" for making good out-group comparison with amniotes, fish groups do not exhibit the fundamental shift in body plan (limbs, rib cage, epidermal scales, neck, narrow tail, etc.) that accompanied evolution into a terrestrial environment and thus are "under-specialized." Nonetheless, especially for looking at permeability and protective functions of the skin, some reference to sarcopterygian fish, and even to actinopterygians, can help discern the range of variability that vertebrates in general exhibit.

As difficult as it is to pinpoint a simple extant out-group to the amniotes, it is even more difficult to find a good extant surrogate for the earliest amniotes–that is, a group whose skin features are relatively unchanged since their origin. Lepidosaurs (lizards, snakes, and *Sphenodon*) exhibit epidermal scales that may or may not have been common in early tetrapod groups–the limited fossil skin evidence makes this difficult to discern. Lepidosaur-type scales, nonetheless, are frequently considered primitive for amniotes and they might make good surrogates for looking at the protective covering function of skin. However, the wide array of aquatic, terrestrial, fossorial and arboreal environments in which lepidosaurs are found makes it difficult to use their skin as a model for the permeability of early amniote skin. Although, again, the predominant environment of the earliest amniotes is uncertain and they may also have inhabited an equally diverse array of environments. As regards the mechanical support role of skin, snakes, being limbless are likely to exhibit a number of extreme specializations. Lizards are generally smaller than the earliest amniotes, but the largest lizards may serve as good models.

Mammals, birds, crocodiles, and turtles, for obvious reasons, are not good surrogates for early amniotes, especially when it comes to studying skin. Feathers, fur, and thick scutes or a carapace are all very derived relative to early amniotes and significantly affect both permeability and protection. In terms of mechanical support and the role of the dermis, these groups may be as akin to early amniotes as lepidosaurs.

Unfortunately then, to get a good sense of whether amniotes exhibit any unique skin specializations, some consideration of all of the amniote subgroups is necessary, with extra focus on lepidosaurs perhaps proving fruitful. By the same token, to get a good grasp on whether any potential amniote specialization is unique relative to their ancestors and thus represents a true evolutionary innovation, some comparison must be made with a diverse array of other vertebrates with some extra focus on lissamphibians, again, perhaps proving fruitful. However, despite these caveats for choosing which vertebrate groups to analyze, a review of skin function does prove valuable for illuminating some aspects of amniote origins. If nothing else, the rather simplistic view (all too frequently espoused in textbooks) that amniotes simply need less permeable skin to handle their dry new "fully terrestrial" environment can be laid to rest. Furthermore, the often overlooked and ever-important mechanical role of the skin can be given its due consideration.

SKIN AS PERMEABILITY BARRIER: DEBUNKING THE MYTH OF A NEED FOR TERRESTRIAL WATER-PROOFING

The permeability of membranes–be they cell membranes, respiratory surfaces, or an integument–is an extremely labile property. Many organisms routinely encounter drastic changes in the concentration gradients across their membranes, be it for oxygen, carbon dioxide, water vapor or dissolved ions. Most organisms accommodate to these changes *physiologically* at least in part by altering the permeability of their membranes. *Evolutionarily*, membrane permeability is also extremely labile. Most subgroups of amniotes (such as lizards) as well as most subgroups of lissamphibians (such as salamanders) each include diverse species that inhabit a wide variety of osmoregulatory niches from extremely saline to fresh water and from very xeric to very humid atmospheres (Feder, 1992; Gans and Dawson, 1976). Thus, to suggest that the evolution of a much larger group of vertebrates, such as the amniotes, is predicated on the establishment of a particular osmoregulatory role for *any* membrane is ludicrous at best. Nonetheless, the notion that a relatively waterproof integument is a fundamental part of the amniote *gestalt* is so deeply embedded in our folklore of vertebrate evolution that a thorough and orderly debunking would seem to be in order.

The basic question is the following: do amniotes exhibit some unique morphological or biochemical feature of their skin that regulates the flow of gases or water in and out of the organism? The two media–air and water–present different problems of analysis and are examined separately below. For each medium, I examine whether theory (based on Fick's diffusion equation) would suggest some unique role for amniote skin in regulating permeability followed by a review of available data on actual permeabilities in a wide diversity of vertebrates.

Gas Exchange

Cutaneous gas exchange is a prevalent part of vertebrate physiology. It occurs in all major vertebrate groups, including amniotes. Feder and Burggren (1985) have elegantly analyzed the potential ways, based on diffusion theory, that skin morphology could be altered to affect gas exchange.

Fick's diffusion equation

$$\frac{M_X}{dt} = \frac{K_X \cdot A \cdot (C_1 - C_2)}{L}$$

gives the mass of substance **x** (M_x) that will diffuse a given distance (**L**) across a given area (**A**) under an initial concentration differential (C_1-C_2). K is a constant, the diffusion coefficient, that depends on the chemical properties of substance **x**. Fick's equation illustrates the limited options for affecting the rate of gas exchange by changes in skin anatomy: (1) alter surface area; (2) alter diffusion length; or (3) alter the diffusion gradient by perfusion of the internal environment or ventilation of the external environment.

Increasing surface area results in an increase in diffusion rates. Although lateral skin folds and other surface area alterations are frequently found in select vertebrate species, amniotes do not exhibit a fundamental shift in overall shape or skin fitting that suggests a significant evolutionary change in external surface area. Microscopic plicae found in conjunction with epidermal scales of lizards and snakes have a significant effect on surface area (Krejsa, 1979). However, no one has analyzed the surface area effects of microscopic folding around epidermal scales versus similar folding found around bony dermal fish scales (Krejsa, 1979). Such an analysis might illuminate a new aspect of control over gas exchange across the skin. However, it is doubtful that it would reveal a unique feature of amniotes relative to other vertebrates.

Following Fick's equation, thicker skin (i.e., a larger diffusion distance) should result in slower diffusion rates. Skin thickness varies considerably within every vertebrate group (Bereiter-Hahn *et al.*, 1986). In general, larger members of the group have thicker skin and, as a result, lower rates of cutaneous gas exchange. Amniotes exhibit a thicker mat of dermal collagen (see Skin as Mechanical Support: From Crossed Fibers to a Dermal Mat) than similarly sized lissamphibians and fish. However, cutaneous capillaries are in the epidermis or at the dermal-epidermal boundary and the thickness of the dermal collagen layer is insignificant to the exchange of gases.

Enhanced perfusion to the skin by ventilation of the surrounding fluid can also significantly affect diffusion rate by raising or lowering the concentration gradient. Amniotes do not exhibit any consistent mode of external skin ventilation (nor does any other vertebrate group, although interesting behaviors are found in select species). Variability in skin perfusion among amniotes is poorly studied at best, but again no consistent unique amniote pattern emerges from the available data (Feder and Burggren, 1985).

In summary, theoretical constraints on cutaneous gas exchange do not reveal any obvious amniote feature that would significantly alter diffusion rates. This lack of a consistent morphological feature that correlates with any of the potential controls on gas exchange is reflected in the empirical data on cutaneous respiration in vertebrates. Amniotes, like lissamphibians and fishes show a wide range of variability in their use of cutaneous gas exchange (Table 1). Contrary to popular perception, cutaneous respiration is widely and variably

Table 1. Range of Cutaneous Gas Exchange in Different Vertebrate groups.

	Uptake/Release across skin	
Group	**O_2(%)**	**CO_2 (%)**
Actinopterygian fish (no air breathers)	6-40	NA
air-breathing fishes	12-48	NA
Lissamphibians (terrestrial forms)	14-82	29-80
Lissamphibians (aquatic forms)	29-82	29-100
Lepidosaurs (breathing air)	2-19	4-33
Lepidosaurs (forcibly submerged)	5-38	NA

Percentages indicate portion of total O_2 uptake or CO_2 release that occur through the skin. For each group, the range of observed values is given with data from sample species at each end of the range. Data are summary from Feder and Burggren's (1985) comprehensive survey. See their work for comprehensive reference to experimental work measuring cutaneous gas exchange rates. NA, not available.

used by vertebrates, independent of their phylogenetic affinities or degree of terrestriality.

Water

As with gases, Fick's diffusion equation describes the movement of water across the skin. The same potential morphological alterations that might affect gas exchange could be seen to affect water diffusion. As the previous discussion indicates, amniotes do not exhibit any unique skin morphologies that affect diffusion across the skin, be it gas or water.

However, water diffusion in the terrestrial environment does present two complications to the simple theoretical model represented by Fick's equation. One complication is that diffusion of water vapor into the aqueous cellular environment involves a phase shift. Fick's equation can be elaborated to account for this phase shift (see Mautz, 1982, for one way of doing this) but it is difficult to relate the theoretical implications of a phase shift to any real morphologies of the membrane involved. However, it is this very phase shift that is instrumental in evaporative cooling. Select amniotes exhibit a number of unique skin features that appear to be adaptive for evaporative cooling (for example, dark coloration, gular folds, highly vascularized ears, and spinal plates). However, again, amniotes as whole do not exhibit any consistent skin features related to evaporative cooling and those specializations that do show up appear to be evolutionary adaptations to a particular type of terrestrial environment, usually an extremely xeric or extremely humid one. Among modern lizards, humid environment species show up to eight times greater rates of water flux than arid environment species (Nagy, 1982).

The second complication that water diffusion across the skin presents is the hydration of the skin itself as water enters the cells. Hydrated cells are thicker resulting in slower diffusion rates across the skin. At the same time, as cells hydrate the diffusion gradient is lowered with the same result–slower diffusion rates. More importantly however, the affinity that the molecules in the epidermis have for water plays an important role in water gain and loss.

All extant amniotes have a keratinized outer region of the epidermis (the stratum corneum) that is variously elaborated into scales, scutes, feathers, or hair. Keratin itself is very hydrophilic (Hill and Wyse, 1989) and the cells of the stratum corneum can quickly becomes hydrated in a very humid or moist environment. However, most terrestrial vertebrates use some sort of lipid coating to limit keratin hydration (Hill and Wyse, 1989). In amniotes, this lipid layer

Table 2. Cutaneous respiratory water loss in a diversity of amniotes. Percent indicates portion of total evaporation that cutaneous loss represents.

Species	Habitat	Temp (oC)	Cutaneous Water Loss mg/(cm2/day)	%	Reference
Crocodiles					
Caiman crocodilus	aquatic	23	32.9	87	2
Turtles					
Pseudemys scripta	aquatic	23	12.2	78	1
Terrapene carolina	temperate forest	23	5.3	76	1
Gopherus agassizii	desert	23	1.5	76	2
Lizards					
Iguana iguana	tropical forest	23	4.8	72	2
Anolis carolinenesis	forest	30	4.6	45	4
Sauromalus obesus	desert	23	1.3	66	2
Uta stansburiana	semi-desert	30	2.4	43	1
Amphibolurus ornatus	semi-desert	35	10	54	3
Amphibolurus inermis	desert	35	3	43	1
Sphenomorphus labillardieri	temperate	30	6	27	5
Gehyra variegata	semi-desert	30	5	57	1
Snakes					
Natrix taxispilota	aquatic	25	16.7	88	7
Pituophis melanoleucus	desert	25	3.7	64	1
Spalerosophis cliffordii	desert	25	4.8	65	6
Cerastes cerastes	desert	25	1.3	57	1

Data compiled by Bentley (1976). References: 1, Bently (1976); 2, Bently and Schmidt Nielsen, (1966); 3, Bradshaw (1970); 4, Clausen (1967); 5, Dawson *et al.*, (1966); 6, Dmi'el (1972); 7, Prange and Schmidt-Nielsen (1969).

is generally found in the spaces between the cells of the *stratum corneum*. The extent to which a hydrophobic lipid coating interferes with diffusion (in *or out* of the organism) and the extent to which hydrophilic keratin encourages this diffusion may play an important part in amniote osmoregulation. However, whereas a good deal of data are available on *how much* water passes through the skin of various terrestrial amniotes (Table 2), the mechanisms of control over water diffusion remain a wide open area of research. The keratin-lipid trade-off may prove to be an important and unique amniote feature. Nonetheless, it must be kept in mind that this trade-off is a means of *regulating* water exchange in different environments. It is *not* simply a water-proofing mechanism evolved in response to the terrestrial environment. The terrestrial environment ranges from extremely xeric conditions in which water loss is the primary problem to extremely humid conditions in which water gain is of paramount concern.

A survey of cutaneous water loss among a diversity of amniotes reveals a very wide range of variability (Table 2). Studies that separate cutaneous water flux from overall evaporation are rare. However, comparisons with overall evaporative rates in other vertebrates indicate that amniotes are not, in any way, unusually resistant to water exchange across the skin (Lillywhite and Maderson, 1982; Mautz, 1982; Shoemaker and Nagy, 1977).

Comparative study of the mechanisms for control of water exchange across the integument is much needed. In amniotes, the distribution and amount of hydrophilic keratin versus hydrophobic lipids may be a unique mechanism for control of water diffusion across the skin. Another way of properly conceptualizing the osmoregulatory challenges that amniotes face is to equate ion concentration gradients faced by aquatic vertebrates with humidity (or water vapor concentration) variability faced by terrestrial vertebrates. In both cases, a wide range of variability is encountered and, both physiologically and evolutionarily, a wide range of skin resistance to water diffusion is observed. Comparative physiologists have paid scant attention to surveying biochemical and morphological mechanisms for varying skin resistance to water diffusion. From what data is available (Bentley, 1976; Mautz, 1982; Shoemaker and Nagy, 1977) it is clear that a wide variety of mechanisms have evolved within amniotes as well as within their sister taxa. The keratin-lipid trade-off mechanism may be a unique amniote feature, but more research is needed to confirm this. It should, however, be kept in mind that keratin is not an amniote invention. A wide diversity of fish exhibit limited regions of keratinization of the epidermis and some teleosts, including the "algae eaters" common in home aquaria, are covered with epidermal tubercles produced by hypertrophy and keratinization of the epidermal cells (Krejsa, 1979). Lipid and polysaccharide mucous coatings are also common among all vertebrate groups (Krejsa, 1979).

SKIN AS PROTECTIVE COVERING: GAINING SPEED BY LOSING PERIPHERAL BONE

A substantial full-body covering of dermal bone in the form of scales, osteoderms, or heavy plates is the ancestral condition for vertebrates as evidenced by its presence in Paleozoic fish (Carroll, 1988; Romer, 1966) and its continued presence in all modern fish groups. Many early tetrapods have been found with bony scales (Clack, 1989; Godfrey, 1989; Godfrey *et al.*, 1981; Holmes, 1984; Hook, 1983; Panchen, 1972) and their absence in some cases is simply an artifact of preservation. Among those Paleozoic tetrapods likely to be in the immediate ancestry of amniotes, bony scales are known to occur (Clack, 1989; Holmes, 1984), especially in the abdominal region (Carroll, 1988; Romer, 1966). However, with a few strange exceptions (such as the dorsal bony scales of caecilians), all extant land vertebrates (including lissamphibians as well as amniotes) have lost a bony dermal covering outside the skull.

The most obvious functional interpretation for a bony dermal covering is protection against trauma. Vertebrate bone, being one of the strongest and stiffest biological materials ever evolved, provides excellent trauma protection. Why, then, is a bony outer covering lost in amniotes? A reasonable answer lies in understanding something about supporting and moving a mass in the terrestrial environment.

That a heavy and bulky object is more difficult to lift or move than a small object of similar mass is intuitively obvious. The reason that bulky objects are more cumbersome lies in the fact that the weight

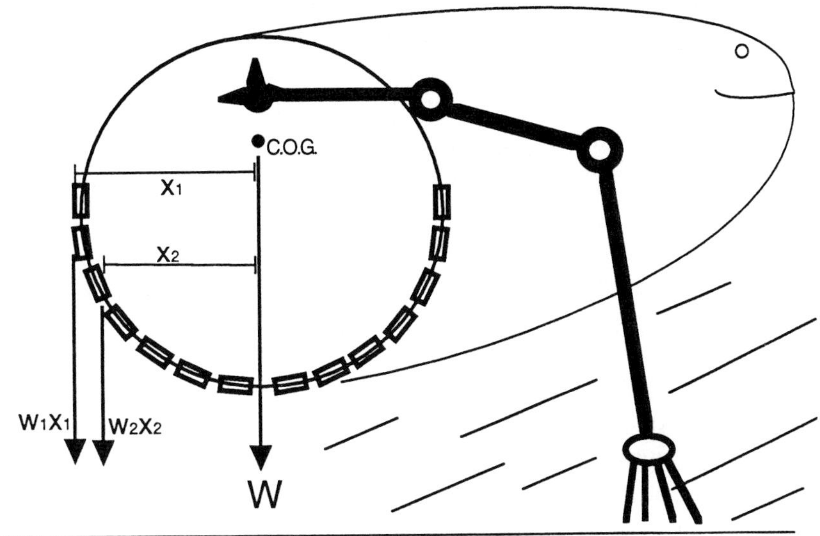

Figure 1. Schematic cross section through a generic tetrapod. Rectangular boxes around the periphery symbolize dermal scales. The overall weight of the animal (**W**) is the sum total of the torques exerted around the center of gravity by all the separate parts of the animal. Each torque is the product of the weight at that point (**w**) and the distance (**x**) from the center of gravity (C.O.G.). Thus, even if $\mathbf{w_1}$ and $\mathbf{w_2}$ are equivalent, the more lateral site will exert a larger torque because $\mathbf{x_1}$ is greater than $\mathbf{x_2}$. This explains why bulky objects are harder to lift and move than smaller objects of the same mass. For vertebrates, dermal scales are relatively dense heavy structures far from the center of mass and thereby they are very destabilizing in the terrestrial environment.

of an object is the sum total of the torques (or moment arms) exerted by each discrete part (recall that the center of gravity is the point at which all these torques cancel each other out–see Fig. 1). The less mass in a position far from the center of gravity, the smaller are the torques. For a stationary object, these torques are relatively insignificant (as long as the object is supported around its center of gravity (otherwise it will not be stationary!). However, put the object in motion and the torques start to be felt.

Bone is very stiff and strong in part by virtue of being dense and heavy. Dermal scales thus add significantly to the overall mass of an animal and their loss will reduce this mass. In addition, because the dermal scales are heavy objects at the periphery of the organism they exert large moment arms about the center of gravity and thus are a double hindrance to speedy locomotion (see Fig. 1).

The earliest land vertebrates retained a full coat of dermal scales and were probably very slow walkers on land. They may well have spent much of their time in the water where gravity does not cause weight torques around the center of gravity because the animal's density is equal to the medium. Dermal scales are gradually lost or reduced in size in early tetrapods. What scant fossil evidence exists (Carroll, 1988; Romer, 1966) suggests that initially dermal scales were lost from the lateral body wall and retained at the midlines, especially the ventral midline (caecilians are the only extant land vertebrates with dermal scales which they have at their dorsal midline). Midline scales exert smaller torques due to a shorter moment arm (Fig. 1). In addition, the scales become much thinner and smaller, sometimes even resembling the thin epidermal scales that show up in amniotes (Romer and Witter, 1941). As dermal scales are reduced and lost, speedier locomotion becomes possible due to the loss of overall mass and, especially important, the reduction of torques that are produced by masses far from the center of gravity.

The loss of dermal scales, although allowing for speedier locomotion, leaves the animal vulnerable to trauma. In lissamphibians, poison glands have evolved as an antidote to the loss of bony protection. In addition, lissamphibians being small and generally arboreal or fossorial are more immune to predation. Amniotes show an even broader diversity of responses to the loss of trauma protection. In general, they exhibit fast locomotion and an endothermic or behaviorally regulated warm metabolism. Keratin in the epidermis may help form a thermal barrier (in the form of fur or feathers, it certainly does). In addition, keratinized scales or scutes might be protection against trauma. In turtles, thick keratinized plates are certainly good protection against trauma, but the mass of the plates

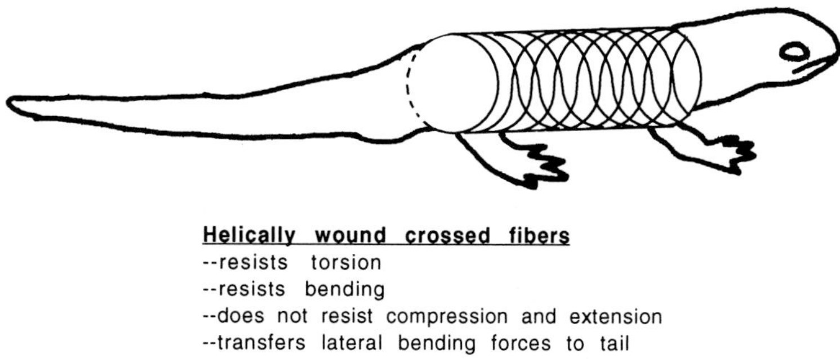

Helically wound crossed fibers
--resists torsion
--resists bending
--does not resist compression and extension
--transfers lateral bending forces to tail

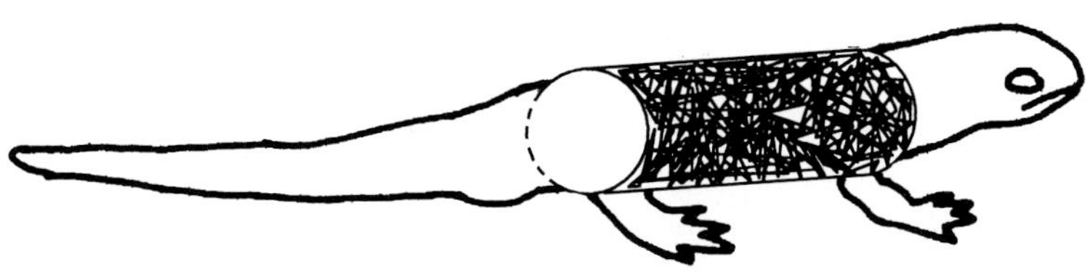

Mat of fibers with no preferential direction
--resists tension in all directions
--withstands traumatic forces

Figure 2. Schematic illustrating helically wound crossed fibers in the body versus a matted dermis with no predominant fiber direction. The two structures have significantly different mechanical properties. A helically wound cylinder works well to transmit lateral bending forces to the tail as well as in resisting torsion and bending. The more matted dermis with no discernible predominant fiber direction resists tension in all directions and may help replace dermal scales as a solid external cylinder for attachment of axial muscles and resistance to trauma.

that of bony dermal armor and turtles are not known for their speedy terrestrial locomotion! Lighter keratinized scales (in lepidosaurs) and fur or feather in birds are probably a good compromise between trauma protection and torque reduction for speedy terrestrial locomotion. Amniotes have evolved a wide diversity of other responses to predator avoidance in the terrestrial environment that are only marginally related to changes in the skin. For example, without the loss of dermal scales (and the subsequent evolution of large keratinized structures such as feathers), the most versatile form of "terrestrial" locomotion–flight–probably would not have evolved.

SKIN AS MECHANICAL SUPPORT: FROM CROSSED FIBERS TO A DERMAL MAT

The functional and evolutionary morphology of the dermal layer of the skin has been particularly neglected by comparative vertebrate biologists. It is widely recognized that the dermis, whether ossified or not, provides structural integrity to the skin (Harkness, 1968). However, biomechanical study of vertebrate support and movement has focused largely on the bony internal skeleton, whereas the mechanics of the external dermis have been largely ignored. In the few cases in which dermal mechanics have been investigated in detail, this largest of organs, not surprisingly, is revealed to play an important role in musculoskeletal function including locomotion (Wainwright *et al.*, 1978) and ventilation (Brainerd *et al.*, 1989).

Most commonly, the dermis of vertebrates takes the form of a fabric with two sets of crossed fibers. This fabric is wrapped helically around the animal, "at a bias" to use the dressmaker's term (see Fig. 2). This helically wound, crossed fiber dermis is likely to be primitive for vertebrates because it occurs early in the ontogeny of every group that has been examined and is retained throughout life in all fish groups (Fig. 3). Developmental repatterning to a more matted dermis is known to occur only in terrestrial vertebrates–birds and mammals among amniotes as well as frogs and salamanders among lissamphibians. Lepidosaurs and turtles, underneath their epidermal scales, appear to retain crossed fibers in the dermis as adults. However, aside from salamanders, dermal fiber patterning in most vertebrate groups is poorly sampled (see Table 3) and the extent of variability within each group is unknown.

The primitive vertebrate dermis can be mechanically modeled as a pressurized cylinder formed from crossed fibers that are helically wound around the animal. This type of structure has several mechanical properties that are important during laterally undulating locomotion (Wainwright, 1988). It resists torsion and bending (common forces during laterally undulating locomotion) and thus helps maintain the structural integrity of the cylinder. In addition, a helically wound dermis serves as an "exotendon" that transmits forces posteriorly to the region of the tail, which is the main propulsive site during laterally undulating swimming (Wainwright *et al.*, 1978). Thus, a helically wound crossed fiber dermis functions to enhance the efficiency of axial lateral undulation, especially during swimming when traveling waves pass axial forces posteriorly to the tail.

Lateral undulation is the primitive mode of axial movement for vertebrates as evidenced by its predominance in all fish groups. Early land vertebrates, including the ancestors of amniotes, were also lateral undulators as evidenced by the morphology of their vertebrae (Edwards, 1977) and the retention of lateral undulation in modern lissamphibians (except frogs) and lepidosaurs.

In amniotes, as well as in lissamphibians, decreasing reliance on traveling waves of lateral undulation has led to the evolution of a thick mat of more randomly oriented dermal fibers (Frolich and Schmid, 1991). Lateral undulation during terrestrial locomotion is often done using standing waves as opposed to the traveling waves most frequently used during swimming (Frolich and Biewener, 1992) and the advantages offered by a helically wound dermis no longer apply. Terrestrial salamanders uniformly exhibit a developmental shift to a more matted dermis, although the few lepidosaurs surveyed retain the helically wound pattern (Table 3). However, axial movement during locomotion in lizards is poorly studied (and they may still use the primitive traveling wave pattern), whereas snakes are well known to regularly use traveling waves of lateral undulation (Jayne, 1986, 1988). Further study of axial locomotion and the role of the dermis is needed for *all* vertebrates and virtually *nothing* has been done in this area on amniotes.

Simply because a structure such as the helically wound crossed fibers of the dermis, is no longer useful is not good reason, evolutionarily, to modify it. Some advantage for evolving an ontogenetic shift to a matted dermis in amniotes (and lissamphibians) is likely to exist. However, the paucity of good data on whole skin mechanical studies from amniotes makes it difficult to do more than speculate on what functional advantage a thick matted dermis provides. Loss of bony dermal scales for speedier locomotion renders

Table 3. Dermal Collagen Fiber Patterns in Chordates.

Species	Fiber Pattern	Reference
Aganathans		
Petromyzon fluviatilis	Crossed	Johnels, 1950
Myxine glutinosa	Embryos crossed	Holmgren, 1946
Elasmobranchs		
Rhizoprionodon terraenovae	Crossed	Motta, 1977
Ginglymostoma cirratum	Crossed	Motta, 1977
Sphyrna lewini	Crossed	Motta, 1977
Mustelus canis	Crossed	Motta, 1977
Mustelus norrisi	Crossed	Wainwright *et al.*, 1978
Negaprion brevirostris	Crossed	Wainwright *et al.*, 1978
Carcharinus acronotus	Crossed	Wainwright *et al.*, 1978
Carcharinus obscurus	Crossed	Wainwright *et al.*, 1978
Carcharinus maculippinis	Crossed	Wainwright *et al.*, 1978
Teleosts		
Fundulus heteroclitus	Crossed	Nado *et al.*, 1969
Hippoglossoides elassodon	Crossed	Brown & Wellings, 1970
Chasmichthyes gulosus	Crossed	Fujii, 1968
Anguilla rostrata	Crossed	Hebrank, 1980
Katsuwonus pelamis	Crossed	Hebrank & Hebrank, 1986
Leiostomus xanthrus	Crossed	Hebrank & Hebrank, 1986
Sturgeons		
Acipenser oxyrhynchus	Crossed	Moss, 1972
Polypterids		
Polypterus senagalus	Crossed	Pearson, 1981
Polypterus bichir lapradei	Crossed	Pearson, 1981
Polypterus endlicheri	Crossed	Pearson, 1981
Erpetoichthyes calabaricus	Crossed	Pearson, 1981
Dipnoans		
Lepidosiren paradoxa	Crossed	Frolich, 1991
Caecilians		
Ichthyophis kohtaoensis	Crossed	Frolich, 1991
Oscaecilia ochrocephala	Crossed	Frolich, 1991
Frogs		
Ascaphis trueii	Tadpole crossed	Frolich, 1991
	Adult matted	Frolich, 1991
Xenopus laevis	Tadpole crossed	Frolich, 1991; Weiss and Ferris, 1954
	Adult crossed	Frolich, 1991
	Adult matted	Fox , 1977

Table 3. Continued.

Species	Fiber Pattern	Reference
Frogs continues		
Scaphious couchii	Tadpole crossed	Frolich, 1991
Rana pipiens	Tadpole crossed	Edds, 1964; Kemp, 1959; Weiss & Ferris, 1954
Rana sylvatica	Tadpole crossed	Kemp, 1959
Rana sp.	Adult matted	Fox, 1977
Salamanders		
Ambystoma maculatum	Larva crossed	Weiss & Ferris, 1954
Ambystoma opacum	Larva crossed	Weiss & Ferris, 1954
11 sp. from six families	Larvae crossed	Frolich, 1991
15 sp. from nine families	Aquatic adults crossed	Frolich, 1991
20 sp. from six families	Terrestrial adults crossed	Frolich, 1991
Mammals		
Sylvilagus sp.	Embryo crossed	Frolich, 1991
	Newborn matted	Frolich, 1991
	Adult matted	Frolich, 1991
Homo sapiens	Adult matted	Millington *et al.,* 1971
Tursiops truncatus	Adult matted*	Huan *et al.*, 1983
Dasypus novemcinctus	Crossed in dermal scales	Moss, 1972
Squamates		
Heloderma horridum	Crossed	Moss, 1972
Egernia kingi	Crossed	Moss, 1972
Iguana iguana	Crossed	Frolich, 1991
Scincus scincus	Crossed	Frolich, 1991
Phyllodactylus siamensis	Crossed	Frolich, 1991
Birds		
Gallus gallus	Embryo crossed	Stuart & Moscona, 1967
	Adult matted	Sengel, 1976

"Crossed" indicates predominant fiber directions that cross at approximately 90°. "Matted" indicates no predominant fiber direction. *Cetaceans have "crossed" fibers in the hypodermis embedded within their subcutaneous fat.

the fluid body of amniotes without good support in the terrestrial environment. Although a helically wound cylinder has some positive attributes (outlined previously), it does not easily resist shortening

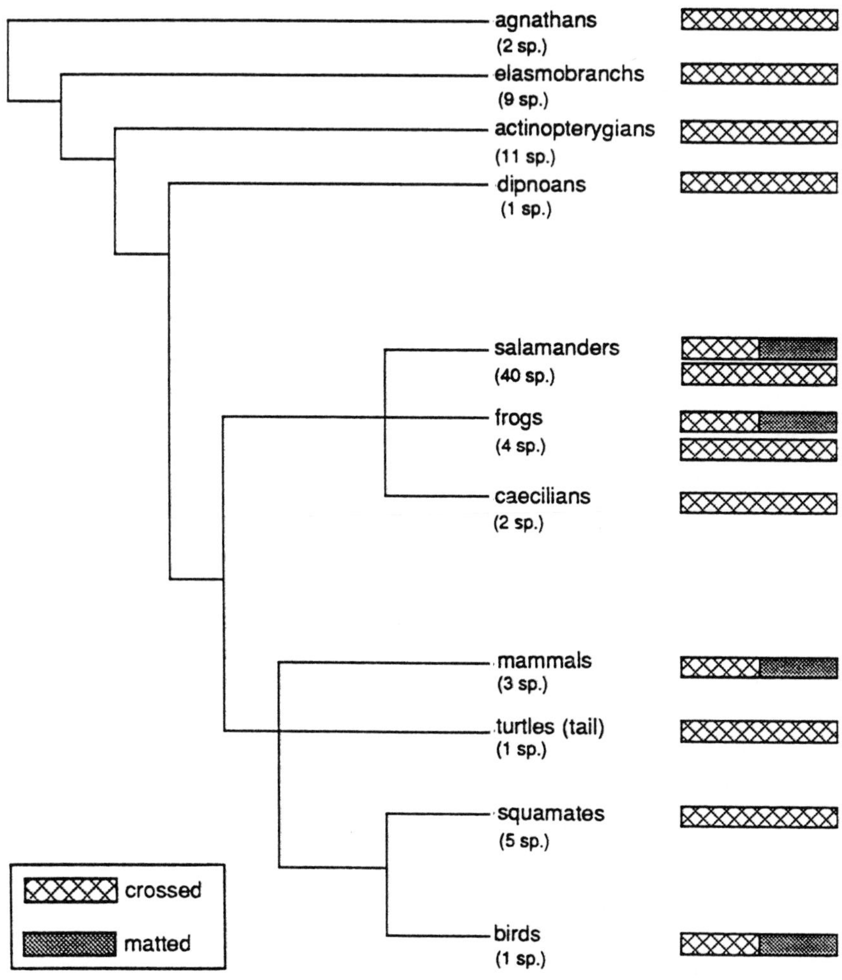

Figure 3. Dermal fiber pattern diversity among chordates. Widely accepted phylogenetic relationships among those taxa for which dermal fiber pattern data are available are shown. The boxes to the right of each taxon indicate dermal fiber pattern ontogenies within each group. Either the crossed fiber pattern is maintained throughout life (as in all fish groups) or initially crossed fibers undergo a developmental transition to a matted pattern (as in salamanders, frogs, mammals, and birds).

(accompanied by an expansion of diameter) and lengthening (accompanied by a narrowing of diameter). For a terrestrial animal trying to control its fluid insides against the forces of gravity and inertia in the relatively sparse medium of air, a solidly woven mat that resists tension in any direction is probably be the most appropriate fiber pattern for its external skeleton. For lepidosaurs and snakes, this is less of an issue if the overlapping keratinized scales of the epidermis create a structure similarly resistant to tension in most directions. Feathers and fur, however, as less interlocked epidermal structures add little to the mechanical integrity of the dermis (and in fact may themselves require the support of a dense underlying mat of fibers).

In conjunction with reorganization of the dermis into a thicker and generally more randomly woven mat of fibers, amniotes show a fundamental reorganization of the axial musculature. Relative to all other vertebrates, amniotes exhibit little evidence of myotomal blocks of axial muscle as adults and the axial muscles are in long strips that can cross several or more vertebral segments. Until more work is done on the comparative physiology of axial muscle function, it is hard to know how important the amniote dermis is as a site of muscle attachment. Unlike fish and lissamphibians, amniotes do not have regular myotomal septae that accept the insertion of large axial bundles along their faces and insert their edges on the dermis. Epaxial muscles in amniotes take advantage of the elaborated vertebral processes that are also a unique amniote feature. However, epaxial muscles also have slips that insert on the dermis and some hypaxial muscles rely solely on the dermis for their origin and insertion. Mechanical and electromyographical study of the axial musculoskeletal system in a wide diversity of amniotes is sorely needed. Until this data is made available, one can only speculate on the function of the many unique properties of the amniote axial system, including the matted dermis. A current resurgence of interest in the developmental patterning of the vertebrate body axis bodes well for understanding how the axial structures, including the dermis come to have their varied forms. A concurrent surge of interest in the mechanical function of the body axis, and especially the dermis as it matures from crossed fibers to a woven mat, will hopefully follow. Then it will become possible to move beyond speculation and emerge with a far more robust picture of the evolution of terrestrial support and locomotion in amniotes.

SUMMARY

The role of the skin in the origin of the amniotes is most commonly associated with the notion that amniotes, relative to other vertebrates, have evolved a waterproof integument. However, a systematic analysis of the environmental constraints on membrane permeability reveals that amniotes, like other vertebrate groups, confront a diverse range of osmoregulatory challenges. The extremes of humidity or salinity in the environment are routinely inhabited by all fish and land vertebrate groups including amniotes. Furthermore, the skin of amniotes has been shown empirically to exhibit a wide range of resistances to both gas and water flux. Finally, amniotes, considered as a whole, do not show any morphological innovations related to osmoregulation. Even the keratinized stratum corneum is not be a unique amniote feature and is only secondarily related to water exchange with the environment. A trade-off between hydrophilic keratin and hydrophobic lipids in the epidermis may be a unique amniote mechanism for either physiologically (during the life cycle of one organism) or evolutionarily (over generations) altering skin resistance to water. However, comparative study of the mechanisms that vertebrates use to regulate water movement across the skin is needed, especially from an integrated biochemical/morphological perspective.

Comparative vertebrate anatomists have neglected analyzing protective and mechanical functions of the whole skin. Early land vertebrates, in both the lissamphibian and amniote lineage, show an evolutionary tendency to lose bony dermal scales culminating in the virtual absence of dermal bone outside the skull in all extant land vertebrates. In amniotes, the protection against trauma provided by bony scales may be coopted by integumental appendages such as scales, scutes, feathers, and fur. Certainly the thick carapace of turtles provides excellent trauma protection.

The loss of dermal bone also reduces overall mass and eliminates the de-stabilizing moment arms that heavy peripheral structures cause. This allows for smaller muscular forces to easily initiate quick changes in the inertia of the whole body leading to speedy terrestrial locomotion. The structural integrity and axial muscle attachment site that dermal bone provides in fish groups is, in adult amniotes, served by a much thickened and matted dermal collagen layer.

These significant morphological innovations in the structures that confer mechanical integrity to the whole skin are more important to the origin of amniotes than any osmoregulatory changes. Without the evolution of a lightweight, yet structurally robust, external skeleton (in the form of a matted dermis), running, flight, bipedalism, and the whole diversity of high-speed amniote locomotor modes might not have evolved. Comparative study of the role of the whole skin as an integrated structural component in the locomotor system is a ripe area for future research.

LITERATURE CITED

Bentley, P.J. 1977. Osmoregulation. Pages 365-412 in: *Biology of the Reptilia, Volume 5, Physiology A.* (C. Gans and W.R. Dawson, eds.). London: Academic Press.

Bentley, P. J. and Schmidt-Nielsen, K. 1966. Cutaneous water loss in reptiles. *Science,* 151:1547-1549.

Benton, M. J. 1991. Amniote phylogeny. Pages 317-330 in: *Origins of the Higher Groups of Tetrapods.* (H. P. Schultze and L. Trueb, eds.). Ithaca, NY: Comstock Publishing Associates.

Bereiter-Hahn, J., Matoltsy, A. G., and K. Sylvia Richards. 1986. *Biology of the Integument: Volume Two, Vertebrates.* Berlin: Springer-Verlag.

Bradshaw, S.D. 1970. Seasonal changes in the water and electrolyte metabolism of *Amphibolurus* lizards in the field. *Comparative Biochemistry and Physiology,* 36:689-718.

Brainerd, E.L., K. F. Liem, and C. T. Samper. 1989. Air ventilation by recoil aspiration in fishes. *Science,* 246:1593-1595.

Brown, G. A., and S. R. Wellings. 1970. Electron microscopy of the skin of the teleost, *Hippoglossoides elassadon. Z. Zellforsch.* 103: 149-169.

Carroll, R. L. 1982. Early evolution of reptiles. *Annual Review of Ecology and Systematics,* 13:87-109.

Carroll, R. L. 1988. *Vertebrate Paleontology and Evolution.* New York: W. H. Freeman and Company.

Carroll, R. L. 1991. The origin of reptiles. Pages 331-353 in: *Origins of the Higher Groups of Tetrapods* (H. P. Schultze and L. Trueb eds.). Ithaca, New York: Comstock Publishing Associates.

Clack, J. A. 1989. Two new specimens of Anthrocosaurus (Amphibia: Anthrocosauria from the Northumberland coal measures. *Palaeontology*, 30:15-26.

Claussen, D. L. 1967. Studies of water loss in two species of lizards. *Comparative Biochemistry and Physiology* 20:115-130.

Crawford, E.C., and G. Kampe. 1971. Physiological responses of the lizard *Sauromalus obesus* to changes in ambient temperature. *American Journal of Physiology* 210:198-210.

Dawson, W.R., V. H. Shoemaker, P. and Licht. 1966. Evaporative water losses of some small Australian lizards. *Ecology*, 47:589-594.

Dillard, A. 1974. *Pilgrim at Tinker Creek.* New York: Harper and Row.

Dmi'el, R. 1972. Effect of activity and temperature on metabolism and water loss in snakes. *American Journal of Physiology*, 223:510-516.

Edds, M. V., Jr. 1964. The basement lamella of developing amphibian skin. Pages 245-250 in: *Small Blood Vessel Involvement in Diabetes Mellitus* (Siperstein, M., ed). Washington, D. C.: American Institute of Biological Science.

Edwards, J. L. 1977. The evolution of terrestrial locomotion. Pages 553-577 in: *Major Patterns in Vertebrate Evolution* (M.K. Hecht, P.C. Goody and B.M. Hecht eds). New York: Plenum Press.

Feder, M.E. 1992. A perspective on environmental physiology of the amphibians. Pages 1-6 in: *Environmental Physiology of the Amphibians* (M. E. Feder and W. W. Burggren, eds.). Chicago: The University of Chicago Press.

Fox, H. 1977. The anuran tadpole skin: changes occurring in it during metamorphosis and some comparisons with that of the adult. *Symposia of the Zoological Society of London*, 39:269-289.

Frolich, L. M., and A. A. Biewener. 1992. Kinematic and electromyographic analysis of the functional role of the body axis during terrestrial and aquatic locomotion in the salamander *Ambystoma tigrinum. Journal of Experimental Biology,* 162:107-130,

Frolich, L. M., and T. M. Schmid. 1991. Collagen type conservation during metamorphic repatterning of the dermal fibers in salamanders. *Journal of Morphology*, 208:99-107.

Fujii, R. 1968. Fine structure of the collagenous lamella underlying the epidermis of the Goby, *Chasmichthys gulosus. Annotationes Zoologicae Japonenses*, 41:95-106.

Gans, C., and W. R. Dawson. 1976. Reptilian physiology: an overview. Pages 1-17 in: *Biology of the Reptilia, Volume 5, Physiology A.* (C. Gans and W. R. Dawson eds.). London: Academic Press.

Godfrey, S. J. 1989. The postcranial skeletal anatomy of the Carboniferous tetrapod *Greererpeton burkemorani* Romer, 1969. *Philosophical Transactions of the Royal Society of London*B, 323:75-133.

Godfrey, S. J., Holmes, R. B., and Laurin, M. 1991. Articulated remains of a Pennsylvanian embolomere (Amphibia: Anthracosauria) from Joggins, Nova Scotia. *Journal of Vertebrate Paleontology,* 11:213-219.

Harkness, R. D. 1968. Mechanical properties of collagenous tissues. Pages 247-310 in: *Treatise on Collagen, Volume 2A.* (Gould, B. S., ed.). New York: Academic Press.

Haun, J. E., E. W. Hendricks, F. R. Borkat, R. W. Kataoka, D. A. Carder, and N. K. Chun. 1983. Dolphin hydrodynamics annual report FY 82. *Naval Ocean Systems Center Technical Report,* 935:7-24.

Hebrank, M. R. 1980. Mechanical properties and locomotor functions of eel skin. *Biological Bulletin,*158:58-68.

Hebrank, M. R., and J. H. Hebrank. 1986. The mechanics of fish skin: Lack of an "external tendon" role in two teleosts. *Biological Bulletin,*171:236-247.

Hill, R. W., and G. A. Wyse. 1989. *Animal Physiology, 2nd edition.* New York: Harper Row.

Holmes, R. B. 1984. The Carboniferous amphibian *Proterogyrinus scheelei* Romer and the early evolution of tetrapods. *Philosophical Transactions of the Royal Society of London* B, 306:431-524.

Holmgren, N. 1946. On two embryos of *Myxine glutinosa. Acta Zoological,* 26:6-14.

Hook, R. W. 1983. *Colosteus scutellatus* (Newberry), a primitive temnospondyl amphibian from the Middle Pennsylvanian of Linton, Ohio. *Novitates,* 2770:1-41.

Jayne, B. C. 1986. Kinematics of terrestrial snake locomotion. *Copeia,* 1986:915-927.

Jayne, B. C. 1988. Muscular mechanisms of snake locomotion: an electromyographic study of the sidewinding and concertina modes of *Crotalus cerastes, Nerodia fasciata,* and *Elaphe obsoleta. Journal of Experimental Biology,* 140:1-33.

Johnels, A. G. 1950. On the dermal connective tissue of the head of *Petromyzon. Acta Zoologica,* 31:177-185.

Kemp, N. E. 1959. Development of the basement lamella of larval anuran skin. *Developmental Biology,* 1:459-476.

Krejsa, R. J. 1979. The comparative anatomy of the integumental skeleton. Pages 112-191 in: *Hyman's Comparative Vertebrate Anatomy, 3rd Edition.* (M.H. Wake ed.). Chicago: The University of Chicago Press.

Lillywhite, H. B., and P. F. A. Maderson. 1982. Pages 398-442 in: *Biology of the Reptilia, Volume 12, Physiology A.* (C. Gans and F. H. Pough eds.). London: Academic Press.

Mautz, W. J. 1982. Patterns of evaporative water loss. Pages 443-481 in: *Biology of the Reptilia, Volume 12, Physiology A.* (C. Gans and F. H. Pough eds.). London: Academic Press.

Millington, P. F., T. Gibson, J. H. Evans, and J. C. Barbenel. 1971. Structural and mechanical aspects of connective tissues. Pages 189-248 in: *Advances in Biomedical Engineering, Volume 1.* (Kenedi, R.M. ed). London: Academic Press.

Moss, M. 1972. The vertebrate dermis and the integumental skeleton. *American Zoologist*, 12:27-34.

Motta, P. J. 1977. Anatomy and functional morphology of dermal collagen fibers in sharks. *Copeia*, 1977:454-464.

Nadol, J. B., Jr., J. R. Gibbins, and K. P. Porter. 1969. A reinterpretation of the structure and development of the basement lamella: an ordered array of collagen in fish skin. *Developmental Biology*, 20:304-331.

Nagy, K. A. 1982. Field studies of water relations. Pages 483-501 in: *Biology of the Reptilia, Volume 12, Physiology A.* (C. Gans and F. H. Pough eds.). London: Academic Press.

Panchen, A. L. 1972. The skull and skeleton of *Eogyrinus attheyi* Watson (Amphibia: Labryinthodontia). *Philosophical Transactions of the Royal Society of London* B, 263:279-326.

Pearson, D. M. 1981. Functional aspects of the integument in polypterid fishes. *Zoological Journal of the Linnean Society*, 72:93-106.

Prange, H. D., and Schmidt-Nielsen, K. 1969. Evaporative water loss in snakes. *Comparative Biochemistry and Physiology*, 28:973-975.

Romer, A. S. 1966. *Vertebrate Paleontology, Third Edition.* Chicago: The University of Chicago Press.

Romer, A. S., and R. V. Witter. 1941. The skin of the rhachitomous amphibian *Eryops*. *American Journal of Science*, 239:822-824.

Schmidt-Nielsen, K., and P. J. Bentley. 1966. Desert tortoise *Gopherus agassizii*: cutaneous water loss. *Science*, 154:911.

Sengel, P. 1976. *Morphogenesis of Skin.* Cambridge: Cambridge University Press.

Shoemaker, V.H., and K. A. Nagy. 1977. Osmoregulation in amphibians and reptiles. Pages 449-472 in: *Annual Review of Physiology, Volume 39.* (E. Knobil, R. P. Sonnenschein, and I. S. Edelman, eds.). Palo Alto, California: Annual Reviews, Incorporated.

Stuart, E. S., and A. A. Moscona. 1967. Embryonic morphogenesis: role of fibrous lattice in the development of feathers and feather patterns. *Science*, 157:947-948.

Wainwright, S. A. 1988. *Axis and Circumference.* Cambridge, Massachusetts: Harvard University Press.

Wainwright, S. A., F. Vosburgh, and J. H. Hebrank. 1978. Shark skin: function during locomotion. *Science*, 202:747-749.

Weiss, P., and W. Ferris. 1954. Electron-microscopic study of the texture of the basement membrane of larval amphibian skin. *Journal of Biophysical and Biochemical Cytology*, 40:528-540.

CHAPTER 11

LOCOMOTOR FEATURES OF TAXA SPANNING THE ORIGIN OF AMNIOTES

Stuart S. Sumida

INTRODUCTION

Analyses of the origin of amniotes has long been tied to advances in "terrestriality." Any analysis involving taxa spanning this transition demands a functional survey of locomotor features. Recently, the study of late Paleozoic tetrapods has entered the realm of experimental analysis and a more rigorous biomechanical approach to osteological remains is becoming more predominant (Lombard and Sumida, 1992). The application of functional interpretations based on experimentally generated models utilizing extant taxa allows more confidence in models of locomotory behavior in basal tetrapods.

Just as important as a functional perspective is a phylogenetic perspective. During the past decade, the use of cladistic analysis in combination with large computer-analyzed databases has provided new hypotheses of relationship for taxa near the origin of amniotes. They provide a phylogenetic framework within which locomotory features may be surveyed. Recent phylogenetic hypotheses have, in some cases, provided novel proposals regarding the arrangement of the relationships of late Paleozoic tetrapod taxa. In other cases, older hypotheses that had been discarded previously have been resurrected with new and (presumably) more reliable methods of analysis. The utilization of rigorously documented phylogenetic hypotheses as a framework for functional analysis of primitive tetrapods is important in two regards: (1) It allows the development of hypotheses regarding the evolution of morphological and behavioral traits such as locomotor features, and (2) it alerts us to taxa that must be included in such a survey, taxa that in some cases might have been otherwise omitted.

Phylogenetic Context

This study does not attempt to generate a new or alternative hypothesis of the relationships of basal amniotes and their presumptive sister taxa. Rather, it attempts to survey the locomotory features of those taxa, with a phylogenetic hypothesis as the guide for the survey. Over the past three decades a succession of phylogenetic hypotheses have centered on the question of the origin and interrelationsips of amniotes (Carroll, 1969a, 1970; Heaton, 1980; Holmes, 1984; Heaton and Reisz, 1986; Reisz, 1980, 1986; Gauthier *et al.*, 1988; Berman *et al.*, 1992). Of these, later analyses have been revolutionized by the application of cladistic methodology to the generation of such hypotheses. Recently, considerable attention has been paid the discovery of a proposed "earliest amniote", *Westlothiana lizziae*, from the Visean of Scotland. Although Smithson (1989) originally designated *Westlothiana* as the earliest known reptile, more detailed subsequent analyses (Smithson *et al.*, 1994; Carroll, 1995) resulted in an amended diagnosis place it in a position intermediate between more traditionally defined anthracosaurs and amniotes.

The final few years of the twentieth century has witnessed the advent of cladistic analyses generated by powerful computer algorithms utilizing extremely large data sets (Gauthier *et al.*, 1988; Berman *et al.*, 1992; Laurin and Reisz, 1994; Carroll, 1995; Lee, 1995). The studies have been influential in shaping recent understanding of the interrelationships of basal amniotes and provided part of the incentive for this review. Figure 1 is an attempt to synthesize features common to these hypotheses. It does not pretend to reconcile all of the interpretive differences of these studies. A few features are noteworthy: (1) The terms Batrachosauria and

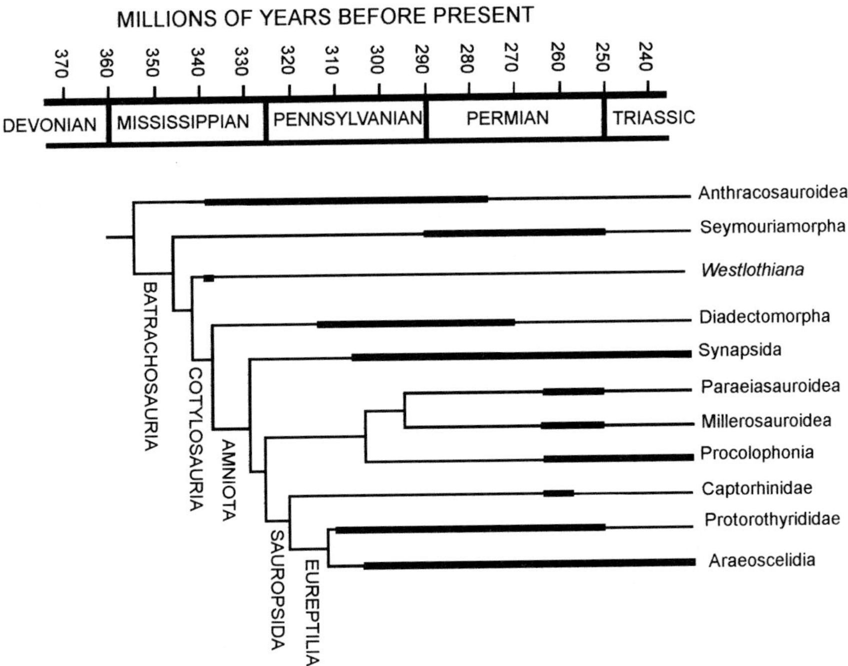

Figure 1. Cladogram providing a phylogenetic framework for the features discussed in this study. The cladogram is a synthesis based primarily on the hypotheses of Bolt and Lombard (1992), Lombard and Sumida (1992), Carroll (1995), Laurin and Reisz (this volume) and Lee and Spencer (this volume).

Cotylosauria are those defined by Laurin and Reisz (this volume); (2) the Amniota are characterized in a manner similar to most of the studies cited, though Berman *et al.* (1992) and Lee and Spencer (this volume) have suggested that there is evidence that Diadectomorpha might be realistically considered as amniotes; (3) the Sauropsida are as defined by Lee and Spencer (this volume); (4) the Eureptilia are characterized *sensu* Gauthier *et al.* (1988); and (5) the Paraeiasaurioidea, Millerosauroidea, and Procolophonia have often been grouped as the "Parareptilia" (Olson, 1947; Gauthier, *et al.*, 1988; Lee, 1995; others), however, this group is undergoing aggressive restudy at the time of this writing, and their interrelationships are not interpreted here in a formal taxonomic sense.

Taxa Considered

Based on the commonalities of recent phylogenetic studies combined into figure 1, members of the following groups are included in this review: representative basal anthracosaurs, seymouriamorphs, diadectomorphs, pelycosaurian-grade synapsids, basal sauropsids including millerittids and procolophonians, captorhinid reptiles, protorothyridid reptiles, and araeoscelidian reptiles (which include basal diapsid reptiles). Both Gauthier *et al.* (1988) and Laurin and Reisz (1995) have noted the position of mesosaurs near the base of the amniote radiation; however, their adaptation to an aquatic life-style precludes their inclusion here.

Schultze (1987) and Hopson (1991) have argued convincingly that defensible phylogenetic analyses must include, whenever possible, all members of a group being considered. When this is not practical, it is imperative that the basal members of any clade be used in the analysis so as to avoid basing hypotheses of transformation between clades on highly derived, terminal taxa. The latter approach is necessarily taken here. The embolomere *Proterogyrinus* is among the most terrestrial and the most thoroughly described of any anthracosaur (Holmes, 1984). *Seymouria* is easily the best known of seymouriamorphs (Berman *et al.*, 1987). All genera of the Diadectomorpha (*Limnoscelis, Tseajaia, Diadectes*) for which adequate postcranial materials are known are included (Berman and Sumida, 1990; Berman et al, 1992; Sumida *et al.*, 1992). Following Reisz *et al.* (1992), the primitive caseosaurian pelycosaurs *Casea* and *Aerosaurus* and basal eupelycosaurian synapsids are considered. The most primitive captorhinids for which complete postcranial material is known are *Eocaptorhinus*, *Captorhinus,* and *Labidosaurus*; information for this group is a synthesis of data from these genera (Holmes, 1977; Dilkes and Reisz, 1986; Heaton and Reisz, 1980; Sumida, 1989a). Basal procolophonids are probably best represented

Figure 2. Left lateral reconstructions of representative late Paleozoic tetrapods. (A) the anthracosauroidian anamniote *Protergyrinus scheeli*, primarily after Holmes (1984); (B) the caseoaurian pelycosaur *Aerosaurus wellesi*, after Langston and Reisz (1981) and Romer and Price (1940); (C) the captorhinid reptile *Eocaptorhinus laticeps*, after Heaton and Reisz (1980); and (D) the araeoscelidian reptile *Araeoscelis gracilis*, after Reisz *et al.* (1984). ➔

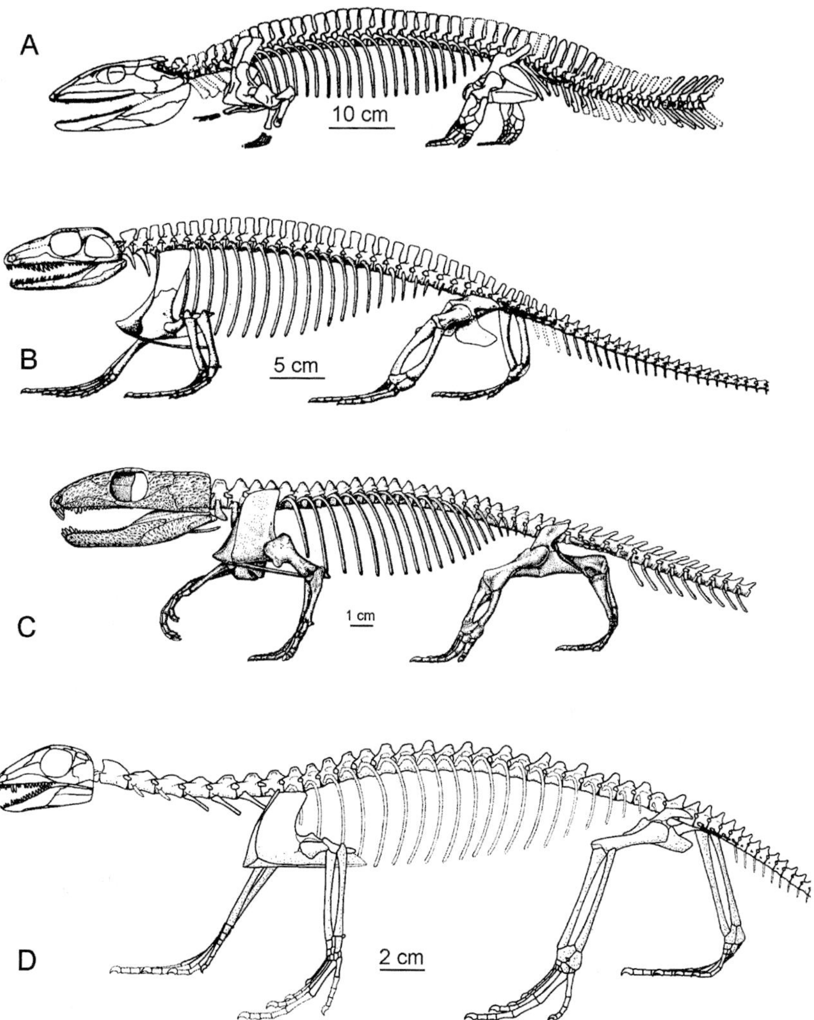
A
10 cm
B
5 cm
C
1 cm
D
2 cm

by *Procolophon* (Lee, 1993). *Parieasaurus* is representative of parieasaurs, though the postcranial skeleton in *Bradysaurus* is also well known and utilized here (Boonstra, 1931; Haughton and Boonstra, 1931; Boonstra, 1934). *Milleretta* (Gow, 1972; Thomasen and Carroll, 1981) is utilized here to represent millerettids. The record for postcranial material of protorothyridids is incomplete at best, so data for the group are presented as a synthesis of the best known genera, *Protorothyris* and *Paleothyris* (Clark and Carroll, 1973; Heaton and Reisz, 1986). Araeoscelidian reptiles are represented by *Petrolacosaurus* and *Araeoscelis* (Vaughn, 1955; Reisz, 1981; Reisz *et al.*, 1984). Figures following illustrate particular features of the taxa ennumerated above. Further, whole body reconstructions of representative taxa are illustrated in figure 2.

Morphology and Methods of Comparison

Wherever possible, description, discussion, and illustrations are based on actual examination of specimens. There are some groups for which this was problematic, particularly the "parareptiles". In these cases, the literature was necessarily utilized.

Whereas many previous surveys of locomotor capabilities (e.g., Romer, 1922; Holmes; 1980) have focused primarily on the appendicular skeleton, it has been noted that the vertebral column plays an important role in the locomotor behavior of late Paleozoic tetrapods (Olson, 1936a; Holmes, 1989; Sumida, 1987, 1990); thus, structure and regional patterns within the vertebral column are presented. The pectoral and pelvic girdles and long bones of the limbs are each considered in turn. The component bones of the manus and pes are not addressed individually; rather, they are treated as functional wholes.

Although in some cases reconstructions of osteological materials are based on composite studies, as much as possible reconstructions are based on the most complete specimens available. Wherever possible, functional interpretations are based on hypotheses that have been generated by experimental models using extant organisms.

AXIAL SKELETON

General Structure

The vertebral column of most late Paleozoic tetrapods is composed of multipartite elements with varying degrees of fusion (Fig. 3). In amniotes and their closest sister taxa the vertebral body is composed of two elements: a pleurocentrum that dominates in size and an anteriorly placed intercentrum which is usually reduced to a smaller, crescent-shaped wedge. Laterally directed transverse processes which provide articular surfaces for the ribs vary in size depending on their position; more anteriorly they may span both the neural arch and the centrum, but they are quite small in the most caudal regions of the body. In many of the taxa surveyed here, the neural arches are conspicuously "expanded" or "swollen" giving the outline of the arch a broadly convex appearance. The "swollen" form of the neural arch results in a broadly flared support for the anterior zygapophyses and transverse processes. A neural spine extends from the dorsal midline of the neural arch and is of variable construction in the taxa surveyed.

Dorsal vertebrae

The primitive condition for the taxa surveyed here is exemplified by *Proterogyrinus* (Fig. 3.). The intercenturm is relatively larger than in any of the other taxa examined. It is a single element, U-shaped in anterior and posterior view, but was quite likely larger in life due to the preesence of cartilage. In mature individuals, the pleurocentrum is notochordal and closed though not fused dorsally; in immature specimens the pleurocentrum is open dorsally. The neural arch articulates with the pleurocentrum primarily. It is fused medially and has a tall, quadrangular neural spine. Zygapophyseal facets are set at an angle of approximately 20° to the horizontal. This would imply that movements more complex than simple lateral bending must have been possible. In the event of lateral bending, the angle of the zygapophyseal articulations would have imposed a coupled rotation. (See Holmes, 1989, for a similar analysis of rhachitomous vertebrae.)

In the remainder of taxa examined here, the pleurocentrum is conspicuously larger than the intercentrum. In most taxa other than *Seymouria*, centra are amphicoelous and the intercentrum is reduced to an almost inconsequential wedge. Although the dorsal vertebrae in most araeoscelidians are generally similar to those of their outgroups, those of the cervical region are conspicuously elongate and help to form a clearly distinguishable "neck" region.

In association with a well-developed pleurocentrum, the neural arch is extremely robust. Seymouriamorphs, diadectomorphs, and most amniotes have a prodigious dorsolateral expansion of the neural arch. On the other hand, *Westlothiana*, protorothyridids, most synapsids, and certain araeoscelidian reptiles do not conform to this pattern. The neural arches in *Westlothiana* are only slightly convex. The neural arches of protorothyridids and some specimens of the the basal diapsid *Petrolacosaurus* are laterally convex. Only one basal synapsid (*Varanosaurus*) exhibits expanded neural arches and this is probably a condition developed convergently (Sumida, 1989b). With this exception, the vertebrae of pelycosaurian-grade synapsids are fairly consistent in their construction. Neural arches are concave, often with depressions or pits that presumably reduced the amount of bone while still retaining adequate structural integrity via an arched construction. Among taxa regarded as "parareptilian", Gow (1972) described the neural arches in millerettids as somewhat broad, but they are not as conspicuously expanded as the neural arches of parieasaurs. Only diadectomorphs have neural arches that are more stoutly and broadly constructed than those of parieasaurs.

Figure 3. Left lateral views of single or paired dorsal vertebrae of representative Late Paleozoic tetrapod taxa. (A) The embolomerous anthracosaurian *Proterogyrinus* (primarily after Holmes, 1984); (B) the seymouriamorph *Seymouria*; (C) *Westlothiana*, a tetrapod of uncertain affinities (after Smithson *et al.*, 1994); (D) the diadectomorph *Limnoscelis*; (E) the diadectomorph *Diadectes*; (F) the caseosaurian pelycosaur *Varanops* (primarily after Williston, 1911); (G) the caseosaurian pelycosaur *Cotylorhynchus* (after Stovall *et al.*, 1966); (H) the millerettid *Milleretta* (partially after Gow, 1972); (I) the parieasaur *Bradysaurus* (partially after Boonstra, 1934; Romer, 1956); (J) the captorhinid reptile *Captorhinus*; (K) the protorothyridid reptile *Protorothyris* (partially after Clark and Carroll, 1973; Heaton and Reisz, 1986); (L) cervical (left) and mid-dorsal (right) vertebra of the araeoscelidian reptile *Araeoscelis* (partially after Vaughn, 1955). All scalebars are equivalent to 1 cm except C where it equals 1 mm. Abbreviations: ic, intercentrum; na, neural arch; ns, neural spine; pc, pleurocentrum. ➔

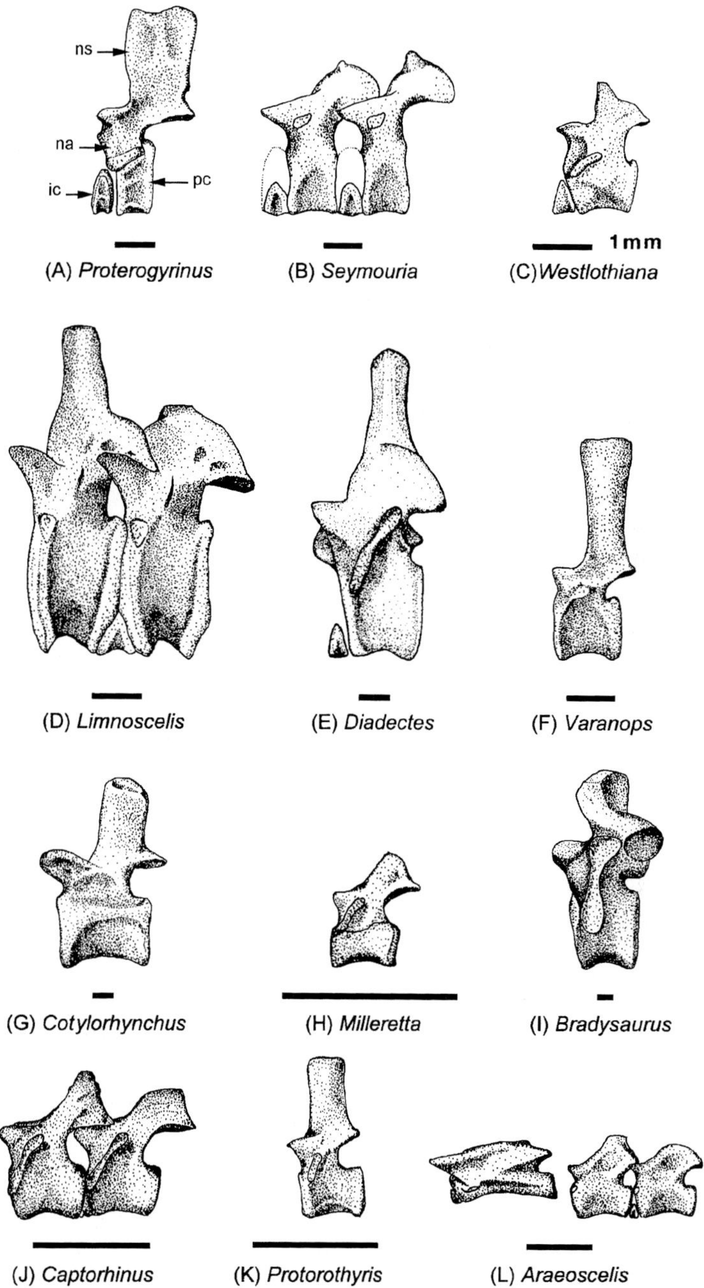
ns
na
ic
pc
1mm
(A) Proterogyrinus
(B) Seymouria
(C) Westlothiana
(D) Limnoscelis
(E) Diadectes
(F) Varanops
(G) Cotylorhynchus
(H) Milleretta
(I) Bradysaurus
(J) Captorhinus
(K) Protorothyris
(L) Araeoscelis

The neural spines in *Proterogyrinus* and most pelycosaurs are laterally compressed and blade like. Those of *Westlothiana*, millerettids, and most of the column in parieasaurs are homogeneously conical. However, in many basal amniotes and their sister groups, there is considerable variability of neural spine structure, both among taxa and within individuals. Neural spines of two general types are found: "tall types" that are approximately conical in shape, and "low types" that are shorter, narrower, and wedge like. In fact, these two spine morphologies often alternate with one another in position through certain regions of the column. In regions of alternation of neural spine height, the zygapophyseal articulations are usually tilted relative to the horizontal plane precluding simple lateral bending.

Neural spine height and width alternate in short stretches of the column of *Seymouria* as well as many specimens of diadectomorphs; however this is never expressed throughout the column. The zygapophyseal articulations are nearly horizontal in most cases; however, they are slightly tilted relative to the horizontal plane in regions of neural spine variability. Representative specimens of the diadectomorphs *Limnoscelis*, *Tseajaia*, and *Diadectes* exhibit alternation of neural spine height. Most diadectomorphs have accessory articulations beyond the central and zygapophyseal contacts.

The neural spines of basal synapsids are uniformly well-developed and quite variable in their morphology among taxa; however, alternation of spine structure is manifested only in the genus *Varanosaurus*. Perhaps notably, *Varanosaurus* is the only pelycosaur that has expanded neural arches and alternation of neural spine morphology as well (Sumida, 1989b). The spines are short in *Aerosaurus* and *Casea*, but show varying degrees of elongation in more derived genera. Unfortunately, vertebral structure is not known for the most basal synapsids, *Oedaleops* and *Eothyris*. Most basal synapsids show a greater degree of zygapophyseal tilt than do seymouriamorphs or diadectomorphs.

The neural spines of millerettids are not particularly well known. Those of pareiasaurs are somewhat better known. They are are stoutly constructed. However, the spines are relatively shorter than those of most diadectids and no examples of alternation of neural spine height are known for any members of the group. One exception to this is in the cervical region in *Bradysaurus* in which the spine of the third cervical is markedly smaller than that of the axis preceding it and the fourth following it. It may be that the large size of both diadectids and parieasaurs supplanted the need for the functional adaptations associated with alternation of neural spine height.

In most reptilian groups (sensu Gauthier *et al.*, 1988) with swollen neural arches, neural spine morphology inevitably varies. Whenever spine height alternates, zygapophyses diverge from a simply horizontal orientation, taking on a dorsolateral angle from 15 to 25° to the horizontal plane. Although neural arches are narrow and spines do not generally alternate in height in protorothyridids, the presence of a single low-spined cervical (as seen in the parieasaur *Bradysaurus*) is common. Alternation between tall, conical spines with low, narrow, wedge-shaped spines is found in all members of the Captorhinidae for which the vertebral column is adequately known as well as in araeoscelidian reptiles. Expanded neural arches are found in most araeoscelids, and with them alternation of neural spine structure. Reisz (1981) noted that the neural arches in *Petrolacosaurus* were markedly narrower than those of other araeoscelidians and of captorhinids. However, although spine height does not alternate in most specimens, neural spine width does. Some vertebral material clearly associated with crania assignable to *Petrolacosaurus* does have expanded neural arches (Sumida, 1990). Notably, these specimens also demonstrate alternation of neural spine height, though not to the degree seen in other taxa.

The common possession of laterally expanded neural arches, tilted zygapophyses, and alternation of neural spine height appears to represent a related functional complex among taxa at or near the origin of amniotes. Perhaps most importantly, whenever the phenomenon is encountered its greatest expression is near the limb girdles. In the captorhinid reptile *Labidosaurus* it occurs throughout the column, but its most extreme expression is near the limb girdles. The phenomenon is not found throughout the column in *Captorhinus* (Vaughn, 1970; Sumida, 1991) or *Eocaptorhinus* (Dilkes and Reisz, 1986; Sumida, 1991) but it is expressed near both limb girdles in the former and near the pelvic girdle of the latter. Alternation is most prominent in small to moderate sized animals that retain a robust build. In very large taxa,

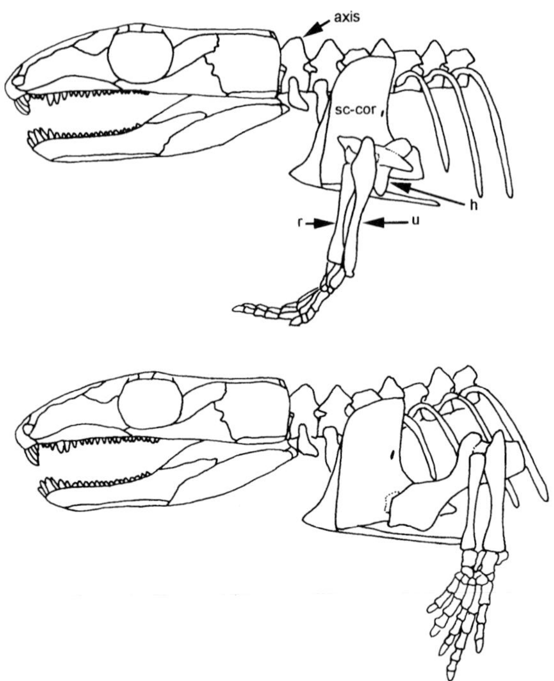

Figure 4. Diagrammatic illustration in left lateral view of the forelimb and anterior vertebral column in a captorhinid reptile. (A) Left forelimb in middle of power stroke; (B) left forelimb just out of contact with adjacent vertebral column in extreme dorsiflexion to aid in stabilization and recovery. Abbreviations: axis, axis vertebra; h, humerus; r, radius; sc-cor, scapulocoracoid; u, ulna.

such as diadectids and parieasaurs, only the extremely hypertrophied neural arches remain of the complex. Extremely gracile taxa show evidence of the terminal stages of having abandoned the entire suite of associated characters. Araeoscelidians provide a transitional example of this trend: they are somewhat more gracile than captorhinid reptiles and their neural spine construction does not contrast to the same degree. However, alternation in neural spine height is expressed whenever swollen neural arches are present.

Sumida (1991) suggested that if muscles spanned the interval between tall-type spines, dorsal and lateral vertebral flexion could have been aided. Tilted zygapophyseal articulations in regions of alternation suggest that lateral flexion would have been necessarily accompanied by axial rotation. The greatest expression of the phenomenon is near limb girdles, so it may have facilitated axial rotation and muscular stabilization of the vertebral column during the recovery stroke of the limb. Figure 4 is a diagrammatic representation of this set of associated movements. Ritter (1993) has demonstrated that epaxial muscles of extant varanid lizards aid in postural stability during locomotion. He further sugests that such action might be characteristic of the Amniota (Ritter, 1995, 1996). The morphological complex associated with alternation of vertebral sturcture could have enhnaced this function significantly in basal amniotes.

Atlas-axis complex

Whereas the dorsal vertebrae of the taxa studied do exhibit some variability, the configuration of the atlas-axis complexes is significantly more conservative (Sumida and Lombard, 1991). The atlas-axis complex (Fig. 5) is a multipartite complex composed of the following elements: paired proatlas, atlantal intercentrum and pleurocentrum, atlantal neural arches, axial intercentrum and pleurocentrum, and axial neural arch. In certain of the taxa the central elements are paired. All elements demonstrate varying degrees of fusion.

The atlantal and axial intercentra abut, precluding exposure of the atlantal pleurocentrum on the ventral side of the axial column. The axial pleurocentrum is a single element that does reach the ventral midline. Among mature seymouriamorphs and all other more derived taxa, the atlantal pleurocentrum is represented by a single ossification. However, it must be noted that immature specimens of *Seymouria* still retain paired pleurocentral components of the atlas (Sumida and Lombard, 1991; Sumida *et al.*, 1992). The atlas-axis complex in *Westlothiana* is not known (Smithson *et al.*, 1994). Diadectomorphs and all of the more traditionally defined amniotes share three features: the possession of atlantal neural spines; fusion of the axial neural arch with a tall, blade-like spine to the axial centrum; and articulation (usually fusion) of the atlantal pleurocentrum with the dorsal aspect of the axial intercentrum.

In addition to adaptations to stabilize the column in association with the recovery stroke of the pectoral limb in some of the taxa, there was likely a need to prevent undue lateral swinging of the head during locomotion. Consolidation of the component elements of the complex and an increase in the stabilizing function of occipital muscles are features of all the taxa considered. This trend is carried to its greatest extreme in diadectomorphs and more traditionally defined amniotes. The fusion of the components of the axis vertebra and the enlargement of its neural spine provided for a larger origin of the obliqus capitus magnus, rectus capitus posterior, and spinalis capitis muscles, as well as the nuchal ligament. Fusion of the atlantal pleurocentrum and axial intercentrum further reduced movement in the complex. The posterior projections of paired atlantal neural arches to either side of the axial neural spine likely restricted movement of the atlas relative to the axis and about the spinal axis in general. This restricted any potential movement to the atlantal pleurocentrum-basioccipital joint. All the muscles originating on the axis inserted onto the skull near the sagittal plane, in a position more suitable for dorso-ventral stabilization than for medio-lateral movement of the atlanto-occipital joint. Contraction of the obliquus capitus magnus would have pulled the skull dorsally and caudally toward the axial neural spine, aiding in holding up the head and stabilizing the joint.

APPENDICULAR SKELETON

With the conspicuous exception of Holmes (1977, 1980), few since the early work of Romer (1922) have concentrated on both the osteological and muscular components of the basal tetrapod limb.

Figure 5. Left lateral views of the atlas-axis complex of taxa near the origin of amniotes. Except where noted, all illustrations are after Sumida *et al.* (1992). (A) The embolomerous anthracosaurian *Proterogyrinus*; (B) the seymouriamorph *Seymouria*; (C) the diadectomorph *Limnoscelis*; (D) the diadectomorph *Diadectes*; (E) the caseosaurian pelycosaur *Cotylorhynchus*; (F) the eupelycosaurian pelycosaur *Ophiacodon*; (G) the millerettid *Milleretta* (partially after Gow, 1972); (H) the parieasaur *Bradysaurus* (partially after Boonstra, 1934); (I) the captorhinid reptile *Captorhinus*; (J) the protorothyridid reptile *Paleothyris*; (K) the araeoscelidian reptile *Petrolacosaurus*. All scalebars are equivalent to 1 cm except G where it equals 0.5 cm. Abbreviations: atic, atlantal intercentrum; atna, atlantal neural arch; atpc, atlantal pleurocentrum; axic, axial intercentrum; axna, axial neural arch; axpc, axial pleurocentrum; pro, proatlas. ➔

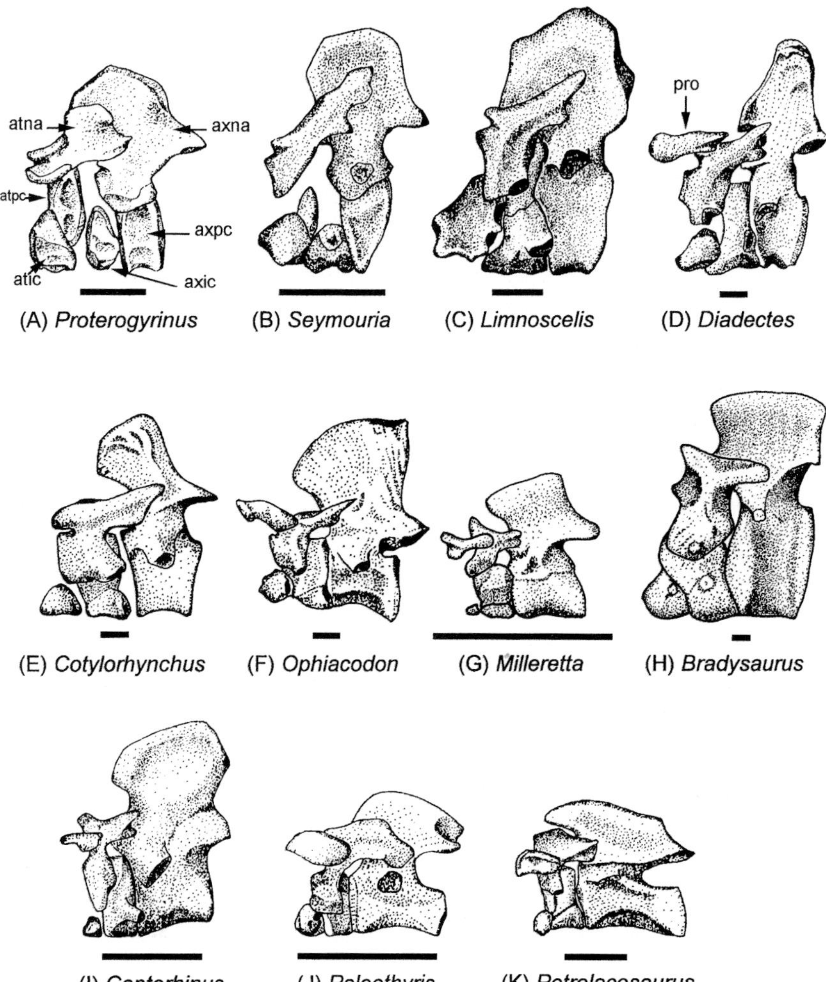
atna
axna
atpc
axpc
atic
axic
pro
(A) Proterogyrinus
(B) Seymouria
(C) Limnoscelis
(D) Diadectes
(E) Cotylorhynchus
(F) Ophiacodon
(G) Milleretta
(H) Bradysaurus
(I) Captorhinus
(J) Paleothyris
(K) Petrolacosaurus

That the appendicular elements supply fewer characters for phylogenetic analyses than do the cranium and lower jaws only compounds this bias. Of the few surveys of tetrapod limb structure that exist, most do not include the more distal antebrachial, crural, and pedal structures; those that do address these areas are generally limited to consideration of single taxa (Romer, 1957; Holmes, 1977, 1980; Sumida, 1989a).

Pectoral Girdle and Limb

Heaton's (1980) inability to produce cladistically useful features of the pectoral girdle in his attempt to distinguish among groups of primitive tetrapods highlights the conservative nature of the complex. The pectoral girdle is a multipartite structure incorporating dermal and endochondral elements (Fig. 6). Over time, the endochondral elements have increased in relative mass in the girdle at the expense of the dermal elements. A median, dermal interclavicle is the only unpaired element of the girdle. Paired dermal clavicles and cleithra form the anterior edge of the complex. On each side, a more robust endochondral scapulocoracoid dominates the remainder of the girdle. Primitively, the scapulocoracoid consisted of only dorsal scapular and ventral coracoid ossifications, with the glenoid cavity spanning their articulation. Two coracoid ossifications is a more derived condition. Unfortunately, the pectoral girdle in the potentially intermediate *Westlothiana* is extremely poorly preserved.

The more basal, distinctly anthrocosaurian condition of the pectoral girdle retains significant development of the dermal elements. Robust clavicular elements are retained in all the taxa surveyed here, and Romer (1922) reconstructed this region as a important area of origin for the clavicular head of the deltoid complex; however, its anterior orientation indicates that it probably provided attachment of

Figure 6. Left lateral views of the pectoral girdle in taxa near the origin of amniotes. (A) The anthracosaurian *Proterogyrinus* (partially after Holmes, 1984; Romer, 1970); (B) the seymouriamorph *Seymouria* (partially after White, 1939); (C) the diadectomorph *Limnoscelis*; (D) the caseosaurian pelycosaur *Varanops* (partially after Langston and Reisz, 1981); (E) the captorhinid reptile *Labidosaurus*; (F) the araeoscelidian reptile *Araeoscelis* (partially after Vaughn, 1955). Abbreviations: ant cor, anterior coracoid; cl, clavicle; cle, cleithrum; cor, coracoid; icl, interclavicle; post cor, posterior coracoid; s, scapula. All scalebars are equivalent to 1 cm. ➔

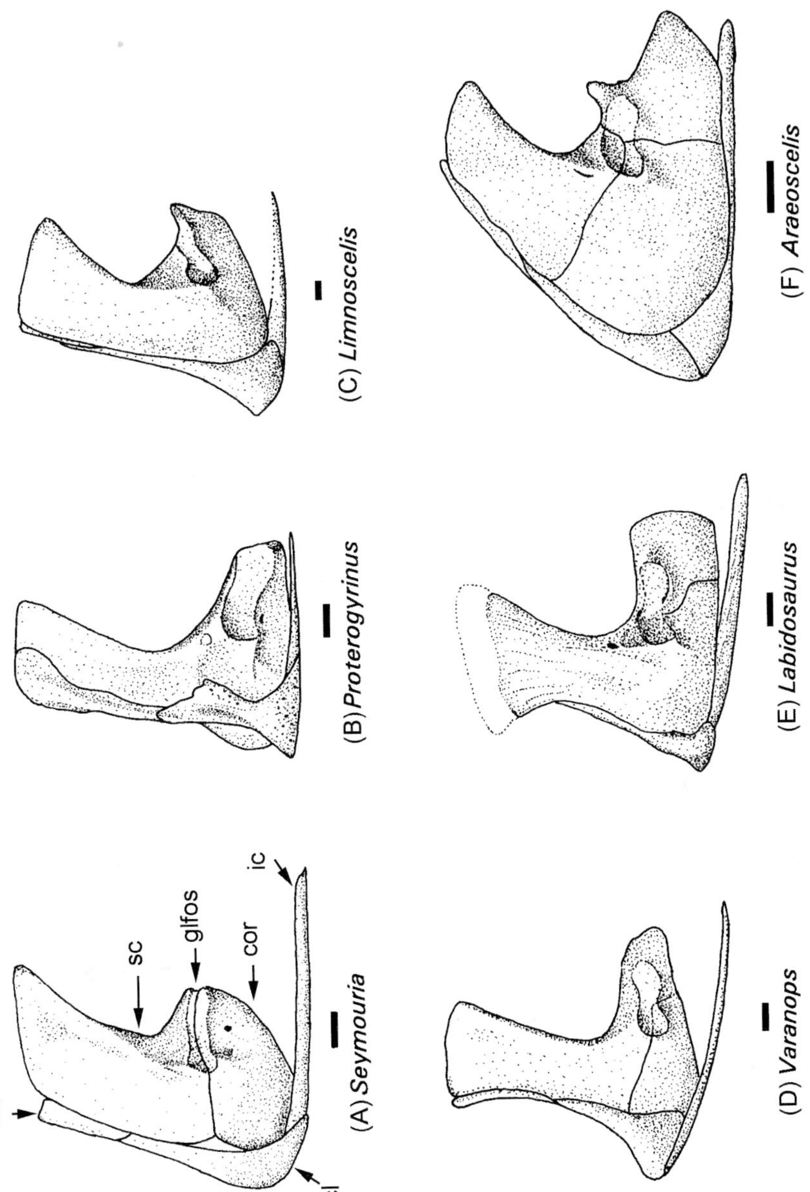
cle
sc
glfos
cor
ic
cl
(A) Seymouria
(B) Proterogyrinus
(C) Limnoscelis
(D) Varanops
(E) Labidosaurus
(F) Araeoscelis

episternal and omohyoid musculature of the neck as well. The cleithrum is reduced to a splint in most amniote taxa. Primitively it is large, extending to the dorsal limit of the girdle in *Proterogyrinus*, *Seymouria*, and diadectomorphs. In amniotes other than synapsids the scapulodeltoid musculature probably originated exclusively from the scapular ossification, whereas in anthroacosaurians it likely spanned the scapula-cleithrum articulation. Pelycosaurs are transitional between these two conditions. *Proterogyrinus*, *Seymouria*, and diadectomorphs share the primitive condition of a broad scapular blade and single coracoid ossification. More derived taxa possess separate anterior and posterior coracoid ossifications.

Whether the coracoid is composed of one or two elements, the configuration of the glenoid fossa is remarkably conservative among the taxa surveyed. It is traditionally described as "screw shaped", wherein the articular surface approximates a partially helical strap; the anterior portion of the fossa faces ventrolaterally, the middle almost directly laterally, and the posterior portion dorsolaterally (Fig. 6). It appears that the cartilaginous covering of the articular surfaces of the glenohumeral joint was rather thin (Holmes, 1977; Sumida, 1989a), allowing reasonable hypotheses of movement to be based on the manipulation of bony elements alone. The articular surfaces of the glenoid and humerus are closely congruent, a condition that likely dictated a highly constrained and stereotyped humeral excursion. In forms as disparate as *Seymouria*, the robust parieasaur *Bradysaurus*,

Figure 7. Reconstructions of left humeri of taxa near the origin of amniotes in distal, ventral aspect. (A) The anthracosaurian *Proterogyrinus* (from Holmes, 1984); (B) the seymouriamorph *Seymouria*; (C) *Westlothiana* (partially after Smithson *et al.*, 1994); (D) the diadectomorph *Limnoscelis*; (E) the diadectomorph *Diadectes* (partially after Romer, 1956); (F) the caseosaurian pelycosaur *Aerosaurus* (partially after Langston and Reisz, 1981); (G) the primitive eupelycosaur *Ophiacodon*; (H) the parieasaur *Bradysaurus* (after Boonstra, 1934); (I) the captorhinid reptile *Captorhinus*; (J) the protorothyridid reptile *Paleothyris* (partially after Heaton and Reisz, 1986); (K) the araeoscelidian reptile *Petrolacosaurus* (after Reisz, 1981); and (L) the araeoscelidian reptile *Araeoscelis* (partially after Vaughn, 1955). All scalebars equal 1 cm. Abbreviations: delt, deltoid process; ect, ectepicondyle; ent, entepicondyle; ent for, entepicondylar foramen; sup, supinator process. ➔

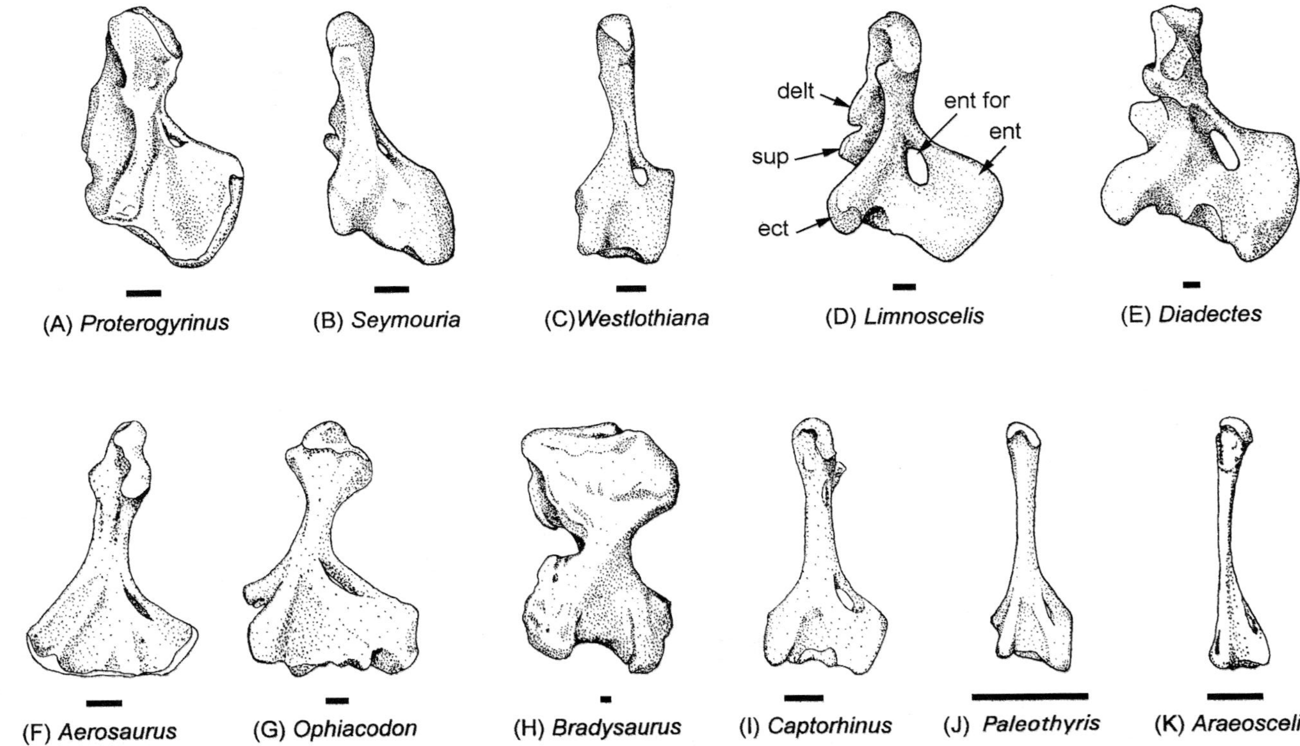
delt
sup
ect
ent for
ent
(A) Proterogyrinus
(B) Seymouria
(C)Westlothiana
(D) Limnoscelis
(E) Diadectes
(F) Aerosaurus
(G) Ophiacodon
(H) Bradysaurus
(I) Captorhinus
(J) Paleothyris
(K) Araeoscelis

and the comparatively gracile araeoscelidian *Petrolacosaurus*, the screw shaped glenoid is retained.

Romer (1922, 1956) characterized the humerus of Paleozoic tetrapods as essentially tetrahedral, an organizational perspective that has been adopted by nearly all subsequent workers. In simplified terms, the axes of the proximal and distal articular surfaces are at nearly right angles to one another. Figure 7 demonstrates these features and significant structural landmarks. In most forms, the proximal articular surface has a high degree of congruence with the glenoid articulation. The anterior region of the proximal articular surface is concave, matching the ventrolaterally directed convexity of the leading surface of the glenoid. Similarly congruent relationships exist for the more posterior surfaces of the glenoid and humeral head. Jenkins (1971) and Holmes (1977) have shown that the complementary articular surfaces constrain a rotation of the humerus along its long axis as it is retracted. At maximal protraction of the humerus, the humeral concavity at the leading margin of the humeral articular head fits snugly into an anterior glenoid convexity, orienting the distal articular surface cranioventrally. Retraction of the humerus prescribes rotation of the humerus on its long axis, resulting in a ventralward orientation of the distal humeral articulation by the middle of the excursion arc and a posteroventral orientation at the end of the arc. Unfortunately, no extant taxa exhibit a gleno-humeral joint similar to the screw-shaped joint of late Paleozoic tetrapods. However, independent analyses of the forelimb in captorhinids and pelycosaurs (Jenkins, 1971; Holmes, 1977; Sumida, 1989a) have all converged on an interpretation similar to that summarized here.

Skeletal indicators of the muscular role critical in postural support dominate the proximal humerus. Although not clearly defined in *Seymouria* or *Westlothiana*, a well developed process for insertion of the latissimus dorsi is extremely well-developed in diadectomorphs, synapsids, and most basal sauropsids (Fig. 7). Its development in tetrapods with a broad range of body sizes is perhaps indicative of increased importance of the retractor musculature in these taxa generally. Extremely broad surfaces of attachment on the dorsal and ventral surfaces of the proximal humeral head are indicative of significant attachments of the scapulohumeralis and coracobrachialis

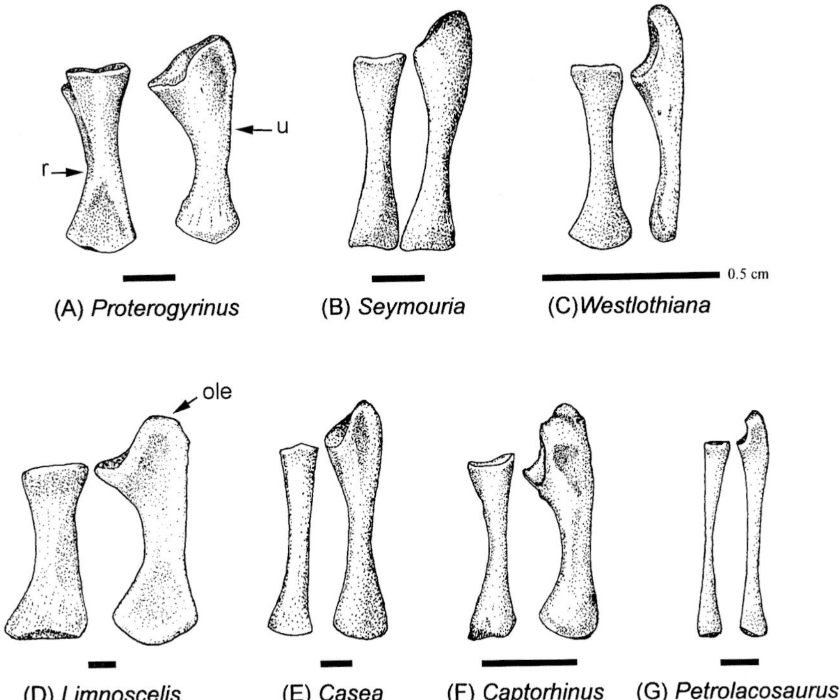

Figure 8. Reconstructions of right antebrachial elements of taxa near the origin of amniotes. (A) The anthracosaurian *Proterogyrinus* (from Holmes, 1984); (B) the seymouriamorph *Seymouria*; (C) *Westlothiana* (partially after Smithson *et al.* 1994); (D) the diadectomorph *Limnoscelis*; (E) the caseosaurian pelycosaur *Casea* (after Romer and Price, 1940); (F) the captorhinid reptile *Captorhinus* (from Holmes, 1977); and (G) the araeoscelidian reptile *Petrolacosaurus* (after Reisz, 1981). Elements are shown slightly disarticulated to facilitate viewing of both the radius (left) and ulna (right). All scale bars equal 1 cm except for C where it equals 0.5 cm. Abbreviations: ole, olecranon process; r, radius; u, ulna.

musculature respectively. The deltopectoral crest is robust in all taxa surveyed.

Whereas the proximal and distal articular heads of the humerus are similarly well developed in all taxa considered, the proportions of the intervening humeral shaft reflect the individual body proportions of each taxon. The humeral shaft is extremely short and almost indistinct in the heavy-bodied *Proterogyrinus*, *Seymouria*, diadectomorphs, parieasaurs, large pelycosaurs, and larger captorhinids. The shaft is slimmer and more elongate in the more gracile *Westlothiana*, protorothyridids, millerettids, protorothyridids, smaller pelycosaurs, small captorhinids, and araeoscelidians. In conjunction with a short humeral shaft, the supinator, entepicondylar, and ectepicondylar processes are robust. The extreme development of these processes provided enormous levers for supinator, flexor, and extensor musculature of the forearm, respectively. Holmes (1977) has suggested that the antebrachium must have been very nearly horizontal at the beginning of the forelimb power stroke, requiring extensive flexor musculature originating from the entepicondyle to maintain postural support during this part of the step cycle. All of the taxa examined here appear to have had such well developed flexor musculature. It is not until diadectomorphs and more derived taxa that the ectepicondyle is better developed, indicating that the contribution of muscular extensors of the antebrachium to propulsion and postural support evolved somewhat later.

Between the entepicondyle and ectepicondyle are an extremely well-developed capitulum and trochlea for radial and ulnar articulations, respectively. Congruence between the distal humerus and antebrachial elements was high, further stereotyping potential movements. The trochlear notch of the ulna is easily recognizable, but the development of a distinct olecranon process that extends well above it is not clearly evident in *Proterogyrinus* or *Seymouria*. An olecranon is present in *Westlothiana* but is not well developed, perhaps because the forelimbs are extremely small size (Smithson *et al.*, 1994). In association with a well-developed olecranon process, the ulnar trochlea is deep and directed anteromedially in diadectomorphs and the Amniota. Manipulation of the components of the elbow joint demonstrates that the anteromedially directed trochlea produces a posterolaterally directed extension of the elbow joint and propulsion. A more efficient lever accommodated the propulsive force delivered by the triceps musculature as the olecranon became more prominent.

In their analysis of the shoulder musculature of varanid lizards Jenkins and Goslow (1983) demonstrated that all three heads of the triceps musculature fired during the middle of propulsive phase. As the firing patterns seen in varanids were broadly similar to those of mammals they characterized as "primitive", they concluded that the activity patterns must have been inherited from some common ancestor of extant reptiles and mammals. Interpolation of this suggestion into the scenario described previously would suggest that in at least some groups of Paleozoic tetrapods similar firing patterns of the triceps musculature may well have occurred. Jenkins and Goslow (1983) further indicated that the vertebral column of varanids exhibited a lateral convexity toward the support phase at the midpoint of the propulsive phase. Similar orientations of the vertebral column may have existed in captorhinid reptiles (Heaton and Reisz, 1980; Sumida, 1989a) and other late Paleozoic tetrapods (Sumida, 1991). Anteromedial force provided by the triceps and directed by the anteromedially oriented trochlear notch of the ulna may have compensated at least partially for the lateral component of vertebral movement near the propulsive forelimb (Figs. 4 and 9).

The presumed attachment of the radius to the ulna via an interosseous membrane and the potentially free rotation of the radius on the capitulum likely indicates that the movement of the radius was primarily limited somewhat by the movement of the ulna on the humerus. A comparison of the potential movements at the distal end of the antebrachium is more difficult, as knowledge of carpal elements is extremely limited in *Proterogyrinus, Seymouria, Westlothiana*, and diadectomorphs. Without knowledge of the manus in these taxa, an understanding of the possible transformations across the transiton to the amniote condition is not possible. However, a brief description of the condition in representative amniotes (Fig. 10) can lend some insight into certain of their locomotory capabilities. The radiale is essentially a distal extension of the radius, whereas the intermedium and ulnare function together as an extension of the ulna. The distal articular surfaces of these two complexes are in separate planes of the manus, precluding a transverse joint at either of their ends. The most reasonable suggestion remains that of Holmes (1977; supported by Sumida, 1989a): The flat articular surfaces of the bones of the manus were all capable of minimal movements at each joint; summed, this allowed for a moderately flexible forefoot. Although the origin of the pisiform is unclear, diadectids, *Tseajaia*, pelycosaurs, captorhinids, protorothyridids, and araeoscelidians all have a pisiform on the ulnar side of the carpus. It is particularly well developed in captorhinids and

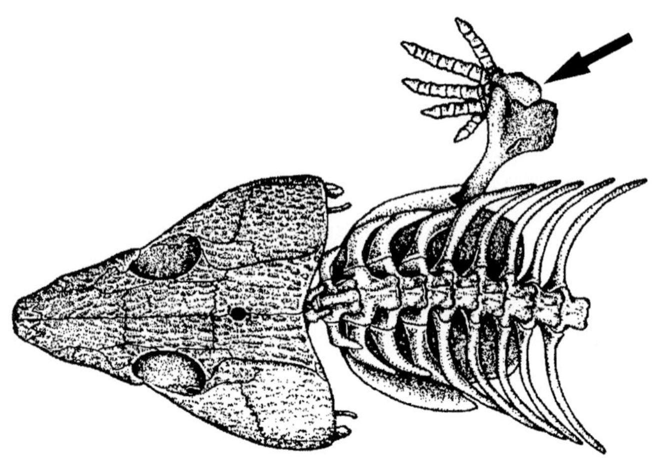

Figure 9. Dorsal view of the anterior portion of the vertebral column and the right forelimb near the middle of the support phase in a captorhinid reptile (based partially on Heaton and Reisz, 1980; Holmes, 1977) to aid in demonstrating the direction of force constrained by the anteromedially directed trochlea of the ulna toward the convexity of the laterally bent vertebral column. Arrow represents the direction of force only and is not meant to imply magnitude.

pelycosaurs, providing a stout lever for the flexor carpi ulnaris and extensor carpi ulnaris. This condition is taken to an extreme in the large captorhinid *Labidosaurus*, which also extends the radiale medially in an analogous fashion on the radial side of the carpus.

In most of the taxa surveyed, the fourth digit of the manus is the largest (Fig. 10), however only protorothyridids (Fig. 10C) and araeoscelidians (Fig. 10D) exhibit a dramatically longer fourth digit (Vaughn, 1955; Carroll, 1969b; Reisz, 1981; Heaton and Reisz, 1986). Thus, models dependent on rotation of the manus and pushoff via the elongate fourth digit (e.g. Rewcastle, 1981) are not applicable to basal amniotes and their presumptive sistergroups. More realistically, the model is applicable only to protorothyridids, araeoscelidians and their derivatives.

Pelvic Girdle and Limb

The structure of the pelvic girdle (Fig. 11) is more conservative than that of the pectoral girdle: Dorsal ilia articulate with sacral

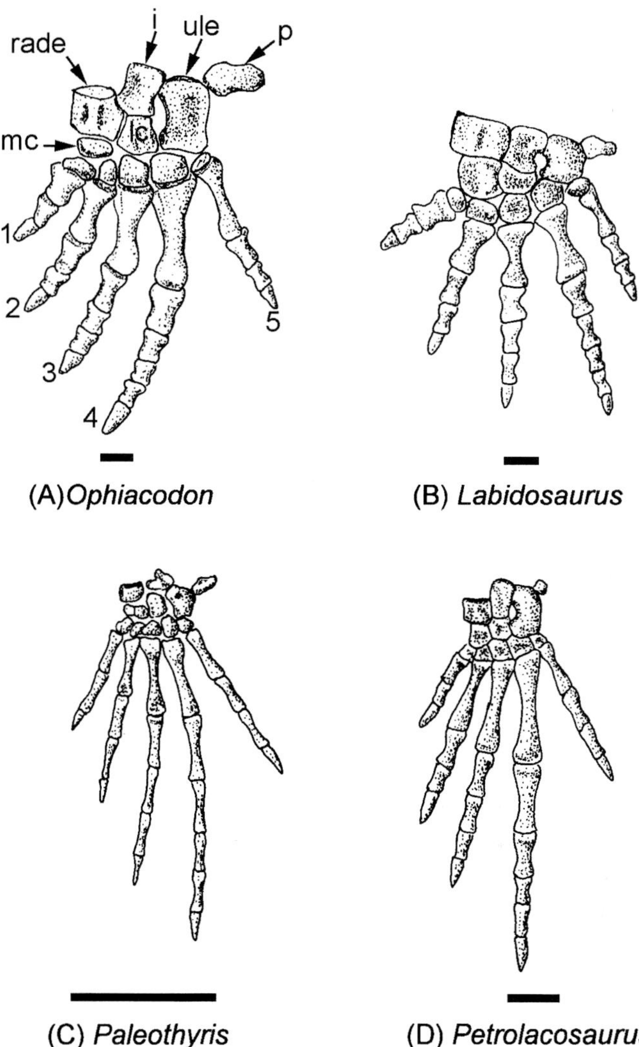

Figure 10. Reconstructions in dorsal view of left manus of basal amniotes. (A) The basal eupelyocosaurian pelycosaur *Ophiacodon* (partially after Romer and Price, 1940); (B) the captorhinid reptile *Labidosaurus*; (C) the protorothyridid reptile *Paleothyris* (after Carroll, 1969b); and (D) the araeoscelidian reptile *Petrolacosaurus* (after Reisz, 1981). In some cases, reconstructions presented here are based on the right manus, but all are presented in left view to facilitate comparison. All scalebars equal 1 cm. Abbreviations: i, intermedium; lc, lateral centrale; mc, medial centrale; p, pisiform; rade, radiale; ule, ulnare; 1-5, first to fifth digits.

vertebrae in varying numbers, often depending on the size of the animal, and quadrangular, plate-like pubic and ischial elements meet one another in the ventral midline. Substantive differences are found primarily in the degree of development of processes of the ilium.

Primitively, as in *Proterogyrinus* and (presumably) *Westlothiana*, the ilium has distinct dorsal and posterior processes. The dorsal process articulated medially with the sacral ribs and laterally served as the origin of the iliofemoralis. Romer (1922, 1956) suggested that the posterior process was a point of origin for tail musculature useful in aquatic locomotion. However, in his detailed analysis of *Proterogyrinus*, Holmes (1984) took issue with this interpretation, pointing out that extant, terrestrial taxa retain similar processes. It is likely that the iliofibularis muscle arose from the posterior process, whereas the iliotibialis probably arose from a ridge running ventral to the junction between the two processes. The processes remain distinguishable and both are robust in *Seymouria*. In diadectomorphs, basal synapsids, parieasaurs, and other reptiles the two processes have essentially consolidated into a single iliac blade. Although this probably did not affect the origins of the iliofemoralis and iliofibularis muscles significantly, it effectively moved the origin of the iliotibialis away from the hip joint, providing it a somewhat greater mechanical advantage in extension of the knee joint. Parieasaurs are unique in having a prodigious anterior extension of the iliac blade, one that is probably associated with the development of as many as four sacral ribs for support in this heavy-bodied group.

Some debate has centered around the function of an externally directed "shelf" of the ilium in diadectomorphs, a feature Heaton (1980) recognized as a synapomorphy of the group. Romer (1922,

---

Figure 11. Reconstructions of the pelvic girdle in taxa near the origin of amniotes. (A) The anthracosaurian *Proterogyrinus* (after Holmes, 1984); (B) the seymouriamorph amphbian *Seymouria* (partially after White, 1939); (C) *Westlothiana* (after Smithson *et al.*, 1994); (D) the diadectomorph *Limnoscelis* (after Berman and Sumida, 1991); (E) the diadectomorph *Diadectes* (after Romer 1922, 1956); (F) the captorhinid reptile *Labidosaurus* (after Sumida, 1989a); and (G) the araeoscelidian reptile *Araeoscelis* (after Vaughn, 1955). All scalebars equal 1 cm. Abbreviations: act, acetabulum; il, ilium; ildpr, dorsal process of ilium; ilpopr, posterior process of ilium; isc, ischium; pu, pubis. ➔

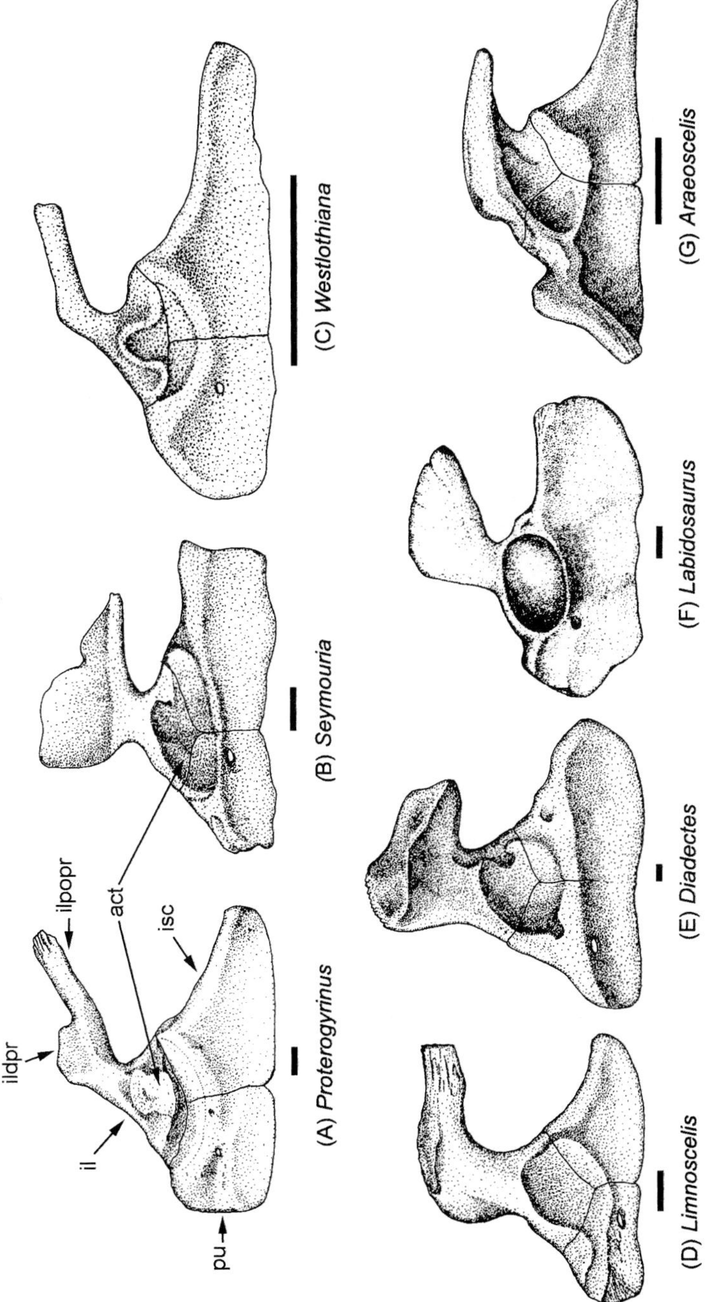
ildpr
ilpopr
il
act
isc
pu
(A) *Proterogyrinus*
(B) *Seymouria*
(C) *Westlothiana*
(D) *Limnoscelis*
(E) *Diadectes*
(F) *Labidosaurus*
(G) *Araeoscelis*

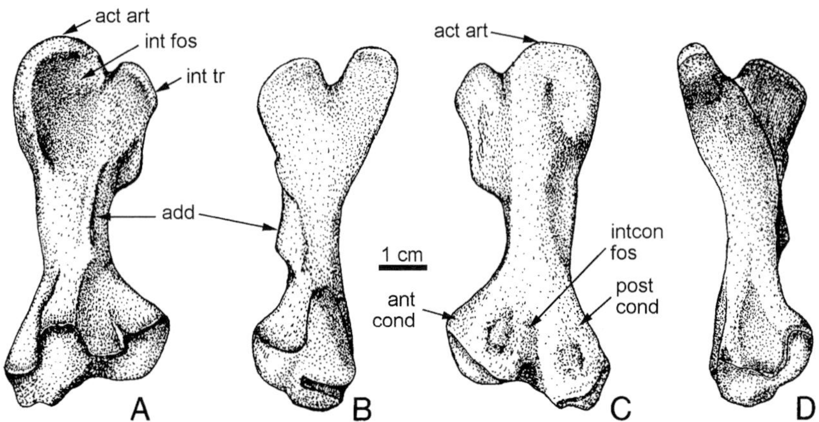

Figure 12. Reconstructions of left femur of the captorhinid reptile *Labidosaurus hamatus* illustrating important processes and landmarks. (A) Ventral view; (B) anterior view; (C) dorsal view; and (D) posterior view. Proximal end of the femur is toward the top of the page in all illustrations. Abbreviations: act art, acetabular articulation; add, adductor crest; ant cond, anterior condyle; intcon fos, intercondylar fossa; int fos, intertrochanteric fossa; int tr, intrnal trochanter; post cond, posterior condyle.

1956) and Olson (1936b) suggested that this represented a "rebuilding" of the iliac blade in diadectomorphs and a transitional stage between amphibians and reptiles in which the epaxial muscles were segregated to the dorsal aspect of the sacroiliac articulation. This implies that the lateral iliac shelf of diadectomorphs is homologous to the iliac blade of more derived tetrapods. Epaxial musculature appears to have been restricted to the medial side of the iliac blade in reptiles (Sumida, 1989a), but the potential form of their insertion into the iliac shelf of diadectomorphs is not clear. In pointing out the existence of a similar shelf in the temnospondyl amphibian *Eryops* Olson (1936b) speculated that it may have served as an expansion of the area of origin for the iliofemoralis muscle (Olson, 1936b).

The acetabulum is markedly conservative in its construction in all taxa surveyed (Fig. 11). It is unremarkable for the most part, smoothly concave and almost circular in outline, with a slight pinching of the recess near its dorsal limit in some taxa. Apparently, it allowed significant potential movement of the femoral head. The proximal head of the femur (Figs. 12 and 13) itself is not nearly as complex as that of the humerus. It is broadly curved and smoothly convex in all taxa examined here. The articular surface is roughly pitted in *Proterogyrinus* and *Seymouria*. Smithson *et al.* (1994) described the proximal articular surface of the femur in *Westlothiana* as more similar to that of protorothyridids than that of *Proterogyrinus*. Smithson *et al.* (1994) illustrate it as more poorly ossified as those of basal reptiles, however this could be an artifact of preservation. In fact, *Westlothiana* is more similar to diadectomorphs and primitive amniotes in the possession of a well developed internal trochanter. The internal trochanter is relatively very large in *Limnoscelis* and *Diadectes*, and distinct in synapsids and basal reptiles. Even the more gracile araeoscelidians display clear evidence of the internal trochanter (Vaughn, 1955; Reisz, 1981).

Although basal anthracosaurs and batrachosaurs surely possessed the puboischiofemoralis internus muscle, the bony markers of its attachment are best seen in *Westlothiana* and more derived taxa. Romer (1922) proposed that the puboischiofemoralis internus in reptiles retains the ancestral functions of protraction and elevation of the femur. All taxa surveyed have an extremely deep intertrochanteric fossa, evidence of the continuing importance of the insertion of the puboischiofemoralis externus as an adductor of the hip joint for postural support. *Seymouria* and heavy-bodied parieasaurs show the most extreme development of the intertrochanteric fossa.

All the taxa surveyed possess a well developed adductor crest on the ventral surface of the femur (Fig. 13). Those with a prominent internal trochanter have refined the crest into a prominent, sharp ridge with a pronounced rugosity for attachment of the caudofemoralis muscle. Contrary to the interpretation of Romer (1922), it appears that the caudofemoralis was a muscle of significant mass in basal amniotes, and possibly their sister groups. Numerous studies of extant reptiles (Snyder, 1954; Rewcastle, 1981; Gatesy, 1990) have confirmed the importance of the caudofemoralis in femoral retraction; given the osteological evidence for its robust presence, the suggestion that it was similarly important in basal amniotes is quite reasonable. In taxa with extremely short femoral shafts the adductor crest runs in a markedly oblique angle from proximal to distal, probably to allow as long an insertion of the adductor musculature as possible. As the shaft becomes more distinct in more derived taxa the adductor crest nearly parallels the long axis of the femur.

The femur in *Araeoscelis* (Fig. 13J) (Vaughn, 1955; Reisz *et al.*, 1984) is distinctly sigmoid, a condition found in other taxa (tentatively) assigned to the family Araeoscelidae [*Kadaliosaurus* (Credner, 1889); *Zarcasaurus* (Brinkman *et al.*, 1984)]. In their studies on extant mammals, Bertram and Biewener (1988) suggested that although bone curvature may augment stress during loading, it serves to greatly increase loading predictability. This requirement may have been met in the strongly sigmoid curvature of the femur in araeoselids, a more lightly built and presumably agile group.

The distal end of the femur is divided into distinct anterior and posterior condyles in all taxa surveyed (Figs. 12 and13). The anterior femoral condyle reaches further distally in strict dorsal or ventral view. The majority of the distal femoral articulation accommodates the tibia (Fig. 15). Holmes (1984) noted that in *Proterogyrinus* there is little ventral exposure of the femoral articulation, whereas the proximal tibial articulation is rather flat, indicative of the need to keep the knee joint somewhat below the level of the acetabulum to allow the crus to be directed toward the substrate. The articular surfaces of the knee joint are not well ossified in *Seymouria* or *Westlothiana*, precluding speculation about its orientation. In diadectomorphs and basal amniotes, a prominent ridge may be seen traversing the proximal head

Figure 13. Reconstructions of the left femur in taxa near the origin of amniotes. (A) the anthracosaurian *Proterogyrinus* (after Holmes, 1984); (B) the seymouriamorph *Seymouria* (partially after White, 1939 and Romer, 1956); (C) *Westlothiana* (partially after Smithson *et al.*, 1994); (D) the diadectomorph *Limnoscelis* (partially after Romer, 1956; Berman and Sumida, 1991); (E) the diadectomorph *Diadectes* (after Romer, 1956); (F) the caseosaurian pelycosaur *Varanops*; (G) the pareiasaurian *Pareiasaurus* (after Romer, 1956); (H) the protorothyridid reptile *Paleothyris* (partially after Carroll, 1969c); (I) the araeoscelidian reptile *Petrolacosaurus* (from Reisz, 1981); and (J) the araeoscelidian reptile *Araeoscelis*, anterior view (after Vaughn, 1955 and Romer, 1956). All scalebars equal 1 cm. ➔

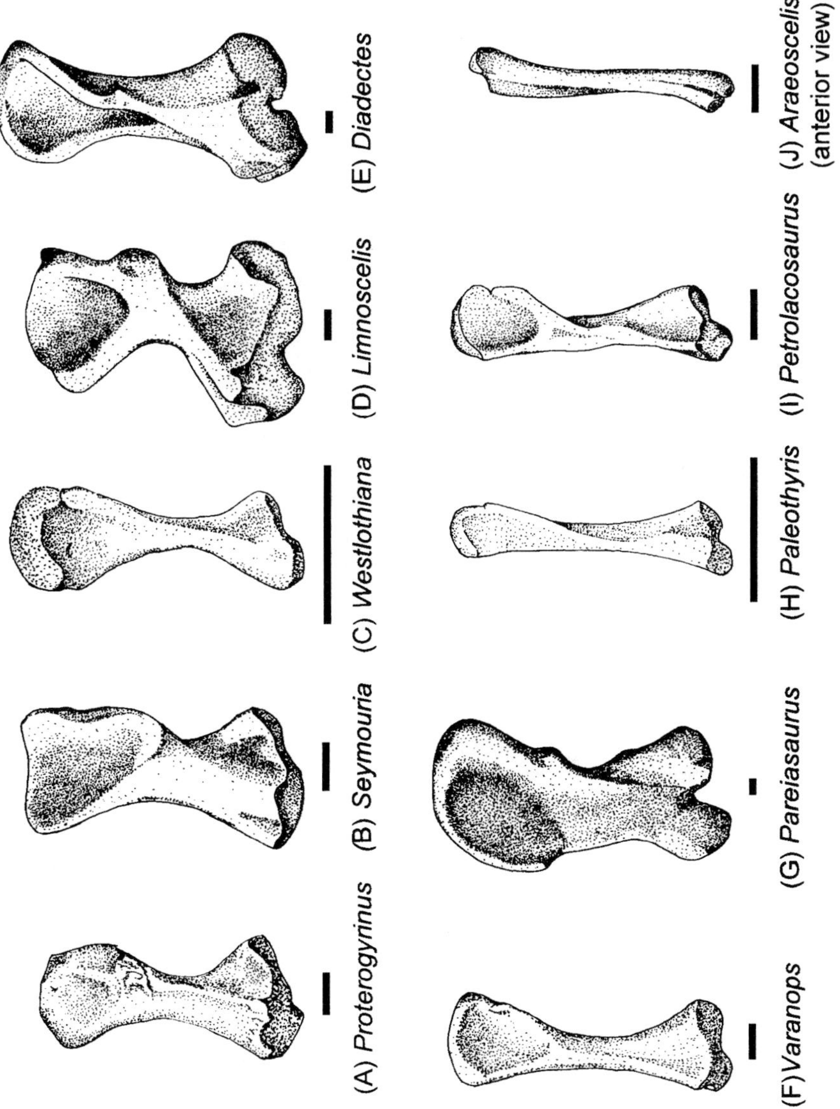
(A) Proterogyrinus
(B) Seymouria
(C) Westlothiana
(D) Limnoscelis
(E) Diadectes
(F) Varanops
(G) Pareiasaurus
(H) Paleothyris
(I) Petrolacosaurus
(J) Araeoscelis (anterior view)

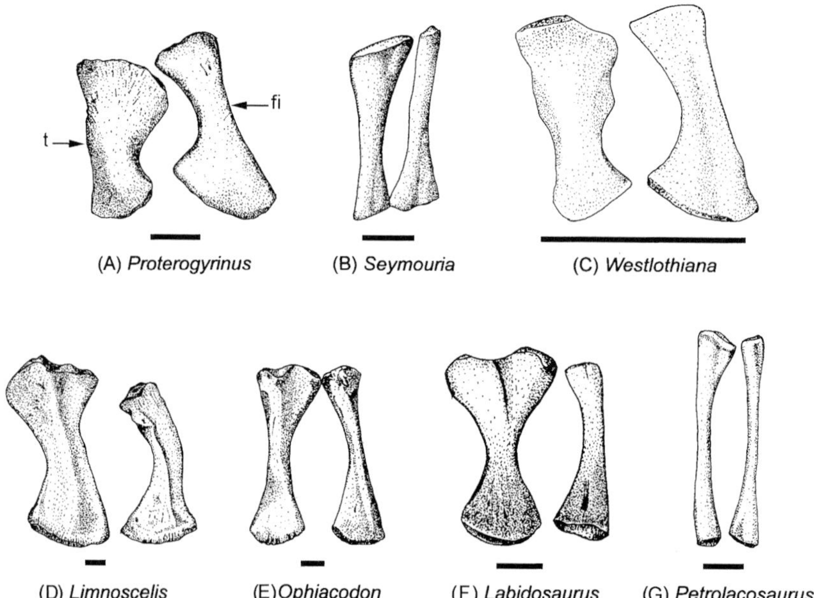

Figure 14. Reconstructions of left crural elements of taxa near the origin of amniotes in dorsal (preaxial) view. (A) The anthracosauroidian *Proterogyrinus* (after Holmes, 1984); (B) the seymouriamorph *Seymouria*; (C) *Westlothiana* (partially after Smithson *et al.* 1994); (D) the diadectomorph *Limnoscelis*; (E) the eupelycosaurian pelycosaur *Ophiacodon* (after Romer and Price, 1940); (F) the captorhinid reptile *Labidosaurus* (after Sumida, 1989a); and (G) the araeoscelidian reptile *Petrolacosaurus* (after Reisz, 1981). Elements are shown slightly disarticulated to facilitate viewing of both the tibia and fibula. All scalebars equal 1 cm. Abbreviations: fi, fibula; t, tibia.

of the tibia, dividing it into a larger anteromedial surface and a somewhat smaller posterolateral surface. Articulation of well preserved materials represented by pelycosaurs and the captorhinids *Labidosaurus* and *Captorhinus* indicates that the ridge lay against the inner edge of the posterolateral femoral condyle and in line with the intercondylar fossa (Fig. 15). Based on this scheme of articulation, the crus is estimated to have laid at an angle of approximately 75 to 80° to the femoral axis (Holmes, 1977; Sumida, 1989a). Because the acetabular articulation was relatively free, it is not possible to determine whether the femur angled slightly ventrally (instead of

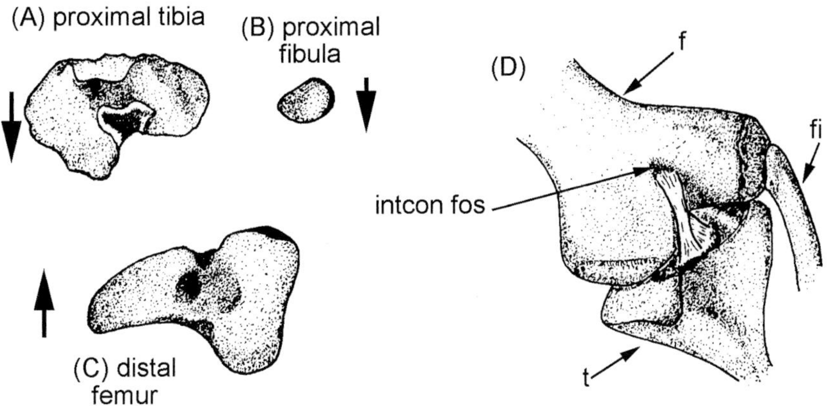

Figure 15. Articular surfaces of the left knee joint of the captorhinid reptile *Labidosaurus hamatus*. (A) Proximal surface of tibia; (B) proximal surface of fibula; (C) distal surface of femur; and (D) reconstruction of hyperflexed knee joint to demonstrate position of reconstructed cruciate ligaments. Large arrows indicate the dorsal direction. Abbreviations as in Figs. 12 and 14.

directly lateral), if the crus was positioned at an angle other than 90° to the substrate, or if there was a combination of the two.

Tibial length is approximately half that of the femur in *Proterogyrinus* and *Westlothiana*. This measure approaches 65 to 75% in *Seymouria* and other more derived taxa except for araeoscelidians, where the measures are approximately subequal. Distinguishing muscle scars on crural elements with confidence is extremely difficult. Some taxa have a well developed cnemial crest on the tibia, but its presence may be a function of size of the animal. In most cases, the tibia is more robust than the fibula. Among diadectomorphs plus amniotes, the tibia is conspicuously more robust and longer than the fibula. However, in clearly anamniote batrachosaurs the fibula approaches the tibia in size and is, in fact, longer than the tibia in *Proterogyrinus* and *Westlothiana*.

The differential lengths of the tibia and fibula (Fig. 14) have clear implications for the cruro-pedal articulation. The tarsus in *Seymouria* is poorly known; however, reconstructions for those in

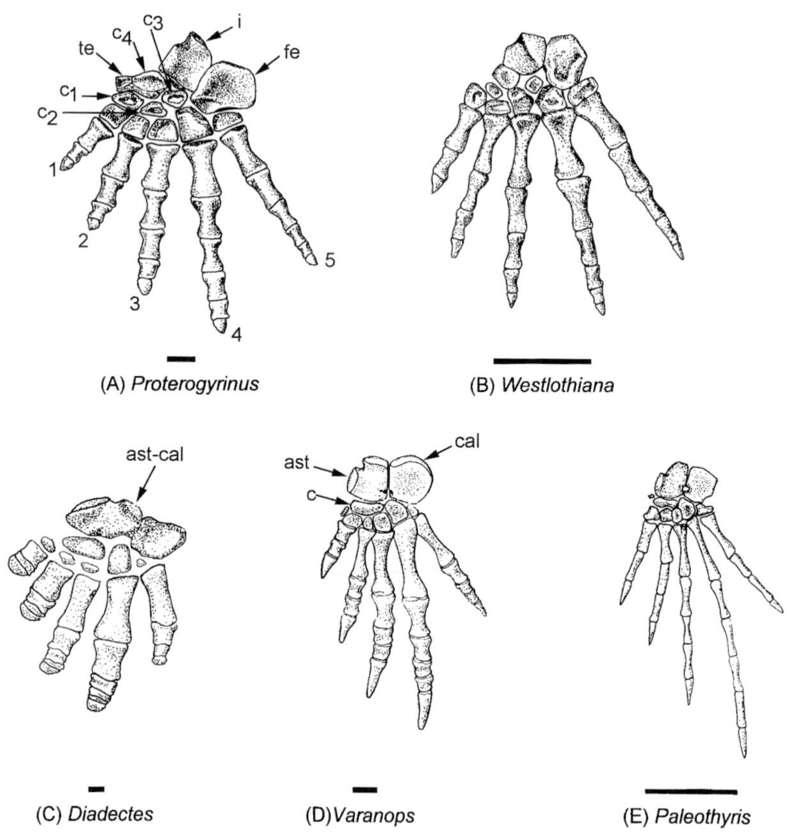

Figure 16. Reconstructions of the left pes in taxa near the origin of amniotes. All are in dorsal view except for B which is in ventral view. (A) The anthracosaurian *Proterogyrinus* (after Holmes, 1984); (B) *Westlothiana* (after Smithson *et al.*, 1994); (C) the diadectomorph *Diadectes* (after Romer and Byrne, 1931; Romer, 1944); (D) the caseosaurian pelycosaur *Varanops*; and (E) the protorothyridid reptile *Paleothyris* (after Carroll, 1969b). All scalebars equal one centimeter. Abbreviations: ast, astragalus; ast-cal, fused astragalus-calcaneus; c, centrale; cal, calcaneus; $c_{1\text{-}5}$; first to fifth centralia; fe, fibulare; i, intermedium; te, tibiale; 1-5, first to fifth digits.

Proterogyrinus and *Westlothiana* are available (Holmes, 1984; Smithson, *et al.*, 1994). More primitively, the fibula articulates with a distinct fibulare and intermedium. It appears that this might be the case in *Seymouria*, but poor preservation makes accurate determination difficult. In *Proterogyrinus* the tibia articulates with the intermedium and an element that Holmes (1984) restored as a fused tibiale plus fourth centrale. Smithson *et al.* (1994) restored the distal articulation of the tibia in *Westlothiana* with the intermedium and tibiale only. In no instance is there an articular plane that completely traverses the foot either at the proximal end of the tarsus or within it. The only complete transverse "joint" of the foot is between distal tarsal elements and the metatarsals.

Considerable argument has surrounded the homologies of the "astragalus" in amniotes. Rieppel (1993) suggested that the amniote astragalus is a neomorph that resulted from ontogenetic repatterning and that it is not homologous to the tibiale plus intermedium (Peabody, 1951). As a group very near the origin of amniotes, the Diadectomorpha would be a logical choice for information regarding this question. However, the group provides only confusing or autapomorphic features. The tarsus is poorly known in *Limnoscelis*. Independent tibiale, intermedium, and fibulare are known but poorly preserved in *Tseajaia*. In some (Romer, 1944; Rieppel, 1993) but not all (Romer and Byrne, 1931) specimens of *Diadectes* (Fig. 16) all three elements appear to be fused into a single "astragalo-calcaneal" element, apparently unique to the genus. Difficulties in understanding the homologies of tarsal elements make a diagnosis of the transformations of the ankle joint difficult, allowing a comparison of potential movements but not encouraging speculation on the homologies of joint articulations.

In amniotes in which these elements can be identified with confidence , the proximal tarsus (Figs. 16 and 17) has consolidated as an L-shaped astragalus and an approximately oval calcaneum. The fibula articulates with both the astragulus and the calcaneum, whereas the tibia articulates with the astragalus only and at a level distinct from that of the fibula. Most araeoscelidians have a unique articulation between the astragalus and tibia in which a distinct ridge of the tibia is set firmly into a trough of the astragalus, essentially creating a "locked" tibiotarsal joint (Vaughn, 1955; Reisz, 1981) (Fig. 17). More generally, the interposition of the fourth distal tarsal prevents the distal astragalo-calcaneal surfaces from defining a midtarsal joint in basal amniotes. As in the anamniotes surveyed in this study, the only

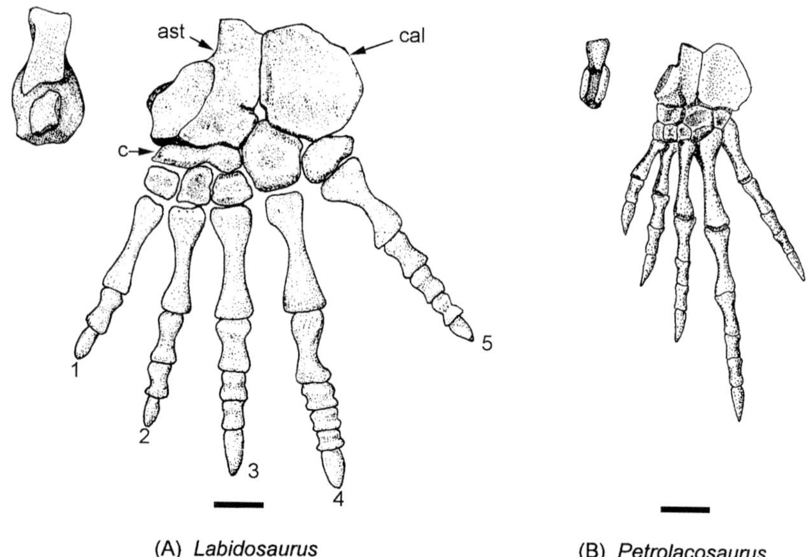

Figure 17. Articular surfaces of the left pes in (A) the captorhinid reptile *Labidosaurus* and (B) the araeoscelidian reptile *Petrolacosaurus*. Both are accompanied by a medial view of the astragalus to demonstrate the unique tongue and groove surface for the locking tibio-talar joint in *Petrolacosaurus* (after Reisz, 1981). Scalebars equal 1 cm. Abbreviations as in Fig. 16.

mesotarsal joint appears to be between distal tarsals and metatarsal elements. Although amniotes retain an intrapedal joint, the proximal tarsus may have been a less flexible mosaic of bones. This is taken to an extreme in araeoscelidians.

Digits of the pes are extensions of the flexible foot. As in the manus, the fourth digit is usually longer than the others in most of the taxa analyzed here but not remarkably so. Schaeffer (1941) suggested that the flexible mosaic of the tarsus in basal reptiles was similar to that in salamanders in which the foot is oriented anteriorly and the entire flexible mosaic was used in a progressive propulsion. Brinkman (1980, 1981) reiterated this view in his examinations of extant iguanid lizards. Rewcastle (1981) proposed that the condition of an extremely elongate fourth digit and the resulting mode of locomotion in iguanid lizards was a reasonable model for "primitive reptilian locomotion". It appears that only the more advanced protorothyridids and araeoscelidians exhibit a phalangeal morphology consistent with Rewcastle's (1981) hypothesis. Thus, Rewcastle's model may be correct, but perhaps at a position within the Reptilia as opposed to a condition characteristic of all amniotes.

STRUCTURAL SUMMARY

The preceding survey demonstrates that there is no clear and distinct transitional point in locomotor features between anamniote and amniote "sides" of the transition. Rather, there is a continuum of changes. Following is a summary of that continuum of changes mapped onto the hybrid hypothesis of relationships presented in Figure 1.

All Taxa Considered

- In all of the taxa considered the pleurocentrum is the dominant element of the vertebral centrum, possibly facilitating stiffening the vertebral column to aid in terrestrial support.
- The abutment of the atlantal intercentrum and axial intercentrum preclude exposure of the atlantal pleurocentrum on the ventral side of the vertebral column.
- The configuration of the gleno-humeral articulation is extremely conservative; the glenoid is screw shaped and the humeral head has high congruence, stereotyping considerably the movements at that joint.
- The humerus may be described generally as tetrahedral, although this feature is considerably streamlined in more derived, gracile forms.
- The manus is a flexible mosaic of bones, though it becomes less so in diadectomorphs and amniotes.
- The pelvic girdle is composed of three fused bones on each side, ilium, ischium, and pubis. The acetabulum is a concave depression at the intersection of these bones and is relatively conservative throughout the taxa included in this survey.
- The concave acetabulum received a smoothly convex femoral head in an articulation that was less constrained that the glenohumeral articulation.

- A deep intertrochanteric fossa of the proximal femoral head underscores the importance of the puboischiofemoralis externus as a postural support muscle.
- A well developed adductor crest underscores the importance of associated postural support muscles to all taxa surveyed here.

Batrachosauria

- In seymouriamorphs and the majority of sister taxa, the neural arches become extremely robust or "swollen," one example of the use of an arched construction to lend structural integrity to the supporting function of the vertebral column. This condition is lost in *Westlothiana* (although it retains convexly bowed neural arches), most synapsids, protorothyridids, and certain araeoscelids.
- The dorsal and posterior processes of the iliac blade no longer exist as separate entities. This may have moved the origin of the iliotibialis muscle further from the hip joint, enhancing its function in extension of the knee.

Westlothiana + *Diadectomorpha* + *Amniota*

- Although the puboischiofemoralis internus muscle certainly existed in more primitive taxa, its insertion on a visible and well-developed internal trochanter is first noted here.
- A pronounced rugosity for the attachment of the caudofemoralis musculature demonstrates the importance of that muscle as the largest retractor of the femur.

Cotylosauria (Diadectomorpha + Traditionally Defined Amniota)

- Variability in the construction of neural spines is present, though relatively irregular and inconsistent in seymouriamorphs. It is first seen as a column-long phenomenon in at least some specimens of diadectomorphs.
- A number of features of the atlas-axis complex point to stabilization of the head: fusion of the axial neural arch and centrum, fusion of the atlantal pleurocentrum to the dorsal aspect of the axial intercentrum, and bracketing of the axial neural spine by posteriorly directed atlantal neural arches.
- A well-developed insertion for the latissimus dorsi muscle is more obvious, although this may be a reflection of size.

- Extreme development of the entepicondyle suggests that flexor musculature of the antebrachium becomes more important in support at the initiation of the contact phase of the forelimb.
- The olecranon process is distinctly larger, suggestive of the increasing importance of the triceps musculature in extension of the elbow and propulsion by the forelimb.
- The anteromedial orientation of the ulna helps to compensate for the lateral component of force generated by the vertebral column as the triceps fires at the midpoint of the propulsive arc of the forelimb.
- A pisiform bone on the ulnar side of the manus may provide for a slightly more efficient lever of the extensor carpi radialis and flexor carpi radialis muscles.
- Tibial length increases beyond half that of the femur, typically measuring 70% or more that of the femur.

Amniota (Synapsida + "Parareptilia" + Eureptilia)

- The scapular blade is narrower than in basal outgroups.
- The humeral shaft becomes markedly more distinct, although the heavy-bodied parieasaurs are an exception. This condition is also found in *Westlothiana*, though it may be convergent due to the extremely small size of its limbs.
- A ridge traversing the proximal tibial head increases the congruence of fit with the distal femoral articular surface.
- A true astragalus develops in association with a consolidation of the tarsus and the tarsus becomes a less flexible mosaic of bones.

Sauropsida ("Parareptilia" + Eureptilia)

- The cleithrum is reduced to a splint, consolidating the scapulodeltoid musculature on the scapular blade; basal pelycosuarian-grade synapsids present an intermediate condition in this regard.

Eureptilia (Captorhinidae + Protorothyrididae + Araeoscelidia)

- Alternation of neural spine height and structure (accompanied by mediolaterally tilted zygapophyses) is refined as a phenomenon most commonly located near the limb girdles. The condition likely allowed muscular control of the vertebral column to aid in stabilization of the trunk and in the recovery stroke of limbs.

Protorothyrididae + Araeoscelidia

- The expanded neural arch diminishes in conjunction with the more lightly built, gracile form of the skeleton. Because some specimens of araeoscelidians retain expanded neural arches, protorothyridids may be autapomorphic for this character, a condition developed convergently with synapsids and *Westlothiana.* In conjunction with this, alternation of neural spine height is not seen in groups lacking expanded neural arches.
- The manus is narrower and more lightly built and the fourth digit is significantly longer than the others. A high degree of rotation of the manus and pushoff via the elongate digit may have evolved in this group.
- As in the manus, the pes is narrower and more lightly built than in outgroups and the fourth digit is considerably longer than the others. Rotation of the pes and pushoff via the elongate digit (as in iguanid lizards) may have evolved in this group.

Araeoscelidia (Basal Diapsida)

- Araeoscelidians are characterized by extremely elongate cervical vertebrae. The head is small and light, resulting in a definitive neck region.
- The femur is slightly sigmoid in shape, allowing greater load predictability.
- In conjunction with the more gracile form of the skeleton, the tibia and femur are nearly equal in length.
- In most, a "locked" tibiotarsal joint lends even more stiffness and structural integrity to the ankle joint.

DISCUSSION AND SPECULATION

As suggested at the beginning of this survey, this study is not intended to define synapomorphies of successively nested taxa. Rather, it has demonstrated that the traditionally accepted notion of a transition to a relatively more "terrestrially adapted" amniote is not appropriate. The anamniote sister groups to the Amniota are also clearly terrestrial taxa.

Anthracosauroids and seymouriamorphs share many locomotor features with amniotes. The primitive tetrapod *Westlothiana* combines a mosaic of primitive and derived features of both the axial and appendicular skeleton. This is not surprising, as its cranial morphology presents a similarly complex admixture of features. In their analysis of its phylogenetic placement, Smithson *et al.* (1994) were forced to redefine the Amniota if it were to be included, or to designate *Westlothiana* as an amniote stem taxon. In either case, the condition of *Westlothiana* further demonstrates that there are few defining boundaries between anamniotes and amniotes, at least based on postcranial features.

The composition and phylogenetic placement of the Diadectomorpha has long been a subject of debate. Most investigators suggest that the Diadectomorpha shares a more recent common ancestor with amniotes than with any other group (e.g. Heaton, 1980; Gauthier t al., 1988; Laurin and Reisz, 1995, also this volume). On the other hand, Berman *et al.* (1992) and Lee and Spencer (this volume) advocate the inclusion of the Diadectomorpha within the Amniota. Hopefully, further study of these groups will resolve these interpretive differences. Meanwhile, it is worthwhile noting that diadectomorphs plus amniotes (Cotylosauria, *sensu* Heaton, 1980) share more locomotor features than any other transitional node in this study. The difficulty in determining where to draw the line between "amphibians" and "amniotes" underscores the blurry nature of the transition between the groups when locomotor features are considered.

One postcranial feature that does appear to support amniote monophyly is the presence of an astragalus. Perhaps significantly, locomotor differences between synapsids and sauropsids appear to minimal. The heavy-bodied pariesaurs are admittedly specialized in their development of proportionally robust supporting structures, but other osteological features are not substantively different. Reptiles on the other hand demonstrate the greatest suite of common features next to the diadectomorphs + amniotes. This may be a result, in part, of the strong emphasis given to that group in recent years. When large thorough studies (e.g. Lee, 1993, 1995; Laurin and Reisz, 1995) of the "parareptilian" (basal sauropsid) groups are integrated into our knowledge of amniote phylogeny, a concomitant understanding of locomotory capabilities may also result.

The close relationship of protorothyridids and araeoscelidians is now strongly supported (Heaton and Reisz, 1986; Berman *et al.*, 1992; Smithson *et al.*, 1994), a close relationship that appears to be reflected in the similarity of locomotor features surveyed here. Together, they will likely provide important out-group data for the analysis of more derived reptilian groups.

ACKNOWLEDGMENTS

This review has benefited significantly from discussions with my close collaborators and advisors during the development of the ideas summarized here. Drs. David Berman, Eric Lombard, and Peter Vaughn have provided help, guidance, wise counsel, and patience. The conclusions presented here have been profoundly influenced by my interaction with them (usually over many cups of coffee); however, any errors of interpretation are my own. Drs. Karen Martin and Kevin Padian reviewed the manuscript and provided comments and suggestions far more numerous and thoughtful than expected of the typical review process. This work has been supported in part by a California State University San Bernardino "minigrant" and Grant #5182-94 from the National Geographic Society. This manuscript is dedicated to the memory of Dr. Everett C. Olson who lives on in the students he produced and the influence his ideas and ideals continue to wield. Of course, my deepest thanks to Dr. Elizabeth Rega.

LITERATURE CITED

Berman, D. S, R. R. Reisz, and D. A. Eberth. 1987. *Seymouria sanjuanensis* (Amphibia, Batrachosauria) from the Lower Permian Cutler Formation of north-central New Mexico and the occurrence of sexual dimorphism in that genus questioned. *Canadian Journal of Earth Sciences*, 24:1769-1784.

Berman, D. S, and S. S. Sumida. 1990. A new species of *Limnoscelis* (Amphibia, Diadectomorpha) from the Late Pennsylvanian Sangre de Cristo Formation of central Colorado. *Annals of Carnegie Museum*, 59:303-341.

Berman, D. S, S. S. Sumida, and R. E. Lombard. 1992. Reinterpretation of the temporal and occipital regions in *Diadectes* and the relationships of diadectomorphs. *Journal of Paleontology*, 66:481-499.

Bertram, J. E. A., and A. A. Biewener. 1988. Bone curvature: sacrificing strength for load predictability? *Journal of Theoretical Biology*, 131:75-92.

Bolt, J. R., and R. E. Lombard. 1992. Nature and quality of the fossil evidence for otic evolution in early tetrapods. Pages 377-403 in: *Evolutionary Biology of Hearing* (A. Popper, D. Webster, and R. R. Fay, eds.). New York: Springer-Verlag.

Boonstra, L. D. 1931. Parieasaurian studies. Part VII.–On the hindlimb of two little-known parieasaurian genera: *Anthodon* and *Parieasaurus*. *Annals of the South African Museum*, 28:429-503.

Boonstra, L. D. 1934. Parieasaurian studies. Part XI.–The vertebral column and ribs. *Annals of the South African Museum*, 31:49-68.

Brinkman, D. 1980. Structural correlates of tarsal and metatarsal functioning in *Iguana* (Lacertilia; Iguanidae) and other lizards. *Canadian Journal of Zoology*, 58:277-289.

Brinkman, D. 1981. The hindlimb step cycle of *Iguana* and primitive reptiles. *Journal of Zoology*, London, 181:91-103.

Brinkman, D., D. S Berman, and D. A. Eberth. 1984. A new araeoscelid reptile, *Zarcasaurus tanyderus*, from the Cutler Formation (Lower Permian) of north-central New Mexico. *New Mexico Geology*, 6:34-39.

Carroll, R. L. 1969a. Origin of reptiles. Pages 1-44 in: *Biology of the Reptilia, Volume 1, Morphology* (C. Gans, A. d'A. Bellairs, and T. S. Parsons, eds.), New York: Academic Press..

Carroll, 1969b. Problems of the origin of reptiles. *Biological Reviews*, 44:393-432.

Carroll, 1969c. A middle Pennsylvanian captorhinomorph, and the inter-relationships of primitive reptiles. *Journal of Paleontology*, 43:151-170.

Carroll, R. L. 1970. The ancestry of reptiles. *Philosophical Transactions of the Royal Society of London* B, 257:267-308.

Carroll, R. L. 1995. Problems of the phylogenetic analysis of paleozoic choanates. *Bulletin of the Museum of Natural Historiy, Paris*, 17:389-445.

Clark, J., and R. L. Carroll. 1973. Romeriid reptiles from the Lower Permian. *Bulletin of the Museum of Comparative Zoology*, 144:353-407.

Credner, H. 1889. Die Stegocephalen und Saurier aus dem Rothliegenden des Plauen'schen Grundes bei Dresden. VIII. Theil. *Kadaliosaurus priscus* Cred. *Zeitschrift deutsche geologische Gesellschaft*, 41:319-342.

Dilkes, D. W., and R. R. Reisz. 1986. The axial skeleton of the Early Permian reptile *Eocaptorhinus laticeps* (Williston). *Canadian Journal of Earth Sciences*, 23:1288-1296.

Gatesy, S. M. 1990. Caudofemoral musculature and the evolution of theropod locomotion. *Paleobiology*, 16:170-186.

Gauthier, J. A., A. G. Kluge, and T. Rowe. 1988. The early evolution of the Amniota. Pages 103-155 in: *The Phylogeny and Classification of the Tetra-pods, Volume 1, Amphibians, Reptiles, Birds*, Systematics Association Special Volume No. 35A (M. J. Benton, ed.). Oxford: Clarendon Press.

Gow, C. E. 1972. The osteology and relationships of the Millerettidae (Reptilia: Cotylosauria). *Journal of Zoology*, London, 167:219-264.

Haughton, S. H., and L.D. Boonstra. 1931. Parieasaurian Studies. Part VI.–The osteology and myology of the locomotor apparatus. A–Hind limb. *Annals of the South African Museum*, 28:297-366.

Heaton, M. J. 1980. The Cotylosauria: a reconsideration of a group of archaic tetrapods. Pages 497-551 *in: The Terrestrial Environment and the Origin of Land Vertebrates*, Systematics Association Special Volume No. 15 (A. L. Panchen, ed.). London: Academic Press.

Heaton, M. J., and R. R. Reisz. 1980. A skeletal reconstruction of the Early Permian captorhinid reptile *Eocaptorhinus laticeps* (Williston). *Journal of Paleontology,* 54:136-143.

Heaton, M. J., and R. R. Reisz. 1986. Phylogenetic relationships of captorhinomorph reptiles. *Canadian Journal of Earth Sciences*, 23:402-418.

Holmes, R. 1977. The osteology and musculature of the pectoral limb of small captorhinids. *Journal of Morphology*, 152:101-140.

Holmes, R. 1980. *Proterogyrinus schelli* and the early evolution of the labyrinthodont pectoral limb. Pages 351-376 in: *The Terrestrial Environment and the Origin of Land Vertebrates* (A. L. Panchen, ed.), Systematics Association Special Volume No. 15, London: Academic Press.

Holmes, R. 1984. The Carboniferous amphibian *Proterogyrinus scheelei* Romer, and the early evolution of tetrapods. *Philosophical Transactions of the Royal Society of London* B, 306:431-527.

Holmes, R. 1989. Functional interpretations of vertebral structure in Paleozoic amphibians. *Historical Biology*, 2:111-124.

Hopson, J. A. 1991. Systematics of the non-mammalian Synapsida and implications of evolution in synapsids. Pages 635-693 *in: The Origin of Higher Groups of Tetrapods: Controversy and Consensus* (H.-P. Schultze and L. Trueb, eds.). Ithaca: Cornell University Press.

Jenkins, F. A. 1971. The postcranial skeleton of African cynodonts. *Bulletin of the Peabody Museum of Natural History*, 36:1-216.

Jenkins, F. A., and G. E. Goslow. 1983. The functional anatomy of the shoulder of the savannah monitor lizard (*Varanus exanthematicus*). *Journal of Morphology*, 175:195-216.

Langston, W., and R. R. Reisz. 1981. *Aerosaurus wellesi*, new species, a varanopseid mammal-like reptile (Synapsida: Pelycosauria) from the Lower Permian of New Mexico. *Journal of Vertebrate Paleontology*, 1:73-96.

Laurin, M., and R. R. Reisz. 1995. A reevaluation of early amniote phylogeny. *Zoological Journal of the Linnean Society*, 113:165-223.

Lee, M. S. Y. 1993. The origin of the turtle body plan: bridging a famous morphological gap. *Science*, 261:1716-1720.

Lee, M. S. Y. 1995. Historical burden in systematics and the interrelationships of 'parareptiles'. *Biological Reviews*, 70:459-547.

Lombard, R. E., and S. S. Sumida. 1992. Recent progress in understanding early tetrapods. *American Zoologist*, 32:609-622.

Olson, E. C. 1936a. The dorsal axial musculature of certain primitive Permian tetrapods. *Journal of Morphology*, 59:265-311.

Olson, E. C. 1936b. The ilio-sacral attachment of *Eryops*. *Journal of Paleontology*, 10:648-651.

Olson, E. C. 1947. The family Diadectidae and its bearing on the classification of reptiles. *Fieldiana: Geology*, 11:3-53.

Peabody, F. 1951. The origin of the astragalus of reptiles. *Evolution*, 5:339-344.

Reisz, R. R. 1980. The Pelycosauria: a review of the phylogenetic relationships. Pages 553-592 in: *The Terrestrial Environment and the Origin of Land Vertebrates*, Systematics Association Special Volume No. 15 (A. L. Panchen, ed.). London: Academic Press.

Reisz, R. R. 1981. A diapsid reptile from the Pennsylvanian of Kansas. *Special Publication of the Museum of Natural History*, No. 7, 74 pages.

Reisz, R. R. 1986. Pelycosauria. *Handbuch der Palaeoherpetologie*, 17:1-102.

Reisz, R. R., D. S Berman, and D. Scott. 1984. The anatomy and relationships of the Lower Permian reptile *Araeoscelis*. *Journal of Vertebrate Paleontology*, 4:57-67.

Reisz, R. R., D. S Berman, and D. Scott. 1992. The cranial anatomy and relationships of *Secodontosaurus*, an unusual mammal-like reptile (Synapsida: Sphenacodontidae) from the Early Permian of Texas. *Zoological Journal of the Linnean Society,* 104:127-184.

Rewcastle, S. C. 1981. Stance and gait in tetrapods: an evolutionary scenario. Pages 239-267 in: *Vertebrate Locomotion* (M. H. Day, ed.), Zoological Society of London Symposium No. 48, London: Academic Press.

Rieppel, O. 1993. Studies on skeleton formation in reptiles. IV. The homology of the reptilian (amniote) astragalus revisited. *Journal of Vertebrate Paleontology*, 13:31-47.

Ritter, D. A. 1993. Epaxial muscle function during lizard locomotion. *American Zoologist*, 33:25A.

Ritter, D. A. 1995. Axial muscle function during lizard locomotion. *American Zoologist*, 35:146A.

Romer, A. S. 1922. The locomotor apparatus of certain primitive and mammal-like reptiles. *Bulletin of the American Museum of Natural History,* 46:517-606.

Romer, A. S. 1944. The Permian cotylosaur *Diadectes tenuitectes*. *American Journal of Science*, 242:139-144.

Romer, A. S. 1956. *The Osteology of the Reptiles*. Chicago: The University of Chicago Press.

Romer, A. S. 1957. The appendicular skeleton of the Permian embolomerous amphibian, *Archeria*. *Contributions, Museum of Geology, University of Michigan*, 13:103-159.

Romer, A. S. 1970. A new anthrocosaurian labyrinthodont, *Proterogyrinus scheelei*, from the Lower Carboniferous. *Kirtlandia*, No. 10, 16. pp.

Romer, A. S., and F. Byrne. 1931. The pes of *Diadectes*: notes on the primitive tetrapod limb. *Palaeobiologica*, 4:25-48.

Romer, A. S., and L. W. Price. 1940. *Review of the Pelycosauria*. Geological Society of America Special Paper, No. 28, 538 pages.

Schaeffer, B. 1941. The morphological and functional evolution of the tarsus in amphibians and reptiles. *Bulletin of the American Museum of Natural History*, 78:395-472.

Schultze, H.-P. 1987. Dipnoans as sarcopterygians. Pages 39-74 in: *The Biology and Evolution of Lungfishes* (W. E. Bemis, W. W. Burggren, and N. E. Kemp, eds.). New York: Alan R. Liss.

Smithson, T. R. 1989. The earliest known reptile. *Nature*, 314:676-678.

Smithson, T. R., R. L. Carroll, R. L. Panchen, and S. M. Andrews. 1994. *Westlothiana lizziae* from the Visean of East Kirkton, West Lothian, Scotland, and the amniote stem. *Transactions of the Royal Society of Edinburgh: Earth Sciences*, 84:383-412.

Snyder, R. C. 1954. The anatomy and function of the pelvic girdle and hindlimb in lizard locomotion. *American Journal of Anatomy*, 95:1-36.

Stovall, J. W., L. I. Price, and A. S. Romer. 1966. The postcranial skeleton of the giant Permian pelycosaur *Cotylorhynchus romeri*. *Bulletin of the Museum of Comparative Zoology*, 135:1-30.

Sumida, S. S. 1987. Two different forms in the vertebral column of *Labidosaurus* (Captorhinomorpha, Captorhinidae). *Journal of Paleontology*, 61:155-167.

Sumida, S. S. 1989a. The appendicular skeleton of the early Permian genus *Labidosaurus* (Reptilia, Captorhinomorpha, Captorhinidae) and the hind limb musculature of captorhinid reptiles. *Journal of Vertebrate Paleontology*, 9:295-313.

Sumida, S. S. 1989b. Reinterpretation of vertebral structure in the early Permian pelycosaur *Varanosaurus acutirostris* (Amniota, Synapsida). *Journal of Vertebrate Paleontology*, 9:419-426.

Sumida, S. S. 1991. Vertebral morphology, alternation of neural spine height, and structure in Permo-Carboniferous tetrapods, and a reappraisal of primitive modes of terrestrial locomotion. *University of California Publications in Zoology*, 122:1-133.

Sumida, S. S., and R. E. Lombard. 1991. The atlas-axis complex in the late Paleozoic genus *Diadectes* and the characteristics of the atlas-axis complex across the amphibian to amniote transition. *Journal of Paleontology*, 65:973-983.

Sumida, S. S., R. E. Lombard, and D. S Berman. 1992. Morphology of the atlas-axis complex of the late Palaeozoic tetrapod suborders Diadectomorpha and Seymouriamorpha. *Philosophical Transactions of the Royal Society of London*, 336:259-273.

Vaughn, P. P. 1955. The Permian reptile *Araeoscelis* restudied. *Bulletin of the Museum of Comparative Zoology*, 113:305-467.

Vaughn, P. P. 1970. Alternation of neural spine height in certain Permian tetrapods. *Bulletin of the Southern California Academy of Sciences*, 164:1-30.

White, T. E. 1939. Osteology of *Seymouria baylorensis* Broili. *Bulletin of the Museum of Comparative Zoology*, 85:325-409.

Williston, S. W. 1911. *American Permian Vertebrates*. Chicago: The University of Chicago Press.

CHAPTER 12

WATER BALANCE AND THE PHYSIOLOGY OF THE AMPHIBIAN TO AMNIOTE TRANSITION

Karen L. M. Martin

Kenneth A. Nagy

INTRODUCTION

Background

Water is the most important substance in an animal's body in terms both of composition and of exchange with the environment. Animals are about 70% water by mass and at least 99% water by number of molecules. Animals, whether aquatic or terrestrial, exchange more water molecules per unit time between themselves and their surroundings than any other kind of molecule. Thus, a change in the water availability of an environment can have a major and rapid impact on an animal.

The challenge of living in a dehydrating environment has been met during the evolution of amniotes by the development of relatively impermeable skin and internalization of respiratory surfaces so that diffusional exchange of water with the environment is much lower in air-breathing than in water-breathing tetrapods. Having a nearly waterproof skin, compared to having a water-permeable skin, is associated with differences in respiratory physiology, osmoregulation, nitrogen metabolism, and thermal relations in the vertebrates. In this chapter, we discuss the differences in the physiology, morphology and ecology of present-day amphibians and reptiles as viewed from the perspective of their water relations. We summarize the general properties of the water budgets of amphibians and reptiles and compare one with the other. Then, we suggest that the patterns seen in these animal groups can provide insight into the transition from water to land that occurred in Paleozoic times, and we suggest ways that amniotes may have evolved from anamniote ancestors.

Assumptions

The transition from anamniotes to amniotes is a soft-tissue transition. The post-cranial skeleton changes little, and the skull changes in small, rather than profound ways. After extensive attention, questions remain as to which of the primitive tetrapods was the first amniote (Carroll, 1970, 1988; Sumida and Lombard, 1991). Romer (1966, p. 102) stated that "primitive Paleozoic reptiles and some of the earliest amphibians were so similar in their skeleton that it is almost impossible to tell when we have crossed the boundary between the two classes." Indeed, the transitional forms assigned to the Diadectomorpha have been variously placed in the Amphibia or Amniota (Sumida *et al.*, 1992; Berman *et al.*, 1992; Romer, 1964; Heaton, 1980; Laurin and Reisz, this volume).

Separation between the major groups of the Lissamphibia (the extant amphibians: Anura, Urodela, and Gymnophiona) and the lineage leading to amniotes occurred at least 250 million years ago, in the late Paleozoic era (Pough, 1983). The suggestion has been made that anurans emerged from the temnospondyl line of Devonian labyrinthodont amphibians wheeas the salamanders and caecilians emerged from the microsaur line of lepospondyl amphibians (Carroll, 1988). If so, the taxon Lissamphibia is polyphyletic and similarities between these groups must be evaluated for possible symplesiomorphies, shared ancestral characters. On the other hand, other scenarios support the monophyly of the Lissamphibia (Laurin and Reisz, this volume). If moist, permeable skin is a symplesiomorphy of the Lissamphibia and not a derived feature for each of the three clades, then presumably the ancestral animal would share this feature. Clearly, there are many morphological differences between paleoamphibians and the Lissamphibia (Romer, 1966; Gans, 1970). Nevertheless they share many aspects of life history and ecology, including a freshwater aquatic larval stage and humid habitats. As in the present, some fossil amphibians remained aquatic throughout life.

Many of the primitive genera of amphibians were heavily scaled (Carroll, 1988). Cutaneous respiration is common among the Lissamphibia but has been discounted for paleoamphibians because of their large body size and the presence of this ossification (Gans, 1970). This has led to the conclusion that the skin of paleoamphibans was not particularly permeable to water, perhaps no more so than the skin of most bony fish (Gans, 1970). However, whether or not large paleoamphibians relied on cutaneous respiration for the majority of their oxygen needs, these scales may not have functioned to prevent water loss. Studies comparing scaled and scale-less snakes show no differences in evaporative water loss rates (Bennett and Licht, 1975; Maderson *et al.*, 1978). Skin permeability is determined in the epidermis, but these scales were probably dermal, as they are in fishes (Bellairs, 1970). The function of these scales may have been to protect the newly terrestrial amphibians from abrasion on the substrate (Frolich, this volume) or from the increased dosagae of ultraviolet light insolation during excursions onto land (Bellairs, 1970).

Because water balance mechanisms have large implications for the physiology of organisms, we propose to consider the possibility that paleoamphibians, although unlike extant amphibians in some ways, may have been similar physiologically. We suggest that there are physiological consequences of water-permeable skin in an aquatic or humid, swampy habitat that may have benefitted the earliest amphibians in their competition with fish. These benefits may have become liabilities in less humid environments, and changes in these systems may have led to the evolutionary transition to primitive reptiles and synapsids. In studying the completion of the vertebrate transition to land, it may be heuristic to examine some aspects of the differences between extant anamniote tetrapods and amniotes, and to use these to illuminate the transition itself.

Selection Pressures

The evolutionary pressures on Paleozoic anamniote tetrapods that led them out of water may have included biological pressures in the form of food source limitations, predation on eggs and larvae, or competition with fish and other amphibians. Physical and ecological factors could have included aquatic hypoxia, poor water quality, and a change to a drier, warmer climate from coal forests and swamps (Zeigler *et al.*, 1979). Animals that emerged onto land more completely during the Paleozoic era may have enjoyed advantages including abundant oxygen in air (Graham *et al.*, 1995), abundant food in the form of insects, little competition for space, and reduced predation on eggs. Terrestrial plants became abundant during the Devonian era and although no adult amphibian, fossil or extant, is herbivorous, fossils indicate that some of the early amniotes may have been (Hotton *et al.*, this volume).

The amniotic egg is the defining feature of the amniote clade, but it is the adult animals that must cope with the demands of true terrestriality. To succeed, the proto-amniote transitional animal had to come to terms with a less forgiving, more variable physical environment. The atmosphere changes far more than an aquatic environment changes, has more extreme conditions of temperature, humidity, and wind levels, and has more rapid fluctuations in these conditions. Wet skin, advantageous as a respiratory surface for amphibians, results in high evaporative water losses in a dry climate. The early amniote maintained hydration out of water, foraged on land, excreted nitrogenous wastes, respired in air, reproduced away from water, produced a cleidoic egg, and moved about terrestrially, among other things. Each of these processes affected, and was affected by, water balance. In addition, each process affected the other processes. Examples of extant adult Lissamphibia can be found that manage one or several of these tasks, in specialized ways. However, no one species of Lissamphibia can do all of these as well as any member of the reptilian grade. This macroevolutionary step involves a suite of adaptations for terrestriality, as contrasted with isolated specializations that permit survival during difficult seasons.

Living on land presents a constant risk of desiccation. In the transition from water to land, water economy must be a key issue. Full terrestriality results in a change from wet, permeable skin requiring frequent access to water to dry, impervious skin that can tolerate wind and insolation. We suggest that this one change affects many aspects of the animal's physiology. To permit full exploitation of a more rapidly fluctuating external environment, we suggest that this evolutionary step hinges on a dramatic increase in homeostatic regulation of the internal environment.

MODERN ANIMALS

Water Budgets

The three parts of an itemized water budget are gain, loss and storage (Shoemaker and Nagy, 1977). Positive storage occurs when an animal retains excess water, as do toads in their urinary bladder, and negative storage is synonymous with dehydration. The most common avenues of water gain are ingestion in the food or by drinking, osmosis through permeable skin, as in frogs and toads, and production of metabolic water *de novo* during oxidation of carbohydrates, fats and protein for energy production. Water leaves animals by evaporation across skin, lungs, or any permeable surface at a rate that is dependent on the temperature and humidity of the air, as well as the permeability of the evaporating surface. The factors that influence the rate of loss may differ, depending on whether evaporation is taking place from the lungs, skin, or eyes. Water is lost as part of the voided urine and feces, and in any glandular secretions released by the animal.

Amphibian Water Budget

Extant amphibians contain more water when fully hydrated, 77 to 83% of body mass, than the 70% of body mass found in extant reptiles and mammals (Bentley, 1966). The gain side of the water budget of a typical amphibian is simplified by the fact that most amphibians do not drink liquid water (Shoemaker *et al.*, 1992). Metabolic water production is determined by the rate of energy metabolism, which in amphibians is relatively low. Typically it ranges from 0.1 to 0.3% of body mass per day, depending on the size of the amphibian. This amount is very small in comparison to the main avenue of water intake, osmotically across the skin. Amphibians can take up water at rates ranging from 30 to 360% of their body mass per day.

Highly permeable skin is a feature shared by all the Lissamphibia, permitting cutaneous respiration (Guimond and Hutchison, 1976); regulation of acid/base balance by CO_2 release (Stiffler *et al.*, 1990); active transport of salt (Stiffler, 1988); release of nitrogenous wastes as ammonia by diffusion into water (Shoemaker, 1987); evaporative cooling, modulating changes in body temperature (Spotila, 1972); and perhaps most important, water absorption through the skin (Shoemaker and Nagy, 1977). "The entire skin of amphibians is a site of seemingly unbridled flux," according to Feder (1992). The skin of Lissamphibia provides almost no barrier to evaporative water loss or gain, enabling the amphibian to rapidly deplete or replenish its stores of water. In some cases the permeable skin is not uniformly distributed over the body but is localized, particularly in a pelvic patch on the ventral surface where it can be pressed against a damp substrate (Stille, 1958).

The largest avenue of water efflux in amphibians is evaporative water loss across the moist and permeable skin. It can range from 3 to 160% of body mass per day (Tracy, 1975), depending on body size, ambient conditions of temperature, wind, and insolation, and on exposure of the animal to these conditions. Water evaporation via lungs and eyes is small relative to skin evaporation. Fecal water losses are determined by the dehydrating capability of the large intestine, which is moderate, and by food consumption, which depends on energy needs. Because amphibians have relatively low energy and food needs, fecal water losses are small relative to evaporative losses. Similarly, water losses via glandular secretions are typically small, but may be large on occasion (Lillywhite, 1971). Water lost as urine can be substantial, ranging from zero to 60% of body mass per day, depending on species and hydration state (Shoemaker and Nagy, 1977). The majority of nitrogenous wastes are excreted via this route. The final composition of the urine is determined by three organs, the kidneys, the cloaca, and the urinary bladder, all of which modify urine composition.

Accumulation of nitrogenous wastes is a constraint on terrestriality in extant Lissamphibia. Aquatic amphibians produce copious dilute urine because water is freely available. They excrete nitrogenous wastes partially through this urine, and partially as ammonia by diffusion across the skin (Shoemaker *et al.*, 1992). Diffusional release of ammonia into air is much lower. Because ammonia is toxic, amphibians switch to forming a less toxic substance, urea, and both aquatic and terrestrial forms generally stop excreting urine when out of water (Shoemaker *et al.*, 1992). In many amphibians, the bladder acts as a fluid store, and water in it is gradually reabsorbed. Meanwhile, urea accumulates in the body and diffuses throughout the body fluids. As a result, amphibians are limited in the duration of their terrestrial excursions by plasma urea levels. This problem becomes particularly acute if the animal is feeding away from water because then additional wastes are produced. As most amphibians typically do not excrete urine on land, they must return to water to rehydrate and only following rehydration is urine formed again and the urea excreted. A few species of frogs which live in arid habitats are able to produce uric acid, and their water requirements are reduced as a result. These same species can have very low rates of evaporative water loss by producing a waxy, waterproofing secretion that is spread over the skin, as in the case of *Phyllomedusa*, and by several layers of dead skin, as in the case of *Chiromantis* (Shoemaker *et al.*, 1992).

The more terrestrial species of lissamphibians are capable of both positive and negative storage of water. Many desert anurans have large urinary bladders, and can store large amounts of water as dilute urine in their bladders. The largest amount of stored urine found to date is 130% of standard (bladder-empty) body mass in burrowed Australian desert frogs, *Cyclorana platycephala* (Shoemaker *et al.*, 1992). When frogs and toads with full bladders are restricted from access to free water, they reabsorb bladder water and maintain normal blood osmotic concentrations while losing water by high evaporation. When bladder water reserves are gone, these anurans continue to survive subsequent dehydration by tolerating additional loss of body water as well as the progressively increasing osmotic concentration in their body fluids. Dehydration tolerance varies with habitat aridity and taxonomic affiliation (Shoemaker *et al.*, 1992), being highest (near 50% of body mass) in desert-dwelling spadefoot toads (McClanahan,

1967). The animal must either return to water, where it can rehydrate by osmosis, or find a hiding place such as a burrow, with a high humidity to limit net evaporation and some moisture in soil or leaf litter for osmotic uptake (McClanahan, 1972). Amphibians that live in seasonally-arid habitats avoid activity during the dry parts of the year, except occasionally foraging at night. They may form cocoons, accumulate urea to maintain a favorable osmotic gradient between themselves and the soil, reduce their already low metabolic rate, and use other physiological and morphological mechanisms to enhance water balance while water availability is low.

In short, most amphibians are built to process water rapidly. Their skin offers little resistance to evaporation; an amphibian away from water will die of dehydration in less than a day or two. Three main points emerge from this analysis. The skin's permeability mandates that it will be a major route of water loss during terrestrial activity away from water. The skin is also the major avenue of water gain, thus behavior, particularly selection of microhabitat, is critically important in achieving water balance in most amphibians while they are active. Third, the more terrestrial and arid-adapted amphibians are able to achieve water balance during seasons when liquid water is unavailable, but usually only by being inactive in burrows or other hiding places.

Reptile Water Budget

Reptiles generally have water requirements that are only about 1 to 5% of those of amphibians and reptiles have much lower rates of water exchange. Reptilian skin has a very high resistance to evaporative water loss (Lillywhite and Maderson, 1988). Thus, those avenues of water flux that were inconsequential in amphibians, such as metabolic water production and fecal water loss, are major aspects of reptilian water budgets.

The main avenue of water intake for many reptiles is in the diet, with intake rates ranging from 0.7 to 2.7% of body mass per day in small arid-habitat species. Metabolic water production in reptiles is determined by the rate of energy metabolism and ranges from 0.1 to 0.5% of body mass per day. This amount accounts for 10 to 20% of total water gain in several small, diurnal, arid-habitat reptiles (Shoemaker and Nagy, 1977). Many species living in temperate habitats can maintain or increase body mass during rainless spring and summer periods when drinking water is not available.

Some reptiles "drink" with the tongue touched to drops of water on vegetation or other substrates. Water gained by drinking may be very important, as in desert tortoises (Nagy and Medica, 1986) and in lizards living in mesic habitats, although actual rates of drinking in the field are largely unknown (Nagy, 1982). Some reptiles have grooves in the skin to channel water droplets of rain or condensation toward the mouth (Withers, 1993).

As with amphibians, evaporation is often the largest avenue of water loss from reptiles. However, the rate can be nearly three orders of magnitude less (Shoemaker and Nagy, 1977; Shoemaker *et al.*, 1992), as little as 0.25% of body mass per day (Mautz, 1982), because of their relatively impermeable skin. Even so, evaporation accounts for 25-75% of total water loss in several species of arid-habitat reptiles (Nagy, 1982). Reptiles living in moist habitats have much higher rates of evaporation, up to 30% of body mass per day in tropical lizards, and 200% per day in dry air in a tropical burrowing snake. In typical reptiles, about half of the total evaporation occurs through the skin, whereas the other half occurs through the respiratory tract. The wet surfaces of eyes can account for 15-65% of total evaporative loss (Mautz, 1982).

Terrestrial reptiles can dehydrate their feces fairly well, as dry as 40% water by mass, and ranging from 20 to 60% of total water loss. Herbivorous reptiles produce more feces per unit of metabolizable energy than carnivorous reptiles, due to the lower digestibility and the lower energy content of plants compared to animal matter. However, if herbivores obtain succulent green plant matter to eat, they ingest much more dietary water than do carnivores (Shoemaker and Nagy, 1977).

Urinary water loss by reptiles can be quite low, amounting to as little as 0.08% of body mass per day in an active desert lizard maintaining a balanced water budget (Shoemaker and Nagy, 1977). They conserve urinary water by excretion of a large portion of their nitrogenous waste as uric acid rather than as urea or ammonia. Uric acid has two beneficial properties. First, it has a low solubility, so it precipitates out of solution at low concentrations and can be eliminated in solid form, saving the water otherwise needed to dissolve urea or ammonia. Neither amphibians nor reptiles can produce urine hyperosmotic to the plasma, so this savings is substantial. Second, uric acid precipitates along with dietary cations such as sodium and potassium, thereby saving the water these ions would otherwise require for their elimination in dissolved form. Thus, the evolution of uricotelism has reduced the water requirements of reptiles, allowing increased independence from free water sources.

Many mesic- and xeric-habitat reptiles have salt glands that produce concentrated salt secretions that eliminate excess dietary salts with little accompanying water. Among desert lizards having nasal salt glands, water losses via nasal secretions are low, 0.1 to 0.3% of body mass per day, but account for about 10% of the small total water loss.

Regarding storage of water, some reptiles are somewhat like amphibians, having the capacity to store up to 30% of their body mass as dilute urine in a urinary bladder. Of course, water-loaded reptiles can produce copious urine while returning their body water volumes to normal, so urinary water losses can exceed 30% of body mass per day (Shoemaker and Nagy, 1977). The desert tortoise can survive up to 50% loss of body mass due to dehydration, after it has exhausted the urinary bladder stores, while the body fluid osmotic concentration rises, as in amphibians. It does not urinate for months and only then after a rainfall allows it to rehydrate (Nagy and Medica, 1986).

To summarize, the ability of many species of reptiles to achieve water balance without drinking is due to their selection of moist foods, and to their low rates of water loss. Reptiles living in dry habitats are relatively waterproof, and most reptiles have water requirements that are only about 1-5% of those of amphibians. The reptiles have nearly impermeable skin, they produce relatively dry feces and may have nasal salt glands that eliminate dietary salts with little water loss. They make urate salts to eliminate waste nitrogen and dietary salts in precipitated form. Desert reptiles may store water in a urinary bladder and tolerate extensive dehydration. These adaptations allow arid-habitat reptiles to be active during dry periods, and to attain water balance using only the succulence of their diet, without needing free water for drinking or any other purpose.

Comparison of Extant Modes of Maintaining Water Balance

Many amphibians can tolerate wide swings in body hydration (Shoemaker *et al.*, 1969; van Buerden, 1984). When full, the mass of water in the urinary bladder may equal the mass of the fully hydrated amphibian with the bladder empty. Water may be reabsorbed into the body from the bladder during terrestrial excursions (Shoemaker, 1964), and after that, the animal may tolerate extreme water deficits before rehydration occurs. In contrast, most reptiles do not experience variation in internal fluid concentration, and maintain body hydration state with low water flux rates. Some species of reptiles, especially those living in very hygric habitats such as wet tropics or moist soil, have water flux rates that are very similar to those of amphibians, including high rates of water gain and loss by evaporation, but these animals do not vary internal fluid concentration substantially.

Semiterrestrial anurans can withstand the loss of about half their body water (400 ml kg^{-1}) and the resultant doubling of body fluid concentrations (Shoemaker and Nagy, 1977). Among amphibians, internal fluids may concentrate to 700 or 800 mOsm, while the desert tortoise, among the most xeric-adapted reptiles, can concentrate its body fluids to 600 mOsm under extreme conditions (Minnich, 1977), although typically in the field concentration goes no higher than 340 mOsm (Minnich, 1979). The time scale in which dehydration occurs in xeric reptiles is also much longer than the brief period that can be tolerated by xeric amphibians.

What an animal can withstand in the laboratory and what it chooses to do in the field may not be the same. Feder and Londos (1984) found that salamanders ceased foraging well before their performance was impaired or their tolerance to desiccation was reached, and they suggested that salamanders voluntarily abandon foraging well before their physiological limits were reached. It is logical to assume that animals will not push themselves to extremes during routine activities such as foraging. This being the case, the time and distance a slow moving, wet animal could spend away from water is quite limited, although it could be extended by the choice of shaded, moist microhabitats, or by moving about during precipitation or at night when water condenses on the soil and leaf litter. Reptiles, with water requirements of only 1-5% that of amphibians, can survive with much less access to water and can use body water more efficiently, permitting much longer terrestrial excursions into much drier habitats.

Heat Balance Implications

One can postulate that once semiterrestrial animals became more restrained in their use of water it was necessary for them to become more deliberate in their temperature regulation. The effects on body temperature when water loss is curtailed through the skin may be very important, as damp skin promotes heat loss, or conversely, minimizes heat gain. A wet-skinned frog in full sun may have a body temperature only a couple of degress higher than a frog in full shade, although double the rate of evaporative water loss (Tracy, 1976). On the other hand, if two ectothermic animals are the same size, but one has wet skin and the other dry, the wet one will be cooler even if both are held in full sun at the same air temperature (Spotila *et al.*, 1992). Increasing environmental temperature will cause body temperatures of both wet and dry animals to increase, but not at the same rate or to the same extent. In the modelling by Spotila *et al.* (1992), dry skin body temperatures increased from 26 to 42.4° C whereas wet-skinned body temperatures increased only from 23.4 to 28.3° C under the identical conditions. In the field, body temperatures of amphibians are generally much lower than those of reptiles (Avery, 1982; Hutchison and Dupre, 1992).

The profound difference in thermal response between a wet-skinned and a dry-skinned animal provides a potential explanation for the contrast in the types of thermoregulation between Lissamphibia and extant reptiles. Lissamphibians generally do not show a preferred body temperature or a specific activity temperature (Hutchison and Dupre, 1992). Thermoregulation in modern amphibians relies mostly on "passive" evaporative cooling and behavioral selection of damp shaded habitats (Hutchison and Dupre, 1992). It is likely that temperature regulation in Lissamphibia is subservient to hydric constraints. In other words, in amphibians, if there is a conflict between hydroregulation and thermoregulation, then thermoregulation takes a lower priority.

Reptiles, by contrast, in general have very narrow ranges of preferred activity temperatures (Brattstrom, 1965). Although ectothermic, many reptiles are able to regulate their body temperature very narrowly with behavioral means (Avery, 1982). The importance of behavioral thermoregulation is shown by the fact that different species of reptiles in the same environment may have different activity temperatures, and reptiles of the same species from different habitats may have the same body temperatures for activity (Bartholomew, 1982). Nocturnal desert geckoes have a lower acitivity temperature than diurnal lizards, but no difference in water flux rates (Nagy and Degen 1988). In addition, the excursion of body temperature between activity and inactivity is much greater in reptiles, as their active temperatures are generally higher than those of amphibians, even for nocturnal reptiles, while body temperatures during inactivity may be substantially lower.

Animals benefit in several ways from a higher body temperature, but it is important to remember that higher body temperatures are enabled or permitted, according to biophysical principles, as a result of the change from wet and permeable to dry and impermeable skin. Permitting a relatively high body temperature reduces the need for loss of evaporative water for cooling in a hot environment. Increased body temperature also increases metabolic rate, which could increase alertness and responsiveness, important for predation or predator avoidance (Bennett, 1991). On the other hand, dry skin creates some new problems for the heat budget of the animal, by increasing sensitivity to a variety of environmental parameters, including solar radiation, wind speed, and heat convection, even though it decreases sensitivity to humidity (Spotila *et al.*, 1992). Body temperature in dry-skinned animals is much more affected by skin reflectance and by body size than is the temperature of a wet-skinned animal. Thus, the reduction in skin permeability to water increases the lability of the body temperature and its sensitivity to some terrestrial conditions.

The wide range of potential body temperatures available to a dry-skinned animal, along with an increased sensitivity to highly variable environmental conditions, could create difficulties for an animal that did not make any effort to control its body temperature. In other words, dry skin not only makes it possible to increase body temperature, it appears to necessitate the inititation of intentional, fairly precise, behavioral temperature selection.

As compared to amphibians, amniotes use more energy on a daily basis. At the same body temperature, the resting metabolic rate of amphibians is only two-thirds that of reptiles of identical mass (Regal, 1983; Pough, 1983). This difference is correlated with the decreased water content as a percentage of total mass in amniotes. A semiterrestrial animal that stays cool as a result of evaporative water loss may be constrained to a low metabolic rate niche. This may explain the absence of herbivory in adult amphibians, as herbivorous reptiles may require high body temperatures for digestion of plant material (Bartholomew, 1966). Reptiles in general operate at a higher metabolic rate than amphibians, and this may be an important part of the physiological distinction between these two groups.

THE FIRST AMNIOTE ANIMALS

Feder (1992) suggests that in amphibians, "fluxes seemingly permit only limited independence from the immediate environment. Accordingly, the internal milieu of amphibians may be far less fixed than that of many other vertebrates." The first amniote, we suggest, accomplished the transition from amphibian anamniote to reptilitan amniote mainly by altering its style of water balance. Restriction of skin permeability permitted greater control over the state of body hydration. Early amphibians remained tied to water for feeding, reproduction, and most of their locomotion, using land probably primarily as a refuge from predators or as an escape route from one drying or degraded aquatic habitat to another, still essentially aquatic, home. Reptiles completed the transition to land by truly exploiting the land's resources as a permanent habitat.

Minimally, the proto-amniote that came to fully exploit the land must have greatly reduced cutaneous evaporative water loss by decreasing skin permeability. However, reducing permeability of the skin also reduced cutaneous reuptake of water, reduced cutaneous respiration, altered acid-base regulation, greatly reduced the utility of ammonia as a waste product, and reduced the cooling effect of evaporation from the body surface, changing heat budget parameters. The simple reduction of water permeability alters nearly all of this animal's systems. Therefore, amniotes have the potential, and probably the necessity, for more refined regulation of blood plasma volume, osmolarity, pH, and body temperature control systems than anamniote tetrapods; in short, the transition to land provides the potential, and possibly the necessity, for increased metabolic homeostasis. To accomplish this, several or all of the following changes were necessary.

Reduced Permeability of the Skin

The permeability of amphibian skin to water is under some hormonal control. Arginine vasotocin (AVT) increases permeability for osmotic uptake of water, but does not affect resistance to evaporation (Shoemaker *et al.*, 1992). This is because the main resistance to evaporation is the boundary layer above the skin, not the permeability itself. In addition, AVT acts as an antidiuretic (Uchiyama and Pang, 1985). During the transition to full terrestriality, AVT secretion may have increased to reduce urinary loss of water, and skin permeability must have been de-coupled from this hormone. Scales are not necessary for reduction of evaporative water loss, as a comparison of scaled and scaleless reptiles has shown (Bennett and Licht, 1975). Heavily keratinized epidermal layers are not particularly necessary in the initial stages, as the "waterproof" frog *Phyllomedusa* has only one layer of stratum corneum, coated in wax and lipids, and yet has greatly reduced permeability to water (Shoemaker *et al.*, 1992). Making the skin permanently and irreversibly impermeable to water solved the problem of water loss but intensified the problem of water gain.

Increased Intake of Dietary Water

An animal eating succulent plants ingests more dietary water than a carnivore does, although plants may be more difficult to digest. Green plants have approximately 1.5 times as much nitrogen as meat or insects per unit of energy and an order of magnitude more water content (Shoemaker and Nagy, 1977). Plants also have higher osmolytes than animal food (Shoemaker and Nagy, 1977). Fossil evidence indicates that the early amniotes were insectivorous (Carroll, 1988), but that herbivory developed early in the amniote lineage (Hotton *et al*, this volume). Insects have only slightly more water per kilocalorie than meat but have more than double the nitrogen, much of which would have to be excreted as wastes. Omnivory would have been ideal for the earliest amniotes in order to maximize benefits and minimize problems of water balance. Increased body temperature would have facilitated increased use of plant matter in the diet.

A further means of adding dietary water would be the use of drinking. This behavioral change is another potential adaptation for life on land, and could have been accomplished initially by the use of tongue flicking, with minimal anatomical or physiological adjustments.

Excreted Nitrogenous Wastes without Losing Excessive Water

The cessation of urine excretion while on land is the extreme response by extant amphibians, necessary to conserve water because of their high cutaneous evaporative water loss . The excretory system of an amniote must excrete nitrogenous wastes terrestrially, in a relatively nontoxic form, using relatively little water, and in addition, it is likely that the excretory system would be required to take an increased role in acid-base regulation. If the animal ate a diet of plants, or a diet of insects or meat and drank water, it could probably meet its dietary needs and have sufficient water to excrete nitrogenous wastes as urea, so it may be that the earliest amniotes did not require uric acid for excretion. *Sphenodon* uses urea, not uric acid (Schmidt-Nielsen and Schmidt, 1973), indicating that this may have been the ancestral condition. Turtle eggs use urea as a nitrogenous excretory product, as well (Packard *et al.*, 1984). The proto-amniote may have continued to use a urinary bladder for storage, and later, reabsorption of water, as do desert tortoises (Nagy and Medica, 1986).

Reproduction on Land without Returning to Fresh Water

Internal fertilization is necessary for the cleidoic egg because of the necessity of a protective shell, secreted before the egg is laid (Packard and Seymour chapter, this volume). Although internal fertilization is necessary for terrestrial reproduction, an intromittent organ is not, as in terrestrial salamanders, *Sphenodon*, and birds. We believe that the presence and structure of extraembryonic membranes are so similar for all amniotes that the amniotic egg probably evolved only once, early on. In its earliest incarnation, it probably had a membranous shell that allowed entry of water and oxygen. A large yolk provided energy. It would probably have been laid in relatively small clutches in a moist environment, perhaps buried in sand or leaf litter. The waste products could have been uric acid, urea, and a little ammonia, although uric acid my not have been necessary in early amniotes. Uricotelism in adults may have preceded evolution of the cleidoic egg (Needham, 1931), because increasing terrestriality of the adult rather than the egg would probably confer an advantage earlier in the evolution of terrestriality. The evolution of the extra-embryonic membranes, in particular the allantois, a ventral out-pocketing of the gut, may have been more an adaptation for sequestering the nitrogenous waste products than a water-conserving measure. On the other hand, the embryo may have tolerated relatively high levels of urea, and the increased osmolarity would have decreased the water potential of the egg and promoted water transport from the soil, as in some turtle eggs (Miller and Packard, 1992).

Increased Behavioral Regulation of Body Temperature

If an ectothermic animal is dry as a result of decreased skin permeability to water, then it must contend with a much more labile body temperature and therefore it is forced either to accept high variability in body temperature, or to control the internal temperature by increased behavioral regulation. The simple solution of permitting body temperature to vary with environment, which worked well in amphibians because the variation was tempered by evaporative cooling and behavioral microhabitat choices, is not as simple for the dry-skinned animal. Because of the greater temperature fluctuations, the enzymes and metabolic functioning of the animal would have to change continuously with temperature over the course of the day. The optimal solution may have been to come upon a specific, rather narrow preferred body temperature range–one that could readily be attained with behavioral means and maintained over most of the period of activity. The reptile *Sphenodon* shows more ancestral traits than most extant reptiles, and has a body temperature during activity that is near the minimum temperature tolerated by squamates voluntarily (Bartholomew, 1982). A higher body temperature would increase metabolism and, in turn, oxygen and food requirements. These increased metabolic needs could potentially lead to increased levels and types of activity, and more complex behavior. Thus, by reducing evaporative water loss, the proto-amniote could concommitantly increase its body temperature and its metabolic rate.

In dry-skinned animals, body size has a greater effect on body temperature than in wet-skinned animals. Even small animals may enjoy a warm body temperature if they have dry skin (Spotila *et al.*, 1992). Early amniotes may have been relatively small (Carroll, 1988); on the other hand, larger animals could maintain a high body temperature longer during the day by inertial homeothermy, although this could increase the time necessary to warm up. Large reptiles in a thermally stable environment may need to spend little effort specifically on thermoregulation (Shine and Madsen, 1996), but they do maintain a relatively constant activity temperature. We suggest that the evolutionary trend during the transition to amniotes is toward a higher and deliberately regulated body temperature, and increased homeostasis, in the presence of increased environmental lability.

Altered Respiratory Mechanisms

With the loss of the skin as an exchange surface, the respiratory system would be forced to increase reliance on the lungs for oxygen consumption and more effective carbon dioxide release. Enclosure of respiratory surfaces in a lung also helps to reduce water loss. The enclosure of the respiratory surfaces and the reliance on a tidal flow of the respiratory medium (Piiper and Scheid, 1975) results in intermittent rather than continuous gas exchange, which alters in particular the release of carbon dioxide. The details of the evolution of lungs have been reviewed extensively (Randall *et al.*, 1981; Little, 1990); suffice it here to emphasize the significance of the enclosure of the respiratory surfaces on other aspects of physiology, including acid-base balance and water balance.

Speculation

Now, we shall set up some hypotheses about the characteristics of the earliest amniote. It seems likely that this tetrapod had a terrestrial diet, probably herbivorous or insectivorous or omnivorous . Its skin was able to resist desiccation in air, probably by thickening and keratinization, and it may have had dermal armor or scales for protection from predators, abrasion, or UV light. Its cleidoic eggs were laid on land in a moist substrate, and had membranous shells permable to water, and true amniotic membranes. The terrestrial hatchlings were not larvae, but fully terrestrial miniature adults. Some of the earliest amniote fossils are those of the Joggins formation of Nova Scotia, in which the reptiles are found associated with tree stumps (Romer, 1966; Carroll, 1988). We speculate that these animals may have sat on tree stumps in basking behavior, similar to extant lizards. Our lines of speculation do not differentiate between particular body sizes, but larger body size could favor herbivory (Pough, 1973; Hotton *et al.*, this volume) unless the animals could select young, tender leaves (Mautz and Nagy, 1987). Carroll (1970, 1988) favors a small size for the transition animal, while Lombard and Sumida (1992) point to a larger animal as the first amniote. Our hypothesis is that the physiological steps from anamniote to amniote may be more definitive than the morphological steps, at least as revealed by the fossil record.

Now, we will try to flesh out the animal a bit more. This putative earliest amniote may have been similar to an extant reptile. The similarities could probably include dry skin and a urinary bladder able to resorb water. This early amniote laid eggs with membranous shells that absorbed water from the environment. It may have had periods of inactivity during the year, and it may have still spent some time in an aquatic habitat, although the skin was relatively impermeable to water when compared to that of amphibians. It may have eaten plants, or insects, or possibly had an omnivorous diet. It basked and behaviorally thermoregulated in order to reach a body temperature that would enable it to be active. What reptile is the most similar to this proto-amniote picture? We suggest that among extant reptiles, turtles in general most closely fit this physiological profile.

What extant amphibian is physiologically most like the earliest amniote? It has relatively impermeable skin, terrestrial reproduction, and forages on land. It tolerates a high body temperature and excretes some of its nitrogenous wastes as uric acid and urea. It lives emerged from water, but it survives seasonal drought and high body temperatures only by becoming inactive. Arboreal "waterproof" frogs in the genera *Phyllomedusa* and *Chiromantis* (Shoemaker *et al.*, 1987) most closely fit this model.

CONCLUSION

Gans and Pough (1982, p. 8) wrote, "We suggest that the absence of a unifying morphological scheme, rather than being an incidental by-product, is an important aspect of the reptilian grade. The lives of Recent reptiles are shaped by a set of shared characteristics that need not produce obvious structural features." The features they suggest as definitive are physiological features: The amniotic egg, ectothermy, low metabolic rate (compared with endotherms), reliance on anaerobiosis for activity, and behavioral temperature regulation. We propose that a similar case may be made for the traits that unite the "amphibian" grade: Great morphological diversity with no obvious unifying morphological characters, but physiological features that set them apart from the bony fishes and from the reptiles. These features include increased variability or relaxed regulation of body hydration levels, high cutaneous permeability to water, relatively low body temperatures for activity, and extremely low metabolic requirments. Pough (1983) described amphibians as being "well known for their generally quiescent and inconspicuous lifestyles and for their low annual use of energy." Bartholomew (1982, p. 344) noted "present-day members of the class Amphibia are morphologically different from the Paleozoic forms which were transitional between fish and reptiles. Nevertheless, they demonstrate clearly the high level of success with which animals are sometimes able to exploit a harsh and demanding environment, despite physiological adjustments that appear at first glance to be modest and ineffective."

Initially we stated that one of the basic issues in comparing amphibians and reptiles is their water budget, and that this affects the whole animal's physiology in a variety of ways. The synergistic effects of providing a dry outer surface are the possibility of warmer body temperatures, higher metabolism, and increased activity and growth rates. This could lead to increased tolerance of high temperatures but perhaps decreased tolerance of low temperatures.

What are the advantages of remaining an amphibian, and not completing the transition to land? The low energy strategy enables amphibians to survive with a very short growing season (Pough,

1983). The geographic range may be limited only by the duration of development from larva to adult for hibernation or aestivaton (Pinder *et al.*, 1992). Some species of the desert or in the arctic may be inactive as much as 6-10 months of the year, a life-style that does hold a certain appeal. With a very low metabolic rate, amphibians can survive on a limited food source. Adult *Rana muscosa* can survive an eight month inactive period with only 4-6% body fat (Bradford, 1983). In addition, adults and larvae do not compete for food; therefore, more members of the population may be supported at once in the same habitat. Temperature selection by habitat is reflected in the biogeography of extant amphibians, which extend to higher altitude and latitude than reptiles; amphibians show greater tolerance to colder climates than reptiles (Pinder *et al.*, 1992). Amphibians can be active at much cooler body temperatures than reptiles (Hutchison and Dupre, 1992); for example, there are observations of frogs swimming under ice (as do some turtles) and salamanders walking in snow. Several species of frogs tolerate being frozen during winter (Storey and Storey, 1986). By becoming inactive, amphibians can tolerate wide fluctuations in internal conditions during drought or very cold conditions.

By releasing the constraints of the hydric niche, the earliest amniotes increased body temperature and metabolic rate, but in so doing may have limited themselves to a more narrow thermal niche. In essence, the reptile by comparison has a higher metabolism and a more narrowly regulated, homeostatic internal milieu that enables it to be more active than an amphibian in a much wider range of terrestrial habitats. In this manner, reptiles have obtained greater independence from the hydric environment by means of a more fixed internal milieu than was the case in anamniote tetrapods.

ACKNOWLEDGMENTS

We are grateful to the following people for helpful comments, helpful arguments, and enlighteningconversations: Brian Henen, Victor Hutchison, Lee Kats, Vaughan Shoemaker, Stuart Sumida, and the UCLA physiological ecology seminar group.

LITERATURE CITED

Avery, R. A. 1982. Field studies of body temperatures and thermoregulation. Pages 93-166 in: *Biology of the Reptilia*, Vol. 12, (C. Gans and F. H. Pough, eds.). New York: Academic Press.

Bartholomew, G. A. 1966. A field study of temperature relations in the Galapagos marine iguana. *Copeia*, 1966:241-250.

Bartholomew, G. A. 1982. Body temperature and energy metabolism. Pages 333-406 in: *Animal Physiology: Principles and Adaptations*, 4th Edition (M. S. Gordon, ed.). New York: MacMillan Publishing.

Bellairs, A. 1970. *The Life of Reptiles*. Volume 1. New York: Universe Books.

Bennett, A. F. 1991. The evolution of aerobic capacity. *Journal of Experimental Biology,* 160:1-23.

Bennett, A. F., and P. Licht. 1975. Evaporative water loss in scaleless snakes. *Comparative Biochemistry and Physiology,* 52A: 213-215.

Bentley, P. J. 1966. The physiology of the urinary bladder of Amphibia. *Biological Reviews,* 41:275-316.

Bradford, D. F. 1983. Winterkill, oxygen relations, and energy metabolism of a submerged dormant amphibian, *Rana muscosa*. *Ecology*, 64:1171-1183.

Brattstrom, B. H. 1965. Body temperatures of reptiles. *American Midland Naturalist,* 73:376-422.

Carroll, R. L. 1970. The ancestry of reptiles. *Philosophical Transactions of the Royal Soceity of London*, B, 257:267-308.

Carroll, R. L. 1988. *Vertebrate Paleontology and Evolution*. New York: Freeman.

Edney, E. B. 1977. *Water Balance In Land Arthropods*. Zoophysiology and Ecology series; Volume 9. Berlin: Springer-Verlag.

Feder, M. E. 1992. A perspective on environmental physiology of the amphibians. Pages 1-6 in: *Environmental Physiology of the Amphibians* (M.E. Feder and W.W. Burggren, eds.). Chicago: The University of Chicago Press.

Feder, M. E., and W. W. Burggren. 1985. Cutaneous gas exchange in vertebrates: Design, patterns, control, and implications. *Biological Reviews,* 60:1-45.

Feder, M. E., and P. L. Londos. 1984. Hydric constraints upon foraging in a terrestrial salamander, *Desmognathus ochrophaeus* (Amphibia: Plethodontidae). *Oecologia*, 64:413-418.

Gans, C. 1970. Respiration in early tetrapods: the frog is a red herring. *Evolution,* 24:723-724.

Gans, C., and F. H. Pough. 1982. Physiological ecology: Its debt to reptilian studies, its value to students of reptiles. Pages 1-13 in: *Biology of the Reptilia*, Vol. 12C (C. Gans and F. H. Pough, eds.). New York: Academic Press.

Graham, J. B., R. Dudley, N. M. Aguilar, and C. Gans. 1995. Implications of the late Paleozoic oxygen pulse for physiology and evolution. *Nature,* 375:117-120.

Guimond, R. W., and V. H. Hutchison. 1976. Gas exchange of the giant salamanders of North America. Pages 313-338 in: *Respiration of Amphibious Vertebrates* (G. M. Hughes, ed.). New York: Academic Press.

Hutchison, V. H., and R. K. Dupre. 1992. Thermoregulation. Pages 206-249 in: *Environmental Physiology of the Amphibians* (M. E. Feder and W.W. Burggren, eds.). Chicago: The University of Chicago Press.

Lillywhite, H. B. 1971. Thermal modulation of cutaneous mucus discharge as a determinant of evaporative water loss in the frog, *Rana catesbiana*. *Comparative Biochemistry and Physiology,* 40A:213-227.

Lillywhite, H. B., and P. F. A. Maderson. 1988. The structure and permeability of integument. *American Zoologist,* 28:945-962.

Little, C. 1990. *The Terrestrial Invasion: An Ecophysiological Approach to the Origins of Land Animals*. Cambridge, UK: Cambridge University Press.

Lombard, R. E., and S. S. Sumida. 1992. Recent progress in understanding early tetrapods. *American Zoologist,* 32:609-622.

Maderson, P. F., W. W. Zucker, and S. I. Roth. 1978. Epidermal regeneration and percutaneous water loss following cellophane stripping of reptile epidermis. *Journal of Experimental Zoology,* 204:11-32.

Mautz, W. J. 1982. Patterns of evaporative water loss. Pages 443-481 in: *Biology of the Reptilia*, Vol. 12C (C. Gans and F. H. Pough, eds.). New York: Academic Press.

Mautz, W. J., and K. A. Nagy. 1987. Ontogenetic changes in diet, field metabolic rate, and water flux in the herbivorous lizard *Dipsosaurus dorsalis*. *Physiological Zoology,* 60:640-658.

McClanahan, L. L. 1967. Adaptations of the spade foot toad, *Scaphiopus couchi,* to desert environments. *Comparative Biochemistry and Physiology,* 20:73-99.

McClanahan, L. L. 1972. Changes in body fluids of burrowed spadefoot toads as a function of soil water potential. *Copeia,* 1972:209-216.

Miller, K., and G. C. Packard. 1992. The influence of substrate water potential during incubation on the metabolism of embryonic snapping turtles (*Chelydra serpentina*). *Physiological Zoology,* 65:172-187.

Minnich, J. E. 1977. Adaptive responses in the water and electrolyte budgets of native and captive desert toroises (*Gopherus agassizi* and *G. polyphemus*). Desert Tortoise Council. Proceedings of the 1977 Symposium: 102-129.

Minnich, J. E. 1979. Reptiles. Pages 391-641 in: *Comparative Physiology of Osmoregulation in Animals* (G. M.. O. Maloiy, ed.). New York: Academic Press.

Nagy. K. A. 1982. Field studies of water relations. Pages 483-501 in: *Biology of the Reptilia*, Vol. 12, (C. Gans and F. H. Pough, eds.). New York: Academic Press.

Nagy, K. A., and A. A. Degen. 1988. Do desert geckoes conserve energy and water by being nocturnal? *Physiological Zoology*, 61:495-499.

Nagy, K. A., and P. A. Medica. 1986. Physiological ecology of desert tortoises in southern Nevada. *Herpetologica,* 42:73-92.

Needham, J. 1931. *Chemical Embryology*. Cambridge, UK: Cambridge University Press.

Packard, G. C., M. J. Packard, and T. J. Boardman. 1984. Influence of hydration of the environment on the pattern of nitrogen excretion by embryonic snapping turtles (*Chelydra serpentina*). *Journal of Experimental Biology*, 108:195-204.

Piiper, J., and P. Scheid. 1975. Gas transfer efficacy of gills, lungs, and skin: theory and experimental data. *Respiration Physiology* 14:115-124.

Pinder, A. W., K. B. Storey, and G. R. Ultsch. 1992. Estivation and hibernation. Pages 250-274 in: *Environmental Physiology of the Amphibians* (M.E. Feder and W.W. Burggren, eds.). Chicago: The University of Chicago Press.

Pough, F. H. 1973. Lizard energetics and diet. *Ecology*, 54:837-844.

Pough, F. H. 1983. Amphibians and reptiles as low-energy systems. Pages 141-188 in: *Behavioral Energetics: The Cost of Survival in Vertebrates* (W. P Aspey and S. I. Lusick, eds.). Columbus: Ohio State University Press.

Randall, D. J., W. W. Burggren, A. P. Farrell, and M. S. Haswell. 1981. *The Evolution of Air Breathing in Vertebrates*. Cambridge, UK: Cambridge University Press.

Regal, P. J. 1983. The adaptive zone and the behavior of lizards. Pages105-118 in: *Lizard Ecology: Studies of a Model Organism* (R. B. Huey, E. R. Pianka, and T. W. Schoener, eds.). Cambridge, Massachusetts: Harvard University Press.

Romer, A. S. 1966. *Vertebrate Paleontology* Third Edition. Chicago: The University of Chicago Press.

Schmidt-Nielsen, K., and D. Schmidt. 1973. Renal function of *Spehnodon punctatum*. *Comparative Biochemistry and Physiology*, 44A:121-129.

Shine, R., and T. Madsen. 1996. Is thermoregulation unimportant for most reptiles? An example using water pythons (*Liasis fuscus*) in tropical Australia. *Physiological Zoology* 69:252-269.

Shoemaker, V. H. 1964. The effects of dehydration on electrolyte concentrations in a toad, *Bufo marinus*. *Comparative Biochemistry and Physiology*, 13:261-271.

Shoemaker, V. H. 1987. Osmoregulation in amphibians. Pages 109-120 in: *Comparative Physiology: Life in Water and on Land* (P. Dejours, L. Bolis, C. R. Taylor, and E. R. Weibel, eds.). Padua: Liviana Press.

Shoemaker, V. H., and K. A. Nagy. 1977. Osmoregulation in amphibians and reptiles. *Annual Review of Physiology*, 39:449-471.

Shoemaker, V. H., S. S. Hillman, S. D. Hillyard, D. C. Jackson, L. L. McClanahan, P. C. Withers, and M. L. Wygoda. 1992. Exchange of water, ions, and respiratory gases in terrestrial amphibians. Pages 125-150 in: *Environmental Physiology of the Amphibians* (M.E. Feder and W.W. Burggren, eds.). Chicago: The University of Chicago Press.

Shoemaker, V. H., L. L. McClanahan, and R. Ruibal. 1969. Seasonal changes in body fluids in a field population of spadefoot toads. *Copeia,* 1969:585-591.

Shoemaker, V. H., L. L. McClanahan, P. C. Withers, S. S. Hillman, and R. C. Drewes. 1987. Thermoregulatory responses to heat in the waterproof frogs *Phyllomedusa* and *Chiromantis*. *Physiological Zoology* 60:365- 372.

Spotila, J. R. 1972. Role of temperature and water in the ecology of lungless salamanders. *Ecological Monographs,* 42:95-125.

Spotila, J. R., M. P. O'Connor, and G. S. Bakken. 1992. Biophysics of heat and mass transfer. Pages 59-80 in: *Environmental Physiology of the Amphibians* (M.E. Feder and W.W. Burggren, eds.). Chicago: The University of Chicago Press.

Stiffler, D. F. 1988. Cutaneous exchange of ions in lower vertebrates. *American Zoologist,* 28:1019-1029.

Stiffler, D. F., M. L. DeRuyter, and C. R. Talbot. 1990. Osmotic and ionic regulation in aquatic and terrestrial caecilians. *Physiological Zoology,* 63:649-668.

Stille, W. T. 1958. The water reabsorption response of an anuran. *Copeia,* 1958:217-218.

Storey, K. B., and J. M. Storey. 1986. Freeze tolerance and intolerance as strategies of winter survival in terrestrially hibernating amphibians. *Comparative Biochemistry and Physiology,* 83A:613-617.

Sumida, S. S., and R. E. Lombard 1991. The atlas-axis complex in the late Paleozoic genus *Diadectes* and the characteristics of the atlas-axis complex across the amphibian to amniote transition. *Journal of Paleontology,* 65: 973-983.

Tracy, C. R. 1975. Water and energy relations of terrestrial amphibians: Insights from mechanistic modelling. Pages 325-346 in: *Perspectives of Biophysical Ecology* (D. M. Gates and R. B. Schmerl, eds.). New York: Springer-Verlag.

Tracy, C. R. 1976. A model of the dynamic exchange of water and energy between a terrestrial amphibian and its environment. *Ecological Monographs,* 46:293-426.

Uchiyama, M., and P. K. T. Pang. 1985. Renal and vascular responses to AVT and its antagonistic analog (KB IV-24) in the lungfish, mudpuppy, and bullfrog. Pages 929-932 in: *Current Trends in Comparative Endcrinology* (B. Lofts and W. N. Holmes, eds.).Hong Kong: Hong Kong University Press.

van Beurden, E. K. 1984. Survival strategies of the Australian water-holding frog, *Cyclorana platycephalus*. Pages 223-234 in: *Arid Australia* (H. G. Cogger and E. E.Cameron, eds.). Sydney: Australian Museum.

Withers, P. 1993. Cutaneous water acquisition by the thorny devil (*Moloch horridus*: Agamidae). *Journal of Herpetology*, 27:265-270.

Zeigler, A. M., C. R. Scotese, W. S. McKerrow, M. E. Johnson, and R. K. Bambach. 1979. Paleozoic paleogeography. *Annual Review of Earth and Planetary Science,* 7:473-502.

CHAPTER 13

RECONSTRUCTING ANCESTRAL TRAIT VALUES USING SQUARED-CHANGE PARSIMONY: PLASMA OSMOLARITY AT THE ORIGIN OF AMNIOTES

Theodore Garland, Jr.

Karen L. M. Martin

Ramon Díaz-Uriarte

INTRODUCTION

Biologists often wonder about the behavioral or physiological characteristics of extinct organisms. Were dinosaurs endothermic? How well could *Archaeopteryx* fly? Unfortunately, fossil information pertaining to behavior or physiology is generally unavailable (but see, for example, Lambert, 1992; Hillenius, 1994; Ruben, 1995; Ruben *et al.*, 1996). In some cases, extant forms exist that are morphologically very similar to ancient forms (Eldredge and Stanley, 1984). If one is willing to assume that the behavior and physiology of these "living fossils," such as the coelacanth, has also experienced relatively little evolutionary change, then a more or less direct window into the past is available (Thomson, 1991; but see Burggren and Bemis, 1990, p. 199).

Amniote Origins

Usually, however, one is limited to a consideration of "ordinary" extant organisms as "models" for extinct forms (e.g., Ruben and Bennett, 1980; Burggren and Bemis, 1990; Ruben, 1991; Ruben and Parrish, 1991; Janis and Wilhelm, 1993; Garland and Carter, 1994; Hackstein and van Alen, 1996; Ruben *et al.*, 1996). The purpose of this chapter is to illustrate some ways in which data on the characteristics of living organisms can be combined with phylogenetic information and recently developed analytical methods to make inferences about the characteristics of hypothetical ancestral organisms.

For illustrative purposes, the phenotypic trait we consider is plasma osmolarity, which previous (nonphylogenetic) studies have shown to vary among phylogenetic lineages (clades) and in relation to ecology or habitat (e.g., freshwater versus saltwater). We have compiled from the literature osmolarity data for a total of 172 vertebrate taxa, including representatives of all major extant lineages (Figs. 1, 2, and Appendix 1). Some living forms are known to be highly derived as compared with their ancestors, and hence constitute "red herrings" with respect to making inferences about extinct forms (Gans, 1970). Given data for a wide variety of extant organisms, therefore, which ones should be used as models for extinct forms–all or a subset? Moreover, how, exactly, should one go about making an inference regarding extinct organisms? Most simply, one could estimate the value of a trait in a hypothetical ancestor as the simple mean of the values observed in its living descendants. Alternatively, one can use additional information on phylogenetic relationships to obtain a better estimate of the ancestral value.

The most common way to use phylogenetic information when making inferences about ancestral trait values is with the general analytical procedure termed "parsimony." Parsimony can be used to "map" or "optimize" a character onto a hypothetical phylogenetic tree, reconstruct values for hypothetical ancestral organisms (internal nodes of the tree), and hence "trace" character evolution (Brooks and McLennan, 1991; Stewart, 1993; Maddison, 1994; Maddison, 1995; Schultz *et al.*, 1996). Appendix 2 discusses the origin and present roles of parsimony analyses in systematic and comparative biology (for some references pertaining to physiological traits, see Burggren and Bemis, 1990; Garland and Carter, 1994, p. 601).

The first and simplest outcome of a parsimony analysis is estimates of where and how many times a particular character state originated in a phylogenetic tree (e.g., Rosenberg, 1996). This allows one to address such questions as: How many times did endothermy arise during vertebrate evolution (Block *et al.*, 1993; Ruben, 1995)? How many times did flight evolve? Estimation of where in a phylogenetic tree a character state first evolved can also allow one to predict which descendant lineages should have it [i.e., before some living taxa have been actually measured (e.g., see Burggren and Bemis, 1990, p. 208 and 220), or for extinct taxa whose physiology or behavior cannot be measured (e.g., Bennett and Ruben, 1986)]. Moreover, parsimony analyses can elucidate the historical pathways of character evolution (e.g., whether a character increased or decreased in size over evolutionary time within a particular lineage), and hence whether the character state in a particular species is derived (sometimes termed "advanced") or similar to the ancestral value (sometimes termed "primitive"). Further, estimates of where, when, and how frequently particular character states evolved can be combined with information on biogeography or environmental changes (e.g., from the geological record) to get at possible *causes* of phenotypic evolution, such as adaptation in response to an altered selective regime [Eastman, 1993; Arnold, 1994: e.g., did changes in habitat aridity precede changes in evaporative water-loss rate of *Coleonyx* lizards? (Dial and Grismer, 1992)]. Attempts to identify "key innovations" also routinely involve parsimony analyses (Nitecki, 1990; Brooks and McLennan, 1991; Eastman, 1993, p. 200). Sometimes inferences about the traits of extinct forms can be combined with mathematical or physical models to "synthesize" extinct forms (e.g., Kingsolver and Koehl, 1985; Tracy *et al.*, 1986; Abler, 1992; Ryan and Rand, 1995).

Although systematists usually deal with discretely valued characters, comparative biologists often study characters that show continuous variation, both within and among species (e.g., body size, metabolic rate, maximal running speed, and home range area). For continuous-valued characters, the procedure termed "squared-change parsimony" is often used to reconstruct values for hypothetical ancestors. In Appendix 3, we explain this procedure with worked examples and present computer simulations that compare its performance with an alternative, nonphylogenetic analysis.

The procedures we discuss are not ways of reconstructing phylogenetic trees. Rather, we use existing hypotheses about phylogenetic relationships (taken from the literature) to construct a composite estimate of phylogeny (as shown in Fig. 1) and then map onto this tree (with 172 tips) the plasma osmolarity data. These physiological data are *not* used to alter the phylogenetic topology that is employed in the analyses. Thus, our procedures are in the tradition of most recent work in "the comparative method," in which the "tip" data being analyzed (information on the phenotypes of a series of terminal taxa, typically species but sometimes populations or averages for genera or higher taxa) are independent of the phylogenetic framework used for analyses (reviews in Brooks and McLennan, 1991; Harvey and Pagel, 1991; Maddison and Maddison, 1992; Eggleton and Vane-Wright, 1994; Garland *et al.*, 1993; Garland and Adolph, 1994; Maddison, 1996; Martins, 1996b,c). Alternatively, one could use the physiological data to influence the estimate of phylogeny used for analysis. Whether the use of phylogenies that are "independent" of the characters under study is the best thing to do is unsettled (review in de Queiroz, 1996); nevertheless, it is often a practical necessity. Formal ways to combine fundamentally different types of data [e.g., DNA sequences, genetic distances (as from DNA hybridization studies), qualitative morphological characters, continuous-valued physiological traits that show broad overlap among the species being analyzed] into a single algorithm for estimation of phylogeny (e.g., Eernisse and Kluge, 1993) are the subject of much current controversy and no simple solution is readily available (de Queiroz *et al.*, 1995; Huelsenbeck *et al.*, 1996a).

PLASMA OSMOLARITY OF VERTEBRATES

For the illustrative purposes of this chapter, we have chosen to analyze plasma osmolarity as a representative physiological trait. No fossil indicator of this trait has ever been proposed, so inferences about its value in extinct organisms must be based on consideration of living forms.

Plasma is the fluid component of blood; it can vary widely in solute concentration. The ability of solutes to cause osmotic pressure and osmosis is measured in terms of osmoles, the osmole being a measure of the total number of particles in a solution. Specifically, one osmole is defined as one gram of nonionizable and nondiffusable substance. *Osmolarity* is the osmolar concentration of a solution, and is expressed as osmoles per liter of solution. As shown in Figures 1 and 2, osmolarity varies widely among species of vertebrates.

Plasma osmolarity is a colligative property, measured as the total number of "osmotically active" particles in solution. If a molecule dissociates into two ions, both may contribute to the osmolarity. Additionally, particle size is unimportant: a small ion such as sodium contributes as equally to osmolarity as a larger macromolecule such as a protein. Thus, two solutions may have the same total osmolarity but differ greatly in chemical composition, as is seen when comparing intracellular and extracellular fluids (Withers, 1992). Physiologically, this means that total osmolarity can be regulated at a particular value concomitantly with some degree of freedom in the actual contents of the body fluids. As discussed below, different vertebrate taxa do indeed exhibit different ways of adjusting total plasma osmolarity.

Although many invertebrates conform to the osmolarity of their surroundings, vertebrates typically regulate the osmolarity of internal fluids at levels different from those found in their environment (Bentley, 1971; Kirschner, 1991; Withers, 1992; Martin and Nagy, this volume). All freshwater vertebrates maintain plasma osmolarities higher than that of freshwater. Most marine vertebrates, on the other hand, maintain values lower than that of saltwater, which is typically about 1000 mOsm. Terrestrial vertebrates ingest freshwater, yet maintain plasma osmolarity at values higher than that of freshwater (freshwater fish do not usually drink their medium). Osmoregulation is an energetically costly process that is maintained presumably because of the biochemical and metabolic benefits of osmotic stability, much as thermoregulation and consequent relative thermal stability confers some benefits (Hochachka and Somero, 1984; Withers, 1992).

Even those marine fishes whose total osmolarity is isosmotic with seawater expend energy to regulate internal ion composition at levels that differ from their environment (Evans, 1979).

Although it is but one component of an animal's overall osmoregulatory scheme, plasma osmolarity offers some practical advantages for a broad-scale comparative study such as we discuss. First, it is a relatively simple trait to measure, so the data set available on vertebrate plasma osmolarity is sufficiently large to be interesting, but not so enormous as to be overwhelming. Second, plasma osmolarity is not dependent on body size or temperature, unlike many other physiological properties, such as metabolic rate. This independence simplifies comparisons among species. We stress that many other physiological properties deserve analysis in the context of amniote origins, including locomotor abilities, preferred body temperature, metabolic rate, blood oxygen carrying capacity, diet, and reproductive or life history traits.

Osmoregulatory requirements or abilities appear to play an important role in restricting most organisms to a particular habitat. The great majority of fish, for instance, are either freshwater or marine, and will quickly die if placed in the "wrong" osmotic environment (Holmes and Donaldson, 1969; Evans, 1979). Some animals, however, show osmoregulatory plasticity. "Anadromous" fish (those that move between fresh and sea water), for example, the true eels *Anguilla* and the salmonids *Onchorhynchus* and *Salmo*, undergo a predictable developmental physiological shift that allows

---

Figure 1; Panel 1. Hypothesis of phylogenetic relationships for 172 extant taxa (mostly different species, but also some subspecies or populations) for which data on plasma osmolarities are available (listed in Appendix 1). The first panel shows tetrapods, and the second shows all remaining vertebrates in our data set. The first column indicates the numerical order from top to bottom. The second column is a two-character identifying code (as required in the computer programs we used). The third column is plasma osmolarity in milliosmoles. The fourth column indicates habitat. Branch lengths in this figure are arbitrary (of the type suggested by Pagel, 1992); for all analyses, however, we assumed that each branch length was equal to one. **AM** indicates the node of interest in this chapter, the ancestor of all amniotes. Note that this tree contains several "soft" polytomies (multifurcations), which reflect uncertainty about relationships; for all of our analyses, however, we have assumed that they are actually hard polytomies (see Discussion). ➔

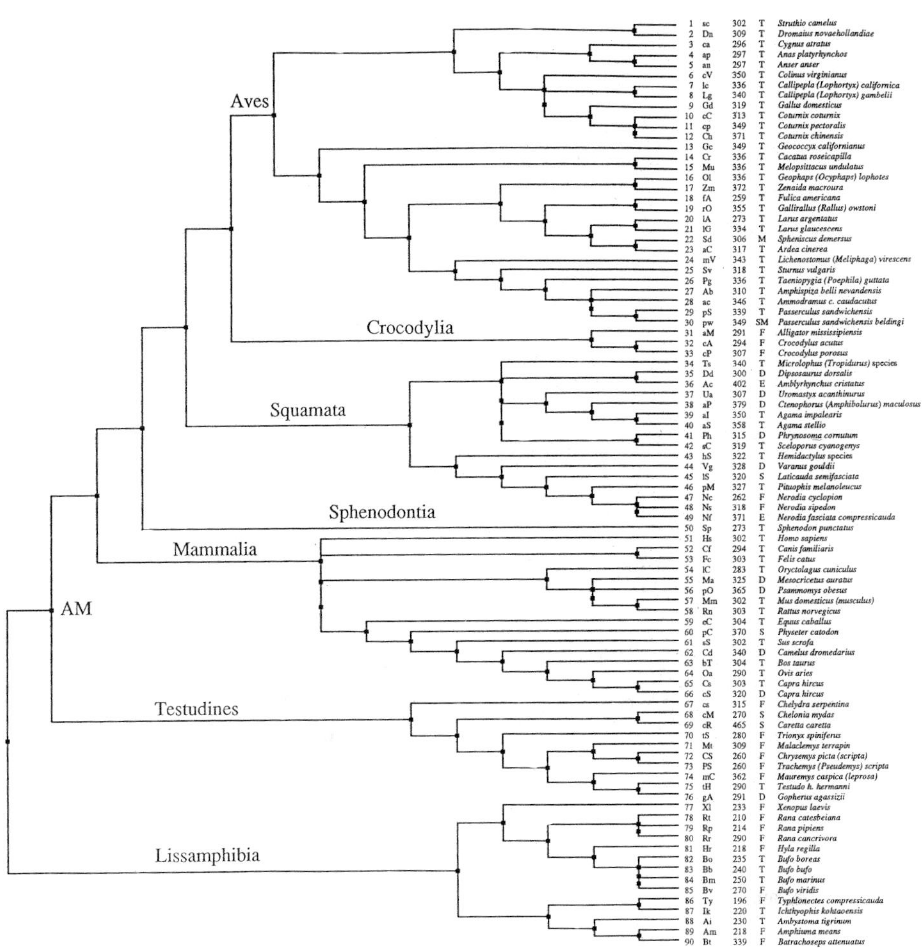
Aves
Crocodylia
Squamata
Sphenodontia
Mammalia
AM
Testudines
Lissamphibia
1 sc 302 T Struthio camelus
2 Dn 309 T Dromaius novaehollandiae
3 ca 296 T Cygnus atratus
4 ap 297 T Anas platyrhynchos
5 an 297 T Anser anser
6 cV 350 T Colinus virginianus
7 lc 336 T Callipepla (Lophortyx) californica
8 Lg 340 T Callipepla (Lophortyx) gambelii
9 Gd 319 T Gallus domesticus
10 cC 313 T Coturnix coturnix
11 cp 349 T Coturnix pectoralis
12 Ch 371 T Coturnix chinensis
13 Gc 349 T Geococcyx californianus
14 Cr 336 T Cacatua roseicapilla
15 Mu 336 T Melopsittacus undulatus
16 Ol 336 T Geophaps (Ocyphaps) lophotes
17 Zm 372 T Zenaida macroura
18 fA 259 T Fulica americana
19 rO 355 T Gallirallus (Rallus) owstoni
20 lA 273 T Larus argentatus
21 lG 334 T Larus glaucescens
22 Sd 306 M Spheniscus demersus
23 aC 317 T Ardea cinerea
24 mV 343 T Lichenostomus (Meliphaga) virescens
25 Sv 318 T Sturnus vulgaris
26 Pg 336 T Taeniopygia (Poephila) guttata
27 Ab 310 T Amphispiza belli nevandensis
28 ac 346 T Ammodramus c. caudacutus
29 pS 339 T Passerculus sandwichensis
30 pw 349 SM Passerculus sandwichensis beldingi
31 aM 291 F Alligator mississipiensis
32 cA 294 F Crocodylus acutus
33 cP 307 F Crocodylus porosus
34 Ts 340 T Microlophus (Tropidurus) species
35 Dd 300 D Dipsosaurus dorsalis
36 Ac 402 E Amblyrhynchus cristatus
37 Ua 307 D Uromastyx acanthinurus
38 aP 379 D Ctenophorus (Amphibolurus) maculosus
39 aI 350 T Agama impalearis
40 aS 358 T Agama stellio
41 Ph 315 D Phrynosoma cornutum
42 sC 319 T Sceloporus cyanogenys
43 hS 322 T Hemidactylus species
44 Vg 328 D Varanus gouldii
45 lS 320 S Laticauda semifasciata
46 pM 327 T Pituophis melanoleucus
47 Nc 262 F Nerodia cyclopion
48 Ns 318 F Nerodia sipedon
49 Nf 371 E Nerodia fasciata compressicauda
50 Sp 273 T Sphenodon punctatus
51 Hs 302 T Homo sapiens
52 Cf 294 T Canis familiaris
53 Fc 303 T Felis catus
54 lC 283 T Oryctolagus cuniculus
55 Ma 325 D Mesocricetus auratus
56 pO 365 D Psammomys obesus
57 Mm 302 T Mus domesticus (musculus)
58 Rn 303 T Rattus norvegicus
59 eC 304 T Equus caballus
60 pC 370 S Physeter catodon
61 sS 302 T Sus scrofa
62 Cd 340 D Camelus dromedarius
63 bT 304 T Bos taurus
64 Oa 290 T Ovis aries
65 Cs 303 T Capra hircus
66 cS 320 D Capra hircus
67 cs 315 F Chelydra serpentina
68 cM 270 S Chelonia mydas
69 cR 465 S Caretta caretta
70 tS 280 F Trionyx spiniferus
71 Mt 309 F Malaclemys terrapin
72 CS 260 F Chrysemys picta (scripta)
73 PS 260 F Trachemys (Pseudemys) scripta
74 mC 362 F Mauremys caspica (leprosa)
75 tH 290 T Testudo h. hermanni
76 gA 291 D Gopherus agassizii
77 Xl 233 F Xenopus laevis
78 Rt 210 F Rana catesbeiana
79 Rp 214 F Rana pipiens
80 Rr 290 F Rana cancrivora
81 Hr 218 F Hyla regilla
82 Bo 235 T Bufo boreas
83 Bb 240 T Bufo bufo
84 Bm 250 T Bufo marinus
85 Bv 270 F Bufo viridis
86 Ty 196 F Typhlonectes compressicauda
87 Ik 220 T Ichthyophis kohtaoensis
88 Ai 230 T Ambystoma tigrinum
89 Am 218 F Amphiuma means
90 Bt 339 F Batrachoseps attenuatus

this transition at a particular life cycle stage (Holmes and Donaldson, 1969; Evans, 1979). Usually, external morphology reflects this shift. Some other fish, for example the salt marsh *Fundulus heteroclitus*, are euryhaline and can move readily between marine and freshwater habitats (Bentley, 1971).

A third type of osmoregulatory plasticity is seen in another estuarine animal, the crab-eating frog *Rana cancrivora* (Gordon *et al.*, 1961; tip 80 in Fig. 1). This frog can live in freshwater or marine habitats as an adult and typically consumes marine invertebrates that are isosmotic with seawater. Plasma osmolarity varies in relation to environmental salinity; for example, at 250 mOsm. in the environment, the plasma osmolarity is 340 mOsm, but at 800 mOsm, the plasma osmolarity is 830 mOsm (Gordon *et al.*, 1961; we use a value of 290 mOsm for this species, which represents animals acclimated to freshwater). The frog does this by increasing or decreasing internal concentrations of one molecule, urea, while regulating the other ions and proteins at constant levels, independent of habitat. A similar mechanism is at work in freshwater elasmobranchs, such as *Carcharhinus leucas* (Thorson *et al.*, 1973; tips 150-151 in Fig. 1) and some skates and rays (Holmes and Donaldson, 1969; Withers, 1992). Note that this isosmolarity with seawater is achieved differently from the method of the hagfishes (tips 171-172 in Fig. 1) and marine invertebrates, which retain an ionic composition much more similar to seawater (Withers, 1992).

Finally, a type of osmotic plasticity is seen among some Lissamphibia (Shoemaker *et al.*, 1992) and the desert tortoise *Gopherus agassizii* (Minnich, 1982). During desiccation, these animals may "relax" osmoregulation, allowing plasma osmolarity to increase. The physiological consequences of this are discussed in detail in the chapter by Martin and Nagy (this volume). Presumably, the ability to tolerate wide variations in plasma osmolarity or body

---

Figure 1; Panel 2. Hypothesis of phylogenetic relationships for 172 extant taxa (mostly different species, but also some subspecies or populations) for which data on plasma osmolarities are available (listed in Appendix 1). This second panel shows all remaining vertebrates in our data set. ➔

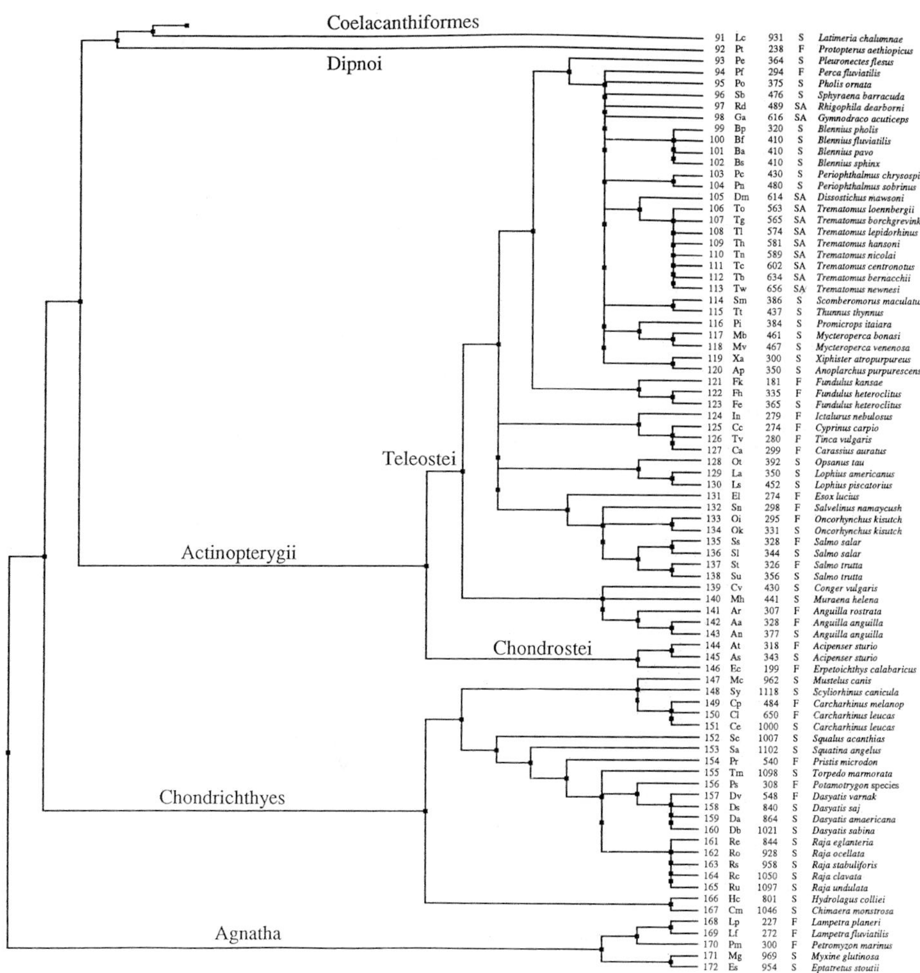
Coelacanthiformes
Dipnoi
Teleostei
Actinopterygii
Chondrostei
Chondrichthyes
Agnatha
91 Lc 931 S Latimeria chalumnae
92 Pt 238 F Protopterus aethiopicus
93 Pe 364 S Pleuronectes flesus
94 Pf 294 F Perca fluviatilis
95 Po 375 S Pholis ornata
96 Sb 476 S Sphyraena barracuda
97 Rd 489 SA Rhigophila dearborni
98 Ga 616 SA Gymnodraco acuticeps
99 Bp 320 S Blennius pholis
100 Bf 410 S Blennius fluviatilis
101 Ba 410 S Blennius pavo
102 Bs 410 S Blennius sphinx
103 Pc 430 S Periophthalmus chrysospilos
104 Pn 480 S Periophthalmus sobrinus
105 Dm 614 SA Dissostichus mawsoni
106 To 563 SA Trematomus loennbergii
107 Tg 565 SA Trematomus borchgrevinki
108 Tl 574 SA Trematomus lepidorhinus
109 Th 581 SA Trematomus hansoni
110 Tn 589 SA Trematomus nicolai
111 Tc 602 SA Trematomus centronotus
112 Tb 634 SA Trematomus bernacchii
113 Tw 656 SA Trematomus newnesi
114 Sm 386 S Scomberomorus maculatus
115 Tt 437 S Thunnus thynnus
116 Pi 384 S Promicrops itaiara
117 Mb 461 S Mycteroperca bonasi
118 Mv 467 S Mycteroperca venenosa
119 Xa 300 S Xiphister atropurpureus
120 Ap 350 S Anoplarchus purpurescens
121 Fk 181 F Fundulus kansae
122 Fh 335 F Fundulus heteroclitus
123 Fe 365 S Fundulus heteroclitus
124 In 279 F Ictalurus nebulosus
125 Cc 274 F Cyprinus carpio
126 Tv 280 F Tinca vulgaris
127 Ca 299 F Carassius auratus
128 Ot 392 S Opsanus tau
129 La 350 S Lophius americanus
130 Ls 452 S Lophius piscatorius
131 El 274 F Esox lucius
132 Sn 298 F Salvelinus namaycush
133 Oi 295 F Oncorhynchus kisutch
134 Ok 331 S Oncorhynchus kisutch
135 Ss 328 F Salmo salar
136 Sl 344 S Salmo salar
137 St 326 F Salmo trutta
138 Su 356 S Salmo trutta
139 Cv 430 S Conger vulgaris
140 Mh 441 S Muraena helena
141 Ar 307 F Anguilla rostrata
142 Aa 328 F Anguilla anguilla
143 An 377 S Anguilla anguilla
144 At 318 F Acipenser sturio
145 As 343 S Acipenser sturio
146 Ec 199 F Erpetoichthys calabaricus
147 Mc 962 S Mustelus canis
148 Sy 1118 S Scyliorhinus canicula
149 Cp 484 F Carcharhinus melanop
150 Cl 650 F Carcharhinus leucas
151 Ce 1000 S Carcharhinus leucas
152 Sc 1007 S Squalus acanthias
153 Sa 1102 S Squatina angelus
154 Pr 540 F Pristis microdon
155 Tm 1098 S Torpedo marmorata
156 Ps 308 F Potamotrygon species
157 Dv 548 F Dasyatis varnak
158 Ds 840 S Dasyatis saj
159 Da 864 S Dasyatis amaericana
160 Db 1021 S Dasyatis sabina
161 Re 844 S Raja eglanteria
162 Ro 928 S Raja ocellata
163 Rs 958 S Raja stabuliforis
164 Rc 1050 S Raja clavata
165 Ru 1097 S Raja undulata
166 Hc 801 S Hydrolagus colliei
167 Cm 1046 S Chimaera monstrosa
168 Lp 227 F Lampetra planeri
169 Lf 272 F Lampetra fluviatilis
170 Pm 300 F Petromyzon marinus
171 Mg 969 S Myxine glutinosa
172 Es 954 S Eptatretus stoutii

hydration has a genetic basis, which is expressed under certain environmental influences.

In summary, plasma osmolarity is a biologically meaningful physiological trait that is regulated by all vertebrates, but in many different ways. As with all aspects of the phenotype, plasma osmolarity of a particular individual (and hence the average value for a set of individuals from a particular species) will depend on both genetic and environmental factors, as well as genotype X environment interaction; it is also affected by short-term acclimation.

Attempts to analyze the evolutionary history of phenotypic traits on a phylogenetic tree presume that the traits are inherited phylogenetically; that is, passed on from ancestors to descendants (Brooks and McLennan, 1991; Harvey and Pagel, 1991; Maddison and Maddison, 1992; Eggleton and Vane-Wright, 1994; Martins, 1996b). Usually we think of genetically-based inheritance (Garland *et al.*, 1992, pp. 29-30), so we implicitly assume that the observed differences among species (e.g., the differences in plasma osmolarities shown in Fig. 1) reflect genetic differences, as opposed to environmental effects. That assumption can only be verified by raising all species under the same environmental conditions prior to measurement of the phenotype. However, a true "common garden" experiment (*sensu* Garland and Adolph, 1991) would be impossible (e.g., raising Antarctic ice fish in a desert environment) for more than a very few, exceptional animals in different clades. Moreover, even short-term acclimation can have strong effects on plasma osmolarities. Thus, in the absence of data from actual common garden experiments, it is unclear whether one should prefer to use plasma osmolarity values taken directly from animals in natural environments, or whether individuals of all species should be acclimated to some set of common conditions before taking measurements. But even acclimation to common conditions would be impossible for the range of species in Figure 1, thus the question is largely academic.

With data that are not derived from animals subject to common rearing and/or acclimation conditions, attempts to reconstruct ancestral plasma osmolarity values may be somewhat misleading. Also, relatively few vertebrates can move between osmotically different habitats (see above), therefore osmoregulatory ability must interact

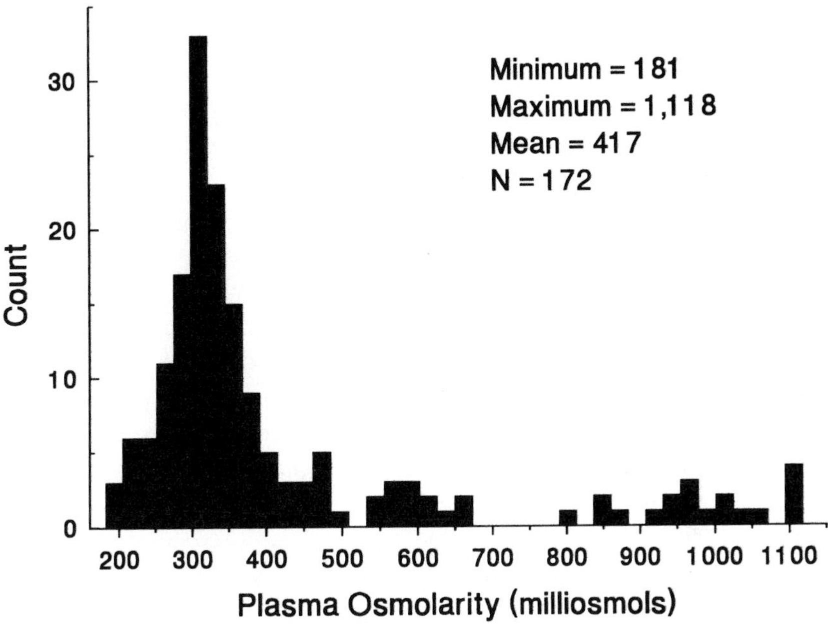

Figure 2. Frequency histogram of plasma osmolarities of 172 vertebrate taxa (data from Fig. 1 and Appendix 1).

with habitat occupancy, which is to some extent a behavioral trait. Also other physiological and biochemical properties can be expected to have coadapted (*sensu* Huey and Bennett, 1987) with plasma osmolarity. That is, natural selection should have favored the correlated evolution of "appropriate" combinations of (1) osmoregulatory "strategies" (e.g., regulating versus conforming), (2) the various physiological properties that are sensitive to plasma osmolarity, and (3) habitat occupancy (including the behavioral aspects of habitat "preference" or "selection"). Indeed, within various clades, plasma osmolarity seems to correlate with habitat type (for some formal statistical tests, see section below on "Choosing Appropriate Models ..."). This is understandable, particularly for aquatic organisms. Body fluids hypertonic to the environment can lead to osmotic water gain, with the need to eliminate large amounts of water while retaining internal solute concentrations; animals with relatively hypotonic body fluids face the opposite problem of water loss and solute gain (Withers, 1992). Thus, one might intuitively expect higher plasma osmolarities in marine organisms than in freshwater organisms, because such a pattern would reduce the energetic cost of osmoregulation.

To reiterate, plasma osmolarity is perhaps best considered as one part of a complex, multivariate phenotype, involving aspects of biochemistry, physiology, morphology, and behavior. If so, then the analyses presented herein may be seriously inadequate; nonetheless, we view them as a useful first step in understanding the evolutionary physiology of osmoregulation. [Another issue that needs further study is the definition and identification of homology with respect to quantitative characters, such as plasma osmolarity (see Burggren and Bemis, 1990; Eggleton and Vane-Wright, 1994; Zelditch *et al.*, 1995; Martins, 1996b; Schultz *et al.*, 1996)].

With the foregoing caveats in mind, we searched the literature for data on plasma osmolarity. When more than one measurement was available for a particular species, usually the most recent measurement was used. Review papers and book chapters were consulted for consensus if multiple values were available. Although we have reported all data as osmolarity, some, particularly the older numbers, were actually measured as freezing point depression, from which we calculated osmolarity. For experiments that altered the internal fluid composition or osmolarity, the control values were used. Euryhaline animals may have more than one value, depending on the habitat; the "normal" or more common habitat for each of these is presented, or, where two habitats are reported, values reflect plasma osmolarities of animals taken directly from, or acclimated to, each habitat.

A caution applies to the higher osmolarity values (greater than 900 mOsm) reported for some of the marine fishes. The standard osmolarity of seawater is 1000 mOsm., but for fishes taken from the field, osmolarity may have been different at that location. Seawater osmolarity was reported as low as 935 and as high as 1108 mOsm in some of these studies. With the exception of hagfish, those marine vertebrates with osmolarity very close to seawater have plasma ion concentrations of most substances that are very similar to those of freshwater animals, along with a larger amount of some osmolyte such as urea or trimethylamine oxide (Withers, 1992); together, these ions and osmolytes allow a match to the environmental osmolarity. Therefore, small differences between the highest values reported in Figure 1 may indicate a conforming response to the immediate marine conditions, rather than an inherent difference in osmoregulation. We have chosen to present the data in the osmolarity units for consistency, although a ratio of the measured plasma osmolarity to the measured seawater from which the fish was taken would in all of these cases be very close to unity.

The "habitat" listed in the fourth column of Figure 1 represents information in the original source, general references, or our knowledge of the animals. In some cases habitat is ambiguous, for example for birds that often occur in deserts but migrate and/or also occur in more mesic situations. Similar caveats apply to some of the lissamphibians, which can be difficult to categorize as terrestrial or freshwater. We should also note that the same habitat, such as freshwater, probably presents very different biological challenges for animals such as snakes (e.g., *Nerodia*) with impermeable integuments as compared with toads (e.g., *Bufo*), with much more permeable integuments or fishes of various types. We would encourage future workers to better define "habitat," including the possibility of describing aspects of habitat or ecology on a quantitative or semi-quantitative scale (see also Garland *et al.*, 1993, pp. 283-284).

PHYLOGENETIC RELATIONSHIPS OF VERTEBRATES

The method for reconstructing the plasma osmolarity of the ancestral amniote that we discuss, squared-change parsimony, requires an estimate of the phylogenetic relationships of all taxa included in the analysis. This same requirement applies to all of the "comparative methods" in current usage (Brooks and McLennan, 1991; Harvey and Pagel, 1991; Maddison and Maddison, 1992; Garland *et al.*, 1993; Eggleton and Vane-Wright, 1994; Garland and Adolph, 1994; Martins, 1996b).

Agreement as to the relationships of all 172 taxa shown in Figure 1 does not exist. We have, therefore, synthesized the available literature in an informal way to provide a composite estimate of the phylogeny (as in the Appendix of Garland *et al.*, 1993). Readers should note that we use the term "composite" specifically because it does not have a formal meaning in systematic biology, unlike such words as "consensus." Composite is used, in its simple dictionary sense, to mean "made up of distinct parts or elements; compounded; not simple in structure" (Oxford English Dictionary, 1971). The term has been used in this way several times in recent papers (e.g., Garland and Janis, 1993; Garland *et al.*, 1993; Purvis, 1995; Pyron, 1996). Our main point in this chapter is to illustrate existing methodology; thus, we invite future researchers to reanalyze the data presented herein as new and better phylogenetic information, data on plasma osmolarities, and analytical methodologies become available. We took phylogenetic relationships of the main vertebrate clades from Ridley (1993), Pough *et al.* (1996), and references therein. Most of these relationships are widely accepted, although the order of the splits within the groups lungfish-coelacanths-tetrapods, and turtles-mammals-squamates, are controversial (e.g., on the position of turtles, see Lee, 1993, 1995 versus Laurin and Reisz, 1995). Under Discussion, we give an example of the effects on our results of changing some of these topological relationships.

Within fishes (Agnatha, Chondrichthyes, Actinopterygii), we followed Nelson's (1994) classification, with additions from de Carvalho (1996) for elasmobranchs. For lissamphibians, we followed Hay *et al.* (1995), Hedges and Maxson (1993), and Larson and Dimmick (1993). For Testudines, we followed Gaffney and Meylan (1988). For squamates, we followed Garland (1994), with additions from Frost and Etheridge (1989) and Greer (1989) for "lizards" and Heise *et al.* (1995) for snakes. For mammals, we followed Garland *et al.* (1993), with additional information from Miyamoto and Goodman (1986), Novacek (1992), and Milinkovitch (1995). For Aves, we followed Sibley and Ahlquist (1990), with additions from Leeton *et al.* (1994) and Bleiweiss *et al.* (1995). For all the tree–in the absence of contrary information–we assumed families and genera (e.g., *Blennius* and *Trematomus*) to be monophyletic, which allowed us to "resolve" some relationships within higher-order groups (see Purvis, 1995). In some cases, we could have resolved particular polytomies with a phenetic criterion, under the assumption that species that are more similarly phenetically (e.g., with respect to plasma osmolarity) are also more closely related (Harvey and Pagel, 1991, p. 157; Pagel, 1992, p. 441). We have not done so because of concern that it might bias our final conclusions (see de Queiroz, 1996).

We used the above references and Barbadillo (1989), Corbet and Hill (1991), Conant and Collins (1991), Frost (1992), Christidis and Boles (1993), Monroe and Sibley (1993), Wilson and Reeder (1993), and Zug (1993) to check species names for accuracy and consistency with the latest taxonomies.

CHOOSING APPROPRIATE MODELS FROM EXTANT FORMS

> "Parsimony and other assumptions are helpful guiding principles; they should not be accepted uncritically, however, but should be combined with our knowledge of the general biology of the animals we study." (Ryan, 1996, p. 7)

In general, one should use all available evidence when tackling a given problem. This notion motivates the "total evidence" approach to phylogenetic inference (e.g., Eernissee and Kluge, 1993). Thus, one could use all available data on plasma osmolarities of vertebrates (as shown in Fig. 1) to infer the value of the ancestral amniote. But doing so would not actually use "all available evidence." For example, Table 1 indicates that plasma osmolarity seems to vary in relation to phylogeny (and habitat–see below). In other words, variation in plasma osmolarity does not seem to be distributed randomly across the tips of the phylogenetic tree. Instead, species within certain clades (lineages) seem to resemble each other more closely than species from some other part of the tree. Members of the Nototheniidae [Antarctic ice fishes (tips 105-113 in Fig. 1)], for instance, have relatively high plasma osmolarities (mean = 597.6, min = 563, max = 656) as compared with most other Actinopterygii.

Resemblance of closely related species is expected (Harvey and Pagel, 1991), even under a simple Brownian motion model of

Table 1. Mean (+/- SE), Range, and Sample Size, of Plasma Osmolarity values (mOsm) of the Vertebrate Clades, Classified by Habitat (Data from Appendix 1; total $N = 172$).

	Saltwater	Saltwater, Antarctic	Freshwater	Estuarine	Terrestrial	Desert
Aves					326 ± 26.6 259-372 30	
Crocodylia			297 ± 8.5 291-307 3			
Squamata	320 1		290 ± 39.6 262-318 2	386 ± 21.9 371-402 2	336 ± 15.9 319-358 6	326 ± 31.5 300-379 5
Sphenodontia (tuatara)					273 1	
Mammalia	370 1				299 ± 7.0 283-304 11	338 ± 20.2 320-365 4
Testudines	368 ± 137.9 270-465 2		298 ± 39.3 260-362 6		290 1	291 1
Lissamphibia			243 ± 47.0 196-339 9		235 ± 11.2 220-250 5	
Coelacanth-iformes	931 1					
Dipnoi (lungfish)			238 1			
Actinopterygii	394 ± 50.4 300-480 27	589 ± 44.3 489-656 11	288 ± 43.4 181-335 16			
Chondrichthyes	983 ± 103.4 801-1118 16		506 ± 125.8 308-650 5			
Cephalaspido-morphi			266 ± 36.8 227-300 3			
Myxini	961 ± 10.6 954-969 2					

Here, the penguin, *Spheniscus demersus*, and the sparrow *Passerculus sandwichensis beldingi*, have been classified as terrestrial. The two animals classified as estuarine are the Galapagos marine iguana (*Amblyrhynchus cristatus*) and the water snake (*Nerodia fasciata compressicauda*). Both of these reptiles may spend time in seawater and fresh water, so may have osmotic loads unlike those of typical terrestrial or freshwater animals. For the squared-change parsimony analyses described in the text, we emphasize the 63 taxa classified here as desert ($N = 10$) or terrestrial ($N = 53$ excluding the penguin).

character evolution (see Appendix 3 section on "Computer Simulations to Test ..."). Thus, statistical tests comparing average values of different clades must incorporate phylogenetic information (Garland *et al.*, 1993). To illustrate that plasma osmolarity is not distributed randomly across the phylogeny for our 172 taxa, we performed a phylogenetic analysis of variance by use of computer-simulated null distributions, as described in Garland *et al.* (1993; see also Martin and Clobert, 1996; Reynolds and Lee, 1996; Harris and Steudel, 1997). For simplicity, we just compared the four clades with a reasonably large number of "Terrestrial" members, as indicated in Table 1 (Aves, N = 30; Squamata, N = 6; Mammalia, N = 11; Lissamphibia, N = 5). Inspection of the summary statistics shown in Table 1 suggests that these four clades differ in average plasma osmolarity. Indeed, a conventional one-way ANOVA yields an F statistic of 28.90, which is highly significant with 3 and 48 degrees of freedom ($P < 0.0001$). This P value, however, cannot be trusted, because it is based on comparison with a conventional F distribution, which effectively assumes that species' plasma osmolarities constitute independent and identically distributed values. In other words, the conventional F distribution is constructed under the assumption of a "star" phylogeny (Garland *et al.*, 1993: see Discussion below). As can be seen from Figure 1, the terrestrial birds (tips 1-30), squamates (subset of the tips 34 - 49), mammals (subset of tips 51-66), and amphibians (subset of tips 77-90) are instead related in a strongly hierarchical fashion.

To account for phylogenetic relationships, we can use Monte Carlo computer simulations to construct a phylogenetically correct distribution of F statistics (Garland *et al.*, 1993). We did so by pruning the tree shown in Figure 1 to include only the 52 taxa of interest and set all branch lengths equal to unity (except for those collapsed in polytomies). We then used the PDSIMUL program to simulate the evolution of plasma osmolarity. We used starting and ending values of 365 mOsm, expected variances at the tips equal to the variance of the real data (1253 mOsm), limits of 181 and 1118 mOsm, and a Brownian motion model. For each of 1000 simulated data sets, we performed a one-way ANOVA with the program PDSINGLE. We then ordered the F values from lowest to highest and determined how they compared with the corresponding F value for the real data set of 52 species. Only 33 of the 1000 F values for simulated data were greater than the real F of 28.90, thus we conclude that the four clades actually do differ significantly (i.e., $P < 0.05$) in average plasma osmolarity.

In addition to variation in relation to clade, Table 1 suggests that plasma osmolarity may vary in relation to ecology or habitat. Within the Actinopterygii, for example, the 27 saltwater forms (mean = 394) tend to have higher values than do the 16 freshwater forms (mean = 288). Similarly, within the Chondrichthyes, saltwater forms (mean = 983, $N = 16$) also tend to have higher values than do freshwater forms (mean = 506, $N = 5$). Within the Mammalia, desert forms (mean = 338, $N = 4$) tend to have higher values than do non-desert ("Terrestrial") forms (mean = 299, $N = 11$).

To illustrate a phylogenetically based statistical test for habitat differences, we can again use computer simulation. Consider the 54 Actinopterygii. One-way ANOVA of the plasma osmolarities of saltwater, saltwater Antarctic, and freshwater forms yields $F = 133.05$, which is nominally highly significant with 2 and 51 df ($P < 0.0001$). We performed simulations using the same parameters as listed above. For the 1000 simulated data sets, F statistics ranged from 0.004 to 34.27. Thus, not one was greater that the F for the real data, and we conclude that habitat differences are statistically significant at $P < 0.001$ (this P value could be even smaller, but we only analyzed 1,000 simulated data sets). We could also perform a two-way analysis of variance, with factors of clade and habitat, but we have not done this for simplicity.

The foregoing differences among cladistically or ecologically defined groups make biological sense. Saltwater forms, for example, must maintain relatively high osmolarities because their aqueous environment contains much higher concentrations of ions than does freshwater. Otherwise, they would tend to lose water to the environment (unless they had impermeable integuments) and hence dehydrate (Withers, 1992). The high plasma osmolarities of the Antarctic ice fishes (tips 105-113 in Fig. 1) prevent freezing in sea water that falls below zero degrees Celsius and contains ice (Dobbs and DeVries, 1975; Withers, 1992; Eastman, 1993). This and some other associations between plasma osmolarity and ecology almost certainly represent evolutionary adaptations, that is, genetically based phenotypes that are the result of past and/or current natural selection.

If the characteristic of interest tends to vary in relation to either phylogeny or ecology, then reconstructing the value of an ancestral form should be facilitated by use of only those forms that are "similar" to the ancestral form. With respect to phylogenetic similarity, this would mean an emphasis on closely related taxa. With respect to ecology, we should focus on forms that live in habitats similar to that of the ancestral form we are trying to reconstruct.

The foregoing two principles seem intuitively obvious. In practice, however, applying them may be complicated. First, consider phylogenetic relatedness. To reconstruct the value of a particular node on a phylogenetic tree, we obviously should begin by considering all available data for descendant taxa. Thus, to reconstruct the plasma osmolarity of the hypothetical organism at the node labeled **AM** in Figure 1 (i.e., the "mother" of all amniotes), we should begin by considering all available data for Testudines, Mammalia, Sphenodontida, Squamata, Crocodylia, and Aves (these being the only amniotes for which data on plasma osmolarity are available). But we should also consider "outgroups" to the amniote clade (see also Huey, 1987). The closest outgroup clade in our data set is the Lissamphibia (see Laurin and Reisz, this volume), for which we have data on 14 species (including some Anura, Gymnophiona, and Urodela). But we can also consider additional outgroups; the next closest cladistically would be the coelacanth and then the lungfish. Should we stop here, or should we include further outgroups? If the latter is the case then we can go to the rest of the bony fishes, then the Chondrichthyes, and finally the jawless fishes (lamprey and hagfish). Should we include all of these outgroups or only a subset?

Now consider habitat type (Table 1). If we knew the habitat of the ancestral form we are trying to reconstruct, then an argument could be made to include in our analyses only those forms that occur in similar habitats. As discussed elsewhere (Berman *et al.*, this volume; Martin and Nagy, this volume), it is likely that the first amniote was fairly terrestrial but had access to freshwater. The adults, we presume, were terrestrial before the eggs were. Fossil evidence for these transitional forms seems to come mostly from terrestrial rather than aquatic sediments (Carroll, 1988; Berman *et al.*, 1992; Lombard and Sumida, 1992; Berman *et al.*, this volume; Hotton *et al.*, this volume). Although the habitat of the ancestral amniote is controversial (see Sumida, this volume), no one has suggested that it lived in Antarctic seawater or was freeze-tolerant. Thus, we might reasonably exclude the nine Nototheniidae (tips 105-113 in Fig. 1) before trying to reconstruct the ancestral amniote's plasma osmolarity.

If a fair amount of data are available, as in the present example of plasma osmolarities (N = 172), then the practitioner probably will be faced with tough decisions about which taxa should be included in attempts to reconstruct an ancestral phenotype. Some simple guidelines are possible (see also Huey, 1987; Burggren and Bemis, 1990). First, one should emphasize close relatives (including some outgroups) and forms thought to be ecologically similar to the extinct form of interest (e.g., based on paleontological information). Second, one should perform several analyses, using the entire data set as well as different subsets, and examine results for consistency: check for taxa that have a large influence on the reconstructed value. An important area for future research on how best to reconstruct ancestral values will be the development of more complicated parsimony-like procedures that allow incorporation of additional information, such as apparent evolutionary shifts in relation to habitat or phylogeny. In their absence, we will, for illustrative purposes, analyze our data set as several different subsets, but concentrate on a subset of 63 terrestrial and desert taxa (see Table 1 and below).

Finally, we emphasize that the original investigators did not collect the present 172 data points for the purpose of reconstructing the plasma osmolarity of the ancestral amniote. Sampling of the vertebrate lineages is obviously very incomplete and uneven, and in no sense should the available data set be considered optimal for our purposes. The bony fishes alone comprise perhaps 20,000 extant species, of which our sample represents but 0.3%. We hope, therefore, that our analyses will prompt other workers to measure additional taxa that would help to refine our estimates of ancestral values and to more thoroughly study covariation between plasma osmolarity and phylogeny or ecology.

Table 2. Effects of Deleting Certain Terminal Taxa on Squared-change Parsimony Reconstructions of Plasma Osmolarity (milliosmols) at Interior Nodes on the Phylogenetic Tree Shown in Figure 1.

Taxa Included	*N*	Simple nonphylogenetic mean	Root node	Ancestor of all amniotes
All vertebrates	172	417	644	365
No "ordinary" saltwater forms	122	358	429	293
Only terrestrial and desert	63	314	263	280#
All amniotes (tips 1-76 in Fig. 1)	76	321	310	310*
Only terrestrial amniotes	58	321	294	294*

For comparison, simple means (i.e., ordinary, nonphylogenetic averages) are also shown. The topology used for analyses was as in Figure 1, but all branch lengths were set equal to unity.

As discussed in the text, we emphasize this subset of 63 taxa as the preferred models for estimation of the plasma osmolarity of the ancestral amniote (node AM in Fig. 1).

* These values are the same as those under root node column because no "outgroups" are present in the reduced phylogeny; thus, this node *is* the root node.

RECONSTRUCTING PLASMA OSMOLARITY AT THE ORIGIN OF AMNIOTES

We now turn to analysis of the data at hand. We use squared-change parsimony to reconstruct the plasma osmolarity of the ancestor of all amniotes, represented by node **AM** in Figure 1. (For comparative purposes, in the next section we also present results for the ancestor of all vertebrates, the basal node in Fig. 1.) For all analyses, we assume that each branch segment is of equal length. These branch lengths are arbitrary, but a justification of their use is presented under Discussion.

Considering all 172 terminal taxa, the squared-change parsimony estimate for the ancestor of all amniotes (node **AM** in Fig. 1) is 365 mOsm. This value is lower than the simple mean (i.e., the ordinary, nonphylogenetic mean, computed in the usual way) of 417

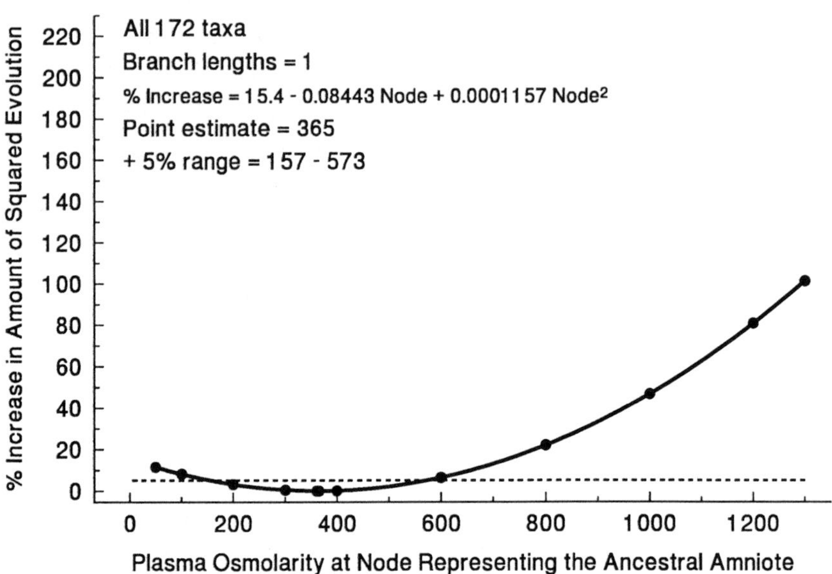

Figure 3. Consequences of fixing nodal value for ancestral amniote at values other than that indicated by unconstrained squared-change parsimony; all 172 taxa. Horizontal dashed line indicates 5% increase in the sum of squared evolutionary changes, compared with the amount required for the unconstrained squared-change parsimony reconstructions.

(Table 2). Figure 3 shows the effect of changing the nodal value for the ancestral amniote to something other than the unconstrained squared-change parsimony reconstruction of 365 mOsm. Any other nodal value requires there to have been more squared change summed over the entire phylogeny. As a heuristic gauge of the uncertainty in this reconstruction (see Appendix 3), we point out the range of nodal values over which this increase is 5%: 157-573 (note that 5% is an arbitrary value).

Effects of deleting various taxa are shown in Table 2. For example, excluding "ordinary" saltwater forms but retaining the Antarctic saltwater forms (leaving N = 122), the point estimate for the

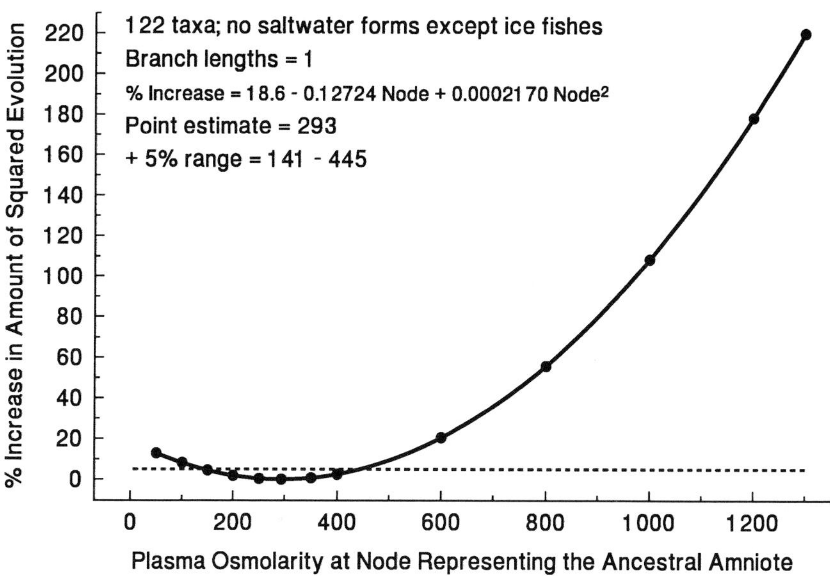

Figure 4. Consequences of fixing nodal value for ancestral amniote at values other than that indicated by unconstrained squared-change parsimony; subset of 122 taxa.

ancestral amniote decreases to 293. The +5% range also decreases to 141-445 (Fig. 4). Considering only the 76 amniotes in our data set, the squared-change parsimony estimate for ancestral amniote is reduced to 310 mOsm, which is only slightly lower than the nonphylogenetic mean of 321 (Table 2).

As discussed in the previous section and in the chapter by Martin and Nagy (this volume), the best "models" for the ancestral amniote may be terrestrial forms. In the remainder of our analyses, therefore, we concentrate on a subset of 63 taxa including only the "Terrestrial" and "Desert" forms of Table 1 (but also excluding the penguin, *Spheniscus demersus*). For this favored subset of 63 terrestrial plus desert taxa, Figure 5 indicates a point estimate of 280 with a +5% range of only 254-306. Thus, as we have restricted the

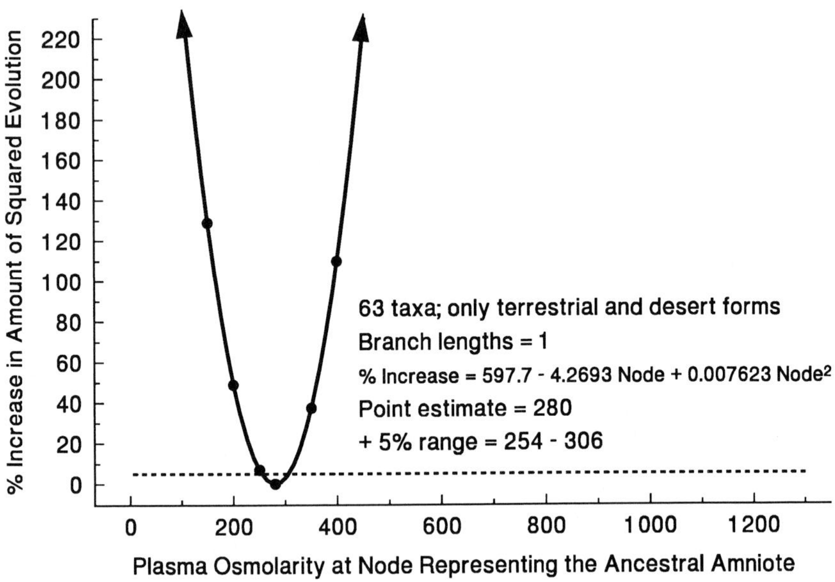

Figure 5. Consequences of fixing nodal value for ancestral amniote at values other than that indicated by unconstrained squared-change parsimony; preferred subset of 63 terrestrial taxa (see text).

taxa under consideration, eliminating those that (as argued above) are not thought to be good models for the ancestral amniote, the point estimate has changed somewhat (decrease of 23.3%), but the main effect has been to decrease dramatically the +5% range (by 87.5%).

Still considering only the 63 taxa, a delete-one jackknife yields an approximate standard error of 6 for the value reconstructed at the ancestral amniote. If the appropriate degrees of freedom were about 60, then we would use a *t*-statistic of 2.0 for $\alpha = 0.05$. This would yield a nominal 95% confidence interval of 268-292. However, because the taxa that we are jackknifing are related in a strongly hierarchical fashion, they cannot be considered as independent data points. Thus, we can only offer this range as a heuristic (see Appendix

3). Still, it is interesting to note that both of these ranges exclude the simple nonphylogenetic mean of 314 mOsm for the 63 terrestrial plus desert taxa.

RECONSTRUCTING PLASMA OSMOLARITY AT THE ORIGIN OF VERTEBRATES

Although estimation of the plasma osmolarity of the ancestor of all vertebrates (i.e., the root node of Fig. 1) is not the focus of this chapter, we will comment briefly on the value reconstructed by squared-change parsimony. As noted in Appendix 3, for the root node and only the root node, this value is identical to the value computed by Felsenstein's (1985) independent contrasts procedure. This is convenient, because a formal analytical procedure for putting a 95% confidence interval on the root node constructed by independent contrasts has recently been developed (T. Garland, Jr., and A. R. Ives, unpublished). Considering all 172 terminal taxa, the squared-change parsimony estimate for the root node of the entire phylogeny is 644 mOsm with a 95% confidence interval of 467 - 821 mOsm [under the assumption, as elsewhere in our computations, that all of the polytomies in Figure 1 are actually hard (see Discussion)]. These values can be compared with the simple (nonphylogenetic) mean of 417 mOsm and a (nonphylogenetic) confidence interval of 383-450 mOsm. Note the lack of overlap of the 95% confidence intervals!

Both squared-change parsimony and independent contrasts (Felsenstein, 1985; see Appendix 3) effectively assume Brownian motion character evolution. This assumption also holds when either method is applied to a star phylogeny, in which case the estimate at the root node would be the same as the nonphylogenetic mean. One component of the Brownian motion assumption is that no directional trends in character evolution have occurred (see Appendix 3). These assumptions, and the others inherent to comparative analyses (see Discussion and Appendix 3), should be kept in mind as we consider reconstructed values and gauges of their uncertainty, including formal confidence intervals.

Inspection of Table 1 or Figure 1 indicates that the only living vertebrates with values of approximately 644 milliosmols are some freshwater Chondrichthyes and the Antarctic ice fishes. To our knowledge, no one has proposed that the ancestor of all vertebrates lived in subzero seawater. Thus, we would reject the inference that the ancestral vertebrate was like extant ice fishes. Eliminating that possibility leaves us with the possible interpretation that the ancestral vertebrate osmoregulated like extant freshwater Chondrichthyes, a clade that is predominately marine. Thorough consideration of this hypothesis is beyond our scope, but is does constitute an example of how the analytical approach we discuss can provide new perspectives on old problems. Existing fossil and morphological/physiological evidence is contradictory as to whether the earliest vertebrates were freshwater or marine (Wake, 1979). If this issue could be settled, then we could instead focus on estimates derived from analyses using only freshwater or marine forms. For example, including only the 45 freshwater taxa, the simple mean is 303 mOsm and the squared-change parsimony reconstruction is 335 mOsm (with all branch lengths set equal to unity). One might also argue that the freshwater Chondrichthyes are specialized (derived) and hence "red herrings" (Gans, 1970). If the five freshwater Chondrichthyes are pruned from the tree (so N = 40), the values are reduced to 277 and 272 mOsm, respectively. All of these are much lower than the corresponding values for the entire 172 taxa (first line of Table 2), and well below the lower bounds of the 95% confidence intervals given above.

In closing, note that we have analyzed the raw plasma osmolarity values. For all 172 taxa, these values show a highly skewed distribution (Fig. 2), although the skew is much less for the subset of 63 taxa that we emphasized in the previous section. When skewness is present, performance of squared-change parsimony reconstructions might be improved by transformation of the data prior to analysis (followed by back-transformation to the original scale of measurement). We see this as an interesting area for future study (see also Díaz-Uriarte and Garland, 1996).

DISCUSSION

Phylogenetic versus Nonphylogenetic Estimates of Ancestral Values

For the subset of 63 terrestrial and desert taxa that we have emphasized as preferred "models" for the ancestral amniote, the squared-change parsimony algorithm reconstructs a value of 280 for the plasma osmolarity of the node corresponding to the ancestor of all amniotes. Two heuristic indicators of uncertainty in this reconstructed value are 254-306 (the range that causes a +5% increase in the sum of squared changes: see Fig. 5) and 268-292 (the nominal 95% confidence interval obtained by jackknifing). The simple (nonphylogenetic) mean for the 63 taxa is 314, which falls outside of either of the foregoing uncertainty ranges. Our limited simulation results (see above) and numerous previous studies of phylogenetically based comparative methods demonstrate that phylogenetic approaches are almost always superior to their nonphylogenetic alternatives (Grafen, 1989; Brooks and McLennan, 1991; Harvey and Pagel, 1991; Martins and Garland, 1991; Maddison and Maddison, 1992; Garland *et al.*, 1993; Eggleton and Vane-Wright, 1994; Garland and Adolph, 1994; Purvis *et al.*, 1994; Díaz-Uriarte and Garland, 1996; Martins, 1996a,b). Thus, although the foregoing ranges are only heuristic, they do suggest that we have gained some degree of accuracy and precision by employing the more complicated phylogenetic estimation procedures.

Let us make clear the relationship between squared-change parsimony reconstructions on a phylogenetic tree and the alternative of ignoring phylogenetic relationships. Imagine that we wished, for whatever perverse reason, to ignore the available phylogenetic information. This is generally equivalent to assuming that the phylogenetic relationships are adequately represented as a single hard polytomy, or "star" phylogeny, with all terminal taxa at the end of equal-length branches that radiate from a single basal node (e.g., Fig. 2 in Garland and Carter, 1994). If we were to apply squared-change parsimony to such a phylogeny, then only a single node (the root) would exist to be estimated, and its value would be estimated as the mean of all adjacent nodes, which would be all of the tips. Thus, the value would be "reconstructed" as the simple mean of the tip values.

Therefore, a conventional analysis, ignoring phylogeny, might estimate the ancestral value as a simple mean of the values for all extant taxa, and this would be exactly equivalent to a squared-change parsimony reconstruction on a star phylogeny. This point has also been made with respect to other phylogenetically based statistical methods (e.g., see Garland *et al.*, 1993; Purvis and Garland, 1993; Garland and Adolph, 1994, p. 822). Analysis of the plasma osmolarity data presented herein (see Table 2), the made-up example in Figure A1, and many other examples in the literature (e.g., Huey and Bennett, 1987; Chevalier, 1991) demonstrate that ancestral values reconstructed with reference to phylogenetic relationships can be quite different from the simple mean of the tip values. Differences are also apparent in the computer simulations described in Appendix 3; for instance, the correlation between the squared-change parsimony value for the root node and the simple mean was only 0.471 under simple Brownian motion and 0.456 under the model simulating an evolutionary trend with limits. The squared-change parsimony reconstructions make use of more information (the estimated phylogenetic relationships) and are arguably "better" in every sense of the word. Thus, they should be used whenever any phylogenetic information is available.

Assumptions of the Squared-Change Parsimony Method

All methods of analyzing comparative data make various assumptions that may be invalid. According to Martins and Hansen (1996, p. 40), "the sum of squared changes algorithm assumes that the distribution of evolutionary changes which gives the smallest sum of squared changes is the most likely to be true"; a second assumption is that "within-species variation does not exist or is negligible in comparison to the level of among-species variability." A third assumption is that the measured differences among the species' mean values (the plasma osmolarities being analyzed) are assumed to reflect genetic differences among the species. This would only be the case if all species had been bred and raised under the same environmental conditions–a common garden experiment (Garland and Adolph, 1991; Garland *et al.*, 1992). This is rarely done for broad-scale comparative studies, let alone for the 172 taxa included in Figure 1. In any case, this third assumption applies to *all* analyses of comparative data, phylogenetic or not.

A fourth assumption is that the phylogeny is known without error. This is unlikely ever to be the case for a real data set involving more than a few taxa. "Ignoring" phylogeny and using conventional statistical estimators does not avoid this problem, however, because it is simply the special case of assuming a star phylogeny with equal-length branches. A star phylogeny is absolutely known to be wrong for most sets of organisms that might be studied; hence, ignoring phylogeny is indefensible. Whatever phylogenetic information is available is likely to be closer to reflecting the truth than is a star, therefore it can and should be used to advantage (e.g., Grafen, 1989; Purvis and Garland, 1993; Losos, 1994; Purvis *et al.*, 1994; Martins, 1996c).

Uncertainty about Phylogenetic Topology

Some ways of dealing with topological uncertainty have been discussed for some other phylogenetic comparative methods (e.g., Garland *et al.*, 1993; Purvis and Garland, 1993; Losos, 1994; Martins, 1996c). Here, we have simply represented obvious topological uncertainty as unresolved nodes (see Fig. 1). For performing the iterative squared-change parsimony algorithm, we have treated these as "hard" polytomies (reflecting true, simultaneous multiway speciation events) rather than "soft" polytomies (reflecting lack of knowledge about what is actually a fully bifurcating topology). In fact, most of the polytomies in Figure 1 actually are soft. However, ways of performing squared-change parsimony computations to account for this uncertainty have not yet been developed. If and when they are developed, the result will almost certainly be to decrease the "confidence" we have in the reconstruction for any particular node.

Some of the phylogenetic relationships depicted in Figure 1 are controversial. An informal way of dealing with this uncertainty would be to redo the analysis with all, or a large number of, the possible alternatives that have been proposed (cf. Losos, 1994). For example, considering the major vertebrate lineages, one could repeat the analysis with alternative arrangements of lungfish-coelacanths-tetrapods and/or testudines-mammals-squamates. Doing so can become quite laborious if many alternatives are considered. For illustrative purposes, we have simply redone our analysis for one such case, switching the position of the lungfish (*Protopterus*) and the coelacanth (*Latimeria*). This particular switch would seem to have the potential for a major effect, because these lineages are only two or three nodes removed from the node we wish to estimate. For all 172 taxa, the topology of Figure 1, and all branch lengths set equal to one, the squared-change parsimony estimate for the ancestral amniote is 365 mOsm, whereas the rearrangement yields a value of 322 mOsm. This effect occurs because the lungfish, which inhabits freshwater, has a much lower plasma osmolarity (238 mOsm) than does the coelacanth (931 mOsm), which inhabits saltwater (Thomson, 1991). Nevertheless, the lower nodal estimate is still well within the +5% range shown in Figure 3 (157-573 mOsm). Moreover, neither of these organisms is retained in our preferred subset of 63 taxa (as used to produce Fig. 5).

Uncertainty About Phylogenetic Branch Lengths

Another assumption of the squared-change parsimony procedure is that the branch lengths are known without error in units of (or proportional to) expected variance of evolutionary change for the character of interest. For expediency, we have simply assumed that all of the branch segments are of equal length, which corresponds to a "speciational" model of character evolution (as in Martins and Garland, 1991; Díaz-Uriarte and Garland, 1996).

As a partial test of the adequacy of our branch lengths, we performed the diagnostic check described by Garland et al. (1992; see also Garland et al., 1991; Díaz-Uriarte and Garland, 1996), which was originally intended for Felsenstein's (1985) method of phylogenetically independent contrasts. Both independent contrasts and squared-change parsimony effectively assume Brownian motion character evolution, thus it seems appropriate to use this diagnostic for the latter as well. For all 172 taxa, the correlation of the absolute values of the standardized contrasts with their standard deviations was -0.056 for all branch lengths set equal. This weak and statistically nonsignificant correlation suggests that the branch lengths are adequate for analyses

Another set of arbitrary branch lengths, corresponding to those depicted in Figure 1, was suggested by Pagel (1992). These branch lengths are constructed by setting all internode branch segments equal to a length of one and then making all branches leading to terminal taxa line up evenly across the top of the tree. These branch lengths are often employed by systematists when presenting line drawings of phylogenetic hypotheses (e.g., as in Fig. 1). For these branch lengths, the correlation of the absolute values of the standardized contrasts with their standard deviations was -0.137. The 2-tailed critical value (α = 0.05) for a correlation coefficient with 170 df is 0.150, therefore these branch lengths seem marginally adequate.

The second set of arbitrary branch lengths that we used corresponds to those suggested by Grafen (1989, his Fig. 2). For these, the correlation of the diagnostic was -0.161 ($P < 0.05$), thus they seem marginally inappropriate for analyses of plasma osmolarity. As discussed elsewhere (Grafen, 1989; Garland et al., 1992; Díaz-Uriarte and Garland, 1996), these arbitrary branch lengths could also be transformed to improve the diagnostic statistic, but in the sake of brevity we have not done so. In any case, a number of examples have now emerged in which branch lengths set equal to unity perform as well as, or better than, others that have been tried (T. Garland, unpublished).

Although the foregoing diagnostic test suggests that branch lengths set equal to unity may be adequate for our purposes, it is still of interest to ask how much our results would change if we used other branch lengths. Table 3 shows the results of using the two other sets of arbitrary branch lengths suggested by Pagel (1992) and by Grafen (1989). For the data set containing only 63 terrestrial plus desert taxa, the value reconstructed for the ancestral amniote increased from 280 to 297 and 294, respectively. These changes are still well within the +5% range shown in Figure 5. For the analysis of all 172 taxa, the three values were 365, 376, and 371, which constitutes a very minor change in relation to the +5% range of 157-573 (Fig. 3). Thus, reconstruction of the plasma osmolarity of the ancestral amniote seems to be relatively insensitive to alterations of phylogenetic branch lengths.

Use of Fossil Information

Our analysis of a physiological trait, plasma osmolarity, has necessarily involved data only for extant forms. For traits such as physiology or behavior, fossil information will rarely be available (Ruben, 1995; Martins, 1996b). If one were studying a morphometric trait, however, such as limb length, then it might be possible to obtain data from both extant and extinct forms. These data could easily be combined in analyses such as we have presented. The only requirement would be that the extinct forms could be placed on the phylogenetic tree; that is, their relationship to extant taxa would need to be known. If branch lengths in units of estimated divergence times were being employed, then extant taxa would line up contemporaneously along the tips of the phylogeny, with extinct forms at the ends of branches terminating below the tips. If one had data for some fossil forms that were closely related to the node of interest (i.e., were topologically close) and occurred not too far away in chronological time (i.e., were at the ends of relatively short branches, assuming that the branch lengths being used for analysis were in units proportional to divergence times), then the extinct forms would have a relatively large effect on the node being reconstructed (because averaging involves adjacent nodes and weighting by the reciprocal of branch length). Thus, inclusion of data for fossil forms could help to narrow the "confidence interval" about the reconstructed nodal value.

Other Uses of Squared-Change Parsimony

In the example we have presented, and in most other published applications of squared-change parsimony procedures (e.g., Huey and Bennett, 1987), a primary goal is to estimate the value of a hypothetical ancestral organism, as represented by an internal node of the phylogenetic tree. Minimization of the sum of the squared changes over the entire phylogenetic tree is taken as the optimality criterion by which to choose a set of internal nodes, including the one(s) of primary interest.

Similar logic could be used to estimate the value of an extant or extinct species for which data were unavailable but which could be placed on the phylogenetic tree; that is, the value for a terminal node rather than an internal node. For example, one might wish to estimate the limb proportions, locomotor abilities, and/or hunting behavior of an extinct mammal that was know only from fragmentary fossils (cf. Garland and Janis, 1993; Janis and Wilhelm, 1993; Harris and Steudel, 1997). In conservation biology, one might wish to estimate the home range area (e.g., Garland *et al.*, 1993) of some (endangered) species that had yet to be studied. Various approaches would be possible,

Table 3. Effects of Different Branch Lengths (All of Which Are Arbitrary) on Squared-change Parsimony Reconstructions of Plasma Osmolarity (mOsm).

	All = 1		Pagel, 1992		Grafen, 1989	
Taxa Included (N)	Root node	Ancestor of all amniotes	Root node	Ancestor of all amniotes	Root node	Ancestor of all amniotes
All vertebrates (172)	644	365	511	376	544	371
All amniotes (76)	310	310*	314	314*	313	313*
No saltwater forms (122)	429	293	398	328	407	314
Only terrestrial (63)	263	280	292	297	289	294
Only terrestrial amniotes (58)	294	294*	309	309*	306	306

* These values are the same as those under Root Node column because no "outgroups" are present in the reduced phylogeny; thus, this node is the root node.

including squared-change parsimony. A range of values for the focal species could be analyzed, and then one would take as the "best" estimate that value which minimized the sum of squared changes over the entire phylogeny. Many of the indices for indexing uncertainty discussed elsewhere in this chapter could also be applied in this context.

Returning to interior nodes, once values have been reconstructed by a parsimony procedure, then simple subtraction can be used to make inferences can be made about the character changes that occurred along particular branch segments. Huey and Bennett (1987), for example, drew inferences about whether particular lineages of lizards had increased or decreased their thermal preferences and tolerances (and whether these changes were associated with biological shifts, such as diurnality versus nocturnality, or nondesert versus desert). In the example presented in this paper, we can also make such inferences. Bear in mind, however, that the usefulness of such inferences depends very directly on the accuracy of the nodal reconstructions. We have presented two ways of indexing uncertainty about nodal reconstructions (see above and Appendix 3); others are discussed below. When one singles out a particular species or lineage to determine if it shows change from the ancestral value, it would make sense first to consider whether its value is outside of such uncertainty intervals (e.g., see Chevalier, 1991, who also incorporated information on within-species variation). (The method of phylogenetically independent contrasts can also be employed to compare single species with a set of others; see Fig. 4 of Garland and Adolph, 1994; McPeek, 1995; Martinez del Rio *et al.*, 1995.)

One example of an evolutionary change in plasma osmolarity is the apparent decrease seen in the lineage leading to the lampreys (Class Cephalaspidomorphi, *Lampetra* and *Petromyzon*: tips 168-170 in Fig. 1). Extant lampreys are anadromous; they have a freshwater larval stage and may enter seawater as adults (Nelson, 1994). One of the species in our data set, *Petromyzon marinus*, is anadromous (Mathers and Beamish, 1974); however, the data for all three species were taken from landlocked freshwater lampreys found in the Great Lakes of North America. Available data for *Petromyzon* caught at sea (cited in Morris, 1971) and for lampreys in 1037 mOsm sea water

(Logan *et al.*, 1980) indicate plasma osmolarities in the range of about 309-361 mOsm. All of these values are much lower than the value of 644 mOsm that is estimated by squared-change parsimony for the ancestor of all 172 vertebrates in our data set (see first row of Table 2). The closest living relatives of the lampreys, the hagfishes (Nelson, 1994: Class Myxini, *Myxine* and *Eptatretus*: tips 171-172) lack a larval stage, are marine, and have much higher plasma osmolarities (954-969 mOsm), values similar to those of saltwater Chondrichthyes. Although the arbitrary branch lengths of Figure 1 do not reflect the fact, it is important to note that lampreys and hagfishes diverged phylogenetically at least 400-500 million years ago (Nelson, 1994); this distant relationship is reflected by their placement in different taxonomic Classes (despite some superficial similarities).

Given that changes along branch segments can be inferred for one trait, then the same can be done for two or more traits. These inferred changes along branch segments can then be correlated. This provides a way to study correlated character evolution (e.g., see Huey, 1987; Huey and Bennett, 1987; Garland *et al.*, 1991; Walton, 1993; Westneat, 1995). Simulation studies indicate that squared-change parsimony estimates of correlated character evolution may be better than some alternatives [e.g., Felsenstein's (1985) phylogenetically independent contrasts] in some cases (Martins and Garland, 1991; Martins, 1996a; see also Pagel, 1993; Bjorklund, 1994). The CMSINGLE program of Martins and Garland (1991) performs the necessary computations. Note, however, that the number of branch segments along which changes can be inferred is greater than the number of original data points (i.e., the data for the tip species). Moreover, the inferred changes are not independent in the statistical sense, because the nodes are computed as averages of surrounding values (Martins and Garland, 1991; Pagel, 1993). Thus, conventional critical values cannot be used for hypothesis testing. Instead, Monte Carlo computer simulations can be used to create empirical null distributions for hypothesis testing (Garland *et al.*, 1991; Martins and Garland, 1991), and computer programs to do so are available from the senior author (CMSINGLE and CMMEANAL of Martins and Garland, 1991; PDSIMUL of Garland *et al.*, 1993). Phylogenetic randomization procedures, as mentioned in the next section, could also be used for this purpose.

With respect to plasma osmolarity, one might test for correlated evolution with habitat shifts. However, the crude categorizations of "habitat" that we have compiled do not constitute a continuous-valued character that ranges simply along a single dimension. Rather, "habitat" encompasses at least one qualitative difference, between aquatic and terrestrial. Phylogenetically based statistical methods for correlating changes in a continuous variable with multiple changes in a complicated, qualitative variables are not well worked out (see Grafen, 1989; Garland *et al.*, 1992, 1993; Garland and Adolph, 1994; McPeek, 1995; Martins, 1996b; Pyron, 1996), so we leave such analyses for the future. Nonetheless, we have shown (see above section on "Choosing Appropriate Models ...") that plasma osmolarity differs significantly among habitat types for the 54 Actinopterygii. This covariation implies that osmolarity has indeed evolved in concert with habitat occupancy.

Future Possibilities for Indexing Uncertainty in Nodal Reconstructions

In closing, we mention some other possibilities for indexing uncertainty in squared-change parsimony reconstructions. Most of these are computer-intensive "resampling" methods of the type that have only recently become popular in ecology and evolutionary biology, although some have a long history (Sokal and Rohlf, 1981; Noreen, 1989; Manly, 1991; Crowley, 1992; Efron and Tibshirani, 1993; Lapointe and Legendre, 1992, 1995; Lapointe *et al.*, 1994). Jackknifing, as we have used above (and see Appendix 3), is one of these methods. Bootstrapping is similar to jackknifing, except that the resampling is done with replacement. The possibility of sampling species with replacement (i.e., a given species' value could appear multiple times in a single resampled data set) does not seem to make much sense. How, for instance, would that species be placed multiple times on a hierarchical phylogeny (as a polytomy?)? Thus, we will not consider bootstrapping. The two other resampling methods that we will consider are Monte Carlo simulations and randomization procedures; the latter can be considered a special case of the former.

Monte Carlo computer simulations have been used to test hypotheses about comparative data (e.g., Garland *et al.*, 1991; Garland *et al.*, 1993; Westneat, 1995; Martin and Clobert, 1996; Reynolds and Lee, 1996; Harris and Steudel, 1997). These procedures involve simulating the evolution of one or more characters along a specified evolutionary tree (topology and branch lengths) under a specified model of evolutionary change (see also Bjorklund, 1994; Díaz-Uriarte and Garland, 1996). For each simulated data set, the statistic of interest is then computed in exactly the same fashion as for the one real data set. The model of evolutionary change is assumed to be representative of the actual processes that resulted in the data about which one wishes to draw inferences (e.g., the average plasma osmolarities of various species of vertebrates). [Simulations can also be used to test the performance of different analytical methods (e.g., Grafen, 1989; Martins and Garland, 1991; Purvis *et al.*, 1994; Martins, 1996a); in Appendix 3, we compare the performance of squared-change parsimony and a conventional nonphylogenetic analysis for estimating the root node of a phylogeny.]

At first thought, Monte Carlo simulations might seem ill-suited to testing hypotheses about values reconstructed for internal nodes. Among other parameters, one must specify the starting value for a simulation, i.e., the value of the character at the root of the phylogenetic tree. If the node of interest *is* the root node, or even one adjacent or almost adjacent to the root, then the results would depend very strongly on a user-specified value, i.e., an assumption of the procedure. This could lead to "inappropriately biasing the analysis" (see de Queiroz, 1996). Nevertheless, further thought suggests the possible utility of computer simulations.

First, if independent fossil (or other) information allowed specification of the starting value (e.g., Garland *et al.*, 1993), then the potential for *inappropriate* bias could be greatly reduced if not eliminated. Second, one could use the simulated data only to gauge variability about the point estimate, but not the point estimate itself. In other words, for the best estimate of the value at the node of interest, one could use the estimate from the real data set (alternatively, see Reynolds and Lee, 1996, p. 740). For the estimate of uncertainty, one could simulate many data sets (e.g., 1000), determine the range of values from the 2.5th to the 97.5th percentiles, divide by two, and use this as an index of a +/- 95% confidence interval about the point estimate from the real data set (cf. Reynolds and Lee, 1996). This "confidence interval" would, of course, be completely predicated on the chosen simulation model (as well as the specified topology and branch lengths). This sort of procedure would be analogous to the "parametric bootstrap" that has recently gained favor in studies of uncertainty in the reconstruction of phylogenetic trees (Huelsenbeck *et al.*, 1996; see also Crowley, 1992, p. 429).

Randomization tests are another example of a computer-intensive resampling method, but these procedures completely avoid the specification of starting values. Randomization tests reshuffle the real data (e.g., the species' mean plasma osmolarities), rather than sampling a subset of it (jackknifing) or creating new data sets by Monte Carlo simulation. Each time the real data are reshuffled, the statistic of interest is recomputed (e.g., the squared-change parsimony value reconstructed for a particular node).

Conventional randomization tests reshuffle the data equiprobably, which, in the context of comparative data, is equivalent to assuming that the character(s) evolved along a star phylogeny with equal branch lengths and by Brownian motion (i.e., it is assumed that nothing like character displacement has occurred). F.-J. Lapointe and the senior author are currently developing a phylogenetic randomization procedure and associated computer programs. This will allow the tip data to be reshuffled in a way that preserves the phylogenetic structure in the data (i.e., closely related species tend to have similar phenotypes). In general, this would mean that values permuted from a given tip are most likely to go back to that tip itself or its sister, next most likely to go to their next closest relative, and so on. As with Monte Carlo simulations, each of the permuted data sets could then be submitted to a squared-change parsimony analysis, a histogram constructed, and the spread of this histogram used to indicate the uncertainty with which we should view the point estimate of the node, i.e., the value reconstructed for the real data set.

Finally, as noted in Appendix 3 (section on "Squared-Change Parsimony and Independent Contrasts"), the root node value reconstructed by squared-change parsimony is the same as that computed by independent contrasts. In principle, independent contrasts solve the problem of the tip data being nonindependent. Thus, for the root node, we believe that jackknifing (applied above and discussed in Appendix 3) may actually have more than heuristic value. This suggestion has not yet been studied. Also for the root node, an analytical procedure based on independent contrasts has been developed (T. Garland, Jr., and A. R. Ives, unpublished results). With the assumption that character evolution can be modeled as Brownian motion, this allows the computation of standard parametric confidence intervals for root nodes (see above section on "Reconstructing Plasma Osmolarity at the Origin of Vertebrates"). Extensions to incorporate within-species variation are also possible (J. Felsenstein, pers. comm.).

The suitability of the resampling methods discussed in this section depends, in part, on how one views interspecific comparative studies in relation to issues of sampling, populations, and the desired nature of the inference. Some of the procedures assume that the available data represent a random sample from some population and/or that the parameters of that population can be fully defined. Exactly what assumptions are made by each method is a complicated subject and beyond the scope of this chapter (Sokal and Rohlf, 1981; Noreen, 1989; Manly, 1991; Crowley, 1992; Efron and Tibshirani, 1993; see also Díaz-Uriarte and Garland, 1996; Martins and Hansen, 1996) But we can mention how some of these issues pertain to our example. The "population" might be considered as all (extant) vertebrates. We cannot fully define that "population," however, because we obviously do not have data for all vertebrates. Moreover, the 172 species in our data set probably should not be considered as a random sample. In general, comparative physiologists do not choose study organisms by reference to a table of random numbers and a list of all extant species (which themselves are probably not a random sample of all species that have ever lived!). Instead (see Garland and Carter, 1994), species are often studied because they are of particular interest [e.g., because they live in extreme environments, such as deserts or the Antarctic (e.g., Eastman, 1993)], because they are particularly suitable for certain physiological measurements, or sometimes just because they happen to be available. And finally, note that we emphasized a still less random subset of 63 taxa for most of our analyses. Future studies will need to consider carefully the optimal application of different computer-intensive resampling methods for making various sorts of inferences from comparative data.

ACKNOWLEDGMENTS

This research was supported by National Science Foundation Grants IBN-9157268 and DEB-9509343 to TG. M. R. de Carvalho kindly provided phylogenetic information and A. R. Ives allowed us to cite unpublished work. We are deeply indebted to C. R. Crumly, R. B. Huey, R. E. Strauss, S. Sumida, and M. L. Zelditch for comments on various versions of the manuscript; we have not always been able to satisfy their concerns nor to reach a consensus on certain issues, and any errors remain our own. J. Felsenstein, A. R. Ives, and F.-J. Lapointe also provided very helpful discussions on related statistical matters.

APPENDIX 1

Complete data set employed in the analyses ($N = 172$). Species are ordered as in the phylogeny shown in Figure 1, and Tip# indicates position of species from top to bottom. Osm. is osmolarity in mOsm. Species names show the most recent taxonomic conventions (see text). Genus or species names in parentheses are the ones used in original sources. Ha, habitat; D, desert; E, estuarine; F, freshwater; M, marine; S, saltwater; SA, saltwater Antarctic; SM, salt marsh; T, terrestrial.

Clade	Species	Tip#	Osm	Ha	Source
Aves	*Struthio camelus*	1	302	T	Altman and Dittmer, 1971
Aves	*Dromaius novaehollandiae*	2	309	T	Skadhauge, 1974
Aves	*Cygnus atratus*	3	296	T	Hughes, 1976
Aves	*Anas platyrhynchos*	4	297	T	Deustch *et al.*, 1979
Aves	*Anser anser*	5	297	T	Zucker *et al.*, 1977
Aves	*Colinus virginianus*	6	350	T	McNabb, 1969
Aves	*Callipepla (Lophortyx) californica*	7	336	T	McNabb, 1969
Aves	*Callipepla (Lophortyx) gambelii*	8	340	T	McNabb, 1969
Aves	*Gallus domesticus*	9	319	T	Skadhauge, 1967
Aves	*Coturnix coturnix*	10	313	T	Osono and Nishimura, 1994
Aves	*Coturnix pectoralis*	11	349	T	Roberts and Baudinette, 1984
Aves	*Coturnix chinensis*	12	371	T	Roberts and Baudinette, 1984
Aves	*Geococcyx californianus*	13	349	T	Ohmart, 1972

Clade	Species	Tip#	Osm	Ha	Source
Aves	*Cacatua roseicapilla*	14	336	T	Skadhauge, 1974
Aves	*Melopsittacus undulatus*	15	336	T	Krag and Skadhauge, 1972
Aves	*Geophaps (Ocyphaps) lophotes*	16	336	T	Skadhauge, 1974
Aves	*Zenaida macroura*	17	372	T	Smyth and Bartholomew, 1966
Aves	*Fulica americana*	18	259	T	Carpenter and Stafford, 1970
Aves	*Gallirallus (Rallus) owstoni*	19	355	T	Carpenter and Stafford, 1970
Aves	*Larus argentatus*	20	273	T	Ensor and Phillips, 1972
Aves	*Larus glaucescens*	21	334	T	Hughes, 1977
Aves	*Spheniscus demersus*	22	306	M	Erasmus, 1978
Aves	*Ardea cinerea*	23	317	T	Lange and Staaland, 1966
Aves	*Lichenostomus (Meliphaga) virescens*	24	343	T	Skadhauge, 1974
Aves	*Sturnus vulgaris*	25	318	T	Braun, 1978
Aves	*Taeniopygia (Poephila) guttata*	26	336	T	Skadhauge and Bradshaw, 1974
Aves	*Amphispiza belli nevandensis*	27	310	T	Moldenhauer and Wiens, 1970
Aves	*Ammodramus (Ammospiza) c. caudacutus*	28	346	T	Poulson, 1969
Aves	*Passerculus sandwichensis*	29	339	T	Goldstein *et al.*, 1990
Aves	*Passerculus sandwichensis beldingi*	30	349	SM	Goldstein *et al.*, 1990
Crocodylia	*Alligator mississipiensis*	31	291	F	Lauren, 1985
Crocodylia	*Crocodylus acutus*	32	294	F	Minnich, 1982; Dill and Edwards, 1931
Crocodylia	*Crocodylus porosus*	33	307	F	Minnich, 1982; Grigg, 1981
Squamata	*Microlophus (Tropidurus)* species	34	340	T	Minnich, 1979; Roberts and Schmidt-Nielsen, 1966
Squamata	*Dipsosaurus dorsalis*	35	300	D	Minnich, 1979; House, 1974
Squamata	*Amblyrhynchus cristatus*	36	402	E	Nagy and Shoemaker, 1984
Squamata	*Uromastyx acanthinurus*	37	307	D	Minnich, 1979; Tercafs and Vassas, 1967
Squamata	*Ctenophorus (Amphibolurus) maculosus*	38	379	D	Minnich, 1982; Braysher, 1976
Squamata	*Agama impalearis*	39	350	T	Minnich, 1979; Tercafs and Vassas, 1967
Squamata	*Agama stellio*	40	358	T	Minnich, 1979; Frenkel and Kraicer, 1971

Clade	Species	Tip#	Osm	Ha	Source
Squamata	*Phrynosoma cornutum*	41	315	D	Minnich, 1979; Roberts and Schmidt-Nielsen, 1966
Squamata	*Sceloporus cyanogenys*	42	319	T	Minnich, 1979; Stolte *et al.*, 1977
Squamata	*Hemidactylus* species	43	322	T	Minnich, 1979; Roberts and Schmidt-Nielsen, 1966
Squamata	*Varanus gouldii*	44	328	D	Minnich, 1979; Green, 1972
Squamata	*Laticauda semifasciata*	45	320	S	Minnich, 1982; Dunson and Taub, 1967
Squamata	*Pituophis melanoleucus*	46	327	T	Minnich, 1979; Komadina and Solomon, 1970
Squamata	*Nerodia cyclopion*	47	262	F	Minnich, 1979; LeBrie and Elizondo, 1969
Squamata	*Nerodia sipedon*	48	318	F	Minnich, 1982; Dessaur, 1970
Squamata	*Nerodia fasciata compressicauda*	49	371	E	Minnich, 1982; Dunson, 1980
Sphenodontia	*Sphenodon punctatus*	50	273	T	Minnich, 1982; Schmidt-Nielsen and Schmidt, 1973
Mammalia	*Homo sapiens*	51	302	T	Guyton, 1991
Mammalia	*Canis familiaris*	52	294	T	Papanek and Raff, 1994
Mammalia	*Felis catus*	53	303	T	Altman and Dittmer, 1971
Mammalia	*Oryctolagus cuniculus*	54	283	T	Keil *et al.*, 1994
Mammalia	*Mesocricetus auratus*	55	325	D	Gottschalk *et al.*, 1963
Mammalia	*Psammomys obesus*	56	365	D	Jamison *et al.*, 1979
Mammalia	*Mus musculus*	57	302	T	Altman and Dittmer, 1971
Mammalia	*Rattus norvegicus*	58	303	T	Ullrich *et al.*, 1963
Mammalia	*Equus caballus*	59	304	T	Altman and Dittmer, 1971
Mammalia	*Physeter catodon*	60	370	S	Gordon, 1982
Mammalia	*Sus scrofa*	61	302	T	Altman and Dittmer, 1971
Mammalia	*Camelus dromedarius*	62	340	D	Schmidt-Nielsen, 1964
Mammalia	*Bos taurus*	63	304	T	Spector, 1956
Mammalia	*Ovis aries*	64	290	T	Dunham *et al.*, 1993
Mammalia	*Capra hircus*	65	303	T	Altman and Dittmer, 1971
Mammalia	*Capra hircus*	66	320	D	Chosniak *et al.*, 1984
Testudines	*Chelydra serpentina*	67	315	F	Minnich, 1982; Dessaur, 1970
Testudines	*Chelonia mydas*	68	270	S	Minnich, 1982; Dessauer, 1970
Testudines	*Caretta caretta*	69	465	S	Minnich, 1982; Schoffeniels and Tercafs, 1965
Testudines	*Trionyx spiniferus*	70	280	F	Minnich, 1982; Dunson and Weymouth, 1965
Testudines	*Malaclemys terrapin*	71	309	F	Minnich, 1979; Gilles-Baillien, 1970
Testudines	*Chrysemys picta (scripta)*	72	260	F	Minnich, 1979; Dantzler and Schmidt-Nielsen, 1966
Testudines	*Trachemys (Pseudemys) scripta*	73	260	F	Minnich, 1982; Platner, 1950
Testudines	*Mauremys caspica (leprosa)*	74	362	F	Minnich, 1982; Schoffeniels and Tercafs, 1965

Clade	Species	Tip#	Osm	Ha	Source
Testudines	*Testudo h. hermanni*	75	290	T	Minnich, 1982; Gilles-Baillien and Schoffeniels, 1965
Testudines	*Gopherus agassizii*	76	291	D	Minnich, 1982; Minnich, 1977
Lissamphibia	*Xenopus laevis*	77	233	F	Shoemaker *et al.*, 1992; McBean and Goldstein, 1970
Lissamphibia	*Rana catesbeiana*	78	210	F	Alvarado, 1979; Yoshimura *et al.*, 1961
Lissamphibia	*Rana pipiens*	79	214	F	Alvarado, 1979; Campbell *et al.*, 1967
Lissamphibia	*Rana cancrivora*	80	290	F	Shoemaker *et al.*, 1992; Gordon *et al.*, 1961
Lissamphibia	*Hyla regilla*	81	218	F	Alvarado, 1979; Mullen, 1974
Lissamphibia	*Bufo boreas*	82	235	T	Alvarado, 1979; Mullen, 1974
Lissamphibia	*Bufo bufo*	83	240	T	Alvarado, 1979; Ferreira and Jesus, 1973
Lissamphibia	*Bufo marinus*	84	250	T	Alvarado, 1979; Middler *et al.*, 1969
Lissamphibia	*Bufo viridis*	85	270	F	Bentley, 1971; Balinsky, 1981
Lissamphibia	*Typhlonectes compressicauda*	86	196	F	Stiffler *et al.*, 1990
Lissamphibia	*Ichthyophis kohtaoensis*	87	220	T	Stiffler *et al.*, 1990
Lissamphibia	*Ambystoma tigrinum*	88	230	T	Alvarado, 1979; Alvarado, 1972
Lissamphibia	*Amphiuma means*	89	218	F	Stanton, 1988
Lissamphibia	*Batrachoseps attenuatus*	90	339	F	Shoemaker *et al.*, 1992; Balinsky, 1981
Coelacanthiformes	*Latimeria chalumnae*	91	931	S	Griffith *et al.*, 1974
Dipnoi	*Protopterus aethiopicus*	92	238	F	Evans, 1979; Smith, 1930
Teleostei	*Pleuronectes flesus*	93	364	S	Bentley, 1971; Lange and Fugelli, 1965
Teleostei	*Perca fluviatilis*	94	294	F	Gordon, 1982; Lutz, 1975
Teleostei	*Pholis ornata*	95	375	S	Bridges, 1993; Barton, 1979
Teleostei	*Sphyraena barracuda*	96	476	S	Evans, 1979; Becker *et al.*, 1958
Teleostei	*Rhigophila dearborni*	97	489	SA	Dobbs and DeVries, 1975
Teleostei	*Gymnodraco acuticeps*	98	616	SA	Dobbs and DeVries, 1975
Teleostei	*Blennius pholis*	99	320	S	Bridges, 1993; House, 1963
Teleostei	*Blennius fluviatilis*	100	410	S	Bridges, 1993; Muller *et al.*, 1973
Teleostei	*Blennius pavo*	101	410	S	Bridges, 1993; Muller *et al.*, 1973
Teleostei	*Blennius sphinx*	102	410	S	Bridges, 1993; Muller *et al.*, 1973
Teleostei	*Periophthalmus chrysospilos*	103	430	S	Bridges, 1993; Lee *et al.*, 1987
Teleostei	*Periophthalmus sobrinus*	104	480	S	Bridges, 1993; Gordon *et al.*, 1965

Clade	Species	Tip#	Osm	Ha	Source
Teleostei	*Dissostichus mawsoni*	105	614	SA	Dobbs and DeVries, 1975
Teleostei	*Trematomus loennbergii*	106	563	SA	Dobbs and DeVries, 1975
Teleostei	*Trematomus borchgrevinki*	107	565	SA	Dobbs and DeVries, 1975
Teleostei	*Trematomus lepidorhinus*	108	574	SA	Dobbs and DeVries, 1975
Teleostei	*Trematomus hansoni*	109	581	SA	Dobbs and DeVries, 1975
Teleostei	*Trematomus nicolai*	110	589	SA	Dobbs and DeVries, 1975
Teleostei	*Trematomus centronotus*	111	602	SA	Dobbs and DeVries, 1975
Teleostei	*Trematomus bernacchii*	112	634	SA	Dobbs and DeVries, 1975
Teleostei	*Trematomus newnesi*	113	656	SA	Dobbs and DeVries, 1975
Teleostei	*Scomberomorus maculatus*	114	386	S	Evans, 1979; Becker *et al.*, 1958
Teleostei	*Thunnus thynnus*	115	437	S	Holmes and Donaldson, 1969; Becker *et al.*, 1958
Teleostei	*Promicrops itaiara*	116	384	S	Holmes and Donaldson, 1969; Becker *et al.*, 1958
Teleostei	*Mycteroperca bonasi*	117	461	S	Holmes and Donaldson, 1969; Becker *et al.*, 1958
Teleostei	*Mycteroperca venenosa*	118	467	S	Holmes and Donaldson, 1969; Becker *et al.*, 1958
Teleostei	*Xiphister atropurpureus*	119	300	S	Bridges, 1993; Evans, 1967
Teleostei	*Anoplarchus purpurescens*	120	350	S	Bridges, 1993; Barton, 1979
Teleostei	*Fundulus kansae*	121	181	F	Evans, 1979; Stanley and Fleming, 1964
Teleostei	*Fundulus heteroclitus*	122	335	F	Bentley, 1971; Pickford *et al.*, 1966
Teleostei	*Fundulus heteroclitus*	123	365	S	Bentley, 1971; Pickford *et al.*, 1966
Teleostei	*Ictalurus nebulosus*	124	279	F	Gordon, 1982; Umminger, 1971
Teleostei	*Cyprinus carpio*	125	274	F	Evans, 1979; Houston and Madden, 1968
Teleostei	*Tinca vulgaris*	126	280	F	Holmes and Donaldson, 1969; Keys and Hill, 1934
Teleostei	*Carassius auratus*	127	299	F	Gordon, 1982; Umminger, 1971
Teleostei	*Opsanus tau*	128	392	S	Bentley, 1971; Lahlou *et al.*, 1969
Teleostei	*Lophius americanus*	129	350	S	Gordon, 1982; Forster and Berglund, 1956
Teleostei	*Lophius piscatorius*	130	452	S	Evans, 1979; Brull and Nizet, 1953
Teleostei	*Esox lucius*	131	274	F	Holmes and Donaldson, 1969

Clade	Species	Tip#	Osm	Ha	Source
Teleostei	*Salvelinus namaycush*	132	298	F	Holmes and Donaldson, 1969; Hoffert and Fromm, 1966
Teleostei	*Oncorhynchus kisutch*	133	295	F	Holmes and Donaldson, 1969; Conte, 1965
Teleostei	*Oncorhynchus kisutch*	134	331	S	Holmes and Donaldson, 1969; Conte, 1965
Teleostei	*Salmo salar*	135	328	F	Evans, 1979; Parry, 1961
Teleostei	*Salmo salar*	136	344	S	Evans, 1979; Parry, 1961
Teleostei	*Salmo trutta*	137	326	F	Evans, 1979; Gordon, 1959
Teleostei	*Salmo trutta*	138	356	S	Evans, 1979; Gordon, 1959
Teleostei	*Conger vulgaris*	139	430	S	Holmes and Donaldson, 1969; Boucher-Firley, 1934
Teleostei	*Muraena helena*	140	441	S	Holmes and Donaldson, 1969; Boucher-Firley, 1934
Teleostei	*Anguilla rostrata*	141	307	F	Holmes and Donaldson, 1969; Butler *et al.*, 1969
Teleostei	*Anguilla anguilla*	142	328	F	Evans, 1979; Sharatt *et al.*, 1964
Teleostei	*Anguilla anguilla*	143	377	S	Evans, 1979; Sharatt *et al.*, 1964
Chondrostei	*Acipenser sturio*	144	318	F	Holmes and Donaldson, 1969; Magnin, 1962
Chondrostei	*Acipenser sturio*	145	343	S	Holmes and Donaldson, 1969; Magnin, 1962
Chondrostei	*Erpetoichthys calabaricus*	146	199	F	Gordon, 1982; Lutz, 1975
Chondrichthyes	*Mustelus canis*	147	962	S	Holmes and Donaldson, 1969; Doolittle *et al.*, 1960
Chondrichthyes	*Scyliorhinus canicula*	148	1118	S	Evans, 1979; Payan and Maetz, 1971
Chondrichthyes	*Carcharhinus melanop*	149	484	F	Holmes and Donaldson, 1969; Smith, 1931
Chondrichthyes	*Carcharhinus leucas*	150	650	F	Gordon, 1982; Thorson *et al.*, 1973
Chondrichthyes	*Carcharhinus leucas*	151	1000	S	Gordon, 1982; Thorson *et al.*, 1973
Chondrichthyes	*Squalus acanthias*	152	1007	S	Holmes and Donaldson, 1969; Murdaugh and Robin, 1967
Chondrichthyes	*Squatina angelus*	153	1102	S	Holmes and Donaldson, 1969; Pora, 1936
Chondrichthyes	*Pristis microdon*	154	540	F	Holmes and Donaldson, 1969; Smith, 1931
Chondrichthyes	*Torpedo marmorata*	155	1098	S	Holmes and Donaldson, 1969; Pora, 1936
Chondrichthyes	*Potamotrygon* species	156	308	F	Withers, 1992; Thorson *et al.*, 1967
Chondrichthyes	*Dasyatis varnak*	157	548	F	Holmes and Donaldson, 1969; Smith, 1931
Chondrichthyes	*Dasyatis saj*	158	840	S	Holmes and Donaldson, 1969

Clade	Species	Tip#	Osm	Ha	Source
Chondrichthyes	*Dasyatis amaericana*	159	864	S	Holmes and Donaldson, 1969; Bernard *et al.*, 1966
Chondrichthyes	*Dasyatis sabina*	160	1021	S	Evans, 1979; deVlaming and Sage, 1973
Chondrichthyes	*Raja eglanteria*	161	844	S	Holmes and Donaldson, 1969; Price and Creaser, 1967
Chondrichthyes	*Raja ocellata*	162	928	S	Holmes and Donaldson, 1969; Maren *et al.*, 1963
Chondrichthyes	*Raja stabuliforis*	163	958	S	Holmes and Donaldson, 1969; Maren *et al.*, 1963
Chondrichthyes	*Raja clavata*	164	1050	S	Bentley, 1971; Murray and Potts, 1961
Chondrichthyes	*Raja undulata*	165	1097	S	Holmes and Donaldson, 1969; Pora, 1936
Chondrichthyes	*Hydrolagus colliei*	166	801	S	Holmes and Donaldson, 1969; Urist, 1966
Chondrichthyes	*Chimaera monstrosa*	167	1046	S	Kirschner, 1991; Robertson, 1976
Cephalaspidomorphi	*Lampetra planeri*	168	227	F	Holmes and Donaldson, 1969; Bull and Morris, 1967
Cephalaspidomorphi	*Lampetra fluviatilis*	169	272	F	Evans, 1979; Pickering and Morris, 1970
Cephalaspidomorphi	*Petromyzon marinus*	170	300	F	Gordon, 1982; Mathers and Beamish, 1974
Myxini	*Myxine glutinosa*	171	969	S	Evans, 1979; Robertson, 1976
Myxini	*Eptatretus stoutii*	172	954	S	Evans, 1979; McFarland and Munz, 1965

APPENDIX 2: PARSIMONY IN SYSTEMATIC AND COMPARATIVE BIOLOGY

The English word "parsimony" is derived from the Latin stem "pars-," which denotes "to spare, save" (Oxford English Dictionary, 1971). As a logical principle, parsimony dictates, in essence, that simple explanations are generally to be preferred. Parsimony is often used as a procedure to infer the phylogenetic relationships of organisms and, as we discuss in this chapter, the evolution of particular characters in the context of an (independent) hypothesis of phylogenetic relationships. Use of parsimony as a philosophical or as an operational principle in systematic biology seems to have at least two origins (reviews in Farris, 1983; Sober, 1988; Edwards, 1996). At present, the justification of parsimony as a procedure for phylogenetic inference remains quite controversial, in part because "parsimony" means different things to different people [cf. Kluge and Wolf (1993) and Sanderson (1995) on the meaning of "cladistics"].

What might be termed "cladistic parsimony" seems to derive from the "Law of parsimony: the logical principle that no more causes or forces should be assumed than are necessary to account for the facts" (Oxford English Dictionary,

1971). To quote Farris (1983, p. 7): "Most phylogeneticists recognize that inferring genealogy rests on the principle of parsimony, that is, choosing genealogical hypotheses so as to minimize requirements for ad hoc hypotheses of homoplasy." "Homoplasy" describes characters or states of characters that are shared by two or more taxa but are not homologous (derived from a common ancestor). "Put simply, homoplasy exists if...two taxa showing the character have a common ancestor that does not have the character" (Wiley, 1981, p. 12). Homoplasy can result from parallel or convergent evolution or from evolutionary reversals.

Cladistic parsimony argues that we should prefer phylogenetic hypotheses that minimize the amount of homoplasy in the characters used in the analysis (e.g., DNA sequences and morphology). For a given data set, this generally translates into a preference for the phylogenetic tree that requires the fewest total number of steps across all characters being analyzed. However, an important point to note here is that, with categorical variables, the actual amount of evolutionary (genetic) change (on some unspecified biological scale) is not necessarily implied to be the same from, say, a coded state of 0 to 1 as from a state of 1 to 2. Thus, minimizing the amount of homoplasy (misinterpreted homologies, representing parallel and/or convergent evolution) does not necessarily equate to finding the phylogenetic tree that implies the least amount of biological change.

The other origin of parsimony procedures in systematic biology hinges on their being, in some cases, an approximation to the maximum-likelihood solution for a model of random evolution (Edwards, 1996). This is a statistical justification of what might be called "distance parsimony," or a preference for topologies that invoke the minimum net amount of evolutionary change (Edwards, 1996). Related to this is the more colloquial idea that, as expressed in a recent textbook in evolutionary biology, "The parsimony principle is reasonable because evolutionary change is improbable. ... it is more likely that a character will be shared by common descent than by independent, convergent evolution. For any set of species, a phylogeny requiring less evolutionary change is more plausible than one requiring more" (Ridley, 1993, pp. 449-450).

In its purest form, "cladistic parsimony" applies only to discretely valued characters (e.g., Stewart, 1993), although an algorithm for continuous-valued characters is available (Kluge and Farris, 1969; Farris, 1970; Huey and Bennett, 1987; Losos, 1990; Maddison and Maddison, 1992; Miles and Dunham, 1996; Butler and Losos, in review). With respect to morphological characters, systematists generally code them into discrete states (e.g., 0 versus 1) and describe each taxon as being characterized by one state or the other, with some characters coded as multistate (e.g., see Brooks and McLennan, 1991). Sometimes this coding is easy to do, as when a particular feature is either present or absent from all individuals that have been examined for a given taxon. Sometimes the categorization is fuzzier, as when a morphological feature is described as "small, medium or large." Categorization is also difficult when taxa show polymorphism (more than one character state) among its members (Wiens, 1995). Finally, categorization can be performed even when the character under consideration is inherently continuous-valued and polygenic (Falconer and Mackay, 1996), such as body size or shape, or metabolic rate (e.g., Dial and Grismer, 1992; Zelditch *et al.*, 1995; Strait *et al.*, 1996). If the phenotypes of the set of taxa under consideration fall cleanly into a small number of non-overlapping categories, then categorization may not be too disputable. When overlap is considerable, however, then it becomes debatable whether such features of organisms can even provide useful information for systematic purposes (see also comments in Garland and Adolph, 1994, pp. 817-821).

Irrespective of how characters are coded, for a given data set, several or even many different topologies may be equally parsimonious, leaving the investigator with ambiguity as to which particular tree should be preferred. This ambiguity can stem from the data (they may contain insufficient information to allow complete resolution of all relationships, or they may imply no unique resolution) and/or from the methods used to draw inferences from the data. Ways of dealing with these sorts of uncertainty are debatable and beyond the scope of this chapter. As well, in addition to parsimony, which itself has multiple variants (Maddison, 1994; Maddison, 1995; Edwards, 1996), many other ways of constructing and/or choosing phylogenetic trees are available, such as maximum likelihood. For all of these issues, we must refer the reader to other literature (e.g., see Wiley, 1981; Friday, 1987; Felsenstein, 1988a,b, 1992; Sober, 1988; Lynch, 1989; Sarich *et al.*, 1989; Springer and Krajewski, 1989a,b; Goldman, 1990; Mayr and Ashlock, 1991; Miyamoto and Cracraft, 1991; Wiley *et al.*, 1991; Lapointe and Legendre, 1992, 1995; Lapointe *et al.*, 1994; de Queiroz *et al.*, 1995; Hillis, 1995; Purvis, 1995; Zelditch *et al.*, 1995; Hillis *et al.*, 1996; Huelsenbeck *et al.*, 1996; Huelsenbeck, Hillis, and Jones, 1996; Lee and Spencer, this volume; Laurin and Reisz, this volume).

As noted in the Introduction, we are here concerned not with inferring phylogenetic relationships but, rather, with mapping characters onto independently-derived hypotheses of phylogenetic relationships (see de Queiroz, 1996). Our mapping of a character (in our case, plasma osmolarity) will not be used to alter the phylogeny itself. The phylogeny that we will use for analysis is taken from the literature, as an informal synthesis of available information from many different sources (as, for example, in Garland *et al.*, 1993; for a more formal type of synthesis, see Purvis, 1995).

Parsimony as a principle can be used to guide how we map characters onto a phylogenetic tree. It would suggest finding that reconstruction (i.e., set of values at interior nodes) that minimizes the amount of change of the character of interest across the entire phylogenetic tree (also termed "character optimization"). For discretely valued characters, this can be termed Manhattan, Wagner, absolute-change, or linear parsimony. For a given character and a given phylogenetic tree, it is often the case that several or many different parsimony reconstructions can be found which result in the same sum of absolute change (Losos, 1990; Dial and Grismer, 1992; Maddison, 1994; Maddison, 1995; Miles and Dunham, 1996; Butler and Losos, in review). Thus, the most parsimonious reconstruction for a given internal node may be a range of character values. Linear parsimony can be used only with dichotomous trees (Maddison and Maddison, 1992, p. 304). Moreover, different types of parsimony exist, including (1) variants that require a priori specification of the character state at the root of the tree or not and (2) variants that allow reversals of character state or not (Maddison and Maddison, 1992; Maddison, 1994).

APPENDIX 3: SQUARED-CHANGE PARSIMONY

For continuous-valued characters, another procedure exists for mapping characters onto specified phylogenetic trees. This algorithm obviates the need to code different taxa as having one of two or more discrete values. It is now most commonly termed "squared-change parsimony," although the terms "minimum evolution" and "minimum squared-change evolution" are also used (Huey and Bennett, 1987; Losos, 1990; Maddison, 1991; Martins and Garland, 1991; Maddison and Maddison, 1992; Walton, 1993; McArdle and Rodrigo, 1994; Westneat, 1995; Miles and Dunham, 1996). Unlike linear parsimony, squared-change parsimony can reconstruct values at internal nodes as having intermediate states not observed in the actual data (the terminal taxa or tips). An interesting area for future research with continuous-valued characters will be to compare squared-change parsimony with (1) linear parsimony and (2) linear parsimony applied after the character has been coded into a small number of discrete categories (cf. Huey and Bennett, 1987; Losos, 1990; Dial and Grismer, 1992; Miles and Dunham, 1996; Butler and Losos, in review).

Squared-change parsimony is relatively easy to implement through an iterative algorithm as used herein, through a recursive algorithm (Maddison, 1991), or by direct computation (McArdle and Rodrigo, 1994). The iterative algorithm works as follows. First, values for the phenotypes of a series of species are placed onto the tips (terminal nodes or taxa) of a phylogenetic tree. Second, arbitrary values are placed (seeded) at each of the internal nodes of the tree, including the root (or basal) node. Typically, the overall mean value of the tips is used to seed the nodes (so that convergence occurs more rapidly), but any arbitrary value can be used with no effect on the final results. Third, working from one side of the tree to the

Figure A1. Illustration of the squared-change parsimony algorithm. (1) An hypothesis of phylogenetic relationships for six extant species, named A-F. Internal nodes on the tree are given the names g, h, i, j, and rt (for root). Note that nodes g and h are effectively a single node in a polytomy, because one of the internode branches has been set to zero length; this "hard" polytomy indicates a multiway speciation event. (2) The hypothetical measured phenotypes for a continuous-valued character (e.g., body size, plasma osmolarity) for each of the six species (shown as integers only for simplicity). The algorithm begins by "seeding" each of the nodes with an arbitrary value for its phenotype; in (2) the simple mean of the six extant species is used. (3) Final results of the iterative algorithm (i.e., after values converge); these values at the nodes are those which minimize the sum of the squared changes (23 in this example) over the entire tree. The branch lengths shown here are arbitrary; for computations, all were set to equal length one. ➔

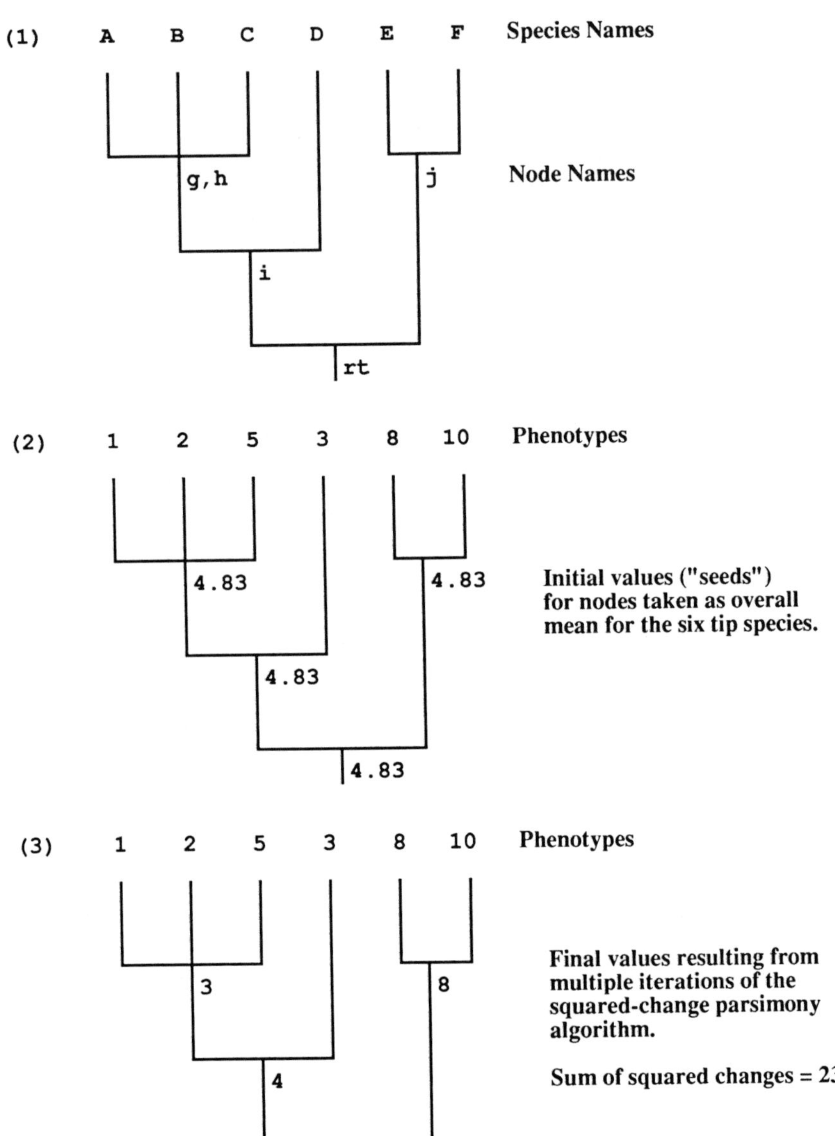
(1)
A B C D E F
Species Names
g,h
j
Node Names
i
rt
(2)
1 2 5 3 8 10
Phenotypes
4.83
4.83
Initial values ("seeds") for nodes taken as overall mean for the six tip species.
4.83
4.83
(3)
1 2 5 3 8 10
Phenotypes
3
8
Final values resulting from multiple iterations of the squared-change parsimony algorithm.
Sum of squared changes = 23.
4
6

other, then back and forth many times, the nodal values are replaced with the average of the values of all adjacent nodes. [If branch segment lengths are not all set equal to one, then this average is computed by weighting by the reciprocal of the branch lengths (see Maddison, 1991; Martins and Garland, 1991).] In a fully bifurcating phylogenetic tree (no polytomies), each node is thus replaced with the average of the three adjacent nodes (which may be tip nodes or internal ones), except for the root node, which is replaced with the average of its two adjacent nodes. At each pass through the tree, the nodal values will change. Eventually, however, the magnitude of the changes becomes smaller and smaller, and the entire set of nodal values converges on the set of values which minimizes the (weighted) sum of the squared changes over the entire tree. These values are unique; no other values for any or all of the nodes would yield a smaller sum of squared changes over the whole tree. This procedure for finding a set of nodal values was hence termed "minimum evolution" by Huey and Bennett (1987; and also by Martins and Garland, 1991; Garland *et al.*, 1991), who first employed it in the present context, but is more accurately termed "squared-change parsimony" (or weighted squared-change parsimony: Maddison, 1991).

Squared-change parsimony also works with "hard" polytomies, reflecting true, simultaneous multiway speciation events. With hard polytomies, a node is again reconstructed as the (weighted) mean of all of its adjacent nodes, which may be more than three (or, for the root node, more than the usual two). A program that computes squared-change parsimony reconstructions with hard polytomies (PDSQCHP) is available from the senior author. As mentioned in the Discussion, formal ways of dealing with the uncertainty represented by "soft" polytomies (which reflect lack of knowledge about the true topology) have yet to be developed.

All parsimony procedures (including those for discrete characters) share the characteristic that they cannot possibly reconstruct values at interior nodes that are outside of the range of values observed at the tips of the tree. This is an unrealistic limitation in some cases, because evolutionary trends may actually result in terminal taxa (e.g., species alive today) that have phenotypes very different from those of their ancestors (e.g., hypothetical example in Fig. 10 of Grafen, 1989; real example in Fig. 2 of Garland *et al.*, 1993). In the context of a given data set, the only way to increase the possible range of ancestral values reconstructed by parsimony procedures is to add additional outgroup taxa to the data set.

A Simple Example of Squared-Change Parsimony

Figure A1 illustrates the squared-change parsimony algorithm with a simple example, including a polytomy (a node with more than two descendants). (For simplicity, we have set all branch segments equal to unity for computations.) Note that the value reconstructed for node **i**, 4, is somewhat different from the simple mean of the six tip species (4.83). The value reconstructed at the root of the tree, 6, differs even more from the simple mean of the tip values.

Justification of Squared-Change Parsimony

Although the procedure has been used many times, the general justification of squared-change parsimony has not been thoroughly considered (see Huey and

Bennett, 1987; Losos, 1990; Harvey and Pagel, 1991; Maddison, 1991; Martins and Garland, 1991; Maddison and Maddison, 1992; Pagel, 1993; McArdle and Rodrigo, 1994; Miles and Dunham, 1996). Under a Brownian motion model of character evolution, squared-change parsimony reconstructions are "similar but not equivalent to a maximum-likelihood estimate" (Huey and Bennett, 1987, p. 1103; see also references in Edwards, 1996). This claim applies if the branch lengths used in computations are in units equal or proportional to expected variance of character evolution (Felsenstein, 1985; Martins and Garland, 1991). In the example shown in Figure A1, all branch lengths were effectively set to one for computations. With variable branch lengths, the algorithm also works, but nodal values are computed as weighted averages, with weighting based on the reciprocal of the branch lengths connecting each node to its adjacent nodes (see Maddison, 1991; Martins and Garland 1991). Brownian motion evolution with all branch segment lengths set equal to unity is equivalent to a "speciational" model of character evolution (termed "punctuational" in Huey and Bennett, 1987; Martins and Garland, 1991) (for more discussion of alternative models for simulating character evolution, see Garland *et al.*, 1993; Bjorklund, 1994; Díaz-Uriarte and Garland, 1996).

According to Maddison (1991, pp. 311 and 312), the squared-change parsimony reconstruction has maximum posterior probability under a Brownian motion model. In other words, minimizing the sum of (weighted) squared changes maximizes the Bayesian posterior probability (i.e., the probability of the reconstructed ancestral character states, given the tip data). (Maximizing the posterior probability is not the same as maximum likelihood estimation: the likelihood of the hypothesis is only one part of the posterior probability, the other two terms being the probability of the hypothesis and the probability of the data.) This, of course, is conditional on the correctness of the model of character evolution (Brownian motion), the topology of the tree, and the branch lengths (which should be in units proportional to expected variance of character evolution).

A closely related view as to the justification of squared-change parsimony stems from the more general context of fitting a model to some empirical data. We have a set of observed data, including the mean phenotypes of a series of species and the relevant phylogenetic topology and branch lengths. To these data we fit a model, i.e., the set of estimates for the phenotypes at all of the internal nodes. We should prefer that model which "best" fits the data (i.e., is most consistent with the data). If our operational definition of "best" is the smallest amount of squared change summed over the whole tree, then we can define "parsimony" in terms of consistency between model and data.

Finally, squared-change parsimony might be justified in any particular application by performing computer (Monte Carlo) simulations and comparing its performance with that of other comparative methods (e.g., Martins and Garland, 1991; Martins, 1996a). This type of justification would, of course, depend on the simulation parameters (including the model of evolution, topology, and branch lengths) being reasonable for the character(s) under study. We have performed one such set of simulations, and the results are described in the next section.

Although they should not be seen as justifications per se, squared-change parsimony also possesses some convenient properties. First, unlike the linear-parsimony alternatives (Butler and Losos, in review), it provides a unique point estimate, a single number, for the value of a continuous-valued character at each of the interior nodes (hypothetical ancestors) on a phylogenetic tree. Indeed, Losos (1990, p. 387) stated that "The major advantage of the squared-change parsimony approach is its analytical tractability." Second, squared-change parsimony is relatively easy to apply. Solutions can be found within minutes or hours, even for hundreds of taxa, with existing computer programs and personal computers. Like it or not, ease of application drives many decisions regarding analytical procedures.

Computer Simulations to Test Performance of Squared-Change Parsimony

We compared how well squared-change parsimony estimates the value at the root node of Figure 1 with the nonphylogenetic alternative of simply computing the mean value for all 172 taxa. We used the PDSIMUL program of Garland *et al.* (1993) to simulate simple Brownian motion character evolution (no limits on how far the phenotype could evolve), with all branch lengths set equal to one (this can be called "speciational" Brownian motion). We started each simulation at a value of 416.55, which is the simple mean of all 172 taxa. For each of 1,000 simulated data sets, we then used the PDERROR program of Díaz-Uriarte and Garland (1996) to compute (1) the simple mean of the simulated tip data and (2) the squared-change parsimony value at the root node. We focused on the root node for convenience: the estimate for the root node is the same for squared-change parsimony and for independent contrasts (see next section), and the latter is much faster to compute (and what PDERROR actually produces).

The simple means of the 1000 simulated data sets ranged from -105 to 955, with an overall mean of 416.10 (+/- standard error = 4.623, variance = 21,371). The squared-change parsimony reconstructions spanned a narrower range, from 167 to 636, with an overall mean of 416.67 (+/- standard error = 2.408, variance = 5,800). Obviously, the means of these distributions do not differ significantly from the true value at the root of the tree (416.55), so both yielded unbiased estimates of the value at the root of the tree. However, comparison of the standard errors (or variances) indicates that the spread of the distributions of the two estimators was very different. The squared-change parsimony reconstruction is much less variable, and hence performs much better.

Figure A2. Worked examples of fixing nodal values when using the squared-change parsimony algorithm. Node **i** (names shown in panel (1) of Fig. A1) is fixed at different values, and the algorithm is applied. When one node is thus fixed, reconstructed values at other nodes are different than when all nodes are free to change (see Fig. A1). Moreover, the sum of squared changes over the whole tree is always greater when any node is fixed at other than its freely-reconstructed value (in Fig. A1, node i was reconstructed as a value of 4, and the sum of squared changes was 23). See Fig. A3 for a graphical representation of this effect. ➔

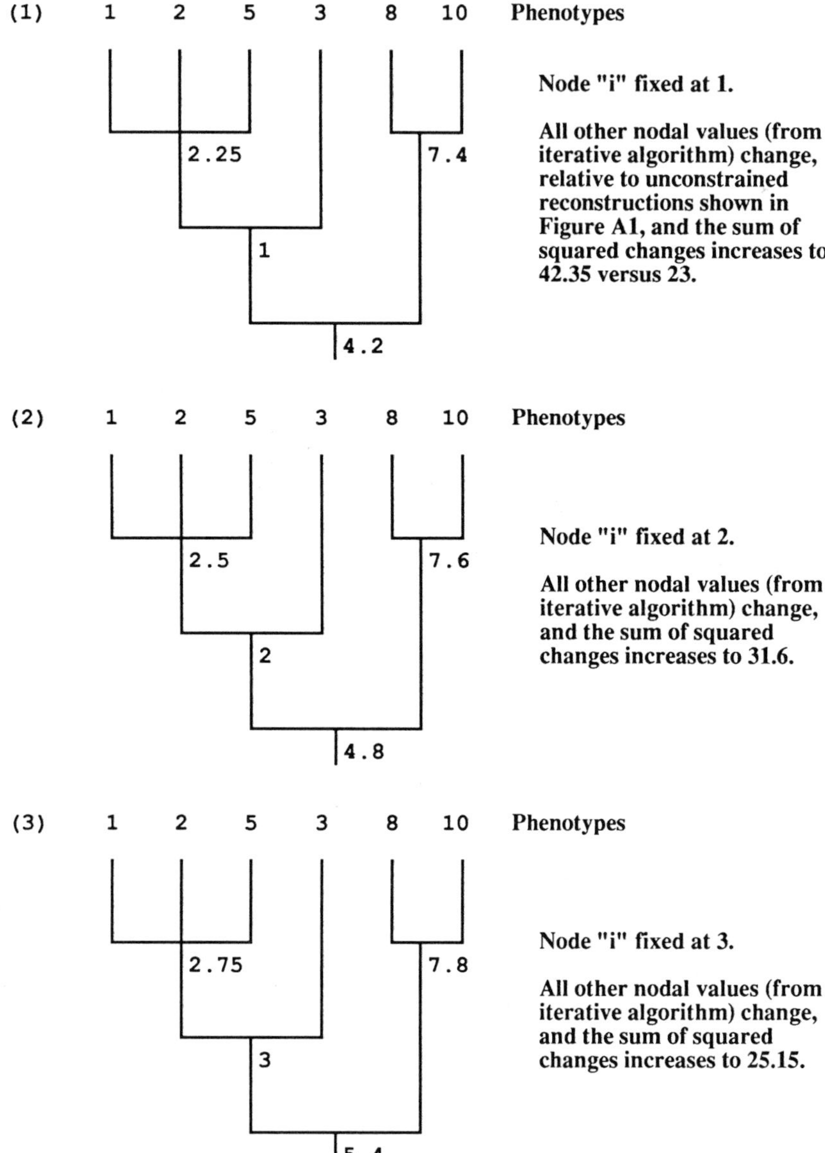
(1)
1 2 5 3 8 10
Phenotypes
2.25
7.4
1
4.2
Node "i" fixed at 1.
All other nodal values (from iterative algorithm) change, relative to unconstrained reconstructions shown in Figure A1, and the sum of squared changes increases to 42.35 versus 23.
(2)
1 2 5 3 8 10
Phenotypes
2.5
7.6
2
4.8
Node "i" fixed at 2.
All other nodal values (from iterative algorithm) change, and the sum of squared changes increases to 31.6.
(3)
1 2 5 3 8 10
Phenotypes
2.75
7.8
3
5.4
Node "i" fixed at 3.
All other nodal values (from iterative algorithm) change, and the sum of squared changes increases to 25.15.

Figure A2; Panel 1.

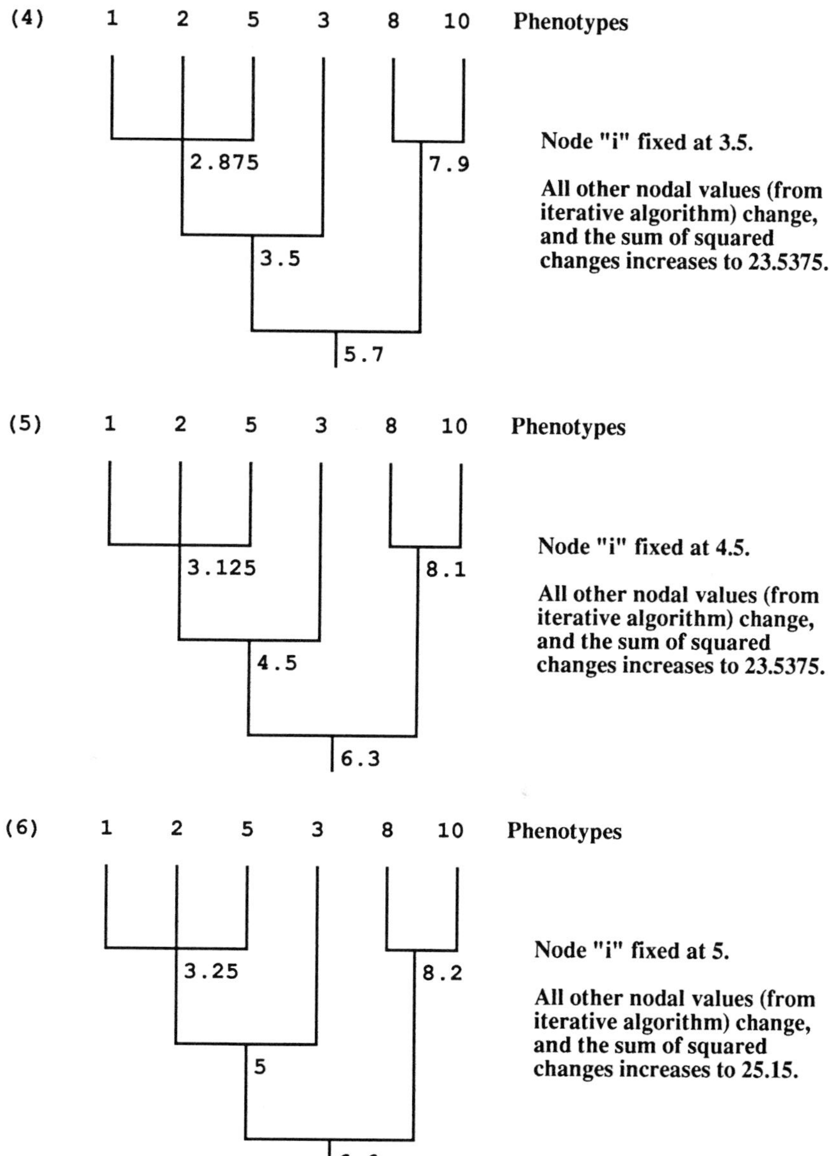
(4)
1 2 5 3 8 10 Phenotypes
2.875
7.9
3.5
5.7
Node "i" fixed at 3.5.
All other nodal values (from iterative algorithm) change, and the sum of squared changes increases to 23.5375.
(5)
1 2 5 3 8 10 Phenotypes
3.125
8.1
4.5
6.3
Node "i" fixed at 4.5.
All other nodal values (from iterative algorithm) change, and the sum of squared changes increases to 23.5375.
(6)
1 2 5 3 8 10 Phenotypes
3.25
8.2
5
6.6
Node "i" fixed at 5.
All other nodal values (from iterative algorithm) change, and the sum of squared changes increases to 25.15.

Figure A2; Panel 2.

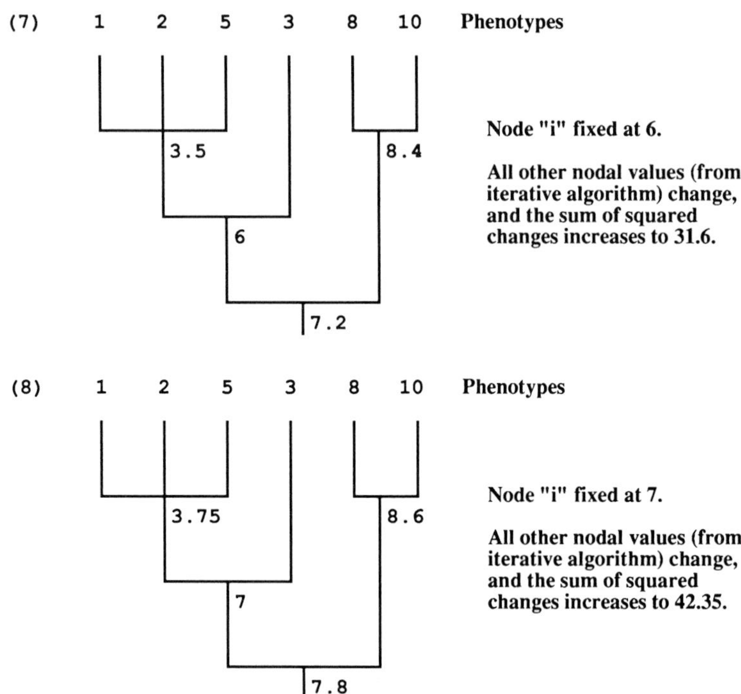

Figure A2; Panel 3.

In the foregoing simulations, simple Brownian motion models a character that is equally likely to increase or decrease in value at any point during its evolution; the magnitude of possible changes is set proportional to the length of the branches on the phylogeny (which can be set to equal length, as we have done, or can be of variable length); the magnitude of possible increases and decreases is also set to be equal, on average, so no overall trends in evolution will occur (except by chance); and no limits are placed on how far the character can evolve. This is the simplest possible model for the evolution of continuous-valued characters, and is the usual starting point in simulation studies. Brownian motion is considered to be a reasonable model for phenotypic evolution by random genetic drift and for evolution in response to certain forms of stochastically changing selection (see Felsenstein, 1985, 1988a).

Obviously, however, Brownian motion may not be a realistic model for the evolution of plasma osmolarities in vertebrates. For example, plasma osmolarity does not evolve without bounds (e.g., see Fig. 2; also note that negative values are physically impossible). Consequently, we also performed a second simulation, intended to be more biologically realistic. Many additional simulations are possible (e.g., see Garland *et al.*, 1993; Díaz-Uriarte and Garland, 1996), but are beyond the scope of this chapter. This second simulation was like the first, except that initial values were set at 365 mOsm (this is the value reconstructed by squared-change parsimony for all 172 taxa; see Fig. 3) and desired final values at 416.55 (the actual simple mean for all 172 taxa), such that a trend for increasing plasma osmolarity was modeled. In addition, we used the "Replace" option of PDSIMUL to put limits on how far the trait could evolve, equal to the range of values in our real data set (181 and 1118; see Fig. 2). (The desired variances across the tips of the simulated data sets were left to be the same as the actual data, which is the default in PDSIMUL.)

With the more complicated model, simulating an evolutionary trend and limits, the squared-change parsimony estimates of the root value were again superior to the simple mean. The simple means ranged from 326 to 804, with an overall mean of 483.69 (+/- standard error = 2.913, variance = 8487). The squared-change parsimony reconstructions spanned a slightly narrower range, from 272 to 693, with an overall mean of 409.01 (+/- standard error = 2.000, variance = 3999). The means of both of these distributions differ significantly from the true value at the root of the tree (365); thus, both estimators were biased and tended to overestimate the value at the root of the tree. The squared-change parsimony estimator was, however, considerably less biased and also had a lower variance.

Squared-Change Parsimony and Independent Contrasts

Squared-change parsimony is not the same as another comparative method used commonly with continuous-valued characters, Felsenstein's (1985) phylogenetically independent contrasts (see Grafen, 1989; Losos, 1990; Martins and Garland, 1991; Garland *et al.*, 1992, 1993; Bjorklund, 1994; McPeek, 1995; Díaz-Uriarte and Garland, 1996; Martins, 1996a; Martins and Hansen, 1996). Squared-change parsimony is a "directional" comparative method (*sensu* Harvey and Pagel, 1991), whereas independent contrasts is a "nondirectional" or "cross-sectional" (Pagel, 1993) method. Directional methods reconstruct ancestral states at each node of a phylogeny and then compute changes between the inferred nodes and other nodes, including the measured tip values. Independent contrasts, on the other hand, uses phylogenetic information (topology and branch lengths) to transform the tip data into a set of N-1 contrasts (worked examples in Garland and Adolph, 1994) that are independent and identically distributed (at least in principle), and which can then be used in many conventional statistical procedures. Independent contrasts also involve calculation of values for interior nodes as an intermediate part of its computations. These nodal values constitute what can be termed "local parsimony" reconstructions, but they do not provide the globally most parsimonious solution because only the information from daughter nodes is used to estimate a given node (Maddison, 1991). Independent contrast nodal values are computed in this way because the goal is to provide independent data points that can then be used in subsequent statistical analyses. The nodal values from squared-change parsimony, on the other hand, are not statistically independent, because the value of each affects the value of all others (for example, Fig. A2). Independent contrast nodal values do not have any of the properties discussed above under the section titled "Justification of Squared-Change Parsimony," therefore they should not be used as estimators of ancestral values per se.

One special case exception exists. The value reconstructed at the root (basal) node is identical for independent contrasts and squared-change parsimony. This makes sense, because the local parsimony of independent contrasts is the same as global parsimony when the entire tree is considered (i.e., from the root node through all descendants). Computer programs for independent contrasts are much faster than those for squared-change parsimony, therefore if one is interested only in the root node, then the former can be used (e.g., PDTREE of Garland *et al.*, 1993; PDERROR program of Díaz-Uriarte and Garland, 1996; others cited in Martins and Hansen, 1996).

Recently, Ryan and Rand (1995; see also Ryan, 1996) have used nodal values from independent contrasts in preference to those from squared-change parsimony, but this practice is unjustified. Ryan and Rand (1995) stated that the two analyses gave similar results in their application, but this will not always be the case.

Indexing Uncertainty in Squared-Change Parsimony Reconstructions

As noted above, linear parsimony reconstructions of discretely valued characters often yield ambiguous values for nodes (Maddison, 1994, 1995). This may seem inconvenient, but it is also useful as an indicator of our incertitude in the reconstructed values (see Appendix 2 in Losos, 1990; Butler and Losos, in review). Recently, Maddison (1995) has discussed ways of calculating probability distributions of ancestral states reconstructed by linear parsimony, given the assumption of a stochastic model of character evolution (see also Schultz *et al.*, 1996). Comparable methods have not yet been fully developed for squared-change parsimony reconstructions, but several heuristic approaches are possible. We explain and then apply two of them, and consider some other possibilities under Discussion. Here we assume that the only data available to the comparative biologist are estimates of the mean phenotype for a series of taxa (the tip values); if estimates of the phenotypic variances for each species are available, then additional procedures are possible (Lynch, 1991; Maddison, 1991; Martins, 1994; McArdle and Rodrigo, 1994; J. Felsenstein, pers. comm.; see also Chevalier, 1991).

Fixing Interior Nodes.–One approach was developed by Huey and Bennett (1987; see also Chevalier, 1991) and is illustrated in Figures A2 and A3. This approach involves fixing a node at an arbitrary value, one other than that which the (unconstrained) squared-change parsimony algorithm delivered. The squared-change parsimony algorithm can then be rerun, but with the value at one (or more) node(s) constrained not to change. The result of this constrained character optimization will always be higher values for the (weighted) sum of the squared changes over the whole tree.

Figures A2 and A3 illustrate that, for node i, fixing it at any value other than 4–the one reconstructed by the unconstrained squared-change parsimony algorithm (Fig. A1)–results in different values at all other internal nodes and a greater amount of squared change over the entire phylogenetic tree. The same principle would hold for any other internal node. If we have reason to believe that

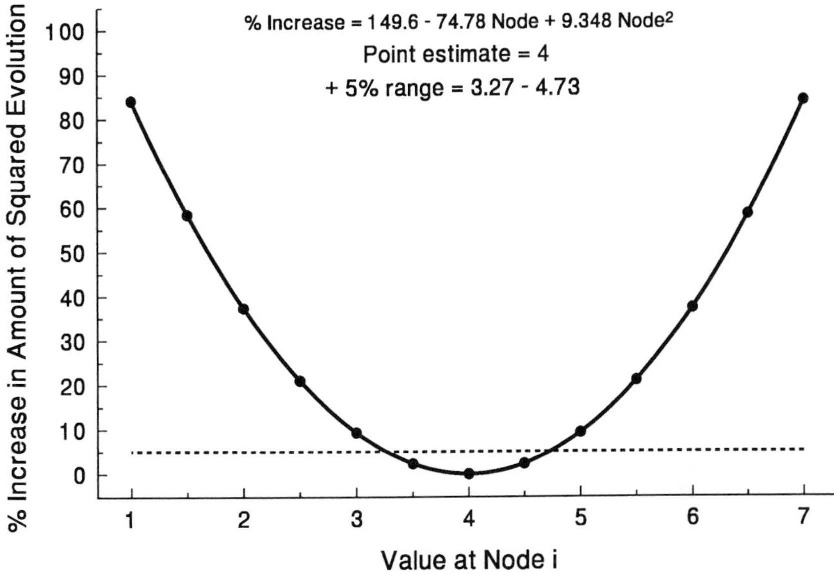

Figure A3. Graphical summary of consequences of fixing nodal values when using the squared-change parsimony algorithm (from Fig. A2).

greater amounts of squared change are less likely, then these other values for node i should themselves be considered less likely to accurately reflect the true (but unknown) value for that ancestral node.

Figure A3 (like Fig. 4 of Huey and Bennett, 1987) shows that over a range of values around the point estimate of 4 the sum of squared changes over the whole phylogeny does not increase very much. For example, in the range 3.27-4.73, the amount of squared evolution increases by only 5%. We caution readers that these intervals cannot be interpreted as "confidence intervals" in the conventional statistical sense.

It is also possible to consider the consequences of fixing a node at a value that is actually outside of the range of values observed for the tips. For example, if we fix node i at a value of 11, the sum of squared change increases to 128.35, or 458% greater than if the value of node i is 4. In general, we can consider any value as a possibility for any ancestral node, but some of these hypothetical values would require there to have been a relatively huge amount of evolution, as compared with the minimum possible, which occurs when all nodes are taken as the values reconstructed by unconstrained squared-change parsimony. In real examples, some values may not be physically (e.g., body mass or plasma osmolarity less than 0) or biologically possible (e.g., vertebrate body temperature greater than about 50 Celsius), and so can be excluded a priori from consideration.

Jackknifing.–Another heuristic way to index uncertainty (or dispersion) in nodal values reconstructed by squared-change parsimony is to use the statistical procedure called jackknifing (Sokal and Rohlf, 1981, pp. 795-799; Manly, 1991; Crowley, 1992; Efron and Tibshirani, 1993). Jackknifing is a way of using the observed data points themselves to gauge the uncertainty in some statistic that can be computed from the data. The general procedure is to first compute the statistic of interest, such as the mean, based on all N data points. As in ordinary statistics, this estimate based on the entire sample is taken as the best estimate of the statistic of interest.

The second step in jackknifing is to recompute the statistic N times, based on deleting each data point from the sample in turn. This is termed resampling. [This "delete-one jackknife" is common and computationally the simplest, but greater numbers of data points can also be deleted and doing so provides better results in some cases (e.g., see Wu, 1986, and discussions following]. The deletion of data points is always done without replacement, unlike the more general method termed "bootstrapping" (see Efron and Tibshirani, 1993).] Thus, if we had a sample of 100 data points, we would recompute the mean 100 times after deleting each of the data points in turn. We would then compute "pseudovalues" for each of these as (N * mean for whole sample) - (N-1) (mean with one data point deleted). The approximate standard error of our original mean is then taken as the square root of (variance of the pseudovalues/N). This approximate standard error can then be used to set confidence intervals by reference to a t-distribution; degrees of freedom are usually assumed to be N-1. In the text, we present the jackknifing computations for the ancestral amniote's plasma osmolarity, as reconstructed by squared-change parsimony. Again, however, this is only a heuristic device, because the degrees of freedom associated with the computations are generally unknown in the face of phylogenetic relatedness and hence nonindependence of the tip values (e.g., see Harvey and Pagel, 1991; Martins and Garland, 1991; Garland et al., 1993; Pagel, 1992, 1993). In addition, the validity of jackknifing needs to be proved analytically or justified by computer simulation for each different type of application (Efron and Tibshirani, 1993).

LITERATURE CITED

Abler, W. L. 1992. The serrated teeth of tyrannosaurid dinosaurs, and biting structures in other animals. *Paleobiology*, 18:161-183.

Altman, P. L., and D. S. Dittmer (eds.). 1971. *Biological Handbooks: Respiration and Circulation.* Bethesda: FASEB Committee on Biological Handbooks.

Alvarado, R. H. 1972. The effects of dehydration on water and electrolytes in *Ambystoma tigrinum*. *Physiological Zoology*, 45:45-53.

Alvarado, R. H. 1979. Amphibians. Pages 261-303 in: *Comparative Physiology of Osmoregulation in Animals, Vol 1* (G. M. O. Maloiy, ed.). New York: Academic Press.

Arnold, E. N. 1994. Investigating the origins of performance advantage: adaptation, exaptation and lineage effects. Pages 123-168 in: *Phylogenetics and Ecology* (P. Eggleton and R. I. Vane-Wright, eds.). Linnean Society Symposium Series Number 17. London: Academic Press.

Balinsky, J. B. 1981. Adaptation of nitrogen metabolism to hyperosmotic environment in Amphibia. *Journal of Experimental Zoology*, 215:335-350.

Barbadillo, L. J. 1989. *La Guía de Incafo de los Anfibios y Reptiles de la Península Iberica, Islas Baleares, y Canarias*. Madrid: Incafo.

Barton, M. 1979. Serum osmoregulation in two species of estuarine blennioid fish, *Anoplarchus purpurescens* and *Pholis ornata*. *Comparative Biochemistry and Physiology*, 643A:305-307.

Becker, E. L., R. Bird, W. Kelly, J. Schilling, S. Soloman, and N. Young. 1958. Physiology of marine teleosts. I. Ionic composition of tissue. *Physiological Zoology*, 31:224-227.

Bennett, A. F., and J. A. Ruben. 1986. The metabolic and thermoregulatory status of therapsids. Pages 207-218 in: *The Ecology and Biology of Mammal-like Reptiles* (N. Hotton, P. D. MacLean, J. J. Roth, and E. C. Roth, eds.). Washington, D.C.: Smithsonian Institution Press.

Bentley, P. J. 1971. *Endocrines and Osmoregulation: A Comparative Account of the Regulation of Salt and Water in Vertebrates*. Heidelberg: Springer-Verlag.

Berman, D. S, S. S. Sumida, and R. E. Lombard. 1992. Reinterpretation of the temporal and occipital regions in *Diadectes* and the relationships of diadectomorphs. *Journal of Paleontology*, 66:481-499.

Bernard, G. R., R. A. Wynn, and G. G.Wynn. 1966. Chemical anatomy of the pericardial and perivisceral fluids of the stingray *Dasyatis americana*. *Biological Bulletin*, 130:18-27.

Bjorklund, M. 1994. The independent contrast method in comparative biology. *Cladistics*, 10:425-433.

Bleiweiss, R., J. A. W. Kirsch, and N. Shafi. 1995. Confirmation of the Sibley-Ahlquist "tapestry." *The Auk*, 112:87-97.

Block, B. A., J. R. Finnerty, A. F. R. Stewart, and J. Kidd. 1993. Evolution of endothermy in fish: mapping physiological traits on a molecular phylogeny. *Science*, 260:210-214.

Boucher-Firley, S. 1934. Sur quelques constitutants chimiques du sang de Congre et de Murene. *Bulletin Institute Oceanographique*, 651:1-6.

Braun, E. J. 1978. Renal response of the starling (*Sturnus vulgaris*) to an interavenous salt load. *American Journal of Physiology*, 234:F270-F278.

Braysher, M. L. 1976. The excretion of hyperosmotic urine and other aspects of the electrolyte balance of the lizard *Amphibolurus maculosus*. *Comparative Biochemistry and Physiology*, 54A:341-345.

Bridges, C. R. 1993. Ecophysiology of intertidal fish. Pages 377-400 in: *Fish Ecophysiology* (J. C. Rankin and F. B. Jensen, eds.). London: Chapman & Hall.

Brooks, D. R., and D. A. McLennan. 1991. *Phylogeny, Ecology, and Behavior. A Research Program in Comparative Biology*. Chicago: The University of Chicago Press.

Brull, L. and E. Nizet. 1953. Blood and urine constituents of *Lophius piscatorius*. *Journal of the Marine Biological Association of the United Kingdom*, 32:321-328.

Bull, J. M. and R. Morris. 1967. Studies on freshwater osmoregulation in th ammocoete larva of *Lampetra planeri* (Blach). 1. Ionic constituents, fluid compartments, ionic compartments and water balance. *Journal of Experimental Biology*, 47:485-494.

Burggren, W. W., and W. E. Bemis. 1990. Studying physiological evolution: paradigms and pitfalls. Pages 191-238 in: *Evolutionary Innovations* (M. H. Nitecki, ed.). Chicago: The University of Chicago Press.

Butler, D. G., W. C. Clarke, E. M. Donaldson, and R. W. Langford. 1969. Surgical adrenalectomy of a teleost fish (*Anguilla rostrata* Lesueur): Effect on plasma cortisol and tissue electrolyte and carbohydrate concentrations. *General and Comparative Endocrinology*, 12:502-514.

Campbell, J. P. , R. M. Aiyawar, E. R. Berry, and E. G. Huff. 1967. Electrolytes in frog skin secretions. *Comparative Biochemistry and Physiology*, 23:213-233.

Carpenter, R.E., and M.A. Stafford. 1970. The secretory rates and the chemical stimulus for secretion of the nasal salt glands in the Rallidae. *Condor*, 72:316-324

Carroll, R. L. 1988. *Vertebrate Paleontology and Evolution*. New York: Freeman.

Chevalier, C. D. 1991. Aspects of thermoregulation and energetics in the Procyonidae (Mammalia: Carnivora). Unpubl. Ph.D. Dissertation, University of California, Irvine.

Chosniak, I., C. Wittenberg, J. Rosenfeld, and A. Shkolnik 1984. Rapid rehydration of and kidney function in the black Bedouin goat. *Physiological Zoology*, 57:573-579.

Christidis, L. and W. E. Boles. 1994. *The Taxonomy and Species of Birds of Australia and Its Territories*. RAOO. Victoria, Australia.

Conant, R., and J. T. Collins. 1991. *A Field Guide to Reptiles and Amphibians: Eastern and Central North America, 3rd edition.* Boston: Houghton Mifflin.

Conte, F. P. 1965. Effects of ionizing radiation on osmoregulation in fish *Oncorhynchus kisutch*. *Comparative Biochemistry and Physiology*, 15:292-302.

Corbet, G. B., and J. E. Hill. 1991. *A World List of Mammalian Species*. Oxford: Oxford University Press.

Crowley, P. H. 1992. Resampling methods for computation-intensive data analysis in ecology and evolution. *Annual Review of Ecology and Systematics*, 23:405-447.

Dantzler, W. H., and B. Schmidt-Nielsen. 1966. Excretion in fresh-water turtle (*Pseudemys scripta*) and desert tortoise (*Gopherus agassizii*). *American Journal of Physiology*, 210:198-210.

de Carvalho, M. R. 1996. Higher-level elasmobranch phylogeny, basal squaleans and paraphyly. In M. L. J. Stiassny, L. R. Parenti, and G. D. Johnson, *The Interrelationships of Fishes*. New York: Academic Press. In press.

de Queiroz, A., M. J. Donoghue, and J. Kim. 1995. Separate versus combined analysis of phylogenetic evidence. *Annual Review of Ecology and Systematics*, 26:657-681.

de Queiroz, K. 1996. Including the characters of interest during tree reconstruction and the problems of circularity and bias in studies of character evolution. *The American Naturalist*. In press.

Dessauer, H. C. 1970. Blood chemistry of reptiles: physiological and evolutionary aspects. Pages 1-72 in: *Biology of the Reptilia, Volume 3: Morphology* (C. Gans and E. Parsons, eds.). New York: Academic Press.

Deutsch H., H. T. Hammel, E. Simon, and C. Simon-Oppermann. 1979. Osmolality and volume factors in salt gland control of pekin ducks after adaptation to chronic salt loading. *Journal of Comparative Physiology*, 129:301-308

deVlaming, V. L., and M. Sage. 1973. Osmoregulation in the euryhaline elasmobranch, *Dasyatis sabina*. *Comparative Biochemistry and Physiology*, 45A:31-44.

Dial, B. E., and L. L. Grismer. 1992. A phylogenetic analysis of physiological-ecological character evolution in the lizard genus *Coleonyx* and its implications for historical biogeographic reconstruction. *Systematic Biology*, 41:178-195.

Díaz-Uriarte, R., and T. Garland, Jr. 1996. Testing hypotheses of correlated evolution using phylogenetically independent contrasts: sensitivity to deviations from Brownian motion. *Systematic Biology*, 45:27-47.

Dill, D. B., and H. T. Edwards. 1931. Physico-chemical properties of crocodile blood (*Crocodylus acutus* Cuvier). *Journal of Biological Chemistry*, 90:515-530.

Dobbs, G. H. III, and A. L. DeVries. 1975. Renal function in Antarctic teleost fishes: serum and urine composition. *Marine Biology*, 29:59-70.

Doolittle, R. F., C. Thomas, and W. Stone, Jr. 1960. Osmotic pressure and aqueous humor formation in dogfish. *Science*, 132:36-37.

Dunham, P. B., J. Klimczak, and P. J. Logue. 1993. Swelling activation of K-Cl co-transport in LK sheep erythrocytes: a three-state process. *Journal of General Physiology*, 101:733-766.

Dunson, W. A. 1980. The relation of sodium and water balance to survival in seawater of estuarine and freshwater races of the snakes *Nerodia fasciata sipodon* and *N. valida*. *Copeia*, 1980:268-280.

Dunson, W. A., and A. M. Taub. 1967. Extra-renal salt excretion in sea snakes (*Laticauda*). *American Journal of Physiology*, 213:975-982.

Dunson, W. A. and R. D. Weymouth. 1965. Active uptake of sodium by softshell turtles (*Trionyx spinifer*). *Science*, 149:67-69.

Eastman, J. T. 1993. *Antarctic Fish Biology: Evolution in a Unique Environment.* San Diego: Academic Press.

Edwards, A. W. F. 1996. The origin and early development of the method of minimum evolution for the reconstruction of phylogenetic trees. *Systematic Biology*, 45:79-91.

Eernisse, D. J., and A. G. Kluge. 1993. Taxonomic congruence versus total evidence, and amniote phylogeny inferred from fossils, molecules, and morphology. *Molecular Biology and Evolution*, 10:1170-1195.

Efron, B., and R. J. Tibshirani. 1993. *An Introduction to the Bootstrap.* New York: Chapman & Hall.

Eggleton, P., and R. I. Vane-Wright, eds. 1994. *Phylogenetics and Ecology*. Linnean Society Symposium Series Number 17. London: Academic Press.

Eldredge, N., and S. M. Stanley (eds.). 1984. *Living Fossils*. Springer Verlag: New York. 291 Pages

Ensor, D. M., and J. G. Phillips. 1972. The effect of dehydration on salt and water balance in gulls (*Larus argentatus* and *L. fuscus*). *Journal of Zoology, London*, 168:127-137

Erasmus, T. 1978. The handling of constant volumes of various concentrations of seawater by the jackass penguin *Spheniscus demersus*. *Zoologica Africana*, 13:71-80.

Evans, D. H. 1967. Sodium, chloride, and water balance of the intertidal teleost, *Xiphister atropurpureus* I. Regulation of plasma concentration and body water content. *Journal of Experimental Biology*, 47:513-517.

Evans, D. H. 1979. Fish. Pages 305-390 in: *Comparative Physiology of Osmoregulation in Animals, Volume 1* (G. M. O. Maloiy, ed.). New York: Academic Press.

Falconer, D. S., and T. F. C. Mackay. 1996. *Introduction to Quantitative Genetics, 4th Edition*. Essex, England: Longman.

Farris, J. S. 1970. Methods for computing Wagner trees. *Systematic Zoology*, 19:83-92.

Farris, J. S. 1983. The logical basis of phylogenetic analysis. Pages 7-36 in: *Advances in Cladistics, Volume 2. Proceedings of the Second Meeting of the Willi Hennig Society* (N. I. Platnick and V. A. Funk, eds.).New York: Columbia University Press.

Felsenstein, J. 1985. Phylogenies and the comparative method. *The American Naturalist,* 125:1-15.

Felsenstein, J. 1988a. Phylogenies and quantitative characters. *Annual Review of Ecology and Systematics*, 19:445-471.

Felsenstein, J. 1988b. Phylogenies from molecular sequences: inference and reliability. *Annual Review of Genetics*, 22:521-565.

Felsenstein, J. 1992. Phylogenies from restriction sites: a maximum-likelihood approach. *Evolution*, 46:159-173.

Ferreira, H. G., and C. H. Jesus. 1973. Salt adaptation in *Bufo bufo*. *Journal of Physiology (London)*, 228:583-600.

Forster, R. P., and F. Berglund. 1956. Osmotic diuresis and its effect on the total electrolyte distribution in plasma and urine of the aglomerular teleost, *Lophius americanus*. *Journal of General Physiology*, 39:349-359.

Frenkel, G. and P. F. Kraicer. 1971. The inhibition of adreno-corticol function in the lizard *Agama stellio*. *General and Comparative Endocrinology*, 17:158-163.

Friday, A. 1987. Models of evolutionary change and the estimation of evolutionary trees. *Oxford Surveys in Evolutionary Biology*, 4:61-88.

Frost, D. R. 1992. Phylogenetic analysis and taxonomy of the *Tropidurus* group of lizards (Iguania: Tropiduridae). *Novitates*, 3033:1-68.

Frost, D. R., and R. Etheridge. 1989. A phylogenetic analysis and taxonomy of iguanian lizards (Reptilia: Squamata). *Miscellanous Publications, University of Kansas Museum of Natural History*, 81:1-65.

Gaffney, E. S., and P. A. Meylan. 1988. A phylogeny of turtles. Pages 157-219 in: *The Phylogeny and Classification of the Tetrapods. Volume 1: Amphibians, Reptiles, Birds* (M. J. Benton, ed.). Oxford: Clarendon Press.

Gans, C. 1970. Respiration in early tetrapods–the frog is a red herring. *Evolution*, 24:223-734.

Garland, T. Jr. 1994. Phylogenetic analyses of lizard endurance capacity in relation to body size and body temperature. In Pages 237-259 in: *Lizard Ecology. Historical and Experimental Perspectives* (L. J. Vitt and E. R. Pianka, eds.). Princeton, New Jersey: Princeton University Press.

Garland, T., Jr., and S. C. Adolph. 1991. Physiological differentiation of vertebrate populations. *Annual Review of Ecology and Systematics*, 22:193-228.

Garland, T., Jr., and S. C. Adolph. 1994. Why not to do two-species comparative studies: limitations on inferring adaptation. *Physiological Zoology*, 67:797-828.

Garland, T., Jr., and P. A. Carter. 1994. Evolutionary physiology. *Annual Review of Physiology*, 56:579-621.

Garland, T., Jr., and C. M. Janis. 1993. Does metatarsal/femur ratio predict maximal running speed in cursorial mammals? *Journal of Zoology, London*, 229:133-151.

Garland, T., Jr., A. W. Dickerman, C. M. Janis, and J. A. Jones. 1993. Phylogenetic analysis of covariance by computer simulation. *Systematic Biology*, 42:265-292.

Garland, T., Jr., P. H. Harvey, and A. R. Ives. 1992. Procedures for the analysis of comparative data using phylogenetically independent contrasts. *Systematic Biology*, 41:18-32.

Garland, T., Jr., R. B. Huey, and A. F. Bennett. 1991. Phylogeny and thermal physiology in lizards: A reanalysis. *Evolution*, 45:1969-1975.

Gilles-Baillien, M. 1970. Urea and osmoregulation in the diamondback terrapin *Malaclemys centrata centrata* (Latreille). *Journal of Experimental Biology*, 52:691-697.

Gilles-Baillien, M. and E. Schoffeniels. 1965. Variations saisonnieres dans la composition du sang de la tortue Greque, *Testudo hermanni* Gmelin. *Annales Societe Royale Zoologique de Belgique,* 95:75-79.

Goldman, N. 1990. Maximum likelihood inference of phylogenetic trees, with special reference to a poisson process model of DNA substitution and to parsimony analyses. *Systematic Zoology*, 39:345-361.

Goldstein, D. L., J. B. Williams, and E. J. Braun. 1990. Osmoregulation in the field by salt-marsh savannah sparrows *Passerculus sandwichensis beldingi*. *Physiological Zoology*, 63:669-682.

Gordon, M. S. 1959. Ionic regulation in the brown trout (*Salmo trutta* L.). *Journal of Experimental Biology*, 36:227-252.

Gordon, M. S. 1982. Water and solute metabolism. Pages 272-332 in: *Animal Physiology: Principles and Adaptations, 8th Edition*,. (M. S. Gordon, G. A. Bartholomew, A. D. Grinnell, C. B. Jorgensen, and F. N. White, eds.). New York: MacMillan.

Gordon, M. S., J. Boetius, I. Boetius, D. H. Evans, R. McCarthy, and L. C. Oglesby. 1965. Salinity adaptation in the mudskipper fish *Periophthalmus sobrinus*. *Hvalradets Skrifter*, 48:85-93.

Gordon, M. S., K. Schmidt-Nielsen, and H. M. Kelly. 1961. Osmotic regulation in the crab-eating frog (*Rana cancrivora*). *Journal of Experimental Biology*, 42:437-445.

Gottschalk, C. W., W. E. Lassiter, M. Mylle, K. J. Ullrich, B. Schmidt-Nielsen, R. O'Dell, and G. Pehling. 1963. Micropuncture study of composition of loop of Henle fluid in desert rodents. *American Journal of Physiology*, 204:532-535.

Grafen, A. 1989. The phylogenetic regression. *Philosophical Transactions of the Royal Society of London, B*, 326:119-157.

Green, B. 1972. Aspects of renal function in the lizard *Varanus gouldii*. *Comparative Biochemistry and Physiology*, 42A:747-756.

Greer, A. E. 1989. *The Biology and Evolution of Australian Lizards*. Chipping North, Australia: Surrey Beatty & Sons. 264 Pages

Griffith, R. W., B. L. Umminger, B. F. Grant, P. K. T. Pang, and G. E. Pickford. 1974. Serum composition of the coelacanth, *Latimeria chalumnae* Smith. *Journal of Experimental Zoology*, 187:87-102.

Grigg, G. C. 1981. Plasma homeostasis and cloacal urine composition in *Crocodylus porosus* caught along a salinity gradient. *Journal of Comparative Physiology*, 144:261-270.

Guyton, A. C. 1991. *Textbook of Medical Physiology, 8th Edition*. Philadelphia: W. B. Saunders Company.

Hackstein, J. H. P., and T. A. van Alen. 1996. Fecal methanogens and vertebrate evolution. *Evolution*, 50:559-572.

Harris, M. A., and K. Steudel. 1997. Ecological correlates of hind limb length in the Carnivora. *Journal of Zoology, London*. In press.

Harvey, P. H., and M. D. Pagel. 1991. *The Comparative Method in Evolutionary Biology*. Oxford: Oxford University Press.

Hay, J. M., I. Ruvinsky, S. B. Hedges, and L. R. Maxson. 1995. Phylogenetic relationships of amphibian families inferred from DNA sequences of mitochondrial 12S and 16S ribosomal RNA genes. *Molecular Biology and Evolution*, 12:928-937.

Hedges, S. B. and L. R. Maxson. 1993. A molecular perspective on lissamphibian phylogeny. *Herpetological Monographs*, 7:27-42.

Heise, P. J., L. R. Maxson, H. G. Dowling, and S. B. Hedges. 1995. Higher-level snake phylogeny inferred from mitochondrial DNA sequences of 12S rRNA and 16S rRNA genes. *Molecular Biology and Evolution*, 12:259-265.

Hillenius, W. L. 1994. Turbinates in therapsids: evidence for Late Permian origin of mammalian endothermy. *Evolution*, 48:207-229.

Hillis, D. M. 1995. Approaches for assessing phylogenetic accuracy. *Systematic Biology*, 44:3-16.

Hillis, D. M., C. Moritz, and B. K. Mable (eds.). 1996. *Molecular Systematics, 2nd edition*. Sunderland, Mass.: Sinauer Associates.

Hochachka, P. W., and G. N. Somero. 1984. *Biochemical Adaptation*. Princeton, New Jersey: Princeton University Press.

Hoffert, J. R., and P. O. Fromm. 1966. Effect of carbonic anhydrase inhibition on aqueous humor and blood bicarbonate ion in the teleost (*Salvelinus namaycush*). *Comparative Biochemistry and Physiology*, 18:333-340.

Holmes, W. N., and E. M. Donaldson. 1969. The body compartments and the distribution of electrolytes. Pages 1-89 in: *Fish Physiology, Volume I* (W. S. Hoar and D. J. Randall, eds.). New York: Academic Press.

House, C. R. 1963. Osmotic regulation in the brackish water teleost, *Blennius pholis*. *Journal of Experimental Biology*, 40:87-104.

House, D. G. 1974. Modification of urine by the cloaca of the desert iguana, *Dipsosaurus dorsalis*. Unpublished Masters thesis, University of Wisconsin, Milwaukee.

Houston, A. H., and J. A. Madden. 1968. Environmental temperature and plasma electrolyte regulation in the carp, *Cyprinus carpio*. *Nature*, 217:969-970.

Huelsenbeck, J. P., J. J. Bull, and C. W. Cunningham. 1996. Combining data in phylogenetic analysis. *Trends in Ecology and Evolution*, 11:152-158.

Huelsenbeck, J. P., D. M. Hillis, and R. Jones. 1996. Parametric bootstrapping in molecular phylogenetics: applications and performance. Pages 19-45 in: *Molecular Zoology* (J. D. Ferraris and S. R. Palumbi, eds.). New York: Wiley-Liss.

Huey, R. B. 1987. Phylogeny, history, and the comparative method. Pages 76-98 in: *New Directions in Ecological Physiology* (M. E. Feder, A. F. Bennett, W. W. Burggren, and R. B. Huey, eds.). New York: Cambridge University Press.

Huey, R. B., and A. F. Bennett. 1987. Phylogenetic studies of coadaptation: preferred temperatures versus optimal performance temperatures of lizards. *Evolution*, 41:1098-1115.

Hughes M. R. 1976. The effects of salt water adaptation on the Australian black swan, *Cygnus atratus* (Latham). *Comparative Biochemistry and Physiology*, 55:271-277.

Hughes M. R. 1977. Observations on osmoregulation in glaucous-winged gulls, *Larus glaucescens*, following removal of the supraorbital salt glands. *Comparative Biochemistry and Physiology*, 57A:281-287.

Jamison, R. L., N. Roinel, and C. de Rouffignat. 1979. Urinary concentrating mechanism in the desert rodent *Psammomys obesus*. *American Journal of Physiology*, 236:F448-F453.

Janis, C. M., and P. B. Wilhelm. 1993. Were there mammalian pursuit predators in the Tertiary? Dances with wolf avatars. *Journal of Mammalian Evolution*, 1:103-125.

Keil, R., R. Gerstberger, and E. Simon. 1994. Hypothalamic thermal stimulation modulates vasopressin release in hyperosmotically stimulated rabbits. *American Journal of Physiology*, 267:R1089-R1097.

Keys, A., and R. M. Hill. 1934. The osmotic pressure of the colloids in fish sera. *Journal of Experimental Biology*, 11:28-33.

Kingsolver, J. G., and M. A. R. Koehl. 1985. Aerodynamics, thermoregulation, and the evolution of insect wings: differential scaling and evolutionary change. *Evolution*, 39:488-504.

Kirschner, L. B. 1991. Water and ions. Pages 13-107 in: *Environmental and Metabolic Animal Physiology: Comparative Animal Physiology, 4th Edition* (C. L. Prosser, ed.). New York: Wiley-Liss.

Kluge, A. G., and J. S. Farris. 1969. Quantitative phyletics and the evolution of anurans. *Systematic Zoology*, 18:1-32.

Kluge, A. G., and A. J. Wolf. 1993. Cladistics: what's in a word? *Cladistics*, 9:183-199.

Komadina, S., and S. Solomon. 1970. Comparison of renal function of bull and water snakes (*Pituophis melanoleucas* and *Natrix sipedon*). *Comparative Biochemistry and Physiology*, 32:333-343.

Krag B., and E. Skadhauge. 1972. Renal salt and water excretion in the budgerigar (*Melopsittacus undulatus*). *Comparative Biochemistry and Physiology*, 41A:667-683.

Lahlou, B., I. W. Henderson, and W. H. Sawyer. 1969. Renal adaptations by *Opsanus tau*, a euryhaline aglomerular teleost, to dilute media. *American Journal of Physiology*, 216:1273-1278.

Lambert, W. D. 1992. The feeding habits of the shovel-tusked gomphotheres: evidence from tusk wear patterns. *Paleobiology*, 18:132-147.

Lange, R., and K. Fugelli. 1965. The osmotic adjustment in the euryhaline teleosts, the flounder *Pleuronectes flesus* L. and the three-spined stickleback *Gasterosteus aculeatus* L. *Comparative Biochemistry and Physiology*, 15:283-292.

Lange, R., and H. Staaland. 1966. Anatomy and physiology of the salt gland in the grey heron, *Ardea cinerea. Nytt Magasin foer Zoology (Oslo)*, 13:5-9.

Lapointe, F. J., and P. Legendre. 1992. Statistical significance of the matrix correlation coefficient for comparing independent phylogenetic trees. *Systematic Zoology*, 41:378-384.

Lapointe, F. J., and P. Legendre. 1995. Comparison tests for dendrograms: a comparative evaluation. *Journal of Classification*, 12:265-282.

Lapointe, F. J., J. A. W. Kirsch, and R. Bleiweiss. 1994. Jackknifing of weighted trees: validation of phylogenies reconstructed from distance matrices. *Molecular Phylogenetics and Evolution*, 3:256-267.

Larson, A., and W. D. Dimmick. 1993. Phylogenetic relationships of the salamander families: an analysis of congruence among morphological and molecular characters. *Herpetological Monographs*, 7:77-93.

Lauren, D. J. 1985. The effect of chronic saline exposure on the electrolyte balance, nitrogen metabolism, and corticosterone titer in the American alligator, *Alligator mississippiensis. Comparative Biochemistry and Physiology*, 81A:217-223.

Laurin, M., and R. R. Reisz. 1995. A reevaluation of early amniote phylogeny. *Zoological Journal of the Linnean Society*, 113:165-223.

LeBrie, S. J., and R. S. Elizondo. 1969. Saline loading and aldosterone in water snakes *Natrix cyclopion. American Journal of Physiology*, 217:659-660.

Lee, C. G., W. P. Low, and Y. K. Ip. 1987. Na+, K+, and volume regulation in the mudskipper *Periophthalmus chrisospilos. Comparative Biochemistry and Physiology*, 87A:439-448.

Lee, M. S. Y. 1993. The origin of the turtle body plan: Bridging a famous morphological gap. *Science*, 261:1716-1720.

Lee, M. S. Y. 1995. Historical burden in systematics and the interrelationships of 'parareptiles' *Biological Reviews*, 70:459-547.

Leeton, P. R. J., L. Christidis, M. Westerman, and W. E. Boles. 1994. Molecular phylogenetic affinities of the night parrot (*Geopsittacus occidentalis*) and the ground parrot (*Pezoporus wallicus*). *The Auk*, 111:833-843.

Logan, A. G., R. Morris, J. C. Rankin. 1980. A micropuncture study of kidney function in the river lamprey *Lampetra fluviatilis* adapted to sea water. *Journal of Experimental Biology*, 88:239-247.

Lombard, R. E., and S. S. Sumida. 1992. Recent progress in understanding early tetrapods. *American Zoologist*, 32:609-622.

Losos, J. B. 1990. Ecomorphology, performance capability, and scaling of West Indian *Anolis* lizards: An evolutionary analysis. *Ecological Monographs*, 60:369-388.

Losos, J. B. 1994. An approach to the analysis of comparative data when a phylogeny is unavailable or incomplete. *Systematic Biology*, 43:117-123.

Lutz, P. L. 1975. Osmotic and ionic composition of the polypteroid *Erpetoichthys calabaricus. Copeia*, 1975:119-123.

Lynch, M. 1989. Phylogenetic hypotheses under the assumption of neutral quantitative-genetic variation. *Evolution*, 43:1-17.

Lynch, M. 1991. Methods for the analysis of comparative data in evolutionary biology. *Evolution*, 45:1065-1080.

Maddison, D. R. 1994. Phylogenetic methods for inferring the evolutionary history and process of change in discretely valued characters. *Annual Review of Entomology*, 39:267-292.

Maddison, W. P. 1991. Squared-change parsimony reconstructions of ancestral states for continuous-valued characters. *Systematic Zoology*, 40:304-314.

Maddison, W. P. 1995. Calculating the probability distributions of ancestral states reconstructed by parsimony on phylogenetic trees. *Systematic Biology*, 44:474-481.

Maddison, W. P. 1996. Molecular approaches and the growth of phylogenetic biology. Pages 47-63 in: *Molecular Zoology* (J. D. Ferraris and S. R. Palumbi, eds.). New York: Wiley-Liss.

Maddison, W. P., and D. R. Maddison. 1992. *MacClade. Analysis of Phylogeny and Character Evolution. Version 3*. Sunderland, Massachusetts: Sinauer Associates.

Magnin, E. 1962. Recherches sur la systematique et la biologie des acepenserides *Acipenser sturio*, *Acipenser oxyrhynchus*, et *Acipenser fulvescens*. Ch.4. Quelques aspects de l'equilibre hydromineral. *Annales Station Centrale d'Hydrobiologie Appliquee*, 9:170-242.

Manly, B. F. J. 1991. *Randomization and Monte Carlo Methods in Biology*. New York: Chapman and Hall.

Maren, T. H., J. A. Rawls, J. W. Burger, and A. C. Myers. 1963. The alkaline (Marshall's) gland of the skate. *Comparative Biochemistry and Physiology*, 10:1-16.

Martin, T. E., and J. Clobert. 1996. Nest predation and avian life history evolution in Europe versus North America: a possible role of humans? *American Naturalist*, 147:1028-1046.

Martinez del Rio, C., K. E. Brugger, J. L. Rios, M. E. Vergara, and M. Witmer. 1995. An experimental and comparative study of dietary modulation of intestinal enzymes in European starlings (*Sturnus vulgaris*). *Physiological Zoology*, 68:490-511.

Martins, E. P. 1994. Estimating the rate of phenotypic evolution from comparative data. *The American Naturalist*, 144:193-209.

Martins, E. P. 1996a. Phylogenies, spatial autoregression and the comparative method: a computer simulation test. *Evolution*. In press.

Martins, E. P. (ed.) 1996b. *Phylogenies and the Comparative Method in Animal Behavior*. Oxford: Oxford University Press.

Martins, E. P. 1996c. Conducting phylogenetic comparative studies when the phylogeny is not known. *Evolution*, 50:12-22.

Martins, E. P., and T. Garland, Jr. 1991. Phylogenetic analyses of the correlated evolution of continuous characters: a simulation study. *Evolution*, 45:534-557.

Martins, E. P., and T. F. Hansen. 1996. The statistical analysis of interspecific data: a review and evaluation of phylogenetic comparative methods. Pages 22-75 in: *Phylogenies and the Comparative Method in Animal Behavior* (E. P. Martins, ed.). Oxford: Oxford University Press.

Mathers, J. S., and F. W. H. Beamish. 1974. Changes in serum osmotic and ionic concentration in the landlocked *Petromyzon marinus*. *Comparative Biochemistry and Physiology*, 49A:677-688.

Mayr, E., and P. D. Ashlock. 1991. *Principles of Systematic Zoology, 2nd Edition*. New York: McGraw-Hill.

McArdle, B., and A. G. Rodrigo. 1994. Estimating the ancestral states of a continuous-valued character using squared-change parsimony: an analytical solution. *Systematic Biology*, 43:573-578.

McBean, R. L. and L. Goldstein. 1970. Renal function during osmotic stress in the aquatic toad *Xenopus laevis*. *American Journal of Physiology*, 219:1115-1123.

McFarland, W. N., and F. W. Munz. 1965. Regulation of body weight and serum composition by hagfish in various media. *Comparative Biochemistry and Physiology*, 14:393-398.

McNabb, F. M. A. 1969. A comparative study of water balance in three species of quail-II. Utilization of saline drinking solutions. *Comparative Biochemistry and Physiology*, 28:1059-1074.

McPeek, M. A. 1995. Testing hypotheses about evolutionary change on single branches of a phylogeny using evolutionary contrasts. *The American Naturalist*, 145:686-703.

Middler, S. A., C. R. Kleeman, C. R. Edwards, D. Brody. 1969. Effect of adenohypophysectomy on salt and water metabolism of the toad *Bufo marinus* with studies of hormonal replacement. *General and Comparative Endocrinology*, 12:290-304.

Miles, D. B., and A. E. Dunham. 1996. The paradox of the phylogeny: character displacement of analyses of body size in island *Anolis*. *Evolution*, 50:594-603.

Milinkovitch, M. C. 1995. Molecular phylogeny of cetaceans prompts revision of morphological transformations. *Trends in Ecology and Evolution*, 10:328-334.

Minnich, J. E. 1977. Adaptive responses in the water and electrolyte budgets of native and captive desert tortoises (*Gopherus agassizi* and *G. polyphemus*). *Desert Tortoise Council Proceedings of the 1977 Symposium*, pages 102-129.

Minnich, J. E. 1979. Reptiles. Pages 391-617 in: *Comparative Physiology of Osmoregulation in Animals, Volume 1* (G. M. O. Maloiy, ed.). New York: Academic Press.

Minnich, J. E. 1982. The use of water. Pages 325-395 in: *Biology of the Reptilia, Volume 12*, (C. Gans, ed.). New York: Academic Press.

Miyamoto, M. M., and J. Cracraft, (eds.) 1991. *Phylogenetic Analysis of DNA Sequences*. Oxford: Oxford University Press.

Miyamoto, M. M., and M. Goodman. 1986. Biomolecular systematics of eutherian mammals: phylogenetic patterns and classification. *Systematic Zoology*, 35:230-240.

Moldenhauer R. R., and J. A. Wiens. 1970 The water economy of the sage sparrow, *Amphispiza belli nevadensis*. *Condor*, 72:265-275.

Monroe, B. L. Jr., and C. G. Sibley. 1993. *A World Checklist of Birds*. New Haven: Yale University Press.

Morris, R. 1971. Osmoregulation. Pages 198-239 in: *The Biology of Lampreys, Volume 2*, (M. W. Hardisty and I. C. Potter, eds.). London: Academic Press.

Mullen, T. L. 1974. Ionic regulation in anurans. Unpublished Ph.D. dissertation, Oregon State University, Corvallis.

Müller, R., K. Boke, U. Martin-Neuhaus, and W. Hanke. 1973. Salzwasseradaption und Stoffwechsel bei Fischen. *Fortschrift Zoologishe*, 32:456-467.

Murdaugh, H. V., and E. D. Robin. 1967. Acid-base metabolism in the dogfish shark. Pages 249-264 in: *Sharks, Skates and Rays* (P. W. Gilbert, R. F. Mathewson, and D. P. Rall, eds.).Baltimore: Johns Hopkins University Press.

Murray, R. W., and W. T. W. Potts. 1961. Composition of the endolymph, perilymph, and other body fluids of elasmobranchs. *Comparative Biochemistry and Physiology*, 2:65-76.

Nagy, K. A., and V. H. Shoemaker. 1984. Osmoregulation in the Galapagos marine iguana, *Amblyrhynchus cristatus*. *Physiological Zoology*, 57:291-300.

Nelson, J. S. 1994. *Fishes of the World, 3rd Edition*. New York: John Wiley and Sons.

Nitecki, M. H. (ed.). 1990. *Evolutionary Innovations*. Chicago: The University of Chicago Press.

Noreen, E. W. 1989. *Computer-Intensive Methods for Testing Hypotheses: an Introduction.* New York: John Wiley and Sons.

Novacek, M. J. 1992. Fossils, topologies, missing data, and the higher level phylogeny of eutherian mammals. *Systematic Biology*, 41:58-73.

Ohmart, R. D. 1972. Physiological and ecological observations concerning the salt-secreting nasal glands of the roadrunner. *Comparative Biochemistry and Physiology*, 43A:311-316

Osono, E., and H. Nishimura. 1994. Control of sodium and chloride transport in the thick ascending limb in the avian nephron. *American Journal of Physiology*, 267:R455-R462.

Oxford English Dictionary, The Compact Edition of the. 1971. Oxford: Oxford University Press.

Pagel, M. D. 1992. A method for the analysis of comparative data. *Journal of Theoretical Biology*, 156:431-442.

Pagel, M. D. 1993. Seeking the evolutionary regression coefficient: an analysis of what comparative methods measure. *Journal of Theoretical Biology*, 164:191-205.

Papanek, P. E., and H. Raff. 1994. Chronic physiological increases in cortisol inhibit the vasopressin response to hypertonicity in conscious dogs. *American Journal of Physiology*, 267:R1342-R1349.

Parry, G. 1961. Osmotic and ionic changes in blood and muscle of migrating salmonids. *Journal of Experimental Biology*, 38:411-427.

Payan, P., and J. Maetz. 1971. Balance hydrique chez les Elasmobranches: argumens en faveur d'un controle endocrinien. *General and Comparative Endocrinology*, 16:535-554.

Pickering, A. D., and R. Morris. 1970. Osmoregulation of *Lampetra fluviatilis* and *Petromyzon marinus* (Cyclostomata) in hypertonic solutions. *Journal of Experimental Biology*, 53:231-243.

Pickford, G. E., P. K. T. Pang, and W. H. Sawyer. 1966. Prolactin and serum osmolality of hypophsysectomized killifish, *Fundulus heteroclitus*, in freshwater. *Nature*, 209:1040-1041.

Platner, W. S. 1950. Effects of low temperature on magnesium content of blood, body fluids, and tissues of goldfish and turtle. *American Journal of Physiology*, 161:399-405.

Pora, E. A. 1936. Sur les differences chimiques et physico-chimiques du sang des deux sexes des Selaciens. *Comptes Rendus des Seances Societe de Biologie et de Ses Filiales et Associees (Paris)*, 121:105-107 and 291-293.

Pough, F. H., J. B. Heiser, and W. N. McFarland. 1996. *Vertebrate Life, 4th Edition.* New Jersey: Prentice Hall.

Poulson, T. L. 1969. Salt and water balance in seaside and sharp-tailed sparrows. *The Auk*, 86:473-489.

Price, K. S., and E. P. Creaser. 1967. Fluctuations in two osmoregulatory components, urea and sodium chloride, of the clearnose skate *Raja eglanteria* Bosc 1802. I. Upon laboratory modification of external salinities. *Comparative Biochemistry and Physiology*, 23:65-76.

Purvis, A. 1995. A composite estimate of primate phylogeny. *Philosophical Transactions of the Royal Society of London. B.*, 348:405-421.

Purvis, A., and T. Garland, Jr. 1993. Polytomies in comparative analyses of continuous characters. *Systematic Biology*, 42:569-575.

Purvis, A., J. L. Gittleman, and H.-K. Luh. 1994. Truth or consequences: effects of phylogenetic accuracy on two comparative methods. *Journal of Theoretical Biology*, 167:293-300.

Pyron, M. 1996. Sexual size dimorphism and phylogeny in North American minnows. *Biological Journal of the Linnean Society*, 57:327-341.

Reynolds, P. S., and R. M. Lee, III. 1996. Phylogenetic analysis of avian energetics: passerines and nonpasserines do not differ. *American Naturalist*, 147:735-759.

Ridley, M. 1993. *Evolution.* Oxford: Blackwell Scientific Publications.

Roberts, J. R., and R. V. Baudinette. 1984. The water economy of stubble quail, *Coturnix pectoralis*, and king quail, *Coturnix chinensis*. *Australian Journal of Zoology*, 32:637-647.

Roberts, J. S., and B. Schmidt-Nielsen. 1966. Renal ultrastructure and excretion of salt and water by three terrestrial lizards. *American Journal of Physiology*, 211:476-486.

Robertson, J. D. 1976. Chemical composition of the body fluids and muscle of the hagfish *Myxine glutinosa* and the rabbit-fish *Chimaera monstrosa. Journal of Zoology, London*, 178:261-277.

Rosenberg, G. 1996. Independent evolution of terrestriality in Atlantic truncatellid gastropods. *Evolution*, 50:682-693.

Ruben, J. A. 1991. Reptilian physiology and the flight capacity of *Archaeopteryx. Evolution*, 45:1-17.

Ruben, J. A. 1995. The evolution of endothermy in mammals and birds: from physiology to fossils. *Annual Review of Physiology*, 57:69-95.

Ruben, J. A., and A. F. Bennett. 1980. The vertebrate pattern of activity metabolism: Its antiquity and possible relation to vertebrate origins. *Nature*, 286:886-888.

Ruben, J. A., and J. K. Parrish. 1991. Antiquity of the chordate pattern of exercise metabolism. *Paleobiology*, 16:355-359.

Ruben, J. A., W. J. Hillenius, N. R. Geist, A. Leitch, T. D. Jones, P. J. Currie, J. R. Horner, and G. Espe III. 1996. The metabolic status of some Late Cretaceous dinosaurs. *Science*, 273:1204-1207.

Ryan, M. J. 1996. Phylogenetics in behavior: some cautions and expectations. Pages 1-21 in: *Phylogenies and the Comparative Method in Animal Behavior* (E. P. Martins, ed.). Oxford: Oxford University Press.

Ryan, M. J., and A. S. Rand. 1995. Female responses to ancestral advertisement calls in Tungara frogs. *Science*, 269:390-392.

Sanderson, M. J. 1995. Objections to bootstrapping phylogenies: A critique. *Systematic Biology*, 44:299-320.

Sarich, V. M., C. W. Schmid, and J. Marks. 1989. DNA hybridization as a guide to phylogenies: A critical analysis. *Cladistics*, 5:3-32.

Schmidt-Nielsen, B., and D. Schmidt. 1973. Renal function of *Sphenodon punctatum. Comparative Biochemistry and Physiology*, 44A:121-129.

Schmidt-Nielsen, K. 1964. *Desert Animals: Physiological Problems of Heat and Water*. London: Oxford University Press.

Schoffeniels, E. and R. R. Tercafs. 1965. Adaptation d'un reptile marin, *Caretta caretta* L., a l'eau douce et d'un reptile d'eau douce, *Clemmys leprosa* L., a l'eau de mer. *Annales Societe Royale Zoologique de Belgique*, 96:1-8.

Schultz, T. R., R. B. Cocroft, and G. A. Churchill. 1996. The reconstruction of ancestral states. *Evolution*, 50:504-511.

Sharratt, B. M., I. Chester Jones, and D. Bellamy. 1964. Water and electrolyte composition of the body and renal funtion of the eel (*Anguilla anguilla*). *Comparative Biochemistry and Physiology*, 11:9-18.

Shoemaker, V. H., S. S. Hillman, S. D. Hillyard, D.C. Jackson, L. L. McClanahan, P. C. Withers, and M. L. Wygoda. 1992. Exchange of water, ions, and respiratory gases in terrestrial amphibians. Pages 125-150 In:

Environmental Physiology of the Amphibians (M. E. Feder and W. W. Burggren, eds.). Chicago: The University of Chicago Press.

Sibley, C. G., and J. E. Ahlquist. 1990. *Phylogeny and Classification of Birds: A Study in Molecular Evolution*. New Haven: Yale University Press.

Skadhauge, E. 1967. In vivo perfusion studies of the water and electrolyte resorption in the cloaca of the fowl (*Gallus domesticus*). *Comparative Biochemistry and Physiology*, 23:483-501.

Skadhauge, E. 1974. Renal concentrating ability in selected West Australian birds. *Journal of Experimental Biology*, 61:269-276.

Skadhauge, E., and S. D. Bradshaw. 1974. Saline drinking and cloacal excretion of salt and water in the zebra finch. *American Journal of Physiology*, 227:1263-1267

Smith, H.W. 1930. Metabolism of the lungfish *Protopterus aethiopicus*. *Journal of Biological Chemistry*, 88:97-130.

Smith, H.W. 1931. The absorption and excretion of water and salts by the elasmobranch fishes. I. Fresh-water elasmobranchs. *American Journal of Physiology*, 98:279-295.

Smyth, M., and G. A. Bartholomew. 1966. Effects of water deprivation and sodium chloride on the blood and urine of the mourning dove. *The Auk*, 83:597-602.

Sober, E. 1988. *Reconstructing the Past: Parsimony, Evolution, and Inference*. Cambridge, Mass.: Massachusetts Institute of Technology Press.

Sokal, R. R., and F. J. Rohlf. 1981. *Biometry, 2nd Edition*. San Francisco: W. H. Freeman and Company.

Spector, W. S. (ed.). 1956. *Handbook of Biological Data*. Philadelphia: W.B. Saunders Company.

Springer, M. S., and C. Krajewski. 1989a. DNA hybridization in animal taxonomy: A critique from first principles. *Quarterly Review of Biology*, 64:291-318.

Springer, M. S., and C. Krajewski. 1989b. Additive distances, rate variation, and the perfect-fit theorem. *Systematic Zoology*, 38:371-375.

Stanley, J. G. and W. R. Fleming. 1964. Excretion of hypertonc urine by a teleost. *Science*, 144:63-64.

Stanton, B. 1988. Electroneutral NaCl transport by distal tubule: evidence for Na+/H+ - Cl-/HCO3- exchange. *American Journal of Physiology*, 254:F80-F86.

Stewart, C.-B. 1993. The powers and pitfalls of parsimony. *Nature*, 361:603-607.

Stiffler, D. F., M. L. deRuyter, and C. R. Talbot. 1990. Osmotic and ionic regulation in aquatic and terrestrial caecilians. *Physiological Zoology*, 63:649-668.

Stolte, H., B. Schmidt-Nielsen, and L. Davis. 1977. Single nephron function in the kidney of the lizard *Sceloporus cyanogenus*. *Zoologische Jahrbuecher*, 81:209-214.

Strait, D. S., M. A. Moniz, and P. T. Strait. 1996. Finite mixture coding: a new approach to coding continuous characters. *Systematic Biology*, 45:67-78.

Tercafs, R. R., and J. M. Vassas. 1967. Comportement osmotique des erythrocytes de lezards. *Archives Internationales de Physiologie et de Biochimie*, 75:667-674.

Thomson, K. S. 1991. *Living Fossil: the Story of the Coelacanth*. New York: W. W. Norton and Company.

Thorson, T. B., C. M. Cowan, and D. E. Watson. 1967. *Potamotrygon* spp: Elasmobranchs with low urea content. *Science*, 158:375-377.

Thorson, T. B., C. M. Cowan, and D. E. Watson. 1973. Body fluid solutes of juveniles and adults of the euryhaline bull shark *Carcharhinus leucas* from freshwater and saline environments. *Physiological Zoology*, 46:29-42.

Tracy, C. R., J. S. Turner, and R. B. Huey. 1986. A biophysical analysis of possible thermoregulatory adaptations in sailed pelycosaurs. Pages 195-206 in: *The Ecology and Biology of Mammal-like Reptiles* (N. Hotton, P. D. MacLean, J. J. Roth, and E. C. Roth, eds.). Washington, D.C.: Smithsonian Institution Press.

Ullrich, K. J., B. Schmidt-Nielsen, R. O'Dell, G. Pehling, C. W. Gottschalk, W. E. Lassiter, and M. Mylle. 1963. Micropuncture study of composition of proximal and distal tubular fluid in rat kidney. *American Journal of Physiology*, 204:527-531.

Umminger, B. L. 1971. Patterns of osmoregulation in freshwater fishes at temperatures near freezing. *Physiological Zoology*, 44:20-37.

Urist, M. R. 1966. Calcium and electrolyte control mechanisms in lower vertebrates. Pages 18-28 in: *Phylogenetic Approaches to Immunity* (R. T. Smith, R. A. Good, and P. A. Meischer, eds.). Gainesville: University of Florida Press.

Wake, M. H. (ed.). 1989. *Hyman's Comparative Vertebrate Anatomy, 3rd edition.* Chicago: The University of Chicago Press.

Walton, M. 1993. Physiology and phylogeny: the evolution of locomotor energetics in hylid frogs. *American Naturalist*, 141:26-50.

Westneat, M. W. 1995. Feeding, function, and phylogeny: analysis of historical biomechanics and ecology in labrid fishes using comparative methods. *Systematic Biology*, 44:361-383.

Wiens, J. J. 1995. Polymorphic characters in phylogenetic systematics. *Systematic Biology*, 44:482-500.

Wiley, E. O. 1981. *Phylogenetics: the Theory and Practice of Phylogenetic Systematics*. New York: John Wiley and Sons.

Wiley, E. O., D. Seigel-Causey, D. R. Brooks, and V. A. Funk. 1991. *The Compleat Cladist: a Primer of Phylogenetic Procedures*. University of Kansas Museum of Natural History Special Publication Number 19,. 158 pages.

Wilson, D. E., and D. M. Reeder. 1993. *Mammal Species of the World.* Washington, D.C.: Smithsonian Institution Press.

Withers, P. C. 1992. *Comparative Animal Physiology*. Fort Worth: Saunders College Publishing.

Wu, C. F. J. 1986. Jackknife, bootstrap and other resampling methods in regression analysis. *Annals of Statistics*, 14:1261-1295.

Yoshimura, H., M. Yata, M. Yuasa, and R. A. Wolbach. 1961. Renal regulation of acid-base balance in the bullfrog. *American Journal of Physiology*, 201:980-986.

Zelditch, M. L., W. L. Fink, and D. L. Swiderski. 1995. Morphometrics, homology, and phylogenetics: Quantified characters as synapomorphies. *Systematic Biology*, 44:179-189.

Zucker, I. H., C. Gilmore, J. Dietz, and J. P. Gilmore. 1977. Effect of volume expansion and veratrine on salt gland secretion in the goose. *American Journal of Physiology*, 232:R185-189.

Zug, G. R. 1993. *Herpetology: an Introductory Biology of Amphibians and Reptiles*. San Diego: Academic Press.

SUBJECT INDEX

—A—

—B—

—C—

—D—

—E—

—X—

—Y—

—Z